NASA SP-4307

Suddenly, Tomorrow Came...

A History of the Johnson Space Center

Henry C. Dethloff

The NASA History Series

National Aeronautics and   1993
Space Administration

Lyndon B. Johnson Space Center

Contents

Illustrations		iv
Foreword		viii
Preface and Acknowledgments		x
1	October 1957	1
2	The Commitment to Space	17
3	Houston - Texas - U.S.A.	35
4	Human Dimensions	53
5	Gemini: On Managing Spaceflight	77
6	The NASA Family	97
7	Precious Human Cargo	117
8	A Contractual Relationship	137
9	The Flight of Apollo	159
10	"After Apollo, What Next?"	187
11	Skylab to Shuttle	209
12	Lead Center	233
13	Space Business and JSC	257
14	Aspects of Shuttle Development	273
15	The Shuttle at Work	285
16	New Initiatives	307
17	Space Station Earth	329
Epilogue		349
MSC/JSC Directors		351
Reference Notes		353
Index		389
The NASA History Series		407

Illustrations

Figures

1.	Interim Facilities Leased as of August 1, 1962	47
2.	Manned Spacecraft Center as of 1962	66-67
3.	A Chronicle of the Last Mercury Flight	68
4.	Launch Vehicles	72
5.	Manned Spacecraft Center Organization as of June 1963	89
6.	Manned Spacecraft Center Organization as of December 1964	90
7.	Apollo Program Government-Industry Functional Matrix	100
8.	The Apollo Stack	145
9.	Flight Operations Division Organization as of December 1962	166
10.	Flight Operations Directorate Organization as of January 1964	167
11.	Flight Control Division Organization as of March 1970	168
12.	Operations-oriented Divisions as of June 1968	170
13.	Johnson Space Center Organization as of March 1973	212-213
14.	Administration and Program Support Directorate Organization as of January 1971	217
15.	Shuttle Management Timeline, 1968 to 1982	228
16.	Orbiter Project Office Organization as of 1973	235
17.	Johnson Space Center Organization as of January 1977	244-245
18.	Civil Service and Support Contractor Employment History	258
19.	Shuttle and Space Station Compared	310
20.	Johnson Space Center Organization as of March 1986	322
21.	Johnson Space Center Organization of April 1987	323
22.	Johnson Space Center Organization as of May 1988	324
23.	Lunar Mission Profile	340
24.	Mars Mission Profile	341
25.	NASA Budget Trends, 1960 to 1990	346
26.	NASA Budget, 1959 to 1989	347

Tables

1. Johnson Space Center Buildings — Construction Costs and Size 50
2. Project Mercury Flight Data Summary 69
3. NASA Astronaut Selections, 1959 to 1969 128
4. Major Apollo Subcontractors 147
5. NASA Budget and Personnel Status 151
6. NASA Budget Requests and Appropriations, FY 1959 to 1971 194
7. Science and Applications Directorate 219
8. Launch Capability, Launch Vehicle Cost and Performance Comparison 236
9. National Launch Vehicles 237
10. Requested and Authorized Apollo and Shuttle Budgets 240
11. Total NASA Awards to Business and Nonprofit Institutions in Texas 259
12. Distribution of JSC Procurements 267
13. Small Business and Minority Business Participation in JSC Procurement Activity 269
14. Geographical Distribution of JSC Procurement Excluding Intragovernmental Actions 270
15. Impact of the NASA Budget on JSC and Houston Area Economy 271
16. Comparative Shuttle and Apollo Launch Vehicle Launch Costs 286
17. Space Shuttle Missions in Brief, 1985 to 1986 290
18. The Shuttle in Flight, 1981 to 1989 302-303

Photographs

President Lyndon B. Johnson 18
The 1000-acre land parcel outside Houston, Texas 40
Congressman Olin E. "Tiger" Teague 41
Building 1, the Administration Building 49
1974 aerial view of Johnson Space Center 51
Mercury-Redstone 3 launch 54
Mercury-Redstone 4 launch 54
Astronaut John Glenn enters the "Friendship" spacecraft during rehearsal exercises ... 55
Scott Carpenter's launch aboard "Aurora 7" 56
Dr. Max A. Faget, Assistant Director for Engineering Development 62
Mercury-Atlas 9 69
Gemini 3, launched March 23, 1965 92

The interior of Gemini 6	93
Astronaut Edward H. White in the microgravity of space	94
Gemini 10, launched July 18, 1966	94
Whitney Darrow, Jr. cartoon	95
AS-201 lift-off, Cape Kennedy, Florida, February 26, 1966	110
Donald K. "Deke" Slayton and George M. Low relax during the AS-202 unmanned flight in August 1966	111
AS-204 astronauts died when an oxygen-enriched fire swept the interior of the spacecraft during preflight tests at Cape Kennedy	112
Astronaut Group I, April 1959	121
The Space Environment Simulation Laboratory	132
Mockup of the command module under construction by North American Aviation at Downey, California	139
Apollo 8 lunar flight marked a giant step toward NASA's lunar landing	171
Apollo 9's Earth-orbital mission of March 1969	172
Apollo 11 launch, July 16, 1969	175
George M. Low watches a television monitor during the lunar surface EVA	175
Flight Director Clifford E. Charlesworth and Eugene Kranz prepare for the change of shifts	176
Artist depictions of the lunar surface and a lunar landing	176
A view of Mission Control Center during Apollo 11 EVA	177
President Richard M. Nixon welcomes the Apollo 11 crew upon the completion of the historic mission	177
Apollo 11's lunar module casts a long shadow upon the Moon and symbolically upon the future of Earth	178
A simple human footprint on the lunar surface	182
Neil Armstrong and Edwin Aldrin raise the flag	183
A gathering of Apollo 14 flight directors at Mission Control Center	199
The LM "Falcon" is photographed against the barren landscape during Apollo 15	200
David R. Scott photographed James B. Irwin as he worked on the lunar roving vehicle	200
Apollo 15 splashdown	201
The lunar surface viewed with a 35 mm stereo close-up camera	202
The Lunar Receiving Laboratory	202
Apollo 16, the fifth lunar landing in the Moon's Descartes area	203
Final Apollo (17) lunar mission, December 13, 1972	203

Apollo 17 splashdown marking the end of the Apollo lunar programs	204
Apollo 17 view of Earth	206
Skylab launch, May 14, 1973	210
Lady Bird Johnson and Center Director Chris Kraft during JSC dedication ceremonies	215
The Soviet Soyuz spacecraft photographed from the American Apollo spacecraft	221
The Apollo-Saturn hardware is transported by barge for display at JSC	222
The anechoic chamber at JSC	241
Astronauts trained to work in the weightless environment of space	250
An astronaut is tested in Chamber A of the Space Environment Simulation Laboratory	251
The Upper Atmosphere Research Satellite in the payload bay of the Earth-orbiting Shuttle *Discovery*	276
The Shuttle made its maiden voyage on April 12, 1981	280
Astronaut George Nelson practices an EVA with a mockup of the MMU	292
Space Shuttle *Atlantis* blasts toward orbit on two powerful SRBs	296
January 28, 1986, *Challenger* lifted off	298
Photograph of the flames developing near the O-Ring on the SRB	298
The crew of STS 51-L	299
The nation deeply mourned the loss of the *Challenger* astronauts	300
Artist conceptions of new ideas: ACRV and Space Station robotics	313
STS-43 soars toward space to begin a 9-day mission	325
Space Center Houston	332
A mockup of the new Space Station	333
Earthrise!	349

Foreword
by Donald K. Slayton

This history of Johnson Space Center (JSC) is a detailed chronicle of the U.S. space program with emphasis on humans in space and on the ground. It realistically balances the role of the highly visible astronaut with the mammoth supporting team who provide the nuts, bolts, and gas to keep the train on the track. It recognizes the early political and technical geniuses who had the vision and ability to create NASA and JSC and keep them expanding at a rapid pace. People like Jim Webb, who was unsurpassed in his ability to create political support and financing, and Bob Gilruth, his counterpart at the technical and operational level, were the real gems in the right place at the right time. They were the true progenitors of manned spaceflight.

This history progresses from when JSC was the Space Task Group, a small cadre of about 300 talented and dedicated ex-NACA and Canadian personnel, to the peak of the Apollo era, when JSC—then called the Manned Spacecraft Center—had thousands of personnel. Yet despite its explosive growth, it never lost its human touch or the "can-do" attitude of its roots.

NASA and JSC became internationally recognized as symbols of excellence both inside and outside government. The image of infallibility grew as we progressed through Mercury and Gemini with major victories and only minor hiccups. Bob Gilruth and his senior aides always knew space travel was risky, but it took the Apollo 1 fire to shock the rest of the world back to reality. I firmly believe that the ultimate total success of the complex Apollo program can be attributed to a large extent to the way the fire dramatically refocused our attention on our goals.

It was inevitable that the post-Apollo reset to near Earth orbit through Skylab, Apollo-Soyuz, and the Space Shuttle program would be anticlimactic for both the players and the spectators. For almost everyone in Houston, these programs, along with a space station, were high on the list of logical consolidation and expansion steps leading to Mars. Unfortunately, none of these logical steps had Apollo's public appeal, so they suffered from disinterest both in the political arena and among the general public. The Shuttle remains a remarkable achievement, but throughout its development it suffered from the lack of a sense of urgency, which led to underfunding. Chris Kraft, Max Faget, Bob Thompson, Aaron Cohen, and other NASA and JSC leaders made this answer to a pilot's prayer a remarkable political achievement when they brought the Shuttle on line with great difficulty.

The Shuttle has brought back some of the public appeal of space travel, primarily because of the size and variety of its crew and the possibility, however remote, that the average citizen might go into space. As usual, the manned aspect has created the catalyst for most forward thinking and planning of future space projects, both national and international. Space officials in what used to be the Soviet Union are enthusiastic about future joint missions to Mars based on the Apollo-Soyuz model. Our many international partners in the next undertaking of NASA and JSC, Space Station Freedom, are enthusiastic about it and dedicated to its success.

World events are catching up with the examples provided by the major manned space programs. Almost every astronaut and cosmonaut who circled planet Earth has observed that from orbit there are no national borders visible on this beautiful globe. All those fortunate enough to view Earth from the Moon were impressed with its similarity to a spacecraft and by its remoteness and insignificance in the Universe. These observations by humans in space have had a profound effect on humans on Earth and provide a strong unifying force for international space exploration. So as tomorrow comes, people of the Earth will inevitably step into the Universe and become true space people—citizens of Mars, the Moon, Venus, and beyond. They will always be building on the achievements of Bob Gilruth and his colleagues at Johnson Space Center.

Preface and Acknowledgments
by Henry C. Dethloff

The history of the Johnson Space Center focuses on an unusual slice of time and human affairs. It has been a time of great changes, the full impact of which are not yet evident. American history and that of humankind has been irrevocably affected by spaceflight. Space has generated new technology, new materials, and a new process of thought about the Earth and the human potential. This book has a beginning and an end, but the story continues, perhaps through all time.

Suddenly, a new tomorrow has come into being. In 1902 H.G. Wells observed that the past, "all that has been and is, is but the twilight of the dawn." Today, because of the American space program, "the world is heavy with the promise of greater things." Indeed, perhaps that day predicted by H.G. Wells has come to be: "when beings, beings who are now latent in our thoughts and hidden in our loins, shall stand upon the Earth as one stands upon a footstool, and shall laugh and reach out their hands amidst the stars."

Each of us have been observers and to some extent participants in the exciting new dimensions of the human experience. As did the African drummers, mentioned in the text, who spread the message that a human was actually walking in space, most of us have heard of or witnessed on television and radio many of the events mentioned. I, for example, then stationed at Jacksonville Naval Air Station, made many flights "downrange" aboard Military Aircraft Transport planes from Cape Canaveral where the Army was testing the new Redstone rocket. I saw Sputnik I, and remember the disbelief, confusion, awe, fear, and wonderment associated with that event. The missile gap, President John F. Kennedy's challenge to go to the Moon, President Lyndon B. Johnson, the Apollo lunar landing, and then Vietnam, the Shuttle, *Challenger*, and the subsequent uncertainties and ambivalence about spaceflight are a past that somehow persists in clinging to the present. I have admittedly written the narrative with a certain sense of involvement; and I expect that the reader will inevitably read with a similar sense of attachment and participation—and that is as it should be. For the most part we have been spectators rather than participants, and those things we have observed have been the externalities and the end product. What we have not been able to observe or understand is how these things came to be.

The story of manned spaceflight is the story of many diverse individuals, and of the collaboration of persons of many backgrounds and persuasions in what became a peacetime mobilization of American human and capital resources. It is a history of science, of engineering, of sacrifice, failures, and great achievement. Johnson Space Center and its personnel are central to the story of the National Aeronautics and Space Administration and manned spaceflight and to the inception of a new epoch in human history. This story seeks to explain how the space voyages, the lunar landing, Mercury, Gemini, Apollo, the Shuttle, and the Space Station came to be, and the role of Johnson Space Center in those developments.

I wish to particularly recognize Oran Nicks and David J. Norton, who provided special insights, direction, and moral support, and reviewed the manuscript as it progressed. Joey

Kuhlman, archivist at the Johnson Space Center History Office, was indispensable as both research assistant and project coordinator. Janet Kovacevich provided continuing counsel and support. Donald L. Hess, JSC History Coordinator and project director through most of the research and writing phase of the book, provided help, support, and direction in ways that I never fully understood or appreciated in that he did so without seeming to impose any constraints on research or content. His participation and style of supervision is greatly appreciated, as are the contributions and assistances of Carol A. Homan who replaced him as JSC History Coordinator.

The entire writing project was characterized by the lack of direction and control by any NASA authority figures, and by the exercise of my complete artistic freedom and professional integrity. It is also characterized by the very professional and critical support of the National Aeronautics and Space Administration Historians, Dr. Sylvia Fries and her successor, Dr. Roger D. Launius. Their breadth of knowledge and technical expertise prevented many pitfalls. They are strong editors and critics.

The quality and precision of the manuscript, to be sure, draws heavily upon the expertise and advice of the JSC History Advisory Committee, specially created to review the draft chapters and offer explanations and advice. Although we met together intermittently, I relied very heavily upon their insights and experience. Joseph P. Loftus, for example, was always available to unravel a knotty problem or explain a seemingly inscrutable technical situation. Henry O. Pohl, Dennis J. Webb, Douglas K. Ward, and Donald E. Robbins constantly tested the mettle of the prose. Daniel A. Nebrig also served on the Advisory Committee. These Advisory Committee members contributed significantly to sharpening my insights and enhancing the accuracy of the manuscript.

I want to thank each one of the many NASA employees or former employees whom I interviewed. They were invariably unstinting in their effort to cooperate, illuminate and explain. They are included in the reference notes. Many who were interviewed then signed on as readers, critics, and advisors. Among these I would like to recognize and thank particularly Paul Purser, Aleck Bond, Bill Kelly, Rod Rose, Chris Kraft, Max Faget, Bob Piland, and John Hodge.

The final product is necessarily mine, and I recognize that the book does not capture the full spirit of the events as they may be recalled by members of the Johnson Space Center and NASA organization. I regret the errors and omissions. The book is an attempt to explain—not so much to those who were directly involved but to those of us on the distant periphery, that is the general public, who watched and simply by virtue of our observations and being became participants in one of the most remarkable stories of modern times—the story of NASA, Lyndon B. Johnson Space Center, and manned spaceflight.

CHAPTER 1: October 1957

"I was at my ranch in Texas," Lyndon Baines Johnson recalled, "when news of Sputnik flashed across the globe . . . and simultaneously a new era of history dawned over the world." Only a few months earlier, in a speech delivered on June 8, Senator Johnson had declared that an intercontinental ballistic missile with a hydrogen warhead was just over the horizon. "It is no longer the disorderly dream of some science fiction writer. We must assume that our country will have no monopoly on this weapon. The Soviets have not matched our achievements in democracy and prosperity; but they have kept pace with us in building the tools of destruction."[1] But those were only words, and Sputnik was a new reality.

Shock, disbelief, denial, and some real consternation became epidemic. The impact of the successful launching of the Soviet satellite on October 4, 1957, on the American psyche was not dissimilar to the news of Pearl Harbor on December 7, 1941. Happily, the consequences of Sputnik were peaceable, but no less far-reaching. The United States had lost the lead in science and technology, its world leadership and preeminence had been brought into question, and even national security appeared to be in jeopardy.

"This is a grim business," Walter Lippman said, not because "the Soviets have such a lead in the armaments business that we may soon be at their mercy," but rather because American society was at a moment of crisis and decision. If it lost "the momentum of its own progress, it will deteriorate and decline, lacking purpose and losing confidence in itself."[2]

According to the U.S. Information Agency's Office of Research and Intelligence, Sputnik's repercussions extended far beyond the United States. Throughout western Europe the "Russian launching of an Earth-satellite was an attention-seizing event of the first magnitude." Within weeks there was a perceptible decline in enthusiasm among the public in West Germany, France and Italy for "siding with the west" and the North Atlantic Treaty Organization (NATO). British-American ties grew perceptibly stronger.[3] Some Americans began to think seriously about building backyard bomb shelters.

That evening after receiving the news, Senator Johnson began calling his aides and colleagues and deliberated a call for the Preparedness Investigating Subcommittee of the Senate Committee on Armed Services to begin an inquiry into American satellite and missile programs. Politically, it was a matter of some delicacy for the Democratic Senate Majority Leader.[4]

Dwight D. Eisenhower was an enormously popular Republican president who had presided over a distinctly prospering nation. He was the warrior president, the victor over the Nazis, and a "father" figure for many Americans. Moreover, race, not space, seemed at the time to be uppermost in the American mind. Governor Orval Faubus of Arkansas had only days before precipitated a confrontation between the Arkansas National Guard and federal authority.

When, at President Eisenhower's personal interdiction, Governor Faubus was reminded that in a confrontation between the state and federal authority there could be only one outcome, the Governor withdrew the Guard only to have extremist mobs prevent the entry of black children into Little Rock High School. Eisenhower thereupon nationalized the National Guard and enforced the decision of the Supreme Court admitting all children, irrespective of race, creed or color, to the public schools.

Finally, Eisenhower had ended the Korean War; he had restored peace in the Middle East following the Israeli invasion of the Sinai peninsula; and in 1955 he had announced the target to launch a man-made satellite into space in celebration of the International Geophysical Year (IGY). And in 1957 the Eisenhower-sponsored interstate highway system was just beginning to have a measurable impact on the lifestyle of Americans.[5] Automobiles were now big, chrome-laden, and sometimes came with air-conditioning and power steering. Homes, too, tended to be big, brick, and sometimes came with air-conditioning and television. There was, however, no question but that the great Eisenhower aura of well-being had been shattered first by recession, then by the confrontations at Little Rock, Arkansas, and now by Sputnik.

The White House commented on October 6, that the launching of Sputnik "did not come as a surprise." Press Secretary James C. Hagerty indicated that the achievement was of great scientific interest and that the American satellite program geared to the IGY "is proceeding satisfactorily according to its scientific objectives"—while President Eisenhower relaxed at his farm. Two days later the Department of Defense concurred that there should be no alarm and that the American scientific satellite program need not be accelerated simply because of the Soviets' initial success. On the ninth, retiring Secretary of Defense Charles E. Wilson termed the Soviet Sputnik "a neat scientific trick" and discounted its military significance.[6]

And that day President Eisenhower announced that the Naval Research Laboratory's Vanguard rocket program, which would launch the IGY satellite into orbit, had been deliberately separated from the military's ballistic missile program in order to accent the scientific nature of the satellite and to avoid interference with top priority missile programs. Had the two programs been combined, he said, the United States could have already orbited a satellite.[7]

Lyndon Johnson, with the approval of Senator Richard B. Russell, Chairman of the Senate Armed Forces Committee, directed the staff of the Preparedness Subcommittee, which he chaired, to begin a preliminary inquiry into the handling of the missile program by the Department of Defense. Independently, Eisenhower met with top military, scientific, and diplomatic advisors and called the National Security Council into session before convening the full cabinet to discuss what could be done to accelerate the United States satellite and guided missile program. The *New York Times* observed that more scientists visited Eisenhower during the 10 days following Sputnik than in the previous 10 months. Neil H. McElroy, who was replacing Charles E. Wilson as Secretary of Defense, and assorted military aides doubted that a speedup of the satellite or missile programs would be feasible given existing technological and monetary limitations. The President for the time concurred that defense spending should be maintained at its then current levels of about $38 billion.[8]

Solis Horwitz, Subcommittee Counsel, reported to Johnson on the 11th that at the preliminary briefing held by the Preparedness Subcommittee staff, Pentagon representatives explained that the Vanguard IGY project and the United States missile program were separate and distinct projects, and that it would be several weeks before they could give an accurate picture of the military significance of the Russian satellite. Moreover, almost everyone had believed the United States would be the first to put up a satellite, and "none of them had given much thought to the military and political repercussions in the event the Soviets were first." At a meeting of the Eighth International Astronautical Federation Congress, the commander and deputy commander of the Redstone Arsenal stated flatly that the United States could have beaten the Russians to space by a year if delays (attributed to the Navy) had not been ordered. McElroy promised to see to it that "bottlenecks" were removed. And retiring Secretary of Defense Charles E. Wilson responded to criticisms that appeals for a faster flow of money for the Vanguard project made between 1955 and 1957 had been "bottled up" in the Secretary's office. Earlier, the press reported that Wilson had an unsympathetic attitude toward basic research, about which he is supposed to have commented: "Basic research is when you don't know what you are doing."[9]

Lyndon Johnson told a Texas audience on October 14 that, "The mere fact that the Soviets can put a satellite into the sky . . . does not alter the world balance of power. But it does mean that they are in a position to alter the balance of power." And Vice President Richard M. Nixon, in his first public address on the subject, told a San Francisco audience that the satellite, by itself, did not make the Soviets "one bit stronger," but it would be a terrible mistake to think of it as a stunt.[10] Sputnik demanded an intelligent and strong response, he said.

The *New York Times* blamed "false economies" by the administration for the Russian technological lead. It reported that the Bureau of the Budget had refused to allow the Atomic Energy Commission to spend $18 million appropriated by Congress on "Project Rover," a nuclear powered rocket research and development program, which "would postpone the time when nuclear power can be used to propel rockets huge distances."[11]

There were scoffers and skeptics, but precious few. The President's advisor on foreign economic affairs called the Soviet satellite "a silly bauble." But by the end of October, the reaction to Sputnik was beginning to take a distinctly different tone. The problem went beyond missiles and defense. It was far more basic. Alan Waterman, Director of the National Science Foundation, submitted a special report to President Eisenhower which indicated that basic research in the United States was seriously underemphasized. The federal government must assume "active leadership" in encouraging and supporting basic research. That same evening Secretary McElroy restored budgetary cuts previously made in arms research. Educators began to insist on greater emphasis on mathematics, physics and chemistry in all levels. Secretary of Health, Education and Welfare Marion Folsom responded that while "more and better science must be taught to all students in secondary schools and colleges," attempts to imitate Soviet education would be "tragic for mankind." Nixon believed that Soviet scientific achievements underscored the need for racial integration in the public schools and elsewhere in the United States. On November 3, a second Soviet triumph in space sorely delimited Folsom's appeal to preserve the tradition of a broad, liberal education. A second much larger and heavier satellite, carrying aboard it a dog named Laika, began Earth orbit.[12]

The next day Johnson, with Richard Russell and Styles Bridges, and all of the Armed Services Committee were briefed at the Pentagon. As Johnson said, "The facts which were brought before us during that briefing gave us no comfort." The next day Johnson decided that the Preparedness Subcommittee should initiate "a full, complete and exhaustive inquiry" into the state of national defenses.[13]

President Eisenhower addressed the Nation on the 7th, telling the people that his scientific friends believed that "one of our greatest and most glaring deficiencies is the failure of us in this country to give high enough priority to scientific education and to the place of science in our national life." He announced the appointment of James Killian, Jr., president of the Massachusetts Institute of Technology, as Special Assistant to the President for Science and Technology, and he elevated the prestigious Science Advisory Committee from Defense to the Executive Office, enlarging its membership from 13 to 18 members. He announced that within the Department of Defense a single individual would receive full authority (over all services) for missile development. Congress, he said, would be presented legislative proposals removing barriers to the exchange of scientific information with friendly nations. The Secretary of State would appoint a science advisor and create science attachés in overseas diplomatic posts. More pointedly, he directed the Secretary of Defense to give the "Army and its German-born rocket experts permission to launch a satellite with a military rocket." Secretary Neil McElroy issued those instructions on November 8.[14]

Eisenhower's initial response to Sputnik emphasized scientific education, basic research, the free exchange of ideas, and the centralization of authority for satellite and missile development outside the prerogative of any single branch of the military services. Although still quite some distance from the conceptualization and organic legislation creating the National Aeronautics and Space Administration, certain parameters for such an organization had become evident in the political and scientific communities by the end of October 1957.[15] But some Americans who had been thinking about bomb shelters began building them.

It was perhaps not inappropriate that Lyndon Johnson compared the Sputnik crisis to Pearl Harbor in his opening remarks for the Preparedness Subcommittee Hearings on November 25:

> A lost battle is not a defeat. It is, instead, a challenge, a call for Americans to respond with the best that is within them. There were no Republicans or Democrats in this country the day after Pearl Harbor. There were no isolationists or internationalists. And, above all, there were no defeatists of any stripe.

But he suggested that Sputnik is an even greater challenge than Pearl Harbor. "In my opinion we do not have as much time as we had after Pearl Harbor," he said.[16] But the subcommittee took the rest of November, December, and most of January to conduct hearings and take counsel on satellite and missile programs.

Distinguished scientists, administrators, and soldiers such as Dr. Edward Teller, "father" of the hydrogen bomb; Dr. Vannevar Bush, president of MIT; General James H. Doolittle, who led the first daring bombing raid over Japan and now presided over the National Advisory Committee for Aeronautics; General Maxwell Taylor, Army Chief of Staff; Dr. Wernher von Braun, Director of the Operations Division of the Army Ballistic Missile Program; Defense Secretary McElroy; dozens of corporate presidents such as Donald W. Douglas with Douglas Aircraft, Robert E. Gross with Lockheed, Roy T. Hurley with Curtis-Wright, Lawrence Hyland (Hughes Aircraft), E. Eugene Root (vice president of Lockheed), S.O. Perry (the chief engineer for Chance-Vought missile program); and flag officers from every service participated in the subcommittee hearings. While "the newspapers have been filled with columns about satellites and guided missiles," Johnson said, "nowhere is there a record that brings together in one place precisely what these things are and exactly what they mean to us."[17] That was the purpose and, to a considerable extent, the accomplishment of Lyndon B. Johnson's hearings. In this, Johnson made a significant contribution to the configuration of the American space program and, at the time unknowingly, to the creation of a space center in Houston, Texas, that would one day bear his name.

Johnson, a Democrat from a then almost overwhelmingly Democratic State, was born near Stonewall, Texas, and received a degree from Southwest Texas State Teachers College in 1933 after teaching at a small Mexican-American school in Cotulla, Texas, and teaching public speaking in the Houston schools. He served as a secretary to Representative Robert M. Kleberg (1932-35), and in 1937 won an election for a vacant seat in Congress caused by the death of the incumbent. In 1938, he was reelected and served four terms in the House before winning his Senate seat in 1948 and again in 1954. He had been a strong partisan of the New Deal and of Franklin Roosevelt and Harry Truman. His elevation to the post of Senate Democratic leader in 1953 and key committee assignments, not to mention his close personal and political relationship with Speaker Sam Rayburn of Texas, afforded Johnson unusual clout and visibility in the Senate. The subcommittee hearings, not wholly innocently it might be added, gave Johnson much greater national visibility. But the truth was that Lyndon Johnson, even in 1957, when it came to satellites and missiles and defense, literally, as he put it in his memoirs, "knew every mile of the road we had traveled."[18]

The subcommittee's first witness, Edward Teller, was born in Budapest, Hungary, in 1908 and educated in Germany, before coming to America in 1935 to serve as professor of physics at George Washington University. He moved to the University of Chicago in 1941, before joining the Los Alamos Scientific Laboratory team, and in 1952 moved to the University of California Radiation Laboratory. Teller attributed America's "missile-gap" to both specific and general situations. Specifically, he said, the United States did not concentrate on missile development because after the war it was not clear how such a missile could be used. More generally, the United States had not committed its money or its talent to the sciences, as had the Soviets. The Soviet achievements, he said, contrary to the popular notion that "their" German scientists are doing the job, must be attributed to the Soviet people and the Soviet scientists. And after considerable discussion and response to questions about national defense, security, and so forth, Teller raised the question: "Shall I tell you why I want to go to the Moon?" And after the laughter subsided he said, "I don't really know. I am just curious."[19]

Vannevar Bush, who received both the bachelor and master of science degrees from Tufts University in 1913 and a doctorate of engineering from Massachusetts Institute of Technology and Harvard University in 1916, was president of the Carnegie Institution before becoming chairman of the corporation of MIT. "Dr. Bush," Johnson addressed him, "for many years Americans have been in the habit of turning to you for good advice and good counsel. It has been a wise habit, and we members of this committee turn to you once again in time of crisis."[20]

In response to questions from Chief Counsel Edwin L. Weisl, Bush explained that the technical problems of the satellite and the ballistic missile are similar. To launch a satellite, very high velocity and effective guidance into orbit are required, and in the case of the intercontinental missile both are necessary, except that one must do "the second one very much better" in order to solve the reentry problem. He advised scattering Strategic Air Command units to make them less vulnerable, and suggested that there was nothing wrong with American scientists, engineers and production. The only problems with the missile and satellite programs, he believed, were organization, planning and past complacency. "We have had a rude awakening," he said, "and now must divest ourselves of our smugness and complacency and get to work." He urged the establishment of a central planning board acting as an advisor to the president and indicated that such an agency had been recommended by the Rockefeller Board in 1953, had received the approval of the Joint Chiefs of Staff, but then had never been implemented.[21]

General James Doolittle received a master and a doctorate of science degree from MIT, and now chaired both the Air Force Scientific Advisory Board and the National Advisory Committee for Aeronautics (NACA). He attributed the current crisis to the fact that the Soviet Union began working intensively on missile development in 1946, while the United States did not begin until 1953. He also said that Soviets worked harder. They had a double incentive system. One is rewarded for excellence—and destroyed if the job was not good, he said. He did not advocate that system. Moreover, he said, the Soviet Union had an "arms" economy and the United States a "butter" economy. About one-fourth of the Soviet Union's gross national product went into the military, while about one-twelfth of America's spending was for defense. And he suggested that the first order in catching up with the Soviet Union would be an overhaul of America's educational system. We need more classrooms and more and better science teachers. Doolittle said that in the Soviet Union the science professor earned roughly 50 times that of the day-laborer, while in the United States "in many cases they do not get as much." We "must give more kudos, more encouragement, more praise, more honor, if you will, to the science students." He believed that Sputnik was a good thing because it alerted Americans to the threat, and the real basis of the threat was Soviet excellence in science and technology.[22]

Undoubtedly one of the witnesses most knowledgeable of missile development was Wernher von Braun. Von Braun began his experiments with liquid fuel rockets in Germany in 1930 as a member of the German Society for Space Travel. It was there that he first encountered one of the three great pioneers in rocketry and space—Hermann J. Oberth.

Oberth was born in 1894 in what is now Hermanstadt, Rumania. When he was 12 years old, his mother gave him a copy of Jules Verne's *De la Terre a la Lune* (From the Earth to the Moon) first published in 1865. That book seems to have provided the common inspiration for the disparate pioneers of space: Robert H. Goddard of the United States, Konstantin E. Tsiolkovsky of Russia, and Oberth of Germany. Oberth designed a long-range liquid fuel rocket in 1917 and completed his doctorate in 1922 with a thesis which became a classic book on the subject of rocketry and space: *Die Rakete zu den Planetenraumen* (The Rocket into Interplanetary Space) published in 1923. The book discussed orbiting space stations, space food, space walks, and possible space missions. He later received a letter from a young German fan who complained that he could not understand Oberth's equations in the book. That young man was Wernher von Braun. Oberth joined the German Rocket Society in 1927, and in 1930 was in Berlin as an advisor for the production of a film entitled *Frau im Mond* (Woman in the Moon). The rocket he constructed for the production never got off the ground, and Oberth turned his talents to the more practical skills of a mechanic and locksmith. Many, many years later in 1955, Oberth joined Von Braun's rocket team at the Redstone Arsenal in Huntsville, Alabama, and so in a sense closed a historic loop that had begun almost 50 years earlier.[23] As early as 1919, Oberth had become aware that a counterpart in the United States was working with rocketry.

The American, Robert H. Goddard, born in 1882, received a doctorate from Clark University and taught physics, but lived and breathed rocketry. Goddard wrote America's first scientific paper on the subject, published by the Smithsonian Institute in 1922 and entitled "A Method of Reaching Extreme Altitudes." It became the subject of some derision in the American press, which labeled Goddard "the Moon rocket man." But Goddard, a technician and tinkerer as well as a theorist, launched the world's first liquid fueled rocket (oxygen and gasoline) from his aunt's homestead in Auburn, Massachusetts, on March 16, 1926. By 1940, Goddard had moved to a ranch in New Mexico and was building rockets 22 feet long, propelled by 250 pounds of liquid oxygen and gasoline and which developed a thrust of 825 pounds. But he worked independently, almost in secret, and without government or institutional support other than for private subsidies from Charles Lindbergh and grants from the Guggenheim fund. Although he died in 1945, long before Sputnik and the reality of space, he had no doubts that space was a part of humanity's future: "for 'aiming at the stars,' " he said, "both literally and figuratively, is a problem to occupy generations, so that no matter how much progress one makes, there is always the thrill of just beginning."[24]

Although recognized only long after his contributions to the theory of space travel, Konstantin E. Tsiolkovsky (1857-1935) was the first to develop the basic theory of rocketry. He prepared an article entitled "Exploration of Cosmic Space by Means of Reaction Devices" in 1898, which was published in 1903. But there seems to have been little application of his theories until much later, and Tsiolkovsky lived most of his life as a deaf and impoverished school teacher. Nevertheless, long after his death he provided inspiration to the Soviet rocket scientists who produced Sputnik.[25] In that moment, German, Russian, and American theory and history joined hands, and they did so perhaps with the metaphysicists and writers of the western world including the ancients who contemplated both their celestial universe and their gods who traversed both the Earth and the heavens, and those more modern dreamers from Leonardo da Vinci to Jules Verne through Edgar Rice Burroughs, Ray Bradbury and Isaac Asimov who made the scientific revolution and man in space a meaningful and popular human experience.

Von Braun's space odyssey began with the production of experimental missiles for the German army's Weapons Department in a program headed by Dr. Ing. H.C. Dornberger, in 1932, prior to Adolf Hitler's elevation to the chancellery. Germany's rejection of the Treaty of Versailles and the rearmament of Germany included the establishment of a permanent missile center at Peenemünde, where the V-2 was developed. It was successfully fired in October of 1942 and began military use in 1944. Finally, by this time some official interest in rocketry was developing in the United States.

A group of scientists at California Institute of Technology, headed by Hungarian-born Dr. Theodore von Karman and including Frank J. Malina, organized a Rocket Research Project in 1939 that focused on design fundamentals of high altitude rockets. In 1944, with military financial support, CalTech reorganized the project as the Jet Propulsion Laboratory which concentrated on jet-assisted aircraft take-off units (JATO). The laboratory also received authorization from Major General G.M. Barnes to proceed with a high altitude rocket project, known officially as Project ORDCIT.[26] As the war's end began to become a reality, military interest in the acquisition of German scientific knowledge, and particularly of V-1 and V-2 weaponry, grew and provided the incentive for what became "Operation Paperclip."

Major General H.J. Knerr, with the Strategic Air Forces, urged General Carl Spaatz to secure established German facilities and personnel before they could be destroyed or dispersed. In early 1945, he also urged Robert A. Lovett, the Assistant Secretary of War for Air, to push for the capture of German war technology, and to allow captured German scientists and their families to immigrate to the United States. Subsequently, on April 26, 1945, the Joint Chiefs of Staff issued an order directing General Dwight Eisenhower to "preserve from destruction and take under your control records, plans, documents, papers, files and scientific, industrial and other information and data belonging to . . . German organizations engaged in military research."[27]

Operation Paperclip, as it was called, became one of the unique finales in the defeat of Nazi Germany. Colonel H.N. Toftoy and Major James P. Hamill masterminded the rocket and missile segment of the project. Toftoy made early contact with a group of scientists, including Von Braun, who opted for capture by the Americans rather than the Russians. Von Braun told the Preparedness Subcommittee that as the Russian Army approached from the east, he and his associates took a vote and unanimously cast their lot with the west. They then somewhat perilously made their way out of Peenemünde and convinced the German navy that they had orders to evacuate with their equipment to a more central location. The group ended up in Bavaria where the American armies found them. During the confusion of Germany's collapse, Colonel Toftoy was unable to get a response from Washington to his request to transfer some 300 German rocket scientists and their families to the United States, and quickly flew to Washington to push his request through. There he secured permission to admit 127 scientists and technicians. The families were to be housed and cared for by United States authorities until they could be transferred at a later date.[28] Von Braun, who had been technical director of the Peenemünde Rocket Center, was one of those 127.

Hamill did more. The Nordhausen V-2 plant, which manufactured the German rockets, was designated to fall within the Soviet occupation zone, and all plans and equipment were to be left for the Soviets. "These orders," Hamill said, "originated at a very high level." But unofficially and off the record, "I was told to remove as much material as I could, without making it obvious we had looted the place." The net result of Operation Paperclip was to bring 300 boxcar loads of materials including plans, manuals, and documents and 100 V-2 rockets to the United States.[29] During his interrogation at Partenkirchen, Germany, in 1945, Von Braun closed with a comment about Moon travel and atomic energy (before the United States dropped its atomic bomb):

> When the art of rockets is developed further, it will be possible to go to other planets, first of all to the Moon. The scientific importance of such trips is obvious. In this connection, we see possibilities in the combination of the work done all over the world in connection with the harnessing of atomic energy together with the development of rockets, the consequence of which cannot yet be fully predicted.[30]

The first contingent of German scientists, including Von Braun and six of his associates, arrived at Fort Strong, Massachusetts, on September 20, 1945. They soon transferred to the Aberdeen Proving Ground in Maryland where they helped process the German guided missile documents. In December, 55 other German rocket specialists were given work at Fort Bliss, Texas, and White Sands Proving Grounds, New Mexico. Von Braun and the men at Aberdeen soon joined the rest at Fort Bliss, and eventually all of the rocket group moved there. Tests with V-2 rockets began in January 1946, and advanced to high altitude experimental tests using V-2 rockets for the Hermes II program. Improved designs and successes led to the search for improved facilities. In 1949, the decision was made to adapt the Huntsville (Alabama) Arsenal, which manufactured chemical mortar and howitzer shells during the war, and the Redstone Ordnance Plant located there, which produced the assembled shells, for the use of the missile team. The Army created the Ordnance Guided Missile Center there in April 1950, at which time Von Braun and about 130 of his associates arrived. The Army team created the Redstone, Jupiter and Juno missiles at the Redstone Arsenal—prior to the launch of Sputnik.[31] In 1951, Von Braun began work on the Army's Redstone missile under the direction of K.T. Keller (who later became president and chairman of the board of Chrysler Corporation). Initially planned for a 400- to 500-mile range, the Redstone soon was adapted to carry a heavier payload over an approximately 175-mile range. In 1955, the longer-range Jupiter rocket program began with the Ballistic Missile Agency under the command of Major General John B. Medaris. The project at first stressed the development of a land-based and sea-based 1500-mile range missile, and the Army Ballistic Missile Agency cooperated with the Navy until the Navy withdrew to develop its own submarine-launched Polaris missile. A single-stage, liquid fueled Jupiter intermediate range ballistic missile (IRBM) was fired on May 31, 1957.[32] Indeed, the Redstone-Jupiter-Juno program and the Polaris program comprised only two of the missile efforts that had been under way in the United States since the close of World War II.

Since 1949, the Naval Research Laboratory had been involved in high altitude rocket research for atmospheric and astrophysics research using liquid and solid rocket propellants in the Viking program. In 1955, the solid fueled Viking held the world altitude record for single-stage rockets. It was from a proposal of the Naval Research Laboratory, in cooperation with the Glenn L. Martin Company, that the launching of the International Geophysical Year satellite was selected by a special advisory board headed by Homer Stewart of the Jet Propulsion Laboratory. The decision, made in August 1955, as Walter McDougall pointed out in . . . *the Heavens and the Earth: A Political History of the Space Age*, stressed both the civilian and the scientific bent of the advisory board and of the project. The decision was supported by the Department of Defense and the administration despite the consensus that the Redstone rocket developed by the Von Braun team "promised a satellite soonest."[33]

Paralleling the missile developments by the Army and Navy, the National Advisory Committee for Aeronautics (NACA) in 1945 began designs for a ramjet-powered aircraft in cooperation (sometimes) with the Army Air Forces and variously Bell and Douglas Aircraft Corporations. Bell began work on the Bell XS-1, while Douglas, working on a proposal for the Navy, began developing the D-558 turbojet. By the mid-1950's, a contract had been awarded to North American Aviation for the X-15, and plans were developing for Project HYWARDS, a successor to the X-15 and a predecessor to the Dyna-Soar, which became a conceptual model for the space shuttle. NACA's Langley Aeronautical Laboratory, specifically the flight research section headed by Robert R. Gilruth, and later the Pilotless Aircraft Research Division (PARD) supported the design efforts of these experimental, rocket-powered aircraft.[34]

Congress founded NACA in 1915, "to supervise and direct the scientific study of the problems of flight with a view to their practical solution." The American Aeronautical Society, founded in 1911, urged the creation of a national aeronautics laboratory, somewhat similar to an earlier but now defunct Langley Aerodynamical Laboratory administered for a few years by the Smithsonian Institution. The proposal generated more controversy and competition than real support, until the outbreak of war in Europe in 1914, and the evident role of aircraft in modern warfare began to stimulate interest in "aeroplanes."[35] The role of aircraft in World War I, before the entry of the United States into that war, captured the attention of the American public much as Sputnik did in 1957. And, as in World War I, the response to the "crisis" was to create a civilian, rather than a military oriented, governmental advisory board. In 1915 the board was NACA. In 1958 the board was a reconstituted NACA, called the National Aeronautics and Space Administration (NASA).

The analogy extended even further. NACA's first chairman, Brigadier General George P. Scriven, explained in the Annual Report for 1915, that while military preparedness seemed to dictate present needs, "when the war is over, there will be found available classes of aircraft and trained personnel for their operation, which will rapidly force aeronautics into commercial fields, involving developments of which today we barely dream."[36] NACA urged and, in August 1916, secured congressional funding for a national civilian aeronautical laboratory. In July 1917, NACA broke ground for the construction of the Langley Aeronautical Laboratory (and a week later Congress approved a $640-million aviation bill).[37]

Just as a reconstituted NACA became the heart of NASA, so Langley's PARD, in a reconstituted form, as the Strategic Task Group, became the nucleus of NASA's man-in-space program. Ultimately, the Strategic Task Group, joined by engineers and specialists from the Canadian subsidiary of Britain's A.V. Roe Corporation, the military services (especially the Air Force) and private industry provided the human resources for the composition of the Manned Spacecraft Center or Lyndon B. Johnson Space Center in Houston, Texas. Among those associated with the Langley Aeronautical Laboratory were Robert R. Gilruth, who became the first director of the Manned Spacecraft Center; Maxime A. Faget, head of the Performance Aerodynamics Branch of PARD and Assistant Director for Research and Development at the Manned Spacecraft Center; and Walter C. Williams, a Langley engineer assigned to supervise flight tests of the Bell XS-1. On a 1947 test flight supervised by Williams, Air Force pilot Charles E. "Chuck" Yeager flew the first manned supersonic flight in history.[38] Williams and Yeager both became key members of the Manned Spacecraft Center team, as did Paul Purser from PARD.

Purser, who worked with Faget on the HYWARDS project and collaborated in the design of the "Little Joe" launch vehicle used in Project Mercury, was an original member of the Strategic Task Group assembled by Gilruth at Langley for the development of a man-in-space program. He served as special assistant to Gilruth during the formative years of the man-in-space program. Christopher C. Kraft, Jr., joined the Langley Laboratory in 1945 and was an original member of Gilruth's Space Task Group (STG), as was Charles W. Mathews who joined the Langley Laboratory in 1943 and had worked on the XS-1 transonic tests. Joseph G. Thibodaux began work at Langley in 1946 heading variously the Materials, Rocket and Model Propulsion Branches. Kenneth S. Kleinknecht joined Gilruth's group at Langley in 1959, after work on the X-15 at the Flight Research Center in California.[39] Thus, variously NACA, the Langley Aeronautical Laboratory, and specifically PARD housed to a considerable extent the people, projects, and aspirations for what would within a year of Sputnik become a defined and institutionalized man-in-space program.

Robert Rowe Gilruth, a 35-year-old aeronautical engineer from Nashwauk, Minnesota, began flight research work at Langley shortly after his graduation with a master of science degree from the University of Minnesota in 1936. In 1945, he organized a research group and conducted transonic and supersonic flight experiments with rocket-powered models, which led to the establishment of PARD. In 1952, Gilruth became Assistant Director of the Langley Aeronautical Laboratory, and in 1958 became director of a new STG organized as a result of the National Aeronautics and Space Act. The early STG, as Paul Purser recalled years later, was something of an ad hoc arrangement, without any official directives or titles established. By 1959, for example, the STG "had never received even as much as a piece of paper from Headquarters establishing the group, and . . . the closest thing to an official pronouncement was the memo that Gilruth himself had written. . . ." Gilruth himself had no official title.[40]

The reality in 1957 was that the United States had diverse and reasonably sophisticated space and missile programs with a relatively long history. The people were in place and had relatively long associations with each other. Moreover, by 1957, the conceptual framework, much of the design, and some of the hardware that would comprise the essential components of America's man-in-space efforts for the next several decades were in place. This analysis, however, was not imminently clear at the conclusion of the extensive hearings conducted by the Preparedness Subcommittee. Other than for Wernher von Braun, relatively few "hands-on" engineers associated with missile or rocket plane development appeared before Congress, although to be sure there were a large number of generals, admirals, and corporate presidents associated with such developments. The hearings were conducted at a much "higher" level and, to an extent, were much more political than technical as might be expected.

The fact that the hearings were political rather than technical, and that the media and the public were truly shocked by that tiny spinning Soviet globe in the sky, led to the institutionalization of an American space program. Senator Johnson released public comments about the hearings from time to time and summarized the work (2313 pages) of the Preparedness Subcommittee after its 3 months of hearings closed. Early in the course of the hearings, one of Johnson's aides commented in a memorandum to Senator Johnson on November 26, 1957, that one clear pattern that emerged from the testimony was the extreme difficulty in pinning down lines of authority for missile and satellite programs. On December 16, Johnson issued a press release saying that "it is apparent that we have the technical skill, the resources and the necessary enthusiasm among our technicians to build any missile that we need and to build it on time. What we have been lacking are hard, firm decisions at high levels."[41] What Johnson was saying was that Sputnik created a chink in Republican political armor and now offered an opportunity for Democratic party leaders.

In December, the Naval Research Laboratory attempted to launch a Vanguard rocket carrying a satellite, but an explosion on the launch pad in front of the press proved only embarrassing. (A second attempt in February 1958 did no better.) The nod then went to Von Braun to launch a Jupiter C (Juno I) carrying a satellite. The successful launch on January 31, 1958, of the Explorer I satellite (weighing 81 pounds) to a maximum altitude of 984 miles considerably bolstered American spirits, but even more significantly the scientific experiments on the Explorer discovered the Van Allen radiation belts surrounding the Earth's atmosphere.[42]

With national confidence bolstered, Congress began moving toward decisions about missiles and space. The Senate approved the creation of a Special Committee on Space and Astronautics (S.R. 256) on February 6, and Senator Carl Hayden as president pro tempore of the Senate called a meeting into session on February 20, where Lyndon Johnson was quickly elected chairman. The committee considered briefly the feasibility of establishing a joint committee with the House but no action was taken. And Johnson outlined what he thought was the primary business of the committee, that being to define who in the executive and legislative branches should have jurisdiction over specific aspects of space and astronautics, how these organizations should be established, and how to deal with the international aspects of space. The committee briefly considered Senator Clinton P. Anderson's memorandum urging a decision on U.S. space objectives as variously a stunt, having to do with military preparedness, or relating to the peacetime uses of space. And he listed options as being to (a) hit the Moon, (b) put a man into space, (c) put an animal into space, or (d) conceivably start thinking about a Mars mission, manned or otherwise.[43] Judging by Senator Anderson's memorandum and the daily press stories relating to space, the country's mood was both feisty and impatient.

It would take a firm hand at the tiller to keep a reasoned course and avoid the pitfalls of unduly hasty decisions. There were a number of such hands, but in retrospect, Lyndon Johnson knew intuitively that space was not simply something "out there," but something intimately associated with the quality of life on Earth. He believed space was the first new physical frontier to be opened since the American West. The Preparedness Subcommittee Hearings continued, and the Special Committee on Space and Astronautics began hearings and independent study. The House of Representatives created a Select Committee on Astronautics and Space Exploration on March 5, under the leadership of John W. McCormack, and began hearings and staff studies. Both the work of the Preparedness Subcommittee and the simple creation as well as the work of the House and Senate Select Committees, "emphasized the importance of a national space program and an agency—preferably independent and civilian—to administer it." Moreover, Johnson's initiative on the Preparedness Subcommittee helped ensure that the decisions relating to space and missile development would occur "in a broad political arena."[44]

Perhaps the strongest incentives and direction leading to the establishment of a national space program under civilian authority came directly from President Eisenhower. Two advisory bodies made similar recommendations to the President. Nelson Rockefeller, who chaired the Rockefeller Brothers Fund which was completing a study of national security, testified before the Preparedness Subcommittee in January that the question as to where the authority for the development of outer space should be housed should be decided by the Secretary of Defense; but by March 5, Lyndon Johnson recalled that: "He changed his mind and recommended to President Eisenhower the establishment of a civilian space agency. The President endorsed his recommendation." Johnson said that in the beginning he had "no firm conviction either way" but by the time the hearings were over, he had been persuaded that the "best hope for peaceful development of outer space rested with a civilian agency."[45] On March 26, President Eisenhower released a document from his Science Advisory Committee entitled "Introduction to Outer Space, An Explanatory Statement. . . ." under his introductory statement which read:

> This is not science fiction. This is a sober, realistic presentation prepared by leading scientists. . . . I have found this statement so informative and interesting that I wish to share it with all the people of America and indeed with all the people of the Earth. . . .These opportunities reinforce my conviction that we and other nations have a great responsibility to promote the peaceful use of space and to utilize the new knowledge obtainable from space science and technology for the benefit of all mankind.
>
> Dwight D. Eisenhower[46]

The Advisory Committee explained that four factors gave "importance, urgency, and inevitability" to the advancement of space technology. Those were "the compelling urge of man to explore and discover" and the necessities of defense, national prestige, and scientific observation and experiment. In very simple language, the report briefly discussed satellites, a manned and unmanned Moon landing, an instrument landing on Mars, a satellite radio network, military applications of space (primarily communications and reconnaissance, specifically rejecting satellites as bomb carriers), costs versus benefits, and finally, a space timetable that concluded with "Human Lunar Exploration and Return" and "much later still" Human Planetary Exploration.[47] To a remarkable extent, the report provided a blueprint for the American space mission over the next several decades.

On April 2, Eisenhower presented a special message and legislation to Congress recommending the creation of the National Aeronautics and Space Agency:

> The new Agency will be based on the present National Advisory Committee for Aeronautics and will continue that agency's well-established programs of aeronautical research. In addition, the new Agency will be responsible for programs concerned with problems of civil spaceflight, space science and space technology.
>
> . . . it is appropriate that a civilian agency of the Government take the lead in those activities related to space which extend beyond the responsibilities customarily considered to be those of a military organization.[48]

The President then instructed the NACA to present full explanations of the proposed legislation to both houses of Congress, and to plan for reorganization as may be required by the legislation. NACA and the Department of Defense were to review programs to decide under which agency they should be placed and what the Department of Defense would need in the future to maintain its military requirements; and NACA was to ensure the participation of the scientific community through discussions with the National Science Foundation and the National Academy of Sciences.[49]

Congress began hearings and study of the proposed legislation immediately, and on July 29 received the President's endorsement of the "National Aeronautics and Space Act of 1958." The act declared that "it is the policy of the United States that activities in space should be devoted to peaceful purposes for the benefit of all mankind." Objectives of American space efforts were to expand human knowledge, to improve the efficiency of aeronautical and space vehicles, and to develop vehicles capable of carrying instruments, equipment and supplies, and living organisms through space. Congress authorized the creation of the National Aeronautics and Space Council (including the Vice President, Secretary of State, Secretary of Defense, the Administrator of NASA, and the Chairman of the Atomic Energy Commission) and NASA which would assume all of the responsibilities, properties, and authority of NACA.[50]

During the deliberations of the proposed legislation, Dr. Hugh L. Dryden, Director of NACA, explained to the House Select Committee that NACA "formally initiated studies of the problems associated with unmanned and manned flight at altitudes from 50 miles up, and at speeds from Mach 10 to the velocity of escape from the Earth's gravity," in 1952. The primary mission of NACA, he stressed, was scientific research for all departments of the government. "In this technological age," he said, "the country that advances most rapidly in science will have the greatest influence on the emotions and imagination of man," and will enjoy the most rapid growth, the highest standard of living, the greatest military potential, and the "respect of the world." There were, in April 1958, 17 unpaid members of NACA appointed by the President who reported directly to the President. The committee established policy and planned research programs conducted by the 8000 scientists, engineers, and supporting personnel who comprised the staff of the agency. NACA's research centers at the time included Langley Aeronautical Laboratory and its associated Pilotless Aircraft Research Station on Wallops Island (with a combined staff of about 3300); Ames Aeronautical Laboratory, California, staffed with 1450 persons; Lewis Flight Propulsion Laboratory, Ohio, with a staff of some 2690; and the High Speed Flight Station at Edwards, California, with a staff of 312.[51]

After the conclusion of the House Select Committee hearings in May, Dryden brought Robert R. Gilruth to Washington to plan a man-in-space program. "There," according to James R. Hansen, "working less than 90 days in one large room on the sixth floor of the NACA building, a small task group of less than 10 men, assembled by Gilruth over the telephone from the staffs of Langley and Lewis laboratories, came up with all of the basic principles of what would become Project Mercury." The plan closely paralleled proposals made by Gilruth's associate, Maxime A. Faget, at a NACA conference on high-speed aerodynamics in March.[52] Thus, before the passage of the act creating NASA, or what became the Johnson Space Center, the United States had a plan and a project group directed toward putting a man in space. It may have been that the creation of the plan by Dryden was consciously or subconsciously directed toward the goal both of preserving a NACA hegemony over space-related activities while at the same time attempting to preserve the essential scientific integrity of NACA programs. It was clear, however, that the new governmental agency for aeronautics and space would be much more operations oriented than had been NACA.

The National Aeronautics and Space Act of 1958 attempted to harness the energies, talents and aspirations of a nation in a bold and exciting new enterprise. The act reflected a remarkable unanimity and commitment by the American people that had perhaps been unmatched in times of peace since the days of Theodore Roosevelt and the construction of the Panama Canal. To be sure, in the minds of many, despite the language of the act, this was not an act of peace but of war, albeit a cold war. Certainly Sputnik was instrumental in the inception and the speedy approval of the Space Act of 1958. America, to be sure, was well on the way to space before Sputnik, and would have been there with or without Soviet competition, but it is most unlikely that the United States would have made the level of commitment to space, in terms of talent, money, organization or popular support, without Sputnik. That extended far beyond space for the United States, and indeed most of the world's peoples began to emerge from Sputnik with a new sense of identity and purpose. Humans were no longer earthbound.

October 1957 was one of those milliseconds in the human experience that marked the beginning of a "giant leap" for all mankind, a leap that might properly be equated to such other moments in history as the discovery of fire, agriculture, the New World, flight, and atomic energy . . . and a leap, to be sure, that is a perilous, difficult, and uneasy one. October 1957 and that October of a year later when NASA officially began functioning were also fundamental to the inception and organization of the Lyndon B. Johnson Space Center in Houston, Texas.

Administrator Thomas Keith Glennan announced that NASA would officially begin functioning on October 1, 1958. On November 3, Robert R. Gilruth, Assistant Director of the Langley Aeronautical Laboratory, announced the formation of a Space Task Group, including himself and 34 other Langley employees. Over the next 3 years this group, which worked together in a seemingly unstructured and almost formless fashion, grew and expanded and developed personal and professional relationships such that when the decision was made to create a NASA "Manned Spacecraft Center," the organization, the experienced personnel and, to a considerable extent, the programs were already in place at Langley and within the NASA community. Thus, October 1957 and October a year later when NASA officially began functioning were critical moments in the inception and organization of what became, after his death, the Lyndon B. Johnson Space Center in Houston, Texas. Johnson helped write and enact the legislation which created NASA. He knew, indeed, every mile of the road America has traveled to space, and he knew intuitively that space was not simply something "out there," but something intimately associated with the quality of life on Earth.[53]

CHAPTER 2: The Commitment to Space

"I can recall watching the sunlight reflect off of Sputnik as it passed over my home on the Chesapeake Bay in Virginia," Dr. Robert R. Gilruth recalled to the audience at the Sixth International History of Astronautics Symposium meeting in Vienna, Austria, in 1972. "It put a new sense of value and urgency on the things we had been doing. When one month later the dog, Laika, was placed in orbit in Sputnik II, I was sure that the Russians were planning for man-in-space."[1] The American response grew from an unusual concatenation of events—a Russian satellite and a dog in orbit, a NACA Pilotless Aircraft Research program, the presence of a large assemblage of German rocket scientists in Huntsville, Alabama, and the sudden unemployment of a Canadian fighter production team. Congress, with NACA/NASA assistance, provided leadership in devising the manned space programs and set the stage for the bold scheme to land an American on the Moon.

In the summer of 1958, as Congress deliberated space legislation, Dr. Hugh Dryden, NACA's Director, called Gilruth and Abe Silverstein, the director of the Lewis Research Center, to Washington to begin formulating a spaceflight program. Silverstein and Gilruth shuttled back and forth from their home offices, usually spending four or five days a week in Washington. For several months, Silverstein noted later, Gilruth's interests had quickly moved in the direction of "manned spaceflight."[2]

Gilruth assembled a small group of associates and advisors, including Max Faget, Paul Purser, Charles W. Mathews, and Charles H. Zimmerman of the Langley Laboratory; Andre Meyer, Scott Simpkinson, and Merritt Preston of the Lewis Laboratory; and many others on an "as needed" basis. He brought in George Low and Warren North from Lewis and Charles Donlan from Langley to help polish the plan in the late summer. The product of these intensive sessions was much more than an organizational format for a work project; it was an engineering design for putting an American in space. As Gilruth said, "we came up with all of the basic principles of Project Mercury," including a pressurized capsule with a blunt face and a conically shaped afterbody containing a contour-shaped couch, to be launched variously by an Atlas or a Redstone, and including a special cluster design proposed by Paul Purser and Max Faget, to be called the "Little Joe," to test an emergency escape device and a water-landing parachute system.[3]

Congress, meanwhile, was deliberating the Eisenhower administration's legislation, introduced by Lyndon B. Johnson and Senator Styles Bridges, calling for the creation of NASA. Hearings were being conducted before the Senate Select Committee on Space and Astronautics, chaired by Johnson, and the House Select Committee on Aeronautics and Space Exploration, chaired by Congressman John W. McCormack.

In July 1958 before final approval of the NASA legislation, Gilruth, with Silverstein and Dryden, presented the concept for manned spaceflight to Dr. James R. Killian (Scientific Advisor to the President) and the President's Scientific Advisory Board. Gilruth and Dryden subsequently appeared before the House Select Committee on Aeronautics and Space Exploration, which began hearings on August 1, and explained the manned spaceflight initiative. Concurrent with the approval of the National Aeronautics and Space Act of 1958, the House created a standing committee on science and astronautics on July 21, headed by Congressman Overton Brooks of Louisiana. Subcommittees included a committee on Scientific Training and Facilities headed by George P. Miller of California, a Subcommittee on Scientific Research and Development headed by Olin E. Teague of Texas, a Subcommittee on International Cooperation chaired by Victor L. Anfuso of New York, and a Subcommittee on Space Problems and Life Sciences under Congressman B.F. Sisk of California. President Dwight D. Eisenhower signed the National Aeronautics and Space Act on July 29. Although the act referred to "manned and unmanned" space vehicles, it by no means specified that the American or NASA "activities in space" necessarily involved placing men or women in space. Not all were convinced (nor would be as the years passed) that a space program and putting humans into space were necessarily synonymous. Nevertheless, in those first weeks following approval of the act, Silverstein and Gilruth urged Dryden to create a special task group to implement a *manned* spaceflight program. [4]

That the American response to Sputnik should literally be to put an "American in space" did not reflect prevailing public opinion or the conventional wisdom of the aeronautical, scientific or military communities. Even among NACA/NASA personnel, many, including senior people, believed that the projected manned spaceflight program was an overreaction at best, a stunt at worst, and necessarily temporary in either event. The "conventional wisdom" was more closely aligned to the idea that manned spaceflight was very premature and could develop only after the technology evolved from unmanned spacecraft. Moreover, many Americans still possessed some innate disaffection for things mechanical, or robotic, that had to do with the further intrusion of machines in the "garden" of American life or, more so, into the "heavens." Flight in any dimension was something some Americans had had difficulty with since the days of the Wright brothers. Despite their reservations and skepticism, Americans had an equally strong, but ambivalent fascination with the "machine." Space vehicles, if such were to be, clearly needed the benign control of the human hand. Although totally unrelated to the in-house NACA/NASA deliberations, a feature article by a prominent political leader in a prominent engineering journal reinforced the arguments in support of manned space vehicles.

In Congress, Senator Lyndon Johnson had become an advocate of a "broader understanding" of the new Space Age. The August edition of the *American Engineer*

Lyndon Johnson knew intuitively that space was not simply something "out there," but something intimately associated with the quality of life on Earth. He believed space was the first new physical frontier to be opened since the American West.

featured an article by Lyndon Johnson, who stressed that America was "badly underestimating the Space Age." Although security had been our first concern, and properly so, Johnson suggested that the overwhelming focus on satellites and missiles missed the point. "The ultimate [purpose] of space vehicles is the transport of man through outer space near or to the Moon, some of the planets, perhaps even to other galaxies. . . . Whatever the date, manned space vehicles will be—when they come—far less of a detail, far more a pinnacle of accomplishment than we now think." The Space Age, Johnson said, will have an impact of the greatest force on how we live and work. "We are underestimating the meaning of this whole new dimension of human experience." We have entered a new frontier, he said, the first new physical frontier to be opened since the American West.[5] Affairs now moved very quickly.

President Dwight D. Eisenhower appointed Dr. T. Keith Glennan as the first Administrator of NASA, and Dr. Hugh L. Dryden, who had headed NACA, to be Deputy Administrator. They assumed their posts on August 19. Glennan, born in Enderlin, North Dakota, in 1905, earned a degree in electrical engineering from Yale University in 1927. His first employment was in the new "sound" movie industry, before joining Electrical Products Research Company, a subsidiary of Western Electric. He became involved primarily in administration rather than research, at times heading divisions of Paramount Pictures, Metro-Goldwyn-Mayer, and Vega Airlines. During World War II, Glennan joined the Columbia University Division of War Research and soon became director of the Navy's Underwater Sound Laboratories at New London, Connecticut. He became president of Case Institute of Technology in 1947 and elevated it into the ranks of the top engineering schools in the Nation. He served as a member of the Atomic Energy Commission between 1950 and 1952. The Space Act declared that "NACA shall cease to exist . . . ," and Glennan announced its close on September 30 and the beginning of NASA on October 1. It is a time of "metamorphosis," he said, ". . . it is an indication of the changes that will occur as we develop our capacity to handle the bigger job that is ahead . . . We have one of the most challenging assignments that has ever been given to modern man."[6]

A few days after NASA became operational, Max Faget, Warren North, Dr. S.A. Batdorf, and Paul Purser went to Huntsville and spent an intensive 2 days discussing with Wernher von Braun and some 30 other engineers and military officers the participation of the Army Ballistic Missile Agency (ABMA) and Redstone in the launch of a manned capsule. On October 7, Glennan, Dryden, and Roy Johnson, Director of the Army's Advanced Research Projects Agency (ARPA), heard Gilruth's final proposal for manned spaceflight that had been approved by a joint NASA/ARPA committee, and which essentially reflected the summer work of Gilruth's task group. "Within two hours," Gilruth said, "we had approval of the plan and a 'go ahead.'" Glennan advised Gilruth to return to Langley and organize a group to manage the project—but to report directly back to Abe Silverstein in the Washington NASA office, rather than to the center director.[7] Not only had a manned spaceflight program been authorized, but the program was to be autonomous and independent of any other NASA center, thus effectually creating the organizational nucleus of what would become the Manned Spacecraft Center or (in 1973) Lyndon B. Johnson Space Center in Houston, Texas. For all practical purposes, the Manned Spacecraft Center existed and operated at the Langley Aeronautical Laboratory for almost 4 years prior to its relocation in Texas. In truth, it may have been that one of the motives for the organization of an autonomous entity to deal with manned spaceflight was to preserve the integrity of the traditional research orientation of the NACA/NASA organization, and possibly even to isolate the project because it was premature or a stunt from the perspective of the mainstream (and presumably more serious) research and scientific efforts of NASA. It could also have been a simple matter of expediency. The establishment of the STG gave the program identity and some protection from agency politics and funding squabbles.

As Glennan explained to the House Committee on Science and Aeronautics in 1959, "To get going, we have had to organize with one hand, while, at the same time, . . .operate with the other." It is not an efficient way to do business, he said, but there was never time to proceed in an orderly fashion. Wesley Hjornevik, who joined the STG as its business and administrative manager, recalled that at what may have been the true moment of inception of the STG, in a meeting with Glennan following the presentation to President Dwight D. Eisenhower and his staff by Gilruth's group, the reality of the manned vehicle project struck. The meeting closed with Glennan's comment, "okay men, let's get on with it." Whereupon Gilruth's mouth "fell open;" he made inquiries about staff, money and facilities. "Glennan," Hjornevik said, "just got red in the face." He had no answers to those questions. He got mad, pounded on the table and repeated, "I said get on with it," and got up and walked out.[8] In a sense, both Hjornevik and Glennan had identified the most prominent and distinctive features of the early manned space effort—its relative spontaneity and organizationally amorphous qualities.

Although the STG was unofficially established on October 8, 1958, it was, as Paul Purser noted later, an ad hoc arrangement, for Gilruth had no written authorization to head the STG or to actively organize and recruit. Gilruth acknowledged that he had been given "a job of tremendous difficulty and responsibility," with no staff and only oral orders to "get on with the job." He credited Floyd Thompson, director of the Langley Center, with not only cooperation but also guidance in establishing the manned space program. And given the fact that Gilruth would be dismembering the Langley Center staff, that was no easy commitment by Thompson. Finally, Gilruth dissipated some of the cloud surrounding the establishment of the STG by announcing in a memorandum dated November 3, 1958, (as suggested by Thompson) that the STG did indeed exist, and that he had the authority to request the transfer of personnel to his group.[9]

That memorandum effectually marked the inception of the Manned Spacecraft Center. The document is significant both for the manner of its promulgation and the fact that it named those who became the "charter members" of the manned spacecraft program. "Recruiting" for the STG began with meetings between Purser, Charles Zimmerman and R.O. House, who agreed to recommend to Floyd Thompson that a proposal be forwarded to NASA Headquarters to create 230 positions on the "space payroll." Of these, 110 were to be directly related to the manned-satellite project, 60 to support groups, and 60 for other space-related projects. Thompson agreed to fund 119 of the positions through Langley, with 36 transfers to be effected immediately. Paul Purser roughed out a "Task Group" memo containing 34 names to which Gilruth added 2.[10]

The next day, November 4, Floyd Thompson scratched a brief approval on the memo saying "This request is okay with the exception of (William J.) Boyer" (whom he wished to retain on his staff).[11]

NASA - Langley
November 3, 1958

MEMORANDUM for Associate Director
Subject: Space Task Group

1. The Administrator of NASA has directed me to organize a space task group to implement a manned satellite project. This group will be located at the Langley Research Center, but in accordance with the instructions of the Administrator, will report directly to NASA Headquarters. In order that this project proceed with the utmost speed, it is proposed to form this Space Task Group around a nucleus of key Langley personnel, many of whom have already worked on this project.

2. It is requested, therefore, that initially the following 36 Langley personnel be transferred to the Space Task Group:

Anderson, Melvin S. (Structures)
Bland, William M., Jr. (PARD)
Bond, Aleck C. (PARD)
Boyer, William J. (IRD)
Chilton, Robert G. (FRD)
Donlan, Charles J. (OAD)
Faget, Maxime A. (PARD)
Fields, Edison M. (PARD)
Gilruth, Robert R. (OAD)
Hammack, Jerome B. (FRD)
Hatley, Shirley (Steno)
Heberlig, Jack C. (PARD)
Hicks, Claiborne R., Jr. (PARD)
Kehlet, Alan B. (PARD)
Kolenkiewicz, Ronald (PARD)
Kraft, Christopher C., Jr. (FRD)
Lauten, William T., Jr. (DLD)
Lee, John B. (PARD)
Livesay, Norma L. (Files)
Lowe, Nancy (Steno)
MacDougall, George F., Jr. (Stability)
Magin, Betsy F. (PARD)
Mathews, Charles W. (FRD)
Mayer, John P. (FRD)
Muhly, William C. (Planning)
Purser, Paul E. (PARD)
Patterson, Herbert G. (PARD)
Ricker, Harry H., Jr. (IRD)
Robert, Frank C. (PARD)
Rollins, Joseph (Files)
Sartor, Ronelda F. (Fiscal)
Stearn, Jacquelyn B. (Steno)
Taylor, Paul D. (FSRD)
Watkins, Julia R. (PARD)
Watkins, Shirley (Files)
Zimmerman, Charles M. (Stability)

(signature)
Robert R. Gilruth
Project Manager

While Gilruth organized his STG at Langley, Abe Silverstein established an office called Manned Space Flight at NASA Headquarters in Washington with George Low as its head. Silverstein, trained as a mechanical engineer, was a veteran flight researcher who joined NACA in 1929. In 1943 George Lewis, who headed the Aircraft Engine Research Laboratory in Cleveland (renamed the Lewis Flight Propulsion Laboratory in 1948), named him to a special committee to coordinate NACA's high-speed aircraft research. Low, who worked with Silverstein in Cleveland and assisted Gilruth's ad hoc committee in planning a spaceflight program, returned with Gilruth to Langley to serve as deputy assistant to Max Faget but was on the job for only a few weeks when Silverstein called him back to

Washington. Low, born in Vienna, Austria, in 1926, left Germany in 1938 and immigrated with his family to the United States. He received the bachelor of aeronautical engineering degree from Rensselaer Polytechnic Institute in 1948, briefly worked for General Dynamics, returned to Rensselaer for a master's degree, and joined NACA as a research scientist at the Lewis Flight Propulsion Laboratory in 1949. He had worked closely with Gilruth in putting together the final plans for Project Mercury in the summer of 1958, and now in Washington with Silverstein, Low considered himself "Bob Gilruth's representative in Washington." He worked very closely with the STG and later the Manned Spacecraft Center until he rejoined Gilruth in Houston in 1964.[12]

Silverstein and Low quickly discovered that while Gilruth's group "had good technical strength," it lacked the personnel and expertise to manage the budgeting, finance, and general administration for a manned satellite program. Low and Silverstein effectually became the personnel and fiscal administrators for the STG, while Gilruth focused on technical management. Low explained later that the STG:

> . . . was a highly technical organization which initially showed little interest in the business management aspects. Personnel management, financial management, etc., were handled on an ad hoc basis. The people were interested in the technical job and had little time for any more than that.[13]

This proved to be both a blessing and a curse. On the one hand, the "manned satellite program," as it was called for a time, was ill-prepared for the rapid physical growth it experienced; and on the other hand, the fluidity of the organization enabled it to do things, as Gilruth observed, that "could only occur in a young organization that had not yet solidified all of its functions and prerogatives."[14] Nevertheless, an administrative crisis would continue to plague the manned spacecraft program through most of its early years. Efforts to deal with the problem led first to an attempt to organize the manned spacecraft program within the administrative structure of Goddard Space Flight Center, being built near Beltsville, Maryland, and finally, to the creation of an autonomous NASA spacecraft center.

Gilruth and his associates plunged ahead with fresh intensity. Silverstein and Low met with Gilruth at Langley weekly; and Gilruth, Paul Purser, or another of the task group went to NASA Headquarters or to another center as often. Ten new members were transferred to the STG from the Lewis Center, including Low, Andre Meyer, Scott Simpkinson, Merritt Preston and Warren North, among others. During the first months of their existence, the group perfected the design and technical specifications for the manned satellite, arranged for launch support with the Air Force's Ballistic Missile Division at Cape Canaveral, worked out test procedures for the capsule and the Redstone rocket, gave intensive attention to the use of Thor versus Jupiter rockets for intermediate-range flights, and resolved many problems relating to trajectory, guidance, astronaut selection and training, recovery, and costs.[15]

The capsule or man-carrying satellite was to have a pressurized breathing atmosphere within a blunt face and conically shaped afterbody. Gilruth attributed the first working design for the capsule to Caldwell (pronounced Cadwell) C. Johnson of the Langley and Wallops Island design group, working closely with others in the STG. Max Faget and Andre Meyer, he said, conceived of the "escape tower" and Faget contributed the contour couch which would protect the occupant from the high g-forces of launch and reentry. The capsule would be launched by an Air Force (Ballistic Missile Division) Atlas rocket, with the Army's Redstone rocket, under development by Von Braun's group in Huntsville, Alabama, used for early test flights. On reentry it would descend by parachute to a water impact. Because it would be America's first manned messenger "to the gods," Abe Silverstein thought the project should be called "Mercury." It was an excellent choice, Gilruth thought, and one that generated great pride. Director Glennan publicly announced the Mercury project on December 17, 1958.[16]

The STG's new project orientation improved both the technical focus of the engineers and the organizational lines of the group. Gilruth, as Director of the STG (and director of Project Mercury), placed Charles Donlan immediately under him as the Associate Director. Upon his graduation from Massachusetts Institute of Technology in 1938, Donlan joined the Langley Aeronautical Laboratory and began work on aircraft spin design criteria. During the war he worked on tests of the Air Force's XS-1 design and became the project engineer for the design and construction of Langley's high-speed (7- by 10-foot) wind tunnel, and subsequently headed the high-speed wind tunnel section. A flight systems division headed by Max Faget, an operations division under Charles Mathews, and a reliability and quality assurance group reported to Gilruth through Donlan. Paul Purser was Special Assistant to Gilruth.[17]

In practice, the association between division heads and the directors—and the staff, wherever they might be—was very informal and collegial. For the most part these were professional engineers who had worked together on various projects in the past and now were joined together to work on another far more exciting and demanding project. Each assumed the tasks they were best suited to perform and critiqued and assisted the others work. And work they did!

They worked holidays, evenings, and weekends. They worked New Year's Day. Gilruth recalled the days of the STG's first year as a time of "the most intensive and dedicated work of a group of people" that he had ever experienced. "None of us," he said, "will ever forget it."[18]

During their first weeks on the job, the STG completed the specifications for the Mercury capsule and placed it, through Langley's procurement officer Sherwood Butler, in the hands of potential contractors who were to return their proposal within approximately 90 days. NASA awarded McDonnell Aircraft Corporation the contract for the construction of the Mercury capsule on January 9, 1959.[19] Thus, the STG early established itself as the design and management team for manned spacecraft programs.

Originally, the manned spacecraft program anticipated considerable in-house design, production, and operations. Gilruth's group, for example, arranged for launch rockets and services through the Air Force and Army Ballistic Missile Agency, and also began work on its own Little Joe rocket to be used for escape system tests at Wallops Island. A group under Scott Simpkinson at the Lewis Laboratory in Cleveland, in cooperation with a small task group under Jack Kinzler at Langley, constructed full-scale Mercury capsule models (called "Big Joe") to be launched aboard Atlas boosters from Cape Canaveral for heat transfer and stability tests.[20] The STG achieved a successful launching of a Mercury prototype vehicle in September 1959, within less than a year of the creation of NASA and the STG.

Gilruth arranged to borrow physicians, flight surgeons, and psychologists from the Army and Navy to advise on the selection of spacecraft crew members. Dryden and Gilruth, in fact, discussed naming such crew members variously "astronauts" or "cosmonauts." Dryden favored the term "cosmonaut," inasmuch as the flights would be made in the cosmos or near space, while the term "astro" or "astral" suggested star flights. "Astronaut," however, became accepted simply by virtue of common usage and preference by team members, and it stuck. The STG medical advisors and psychologists urged the selection of astronauts from the more dangerous professions, such as race car drivers, mountain climbers, scuba divers, or test pilots. Whether it was judiciously, fortuitously, or both, it was President Eisenhower who decided that astronauts should be selected from a pool of military test pilots. And they all breathed a sigh of relief, Gilruth recalled, because it "allowed the delegation of flight control and command functions to the pilot of the satellite."[21]

The new year, 1959, dawned with still only a small group assigned to manned spacecraft projects. The original 35 in the STG had been joined by 10 engineers from Lewis, and another 12 Langley personnel had been shifted to STG projects. Other individuals had been recruited from the Army and the Air Force, but staffing quickly became a serious problem. Floyd Thompson, who cooperated fully with Gilruth's constant requests for personnel from the ranks of Langley staff, finally slowed Gilruth's "raids," which left his own staff so terribly imbalanced, by telling him: "Bob, I don't mind letting you have as many good people from Langley as you need, but from now on I am going to insist that for each man you want to take, you must also take one that I want you to take."[22] The problem with staffing was compounded by the reality that the United States had only a limited supply of aerospace engineers, fewer still with the credentials that would be useful to the STG. Moreover, the postwar aerospace market was a terribly competitive one such that the government had the greatest difficulty competing in the marketplace. This market situation contributed in the long run to greater and greater dependency on contractors for goods and services, but NASA Administrator James Webb believed that greater reliance on private contractors would help build a stronger constituency for NASA programs. Moreover, President Eisenhower abhorred the creation of large federal establishments, particularly those that might compete with private enterprise. But an unusual and highly fortuitous circumstance enabled Gilruth to obtain a new cadre of aerospace engineers which greatly alleviated his recruiting problems and proved extremely important to the American space program over the next several decades.

On February 20, 1959, AVRO Aircraft, Ltd. of Canada, a subsidiary of Britain's A.V. Roe Corporation, closed its doors and terminated about 13,000 employees in response to a decision by the Government of Canada to scrap its plans to build an air defense force centered on the Arrow (CF105) fighter, then reputed to be one of the best designed high-performance aircraft on the drawing board. The AVRO CF100 was in production, and a jet liner, similar to a Learjet, was ready for production. Development of a "state-of-the-art" fighter, however, proved perhaps overly ambitious for Canada and terribly costly and the then highly touted American Bomarc defense system seemingly reduced the necessity for fighters. The result was simply a decision by Prime Minister John Deifenbaker's government to suspend the program. Company officials, hoping to demonstrate the economic impact of such a decision, elected to dramatize their plight by terminating all employees at once.[23] The government, however, was unmoved.

A huge pool of highly qualified aerospace engineers suddenly became available. Among these, for example, were Jim Chamberlin, R. Bryan Erb, Rodney Rose, and others. Erb, who was born in Calgary, was first led to his interests in space by an explorer who visited his fifth grade elementary class, and predicted that one day man would fly to the Moon. That, Erb recalled, caught his attention. He later received a C.E. degree in fluid dynamics at the University of Alberta, and then a master's at the College of Aeronautics in Cranfield, England. At Cranfield, Erb's interest in space was reinvigorated by the visit of science fiction author Arthur C. Clarke, and by the intense interest of members of the British Interplanetary Society. He joined AVRO Aircraft Ltd. in Toronto, for work in thermodynamics in 1955, only to receive a notice one morning that as of the end of the day, on Friday, February 20, 1959, he was unemployed. Similarly, Rod Rose, who was born in Cambridge, England, obtained a fellowship at the Cranfield Institute of Technology after a "Gentleman Apprenticeship" with A.V. Roe in Manchester. He worked for Vickers Supermarine for a time on a Swift transonic airplane before emigrating to Canada in 1957 to work with AVRO Aircraft, Ltd. Rose attributes the demise of the Arrow project largely to politics.[24]

He recalls reporting to work as usual on Friday, February 20, and that about "elevenish" an announcement was made on the speaker system that a serious announcement would be made later in the day. Shortly after 3 p.m., he said, an announcement was made that as of the close of work, all employees were terminated, and would be able to return Monday morning to pick up their belongings. One of the people working with him, Rose recalled, had just arrived from England, was living in a hotel with his wife and child, had received no pay, and had no money. Some 20,000 people, he estimated, were directly affected by the lay-off, and another 100,000 who provided various services to the project were probably put out of work. The major problem, he believed, was that the Arrow project and AVRO were creatures of the Liberal government, and with the return of the Conservative Party to power came a purge of all things associated with the past Liberal Party regime. The purge was so complete, he added, that plans, models, specifications, and designs of the Arrow fighter, engine components, and tests were methodically and deliberately destroyed. It was, he believed, a tragic loss for Canada and the world aerospace industry, for the Arrow CF105 was far ahead of its time.[25]

The expertise developed in work on the Arrow (which had been designed with a Mach 2 performance ability), however, became an invaluable part of the NASA manned spacecraft effort. Rose believed that AVRO expertise including operations experience, real-time telemetry, and "fly-by-wire" [where controls operated through a computer system] knowhow plus Arrow advances in thermodynamics, materials and structures, among other things, greatly facilitated the development of the American manned spacecraft effort.[26]

In this context, Jim Chamberlin, whom Rose described as a brilliant engineer and who would become a key person in the design of the Mercury project, contacted Gilruth, with whom he had close personal and professional associations, and asked if the STG might be interested in the AVRO people.[27] It was an undisguised opportunity, and Gilruth acted immediately.

He, Charles Donlan, Charles Mathews, Paul Purser, and Kimble Johnson promptly flew to Toronto, interviewed about 100 applicants for jobs with the STG, within 10 days extended offers to about 50 AVRO engineers, and received acceptances from 25. Among the 25 was Bryan Erb, whose American connections dated back seven generations to Captain Henry Erb (who threw his lot with the Loyalists in the American Revolution and left the United States for Canada in 1783). Erb, in a sense, had returned home. Another was Rod Rose, who confessed that he had required a bit of persuasion from Jim Chamberlin.[28]

By May most of the 25 AVRO engineers were intensely involved in Project Mercury, most of them in middle-management technical positions, and a few such as Chamberlin and Rod Rose soon in senior level positions. It was, Rose recalled, an instant meshing marred perhaps only by the fact that his immediate supervisor, Jerry Hammack, spoke "Georgia" and constantly chided Rose about his inability to speak "good English." By the end of 1960, six additional former AVRO employees joined the NASA contingent, a few of whom went directly to the Goddard Space Flight Center and to NASA Headquarters. About half of the 31 employees from Canada were born in Canada, half were from England, and one (Tec Roberts) came from Wales. The AVRO/NASA roster included:

Pete Armitage	Bryan Erb	Dave Ewart	Dennis Fielder
Morris Jenkins	Rod Rose	Dick Carley	Tom Chambers
Norm Farmer	John Meson	Bruce Aikenhead	Frank Chalmers
Jack Cohen	Stan Cohn	Gene Duret	Joe Farbridge
John Hodge	Fred Mathews	Owen Maynard	John Shoosmith
George Watts	Stan H. Galezowski	Tec Roberts	George Harris
Dave Brown	Les St. Leger	Burt Cour-Palais	Jim Chamberlin
Len Packham	Bob Vale	Bob Lindley	

The "AVRO connection," as Rod Rose called it, swelled the ranks of the manned spacecraft personnel force from about 135 persons to about 160 by April 1959 and, more importantly, provided engineering talents and expertise which simply were unavailable in the United States. At the time, even qualified aeronautical engineers were hesitant to apply for a position in the STG in the belief that it was temporary at best and "Mickey Mouse" at worst.[29]

Gilruth's needs for additional personnel reflected only one aspect of NASA growth pressures. The STG was a new and still relatively small part of the NASA complex of centers and programs. Abe Silverstein began arrangements for the transfer to NASA of approximately 250 members of the naval research staff who had worked under Dr. Homer Newell on upper atmospheric research and under Dr. John P. Hagan on the Naval Research Laboratory's Vanguard satellite program. Many of these people worked in and around the Washington area, and Silverstein wanted to provide them facilities in the area. When he asked Dryden about possibilities, Dryden commented that "just the day before at a meeting of the National Geographic Society he had been asked by a representative from the Agricultural Department if NASA needed any land in the Washington area for a lab site and that they would welcome NASA's use of land at the Beltsville site." Silverstein followed up and received approval for the transfer of 500 acres. It was, he said, the beginning of the Goddard Space Flight Center, which he named in honor of America's rocket pioneer, Robert H. Goddard. The center was officially created on May 1, 1959.[30]

Because the STG was a "highly technical organization" whose personnel had little time for administration, Silverstein decided to incorporate the STG under the mantle of the new Goddard Space Flight Center. Silverstein arranged the appointment of Harry Goett from the Ames Research Center to head the Goddard Center, with Gilruth to be Deputy Director. Gilruth and the STG, however, would physically remain at Langley until the completion of Project Mercury. In theory, Goett would provide administrative control and

Gilruth technical direction, while Silverstein could provide policy direction and control from Washington.

Several things went wrong with this plan. Once Goett became a director, his formerly warm relationship with Silverstein cooled and cooperation became difficult. And instead of improving the business management of the STG and Project Mercury, the 200 miles separating Gilruth's operations from the administrative center only aggravated management difficulties. Moreover, Gilruth, who once reported directly to Silverstein as an autonomous director, now reported to Silverstein through Goett. It began, as George Low concluded, "a serious rift between Silverstein and Gilruth." This "Goddard interlude" reinforced the perception which was growing that the manned spaceflight initiative needed to be a separate task group, center or entity of some kind.[31] Goddard was only one of the new centers being added to the NASA collection.

In October 1959, President Eisenhower announced the transfer of the Army Ballistic Missile Agency's Development Operation Division in Huntsville, Alabama, and the launch facilities at Cape Canaveral, Florida, to NASA. With congressional approval effective March 14, President Eisenhower, by Executive Order, renamed the Huntsville facility the George C. Marshall Space Flight Center on the following day. On July 1, 1960, Dr. Wernher von Braun became the director of the facility whose primary mission would be to develop "high thrust space vehicles," and more precisely for the moment, the Redstone, Centaur and Saturn rockets.[32]

Von Braun and every center director, including Thompson at Langley and Goett at Goddard, were competitors for the limited supply of men and money in the face of burgeoning programs and responsibilities. Moreover, Von Braun, who had previously been "completely responsive" to NASA (and STG) requirements, now was within the NASA organization an administrator of higher rank than Gilruth and enjoyed greater public recognition. Gilruth's lack of rank within the system was partly alleviated in January 1961 when the STG was broken out of the Goddard organization and restored to its original autonomy with a direct reporting line to Silverstein.

The real issue involved delineating responsibility for the manned space program as an effort distinct from other NASA programs and projects. Silverstein said that with the growth of new projects and the full realization of the scale of the manned effort within the NASA program, "it became clear to Drs. Glennan and Dryden and me that perhaps the concept of using Goddard as a place to house the manned program was wrong and that Goddard should direct the unmanned satellite program and a wholly new center be created for the manned spaceflight program."[33] The general public and Congress, to some extent, were generally oblivious to all of these problems. If it had not been true before, the elections of 1960, which brought John F. Kennedy to the White House, focused national attention on the "missile gap," the "space race," and the "red menace."

Americans were aware that the Soviets had launched Luna I, the first spacecraft into interplanetary space, in January 1959 followed shortly by Luna II, which impacted on the Moon in September, and Luna III, which flew behind the Moon in October. The latter coincided with Premier Nikita Kruschev's visit with President Eisenhower at Camp David. The elections in November were tightly contested by Eisenhower's Vice President, Richard M. Nixon, and the Democratic candidate, John F. Kennedy, who stressed that the all-too-obvious "missile gap" was the product of a past Republican administration which had become too complacent about America's position of power and wealth in the world, and so uncaring that many Americans, particularly minorities, failed to share in the affluent society. The apparent missile gap, accentuated by the Soviet Moon rocket launches, provided a critical edge in the election. Kennedy very narrowly defeated Nixon.

The election returns, however, had not convinced President Eisenhower that a missile gap existed, nor that manned spaceflight could be justified beyond Project Mercury. When his Science Advisory Committee submitted a report, prepared by a panel headed by Dr. Donald Hornig of Brown University, of projected costs of prospective manned space programs, he was understandably concerned. Project Mercury could cost a projected $350 million, an Earth and lunar orbital mission an additional $8 million, and a lunar landing an estimated $26 to $38 million more. When he asked why a lunar landing should be undertaken, the mission was likened by one of the staff to Columbus' voyage to the New World. Eisenhower snorted in response: "I'm not about to hock my jewels." And in the 1962 budget sent to Congress in January 1961, the President questioned the validity of extending manned spaceflight beyond the Mercury project.[34]

Eisenhower was not alone in his perception of the viability of continuing manned space missions. NASA Director Keith Glennan confided to Oran Nicks, who directed NASA's Lunar and Planetary Programs between 1961 and 1968 before becoming an Associate Administrator at Headquarters and then Deputy Director at Langley (1970 to 1980), that his real interest throughout his administration of NASA was *other* than manned spaceflight. Congress, however, was much bolder. In February 1959, the House Select Committee on Astronautics and Space Exploration advised creating programs that would lead to the "manned exploration of the Moon and nearby planets with eventual establishment of scientific bases on these bodies." In July 1960, Congress urged as a high priority program "a manned expedition on the Moon in this decade."[35] During the early years of the manned space program, Congress rather than the executive branch tended to exercise leadership and take the initiative in space program planning. Congress also anticipated President John F. Kennedy's bold initiative for a lunar landing within the decade.

Although President-elect Kennedy had urged a stronger effort in space, he had been and remained ambivalent about "man-in-space." Shortly before Christmas, John Kennedy invited Lyndon Johnson to join him at Palm Beach, Florida, while he was vacationing and recuperating from the vigorous election campaign. Johnson prepared for the meeting by investigating, among other things, the status of the space program. He was informed by his staff that the Nation did not have a comprehensive or centrally coordinated space program, and that at NASA "there has been a continuing lack of leadership and competence, basically in administration but not excluding the scientific field." The Space Council created by the NASA enabling act was moribund, despite Johnson's earlier personal understanding with President Eisenhower to have the President serve as its lead. And he was advised that NASA needed a tough and competent new administrator. The Mercury program, he was informed, had suffered "slippage," and other programs including Saturn, communications and weather satellites, and scientific probes were showing "slippage and failure." Moreover, bitter controversy existed between the Army, Navy and Air Force over roles and missions related to space. The Air Force wanted responsibility for the entire program and would relegate NASA to a strictly advisory role.[36]

At their meeting Kennedy asked Johnson to head the administration's initiatives in prohibiting discrimination against minorities doing business with the government. Johnson agreed, and then was asked by Kennedy what else he would like to do. Johnson replied that he would like to continue his contact with space activities. Kennedy agreed, and issued a press release indicating that he would rely on the Vice President for space leadership. As Johnson recalled, "Every president brings to the office his own special concerns, which are the result of his interests and experience. Space was not one of President Kennedy's primary concerns at that time."[37]

Edward C. Welsh, who became Executive Secretary of the National Aeronautics and Space Council, began drafting amendments to the NASA legislation making the Vice President, rather than the President, a member and the chairman of the Space Council, which was approved by Congress in April. When Kennedy suggested that General James M. Gavin head NASA, Johnson responded that "it would be a serious mistake to appoint any military man to head the organization." And Kennedy responded, "All right, find another administrator." Johnson did. He personally interviewed some 20 prospective candidates and selected James E. Webb, former Director of the Bureau of the Budget and Under Secretary of State during the Truman administration.[38]

NASA Administrator Keith Glennan resigned on January 20, 1961, the last day of President Eisenhower's administration, without having received any statement from the President-elect as to his intentions regarding NASA. Webb, who was formally sworn into office on February 14, asked that Hugh Dryden be retained as Deputy Administrator. Meanwhile, a "lunar flight feasibility committee" chaired by George Low, and including Oran Nicks, Max Faget, and others, prepared a paper for the Vice President which offered a brief technical justification for a lunar landing.[39]

At the end of March, President Kennedy met with Johnson, Budget Director David Bell, and science advisor Jerome B. Wiesner, and others to discuss space matters. One consensus of the meeting was that the United States needed to develop more powerful rocket engines. Johnson advised setting a goal, "a bold and understandable challenge," to move America forward. Johnson said that he continued to discuss "this concept with the President at some length over the next few weeks."[40] The President's and the Nation's problems were soon exacerbated by another spectacular Soviet space achievement and an American-backed military debacle in Cuba.

On April 12, 1961, Major Yuri Gagarin became the first human to "leave this planet, enter the void of space, and return." Public dismay at this new evidence of Soviet space prowess rivaled that of Sputnik 4 years earlier. President Kennedy and Johnson conferred at length on the 19th, and on the 20th Kennedy directed Johnson to head a Space Council inquiry to see "where we stand in space." He asked:

> Do we have a chance of beating the Soviets by putting a laboratory in space, or by a trip around the Moon, or by a rocket to land on the Moon, or by a rocket to go to the Moon and back with a man? Is there any other space program which promises dramatic results in which we could win?[41]

Although the President's memorandum, Johnson recalled, came to him only 3 days after "the disastrous failure at the Bay of Pigs," Kennedy was not trying to use space to divert attention from the debacle in Cuba. Edwin C. Welsh, Executive Secretary to the Space Council, concurred that the collapse of the Cuban invasion did not encourage a space venture, but if anything was a deterrent in that the administration could not afford a failure. On the same day that President Kennedy addressed his memorandum to Johnson, Congress approved an amendment to the Space Act making the Vice President, instead of the President, chairman of the National Aeronautics and Space Council.[42]

Space Council meetings began on April 22. Consultation and advice came from James Webb, who of course was a member of the Council. Johnson invited Frank Stanton, president of Columbia Broadcasting System; George R. Brown, president of the Houston-based Brown & Root Construction firm; and Donald C. Cook, executive Vice President of American Electric Power Company, to meet with the council. Hugh Dryden, Wernher von Braun, Admiral John T. Hayward (Assistant Chief of Naval Operations for Research and Development), and General Bernard A. Schriever were among those consulted. The Space Council reported to the President on April 28 that at the moment neither the United States nor the Soviet Union were known to have the capability of circumnavigating the Moon or landing a man on it, but that "with a strong effort the United States could conceivably be first in those accomplishments by 1966 or 1967."[43]

The scientific community and medical community, and indeed NASA Administrator James Webb, counseled a more moderate approach to the "space problem." An ad hoc Committee on Space headed by Jerome Wiesner, who became Kennedy's Science Advisor, stressed the accumulation of scientific data from unmanned probes. Another special panel of the Science Advisory Committee chaired by Dr. Donald Hornig urged more experiments with animals before men were committed to spaceflight, and gave only "lukewarm" endorsement to Project Mercury. Although Webb sought an expanded space program, he sought a "balanced" program and was uncertain about the costs and propriety of a manned lunar expedition.[44]

Although the debate continues as to whether the manned lunar expedition was inherently a political decision or a scientific decision, the political climate at the time strongly influenced the administration's decision. Views within the technical/scientific community were not clear. Technical people, engineers, test pilots, and life scientists looked at the problem of manned spaceflight from different perspectives. The public both feared the Soviet Union and the risks of an arms/space race. No one understood the extent of real costs involved or could estimate benefits or economic returns. The decision to attempt a manned lunar landing would require a substantial commitment of personnel, talent, and money and would affect the whole society. Leadership in American space initiatives now shifted dramatically from Congress to the White House. On May 25, John Kennedy addressed Congress and the American people:

> With the advice of the Vice President, who is Chairman of the National Space Council, we have examined where we are strong and where we are not, where we may succeed and where we may not. Now it is time to take longer strides—time for a great new American enterprise—time for this nation to take a clearly leading role in space achievement, which in many ways may hold the key to our future on Earth.[45]

I believe that this nation should commit itself to achieving the goal, before the decade is out, of landing a man on the Moon and returning him safely to Earth.[46]

Congress turned to the task of defining and funding the President's new space policy with enthusiasm. Hearings in the House and Senate closed with the approval of approximately $1.7 billion in funding for space, and the promise of an additional $40 to $70 billion expenditure in the decade of the 1960's. A special report released by the House Committee on Science and Astronautics in August explained the "Practical Values of Space Exploration" as the generation of "new knowledge," the enhancement of America's international prestige and stature, and interestingly, the suggestion that space exploration might be a substitute for war. The economic benefits of the space program "spread across the entire industrial spectrum—electronics, metals, fuels, ceramics, machinery, plastics, instruments, textiles, thermals, cryogenics, and a thousand other areas." Space research should generate new industries, new power sources, progress in "human engineering," advanced communication systems, weather prediction and control mechanisms, the development of high-speed lightweight computers, advances in solid state physics, new economic alliances and private enterprises and jobs related to space.[47]

In some respects this new project was thrust upon NASA and its components, including the STG. But in most respects it was a project invited, planned for, dreamed of and enthusiastically entered into by the NASA community. The inception and design of a lunar mission actually pre-dated President Kennedy's announcement by almost 2 years. As the initial flights of Mercury developed, meetings between Silverstein and Gilruth's staff and personnel generated a program that would go beyond Mercury's limited spaceflight and which in three stages (A, B, and C) projected an Earth orbit, a lunar orbit, and a lunar landing. Silverstein prophetically named the project Apollo, for the Greco-Roman god of the Sun and prophecy. In 1960, some STG personnel actually began work on Apollo-related projects. "Gemini," according to Abe Silverstein, "was created as a filler between the Mercury and Apollo programs since it was recognized that the flight operations in Mercury would be terminated long before Apollo hardware would be ready to fly." It was believed that too lengthy an interval without flight would destroy the capability of flight operations and the astronauts.[48]

The administration's endorsement of a program to put an American on the Moon shifted NASA's technological and fiscal focus more fully on its manned spaceflight program, and prominently upon Bob Gilruth and the STG. The lunar landing was to be a NASA objective, and all centers would contribute to its accomplishment. But the new lunar mission seemed to mandate that the manned spacecraft program be established as a separate center, rather than remain under the administrative auspices of Goddard Space Flight Center or the Langley Research Center.

The greatly expanded NASA mission also required an administrative reorganization to accomplish an engineering, scientific and production feat which far exceeded anything the United States previously had entered into and before which (in terms of technical complexity, costs, and, as it turned out, time) those great feats of transcontinental railroads and the Panama Canal paled.

Even as NASA Headquarters and other branches of government began to contemplate moving the space task program to its own site, Webb began to address the new organizational problems relating to a far more massive operations and production effort by NASA. He created the Office of Manned Space Flight through which all programs relating to the lunar landing (and Mercury and Gemini) could be orchestrated. Program offices were also established for Space Science, Applications, and Advanced Research and Technology. All program offices reported through the Associate Administrator, Robert C. Seamans, Jr. (an MIT graduate and RCA engineer), and the Deputy Administrator, Hugh Dryden, to Webb. Homer Newell came from the Vanguard program to be Deputy Director for Space Science; Morton Stoller covered Applications, and Ira H. Abbott—Research and Technology. D. Brainerd Holmes was selected to head the Office of Manned Space Flight because of his experience with RCA in handling "large scale endeavors." As project manager of the Ballistic Missile Early Warning System, he was known as an organization man and credited with being a tough program manager. Holmes brought in Joe Shea from Bell Laboratories to head systems engineering. Bell Laboratories organized a management company called BellComm specifically to provide management assistance for NASA. Abe Silverstein left NASA Headquarters to become the Director of the Lewis Research Center, and Dryden remained the anchor man amidst all of the turnover.[49] New administrators, new organizations, and rapid expansion began to create personnel and management problems at a very critical moment in the life of the manned spacecraft program. These problems were generally sublimated to the great opportunities and excitement and the sheer hard labor involved in the existing programs and the new. To add to the confusion, by mid-1961 the decision was made to relocate the manned spacecraft program onto its own center.

The "slippage" in the space program reported to Lyndon Johnson seems to have faded by March, when the Kennedy administration assumed office. Real progress had in fact been made over the past several years and in 1961 much of the hard work began to bear fruit. The first team of astronauts was selected in 1959. In May of that year, Able, a rhesus monkey, and Baker, a squirrel monkey, were lofted to an altitude of 300 miles and 1500 miles downrange over the British West Indies. On December 4, 1959, and on January 21, 1960, Sam and Miss Sam made successful flights from Wallops Island; while on January 31, 1961, Ham, a chimpanzee, made a full dress suborbital flight in a Mercury capsule launched from Cape Canaveral, Florida, and ended up with "wet pants" when his capsule landed 150 miles beyond the recovery point with a collapsed heat shield which had punctured the capsule. Work on this problem, which was solved by placing impact absorbing metal honeycomb on the aft bulkhead and a cable and spring system between the heat shield and the capsule, enabled the launch of America's first manned flight to proceed. In the interim, Yuri A. Gagarin, a Soviet cosmonaut, made man's first journey into space in a 108-minute orbit of the Earth aboard the 5-ton Russian Vostok spacecraft.[50]

Although American consternation over this latest Soviet triumph led to the perception that Gagarin's flight hastened the launch of America's first astronaut, the fact was that an American launch had been imminent. Within the month, the STG successfully launched Alan Shepard aboard "Freedom 7" on May 5, 1961, for a 15-minute flight downrange. President Kennedy, who offered Shepard his personal congratulations by radio-telephone when he arrived aboard the pick-up carrier, hailed the flight as a "historic milestone," but urged America to "work with the utmost speed and vigor in the further development of our space program." Although unrelated to the Shepard flight, on May 16 a site selection team visited Houston, Texas, one of the many locales being considered as a possible home for a new manned spacecraft center. And then, on May 25, President Kennedy announced the lunar landing initiative. May 1961, was, as Gilruth turned the phrase, "the end of the beginning" for America's manned space program.[51]

On September 19, 1961, NASA announced that its new "spaceflight laboratory" would be located in Houston, Texas, on 1000 acres of land made available to the government by Rice University.[52] By the end of the year and throughout 1962, first hundreds and then thousands of manned spacecraft personnel, contractors, support groups, and their families were making their way to the flat, seemingly hurricane-ridden coastal prairies south of Houston, until then the exclusive habitat of Texas cattle, oil derricks, rice fields, fish, ducks, some alligators, lots of mosquitoes, and a most enthusiastic and receptive local population.

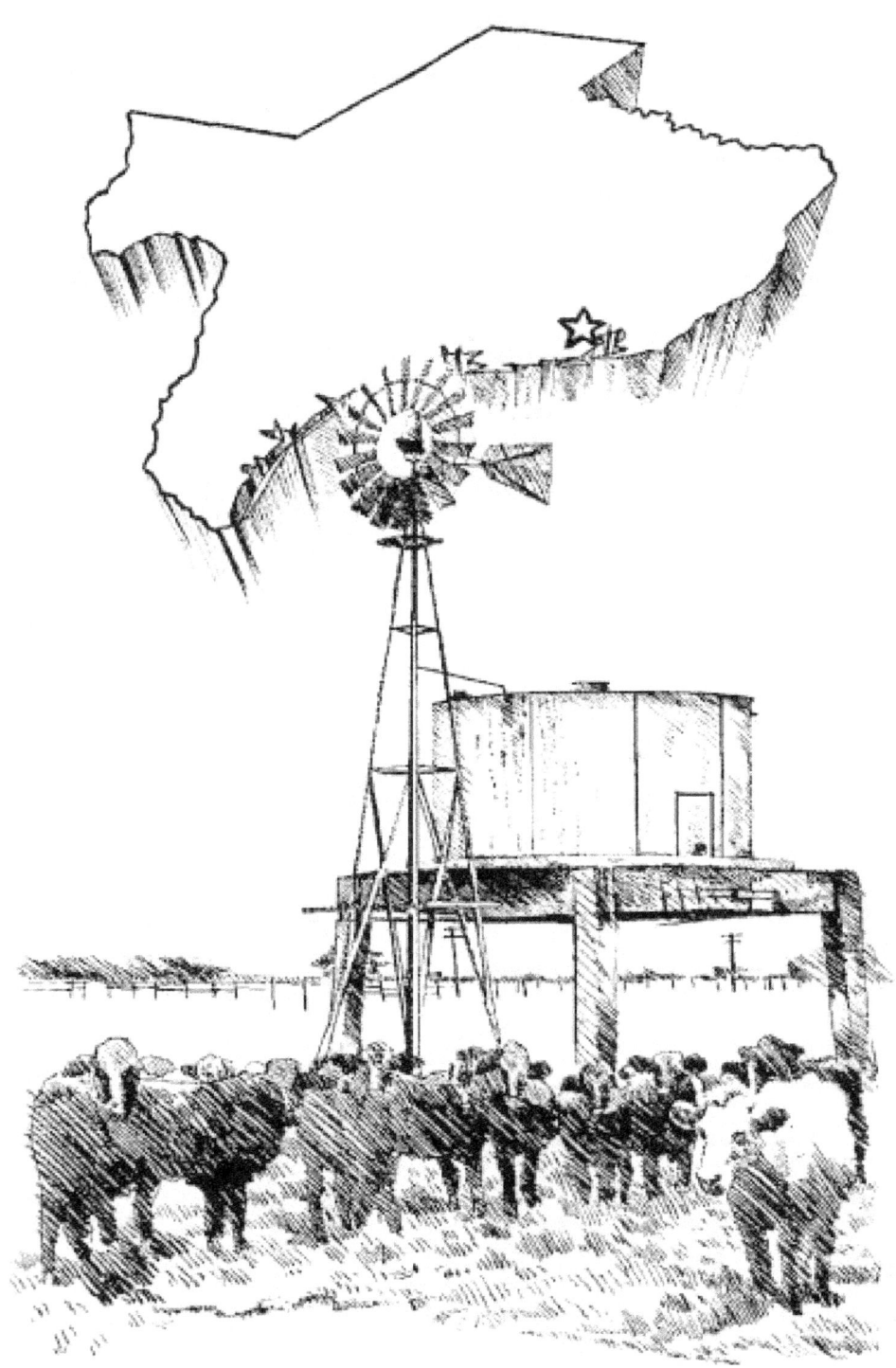

CHAPTER 3: Houston - Texas - U.S.A.

The Space Task Group began as a semiautonomous field unit, an essentially technical engineering organization highly dependent on the Langley Research Center and NASA Headquarters for administrative management and control. Like some great nova it had a seemingly spontaneous birth, conceived by a wholly external cosmic event—the orbiting of Sputnik I. As is often true with such stellar events, the developing American space program had a cloudy and possibly transient future. Yet the new NASA contained a considerable history of research and development in hypersonic flight and rocketry which became the intellectual and sustaining force behind the space program. Those who participated in the manned space program never exhibited any sense of uncertainty or confusion. A "can-do" attitude and determination carried the STG over many bumps, with the questions of how the program should be operated and where it should be located being among the more divisive.

Problems relating to the physical location of the STG and its management developed almost as soon as it came into being. On December 4, 1958, Paul Purser discussed merging people from the Lewis Research Center with STG people at Langley. G. Merritt Preston doubted that Lewis researchers would want to move to Langley, and Purser foresaw "similar problems in the other direction." But everyone agreed that the important thing was to get the job done.[1]

Abe Silverstein promoted the organization of the Goddard Space Flight Center, which came into being in May 1959, as a mechanism "to provide greater autonomy for the manned spaceflight operation, and in recognition that this new center might be a location where the manned space operations could develop to its appropriate stature." Silverstein served as acting director of Goddard during its first few months of operation until Harry Goett, an engineering manager at Ames who headed a "Research Steering Committee for Manned Space Flight," received the appointment. Robert Gilruth became "assistant director for manned satellites." Because Project Mercury was in full swing and because there were no facilities available to receive personnel, no plans were made to physically transfer the STG. Gilruth, in fact, reported to his staff in February 1960 that "no major move of STG personnel is anticipated during the next two to two and one-half years."[2]

It soon became clear to Silverstein, Glennan, and Dryden, however, "that the use of Goddard to house the manned spacecraft program was wrong" and that the existing management structure could jeopardize the program. Wesley Hjornevik, then Glennan's personal assistant, recalls that the decision for an independent location for the STG came, in fact, largely from Glennan. Glennan, with counsel from Hugh Dryden, decided in late 1960 that the manned flight effort should be separated from Goddard, as it became evident that the idea of manned flight was becoming more popular and the perception of it being a stunt began to disappear. There were also more pressing concerns. The attachment of the STG to Goddard placed Gilruth lower in the management chain beneath the Goddard director, Harry Goett, instead of reporting directly to Headquarters. Because of this and other factors, the Gilruth-Goett-Silverstein linkage had become strained. In addition, having a major program located at one center but managed by another created stress. Finally, Glennan and Dryden believed that a continuation and enlargement of the spaceflight program beyond Mercury would result in such massive expansion of the STG that its physical association would result in the Goddard scientific and research programs being overwhelmed. Separating the STG from Goddard would not only protect the integrity of scientific and research programs, but would also help resolve some of the management conflicts that had developed between Gilruth on the one hand and Silverstein and Goett on the other.[3]

Independently, George Low, whose "Manned Lunar Landing Task Group" was studying the possibilities for a manned lunar landing, had come to the realization that the manned spaceflight program should be separated from all other NASA centers and had been quietly urging this course. Silverstein completed a review of the Goddard-STG management problem in November and concurred that the STG should report directly to the Office of Space Flight Programs. On January 1, 1961 (before leaving office on January 20), Glennan issued an order separating the STG from Goddard and restoring it to its original semiautonomous status, and he also left a memorandum for his successor, not yet named, explaining why he had issued the order and recommending that the STG not be collocated at Langley Research Center, Marshall Space Flight Center, Lewis Research Center, or the High Speed Flight Station at Wallops Island. He left open the options of placing the program at Ames or at a new center, and Hjornevik believed that he preferred a new center. Glennan felt, however, that his successor should be the one to determine the location of the new center. But he did appoint a committee headed by Bruce T. Lundin at the Lewis Research Center to investigate the possibilities for relocating the manned flight program.[4]

Glennan indicated that the parameters for relocation should include a preference for a site close to an existing NASA installation, a site that would allow for the development of a life sciences center adjacent to it, that a move should not disrupt the Mercury program, and that contractors participate to a greater extent than they had under the Mercury program. Lundin, with Wesley Hjornevik, Ernest O. Pearson, Jr., and Addison M. Rothrock found a general consensus that the manned spaceflight program required a center of its own, but could get little agreement on where such a center should be located. The committee finally recommended that the STG be relocated at Ames Research Center in California.[5]

But as of April 12, when Yuri Gagarin made his spectacular Earth-orbital flight, a firm decision to create a separate center for the American manned space effort had not been made by the new NASA director, James Webb. Just as Sputnik I precipitated the organization of the STG, Gagarin's flight seemed to mandate the separation and independence of the American manned spaceflight program. The Lundin Committee recommendations were forgotten. In late April, Abe Silverstein, Al Seipert (Associate Director for Administration), and Wesley Hjornevik (who had become Seipert's Deputy Director for Administration) were called into Webb's office to discuss a variety of questions, one of which had to do with projected costs and personnel numbers required for the creation of a new center. Numbers tossed out at that meeting would later become very critical in the formation of a manned spacecraft center. Abe Silverstein believed that 3000 personnel should staff a center, Hjornevik suggested that construction would cost in the realm of $50 million, and Seipert decided that number should be bumped to $60 million. The latter number soon appeared in the NASA appropriations bill. On May 1, Silverstein's office completed a draft for "Organizational Concepts and Staffing

Requirements" of an "independent NASA field center responsible for the conduct of programs for manned spacecraft."[6]

On another occasion in Silverstein's office, Silverstein, Low, and Hjornevik reviewed the possibilities for relocating the manned spacecraft program, and it was at this meeting that Houston, Texas, first came up for consideration. Low recalls that Silverstein impulsively asked the question, "I wonder where Albert Thomas' district is?" It was not a wholly innocent question. Thomas happened to be chairman of the House Appropriations Subcommittee which had responsibility for NASA appropriations. Moreover, years earlier, in October 1958 to be precise, Thomas had urged NASA Director T. Keith Glennan by letter and by telephone to consider Houston as a possible site for a NASA "laboratory." Silverstein and others were aware of these inquiries. Wesley Hjornevik reminded Silverstein that Thomas' district included Houston. They looked at an atlas and noticed that Houston was also the location of Ellington Air Force Base, which had become essentially deactivated since World War II. They also perused possible sites in Florida and California.[7]

Silverstein sent Philip Miller, Chief of the Facilities Engineering Division for Goddard, and John M. Parsons, Associate Director for the Ames Research Center, to Houston to look into location possibilities. Miller and Parsons were met at the airport in Houston on May 16 by George Brown, of the Houston-based Brown & Root construction company, who the previous month met with Johnson's Space Council. With him was Ed Redding representing the Houston Chamber of Commerce. They first went to Rice University to visit with acting president Dr. Carey Croneis and then to Ellington Air Force Base to meet with the base commander Brigadier General Russell F. Gustke. "The General was very cooperative," Miller recalled, "and had been briefed prior to our arrival by Congressman Thomas and indicated an alertness to the confidential nature of the visit." From Ellington, the travelers drove south through open coastal prairie through a large tract of land identified as the West Estate which had recently been donated to Rice University by Humble Oil Company. George Brown indicated that Rice University would be favorably disposed to making the land available to the government for a research center installation. Parsons and Miller went through the 20,000 square foot West Mansion. They also found that barge traffic could navigate Clear Lake (with access to the Houston ship channel) to the property. They later viewed the inoperative Dixon Gun Plant and various industrial facilities in the Houston area and flew back with a report for Silverstein.[8]

Also in May a number of meetings were held which included at various times Bob Gilruth, Wesley Hjornevik, Abe Silverstein, Paul Purser, Paul Dembling (head of the Policy Planning Board) and occasionally James Webb, among others. These discussions resulted in a proposed plan for the organization, prerequisite physical facilities, general criteria for a location, and probable staffing needs of a new center. Another memorandum drafted by Dembling established specific criteria for locating a manned spacecraft center and was intended for circulation to Congress and prospective communities. Silverstein, Max Faget recalled, insisted that the projected center would operate directly under his authority at NASA Headquarters. When he failed to receive support on this issue, he elected to leave Headquarters and return to Lewis Research Center. Dr. Brainerd Holmes was appointed to head the newly created Office of Manned Space Flight, and the decision was made to first create the organization and then decide how much authority would be vested in Headquarters and how much in the center.[9]

The search for a location for the manned spaceflight laboratory now began in earnest amidst wild rumors and during a complete overhaul of the NASA Headquarters staff and organization. Since the summer of 1960, the national press had been generally critical of the progress, or "lack of progress," in the Mercury program, so much so that in November, following two different launch failures, Glennan sent the following message to Gilruth:

> ... I know how discouraging these troubles are to you and your fine staff. Please try to close your ears to the press comments and know that there is no lack of faith in your ability to succeed in this effort. Now is the time for real driving leadership so grit your teeth and dig in. We are solidly behind you and your outfit. [10]

Gilruth and the STG had little time to consider moving anywhere, but concentrated instead on some very tough technical problems and test failures.

Finally, on December 19, a successful firing launched a test capsule to an altitude of 117 nautical miles; and on January 31, 1961, the chimpanzee Ham reached an altitude of 135 nautical miles and landed some 364 miles downrange. A successful test firing in February led to a marginally successful flight test on March 18. Then a bad launch forced a booster destruction order on April 25. A successful test using a Little Joe booster on April 28 preceded the May 5 launch of Mercury Astronaut Alan B. Shepard, Jr., using a Redstone missile on a ballistic flight path to an altitude of 116 miles and a downrange distance of 302 miles.[11] Shepard's 15-minute suborbital flight was tremendously important, but its significance was sorely diminished by Gagarin's 25,000-mile orbital flight aboard Vostok I, a capsule weighing five times as much as the Mercury "Freedom 7." "Getting the job done!" became increasingly important to the STG as technical problems and external pressures mounted. The STG now began focusing on the launch of Mercury-Atlas 1 scheduled for July.

Congress began hearings on the $1.7 billion NASA appropriations bill which included $60 million for the manned spaceflight laboratory. A progress report released by the House Committee on Science and Astronautics on December 30, 1960, indicated some problems and malfunctions in the Mercury program, but explained that the "implementation of a project such as Mercury demands on a continuing basis, boundless energy, enthusiasm and determination. . . . Work on Project Mercury . . . is proceeding on a three shift, seven-day-a-week basis."[12] Overall the report was wholly supportive, and anticipated, in a sense, a favorable response by Congress to the 1962 NASA appropriations bill, which included the allotment for a manned spaceflight laboratory. NASA began the search for a new center location in earnest.

Specific site criteria, made available to Congress and the general public, greatly facilitated the search. The site required access to water transportation by large barges, a moderate climate, availability of all-weather commercial jet service, a well established industrial complex with supporting technical facilities and labor, close proximity to a culturally attractive community in the vicinity of an institution of higher education, a strong electric utility and water supply, at least 1000 acres of land, and certain specified cost parameters. By June, Congressmen, such as Olin Teague, were being inundated with applicants for the spaceflight laboratory. Most simply could not qualify, as Teague explained to some of his Texas constituents, because only a large industrial area would meet the specifications. Houston was "probably the only Texas area being considered at the moment," Teague said.[13]

Webb appointed a site selection team in August chaired by John F. Parsons and including Philip Miller, Wesley Hjornevik, and I. Edward Campagna, the construction engineer for the STG. Hjornevik became ill and was replaced by Martin A. Byrnes. First, a list of 22 cities which met the essential criteria of water and weather was established. This was reduced to nine areas, most of which included some federal facility. They were:

Jacksonville, Florida (Green Cove Springs Naval Station)
Tampa, Florida (MacDill Air Force Base)
Baton Rouge, Louisiana
Shreveport, Louisiana (Barksdale Air Force Base)
Houston, Texas (San Jacinto Ordnance Depot)
Victoria, Texas (FAA Airport)
Corpus Christi, Texas (Naval Air Station)
San Diego, California (Camp Elliott)
San Francisco, California (Benicia Ordnance Depot)

Additional sites were soon identified, bringing the total to 23. Four of the added sites were in the vicinity of St. Louis, Missouri; two additional sites were identified in Houston (including one offered by Rice University and another by the University of Houston). And other sites were variously in Bogalusa, Louisiana; Liberty, Beaumont, and Harlingen, Texas; and Berkeley, Richmond, and Moffett Field, California.[14]

Between August 21 and September 7, the team visited 23 cities, beginning in Jacksonville, Florida, and ending in Palo Alto, California. The routine at each stop involved an afternoon arrival and a greeting by State and local dignitaries, a trip to the hotel where the visitation team explained the selection criteria, a breakfast meeting with townspeople, a visit to the proposed site and the nearby college or university, and a late afternoon departure for the next city on the agenda.[15]

During the visitation, particularly strong political pressure developed from a Massachusetts delegation headed by Governor Volpe and Senator Margaret Chase Smith, which produced a personal inquiry to Webb from President Kennedy. Missouri directed its case through Senator Stuart Symington. California's Congressman George Miller, then acting head of the House Committee on Science and Astronautics, championed the case for his State. Proponents of sites in Boston, Massachusetts, Rhode Island, and Norfolk, Virginia, made separate presentations to Webb and the Headquarters staff, and these additional sites were added to the final review. By the close of the visitation period, the site selection team had identified MacDill Air Force Base at Tampa, Florida, as the preferred site, largely because the Air Force planned to close down its Strategic Air Command operations at that base. A Houston site offered by Rice University was second, and the Benicia Ordnance Depot in the San Francisco Bay Area was third. Before a decision could be made, however, the Air Force decided not to close MacDill, omitting it from consideration.[16]

Houston moved into first place. Webb, now in close contact with President Kennedy on the matter, informed the President on September 14 of the decision made by him and Hugh Dryden. On that date Webb replied in two separate memoranda to President Kennedy's inquiry reviewing for him criteria and procedures for the site selection. One memorandum reviewed procedures, and the other reported that: "Our decision is that this

Suddenly, Tomorrow Came . . .

Texas cattle grazed the 1000-acre land parcel outside Houston at the time its selection as site of the Manned Spacecraft Center was announced. Soon they were herded out to make room for spacecraft, astronauts, and engineers.

laboratory should be located in Houston, Texas, in close association with Rice University and the other educational institutions there and in that region."[17] After advance notifications of the award were made by the Executive Office and from NASA, the public announcement of the location followed on September 19, 1961.

NASA announced that the $60 million manned spaceflight laboratory would be located "in Houston, Texas, on a thousand acres of land to be made available to the government by Rice University. The land, in Harris County, borders on Clear Lake and on the Houston Light and Power Company salt water canal." The laboratory would be "the command center for the manned lunar landing mission and all follow-on manned spaceflight missions." Under the 1962 budget, appropriations for construction included $12.1 million for a Flight Project Facility, $13.2 million for an Equipment Evaluation Laboratory and Support Facility, $3.6 million for a Flight Operations Facility, $26.5 million for an Environmental Testing Laboratory, and $4.5 million for site development and utilities. Webb emphasized that the Houston location would provide an integrated facilities system interconnected by deep water transportation with the expanded lunar launch facilities at Cape Canaveral and the Michoud Plant on the Mississippi River near New Orleans, where the space vehicles were to be fabricated.[18] The acquisition of these facilities in the summer of 1961 completed the assemblage of NASA centers.

The reaction to the Houston location among STG people, including Gilruth, was less than enthusiastic. Gilruth and many others were reluctant to leave the Virginia area which had been home for many years. Texas was not known to be a particularly hospitable place. Not only was he not enamored with the selection of Houston, but his first visit to Houston, did not help his perception of things. Hurricane Carla arrived in the Houston area just before Gilruth.[19]

Martin Byrnes, Gilruth, Walter Williams (the new Associate Director of the Manned Spacecraft Center), John Powers, and Ralph E. Ulmer (a facilities specialist at NASA Headquarters) arrived in Houston on September 22, 1961, as the first official NASA delegation. As the delegation made its way to a motel, "the night was humid," Byrnes recalled, "air . . . heavy with the odor of the Houston channel, [and] industries blowing downwind from petroleum and chemical facilities and the paper mill in that general area" generated some very grim comments. The next day Ed Campagna joined them and they toured the countryside in the vicinity of the new center.[20]

The scene was one of devastation. Telephone lines and debris littered the roadway. Along Farm Road 146 and 528 leading to what would soon be the main entrance to the Manned Spacecraft Center, boats had been hurled into the highway, pieces of houses and buildings lay in the fields, trees were flattened, and fields and pastures were still flooded or sodden with the heavy rains from Carla. Ellington Field, which would provide temporary quarters for the STG, offered dreary wartime housing with peeling paint and a sense of high disrepair. It was altogether uninviting. Early Sunday morning everyone except Byrnes, Powers and Ed Campagna returned to Washington or Langley.[21] Meanwhile the local newspapers and national media carried the story of the selection of Houston as the home for the spaceflight laboratory and the phone never quit ringing.

The local press attributed the selection variously to Rice University, Congressmen Albert Thomas and Bob Casey, President Kennedy, Vice President Johnson, NASA Administrators James Webb and Hugh Dryden, Rice University's new President Kenneth Pitzer and Chancellor Carey Croneis, Rice University's Board Chairman George R. Brown, Humble's Board Chairman Morgan Davis, and the general "can-do" attitude of the Houston community. Although the latter may have been as important as the diligent efforts of Congressman Albert Thomas, the Houston site, as Thomas carefully reiterated, met the requirements of the Moon shot program better than any other.[22]

Although the Houston site neatly fit the criteria required for the new center, Texas undoubtedly exerted an enormous political influence on such a decision.

Congressman Olin E. "Tiger" Teague, member of the House Committee on Science and Astronautics and head of the Subcommittee on Manned Space Flight, is shown here during a tour of Manned Spacecraft Center testing facilities. Congressman Teague was one of the architects of the American space program and an early partisan for the location of a NASA center in Texas.

Lyndon B. Johnson was Vice President and head of the Space Council, Albert Thomas headed the House Appropriations Committee, Bob Casey and Olin E. Teague were members of the House Committee on Science and Astronautics, and Teague headed the Subcommittee on Manned Space Flight. Finally, Sam Rayburn was Speaker of the House of Representatives. In the long run, the resources of Texas and the general enthusiasm of the Houston people combined to win not only the location of the Manned Spacecraft Center but also the hearts of the people who would soon be migrating to this strange new land.

Local enthusiasm and support began to be felt immediately and had a large role in making the new Manned Spacecraft Center really happen. The "can-do" attitude, infused by the excitement and drama of manned spaceflight, suddenly infected the Houston community and in turn rejuvenated a somewhat despondent STG and their families. Wesley Hjornevik immediately sent a small group to join Martin Byrnes (Site Manager) in Houston to begin making arrangements for the move. Among these were Stuart Clark (head of personnel), Burney Goodwin, Eugene Horton, W.A. Parker (Site Procurement Chief), John Vincent, Jeff Davis, Luther Turner, and Robert Peck. We arrived as "heroes," Parker recalled, although we were strictly lower to middle management. "The keys to the city were just all but given us," he said. The group met with representatives of the Chamber of Commerce, Small Business Administration, Better Business Bureau, and civic and social organizations. "The city was ours," Parker recalled. "We had police escort from meeting to meeting because everyone was interested in talking with us. . . . We had a constant audience." He and Marty Byrnes woke up at their motel one morning, he said, to find a letter from the president of Joske's Department Store and a new Stetson hat for each. After a meeting with the Bay Area realtors, the realtors devised a system for meeting every newly arrived STG employee at the airport, and provided each with a tour of the city and housing possibilities. The banks were outstanding. If a NASA employee arrived short of cash, the banks would open for them and provide funds any hour of the day or night. And the Chamber of Commerce produced tickets to the football games, theater, parties, or anything anyone desired. "We were on stage, we were the NASA people, and it was a glorious experience while it lasted."[23] Byrnes, heading the Houston advance party, coordinated the "promotional" aspects of the move with John A. "Shorty" Powers at Langley.

Powers, who made the initial visit with Gilruth in September, had been tremendously impressed with the friendliness and support of the Houston people. According to George Low, it was Powers who lessened the apprehension felt by task group personnel at Langley upon news of the forthcoming move to Houston. He put out an announcement that "Houston was a great place to live," and his theme that we should all go to Houston "and build a new center there" began to change the mood of gloom to one approaching enthusiasm.[24] Houston, albeit with some effort, could be envisioned as the promised land for the STG.

Powers launched a campaign at Langley, in cooperation with the Houston Chamber of Commerce, to make the move not only palatable but attractive. He posted signs all over Langley saying that "Houston is a good place to live!" His office presented slide shows and provided brochures. Ben Gillespie came from Houston to show a movie on the City of Houston and the new site. Powers held open meetings in the Langley cafeteria, and "shot down" the rumors that Houston had a hurricane every year and that hundreds of snakes crawled around the streets. His vigorous campaign closed with special flights being made available to some husbands and wives to fly to Houston and see the city firsthand.[25] It was an admirable and effective promotional campaign. To be sure, many came to Houston resigned to the fact that such a move "went with the job." For some families, the move to Texas was a long and difficult transition. Happily, Houstonians really wanted and welcomed the new NASA center and all of the people with it.

No single person so reflected the sincere warmth, support, and helpfulness of the Houston community as did Mrs. Grace Winn, who seems to have personally touched the lives of most of the STG moving to Houston. By sheer coincidence Grace Winn happened to be spending an extended visit in Washington, D.C. (recovering from a whiplash suffered in an automobile accident), and dropped by to see her old friend Olin E. "Tiger" Teague at the Congressman's office. Teague "was sitting in his office with his coat off in front of his television listening very intensely." "When he saw me," Grace Winn said, "he motioned me to come in, sit down, and listen to the program he was watching." Winn sat down without having the slightest idea what she was hearing, and in a few moments Teague excitedly explained to her that NASA was going to Houston. Grace recalled that she had no idea what NASA was, but that "Tiger" Teague said that since she knew the city and had been a part of it for a long time, she should go and introduce these people to Houston. He then picked up the phone and called Franklin Phillips, assistant to Administrator James Webb, and told him that NASA should send her down there. Phillips asked her to come see him.[26]

The next morning after their conversation, Phillips called Stuart Clarke, Chief of Personnel for the STG at Langley. Clarke asked Grace to come down to Langley the next morning and made reservations for her on a 7:00 a.m. flight. At Langley, Grace visited with Clarke and "Shorty" Powers, who asked her to join the STG in the Public Affairs Office. Grace accepted and on Monday, November 13, reported for work—the thirteenth person with the STG, she recalled, to report for permanent duty in Houston.[27]

Grace went to her office in the Gulfgate Shopping Center, picked up an already massive pile of mail, and was told that she would be moving her office to the Rich Building. Her job was to head the relocation office and facilitate the move of Langley people to Houston. Her approach, she decided, would be to treat people the way she would want to be treated if she were going through a similar move. So she decided the best thing to do would be to meet the new people at the airplane as they arrived. Her first group was to arrive at Houston International Airport (now Hobby Airport), quite some distance from the temporary offices south of Houston. She spent a rainy afternoon waiting for a plane which never came, but which was deflected to Little Rock, Arkansas, because of bad weather. When it did arrive the next day, everyone aboard was "tired, worn out, disgusted and discouraged." Grace took them all to her country club to lift their spirits and gave them a warm, personal welcome. Then she contacted realtors and suggested to them that because of the limited amount of time the families would be in Houston, they all should cooperate and show not only the homes they had listed, but also those listed by their competitors. And they did—thereafter![28] And Grace Winn thereafter personally met most of the newly arriving NASA families at the airport. In a short time she became legend.

She talked to newcomers about schools, homes, children, taxes, and even whether they should leave their comfortable homes and lives in Hampton or Newport News, Virginia, and come to Houston at all. After talking to Grace, many came willingly. She kept cards to send to members of families who were ill, and sometimes took them food or flowers. She knew,

Suddenly, Tomorrow Came . . .

she said, how lonely it is to be sick in a strange town. Max Faget said that Grace Winn wanted everyone to buy a home near her in Memorial Forest, a place, Faget said, that unfortunately was beyond the reach of most of their pocketbooks, and quite distant from the Manned Spacecraft Center.[29]

Grace, officially described in a Manned Spacecraft Center announcement as a "gracious and talented lady," arranged to have manuals for Texas driving tests on hand for the newcomers, rental cars provided at flat minimum rates, brochures about houses and apartments, information from the Better Business Bureau about buying automobiles and property, information on home and deed insurance, and lists of doctors, dentists, veterinarians, baby-sitters, and dealers who sold boats and fishing equipment. She provided maps, books and charts about weather. She also stocked books about local insects and snakes for the wary new arrivals. "We tried," she said, "to think of everything possible." The Retail Credit Association facilitated the hookup of telephones and utilities and opening charge accounts. In 1962 Joan Pesek and Linda Sauter provided her "gracious and patient" assistance in the relocation office. Despite the personal touch and the genuine sense of welcome, moving to Houston was a difficult business. There were few places to rent at a reasonable price anywhere, and few suitable homes for sale in the area. What would become the NASA community was amidst a then desolate and remote farming and fishing area south of Houston.[30]

Grace remembered the excitement when, after a press conference, astronauts Ed White, Jim McDivitt, and Frank Borman decided to stay and look for a home before they returned to Langley. Grace helped them in their search and, at the end, as Ed White negotiated for his house, neighborhood children came up to Grace who had stepped out front and asked her if those men were astronauts. Grace told them "yes," and the kids ran up and down the street yelling "astronauts are in the house!" When White and McDivitt came out, the children asked for their autographs. After obliging, both expressed surprise to Grace that anyone would want their autograph, because they hadn't done anything. She told them that this was Houston's first experience of this type and they would soon know what it really meant to be asked for autographs. Wesley Hjornevik believed that what "really made the difference was the welcome that was given us by everyone NASA people bumped into whether the guy at the gas station, the grocer, or their new neighbors."[31]

The first edition of the *Space News Roundup,* published on November 1, 1961, announced that the old STG had ceased to exist and in its place stood the new Manned Spacecraft Center. Moreover, this center had a new home, in Houston, Texas, where "the people . . . have literally welcomed MSC personnel with open arms."[32] And it was true that Houston had an affair going for the NASA contingent—at least for awhile. After a year or so some of the novelty and maybe even the affection began to wear a bit, but then Houston was so busy growing one could not spend too much time concentrating on any one aspect of that growth. Public demonstrations of support declined after awhile, but Houston's adoption of the MSC personnel and families was every bit genuine.

Although people such as Mrs. Winn helped alleviate the trauma of a distant move, efforts to continue scheduled operations while setting up business in a new city, all the while building a new $60 million center and making massive additions to personnel and programs in a distant place, placed heavy burdens on all personnel and their families. It required a superhuman effort and enormous cooperation from many sectors of public and private life.

Help came from unexpected places. A few days after Hjornevik's advance party arrived, Wes Hjornevik called to ask them to set up offices in Houston, which had not been a part of their assignment. He called Wednesday evening and wanted spaces ready that Friday. By chance, when the call came in, Marvin Kaplan, the manager of the Gulfgate Shopping Center, was with the group. He said he had two empty dress shops that they might use. The six people on hand promptly walked across the street to the shopping center and decided that indeed the Gulfgate Shopping Center would be fine for temporary quarters. Fingers Furniture Company in Houston offered furniture free of charge; Southwestern Bell installed telephones promptly without a purchase order; Joske's Department Store provided the drapes without charge; a leasing car company provided complimentary automobiles until government vehicles became available; and Continental Airlines offered their hostesses to serve as receptionists until regular government staff could be hired. On Friday the General Services Administration (GSA) delivered a load of office furniture from Dallas at the same time a load of furniture arrived from Fingers. Fingers Furniture stored the GSA equipment for later use.[33] By Friday afternoon, the new MSC administrative office was ready for use, but the NASA "outfitters" soon realized that such charitable goodwill could create legal and ethical entanglements and took precautions to legally document all aids, assistances, and purchases. No problems ever resulted from Houston's generosity.

The six-man advance party received their permanent assignment to MSC about 4 weeks after their arrival and thus became the first full-time NASA employees. They spent most of their time providing orientation and "education" for the local citizens. Local businessmen needed to know how to do business with the government. They had difficulty understanding why they could not take their prospective government customers to a ball game, give them gifts, or take them on hunting expeditions. Chambers of Commerce were alerted to the hazards of a "boomtown" development. School boards were informed of the prospective enrollment growths and interest in quality programs.[34] Slowly at first, but at a rising rate, a nascent NASA community began to take form on the southern reaches of greater metropolitan Houston.

In the early days, life tended to be haphazard both at the Langley Center, where new STG personnel were gathering, and in the community adjacent to the new Houston center. Many STG personnel arrived at Langley expecting to be permanently stationed there. Others hired on as STG employees knowing there would be a move, but not knowing where. The transition was difficult on employees and their families. Housing and office space were scarce at Langley and in Texas. Wesley Hjornevik, who arrived at Langley in October 1959, managed to rent a large home for his family who would join him later. He became Grace Winn's counterpart at Langley—but in an official capacity. He managed most of the new hire and transfer operations for the STG. In fact, some new employees such as Bill Parker, Ed Campagna, Floyd Brandon, Stu Clark, and a few others "batched" at Hjornevik's home until "Mrs. Hjornevik finally arrived and booted us out." Clark and Parker then rented three houses on Lighthouse Road on Chesapeake Bay, provided maid service, linens, and utilities, and charged $35 a month to single employees or spouses who needed temporary quarters pending the arrival of their families. At times as many as 25 employees lived in the three homes. Some "dissidents" broke away from the Chesapeake fraternity and began their own enclave elsewhere. In 1962, the Hjornevik's and many of the STG employees made their permanent move to Houston, but the new center was not ready for final occupancy until 1964.[35] Getting temporary office quarters for center operations was about as hectic as acquiring living space.

The Houston Chamber of Commerce provided Martin Byrnes a temporary office in its suite, and Chamber members and employees, including Pat Patillo, Ben Gillespie, Marvin Hurley, and Gordon Turrentine helped assemble lists of spaces that might be leased as temporary quarters. With the help of the GSA field office manager in Houston, Byrnes reviewed the possibilities and visited selected buildings. He brought his recommendations with photographs, floor plans, and descriptions back to Langley where he, Gilruth, Walter Williams, and staff people in various facilities decided to seek four specific buildings located in the vicinity of the Gulf Freeway and the Old Spanish Trail in south Houston. Upon his return to Houston, Byrnes and the Houston GSA office manager went to see the GSA Regional Commissioner in Dallas.[36]

The GSA staff wanted specifications rather than specific buildings, and indicated that staffers could not go to Houston to begin the work for at least 2 weeks, property assessments would take another month, the contract would take 90 days to prepare, and possibly in 6 months the buildings could be under contract. Byrnes responded that he needed them in 10 days, and if it required more time, then NASA would take care of the leasing itself. About this time the commissioner came into the room and said that instructions from the Washington office were to meet the schedules established by NASA. In approximately 3 weeks the Rich Fan Company building, the Houston Petroleum Center, the Farnsworth-Chambers building, and (somewhat later) the Lane-Wells building were under government lease. Dr. Stan White's Life Sciences Division moved to the Lane-Wells building, Max Faget's Space Flight Office took the Rich Building on Telephone Road, while headquarters occupied the Farnsworth-Chambers building, and the Houston Petroleum Center absorbed other offices (figure 1).[37]

Other buildings leased included one formerly used by the University of Houston as a television-radio station, which MSC converted to a computer facility. A former Canada Dry Bottling building became a machine shop; the unused Minneapolis-Honeywell building housed the photographic labs of the Public Affairs Office; apartments were leased; a former bank building became the personnel office; and later a vacated Veterans Administration building in downtown Houston became available.[38] But a large number of MSC personnel found office space at Ellington Field.

Ellington Field, established as a World War I training base was reactivated during World War II. By 1961 the facility had been inactivated to serve primarily as a reserve training facility. The buildings available to NASA were mostly World War II Wherry-type barrack structures with wooden walls and no air-conditioning. The light construction and wooden raised flooring made them unsuitable for labs or shops, but with some renovation (at $6 to $7 per square foot) adequate for office space. Before construction was completed on the MSC site, some 1500 personnel worked at Ellington Air Force Base.[39]

Preliminary steps leading to the construction of the new center in Houston actually began before the acquisition of temporary facilities. Arrangements had been made at the Washington level for construction of the new center to be managed by the Corps of Engineers. On Monday, September 24, Ed Campagna, James M. Bayne and Marty Byrnes flew to Dallas to meet with Colonel Paul West, the District Engineer for the Fort Worth

Houston - Texas - U.S.A.

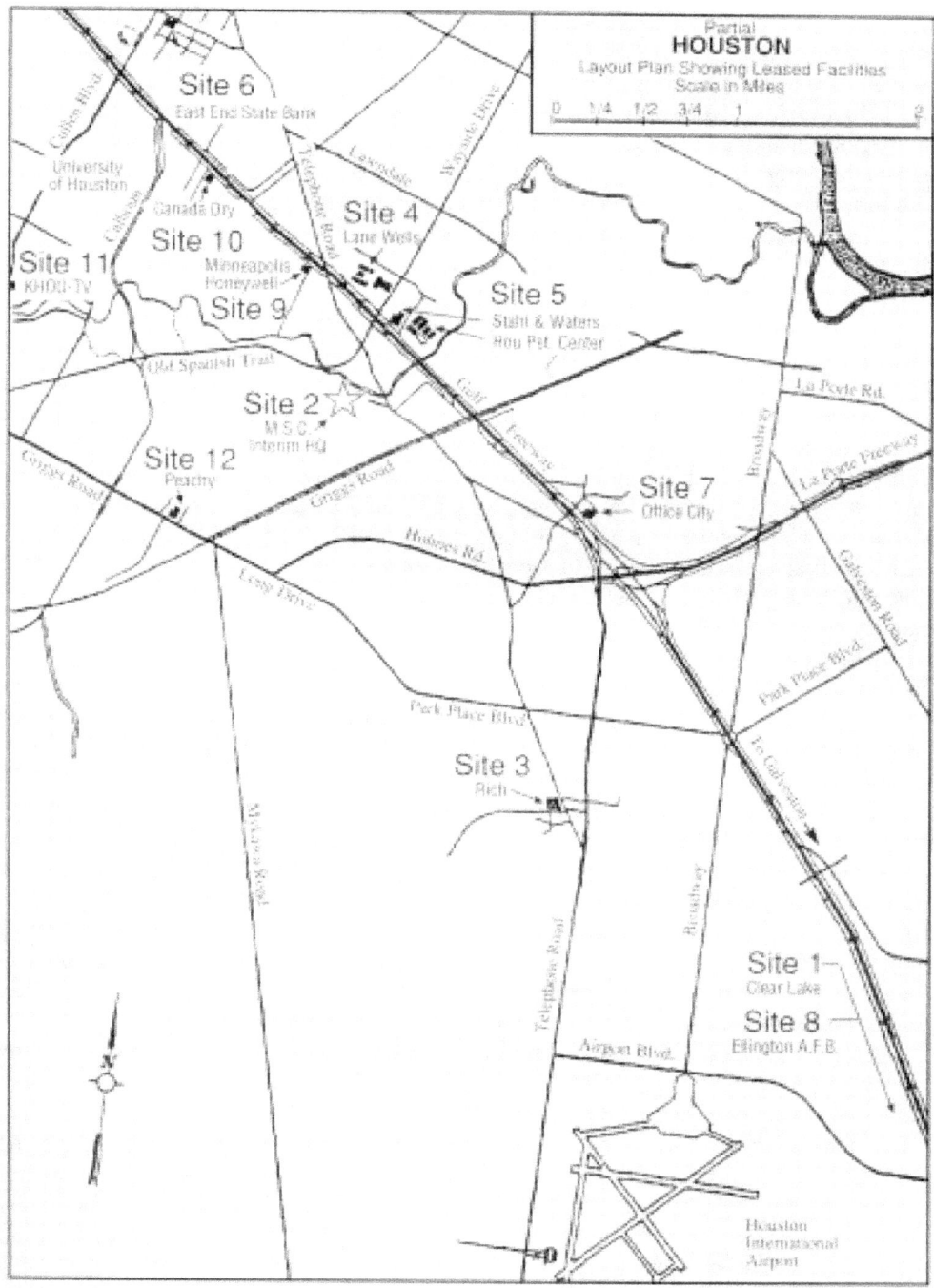

FIGURE 1. Interim Facilities Leased as of August 1, 1962

District of the Corps of Engineers who, with his staff, would supervise design and construction of facilities. Byrnes and Campagna explained that they required a center which would house possibly 5000 people, cost less than $60 million, be designed within 6 months, and be built within 18 months. The reaction from the Corps staff was that it simply could not be done, and that the design time and construction time were wholly unrealistic. "In the midst of a pretty uncontrolled meeting," Byrnes said, Colonel West came into the room and said that he had just talked to the Chief of Engineers in Washington and had been told that the NASA schedule as it had been described "is the way things were going to be."[40] Again the "can-do" attitude, which presumably emanated from the highest levels of the government, prevailed against the impossible.

During 1961 the federal government acquired title to the 1000-acre donation from Rice University and purchased an additional 600 acres needed to give the site frontage on the highway. A 20-acre reserve drilling site fell within NASA's total 1620-acre site. The State legislature authorized and funded the construction of NASA Road 1, a unique and distinctive highway category, for which there was no precedent and has been no sequel. The Corps of Engineers opened a project office in Houston. Design work was underway in January 1962, and construction on the underground utility systems and roadways began in March. Gilruth transferred his headquarters to Houston effective March 1; thus, on that date the Manned Spacecraft Center in Houston, Texas, became a fully operational NASA center, although Project Mercury offices, under the authority of Walter C. Williams, remained at Langley.[41]

The Corps of Engineers had full responsibility for the selection of the Architect/Engineering construction firms, and MSC, represented largely by Ed Campagna and Jim Bayne, served in a consulting capacity. The design process began with the selection of 20 firms considered eligible for a project of the size contemplated. The list included Brown & Root Construction of Houston. When the list was cut to 15 firms, Brown & Root was omitted, but then reinstated at the direction of the Chief of the Corps of Engineers. The list was subsequently whittled to eight firms, and when Brown & Root failed to make the list, Corps Headquarters asked that the company be added. Another cut left five names, again excluding Brown & Root, and after the reinsertion of Brown & Root to the list, that Texas firm, acting on behalf of a consortium of contractors including Charles Luckman of Los Angeles as the "designer," received the $1.5 million design contract for the center.[42] Although it may not have been evidence of a "can-do" attitude, the contractor selection process possibly also felt the invisible hand of the highest levels of government.

Corps supervision of construction also created problems for the developing Manned Spacecraft Center. Corps construction expertise lay in the area of dams and major public works, rather than buildings. Although speed of construction was necessary, speed contributed to errors and oversight. Unilateral decisions made by the Corps during construction affected the operation of MSC. Future MSC operations required sound engineering and quality controls, and the Corps, of course, was charged simply with getting the job done and moving on. A Change Order Board, established by NASA, required that all changes had to be approved by NASA consultants, but this did not close all of the loopholes.[43] Despite difficulties and delays, the new center developed largely according to schedule, and included some real construction and developmental achievements and only a few failures.

Building 1, the Administration Building.

Ed Campagna, for example, listed as among the outstanding construction achievements the overall design concept of using prefabricated exterior building panels, which has since become common practice, but at the time was a pioneering concept. It was a major element in bringing construction to the required speed and keeping costs contained. The center employed a cost-efficient continuous-loop utility services system. A data acquisition center monitored heating and air pumps, fans, valves and electrical systems providing efficient maintenance and operation. Interior office modules, also at the time a novel feature, lowered costs and made for more efficient utilization of interior spaces. On the other hand, the original administrative building design proved nonfunctional, far exceeded costs, and had to be redrafted; the original design and specifications for the environmental chambers totally failed; and utility relays and stations had to be added later to accommodate expansion. Overall, Gilruth and most of the MSC staff believed that the physical facilities filled the requirements of the program and were built with taxpayer cost consciousness in mind. Gilruth attributed much of the efficiency of the plant to the design and planning work of Max Faget, Aleck Bond, and Dick Johnston.[44]

A joint report on the project by the Facilities Division of MSC and the Corps of Engineers for the Fort Worth District concluded:

> Based on economical cost of construction, speed and ease of erection, and general appearance, the architectural concept established by the design of the facilities . . . is considered well suited to the NASA mission and NASA needs. It is functional; it has the clean lines associated with the space age look; it reflects space science in an architectural manner without being ostentatious.[45]

Numerous state and national contractors and suppliers participated in the construction process. Contracts for the first 11 buildings were awarded in December 1962, and within 12 months, by January 1964, 2100 employees were readying for the move to the site with the remaining 600 personnel to be on site by July.[46]

Building a new center, of course, meant much more than building buildings. As Paul Purser carefully explained in an institutional planning study, "Centers are people and competence, not numbers of personnel and facilities,"[47] and it is with that admonition in mind that one must review the lists of buildings built and people hired. And to be sure, there were a lot of both. Between July 1961 and July 1962 the number of MSC employees more than doubled from about 750 to almost 1600 and would almost double again within the year.

That July 4 (1962), Houston celebrated MSC with a truly Texas-style, Texas-sized barbecue. A parade featured in a 60-car motorcade Vice President Lyndon Johnson and Governor Price Daniel of Texas, Senators Ralph Yarborough and John Tower of Texas, Senator Robert S. Kerr of Oklahoma, Congressman George Miller of California (who chaired the House Committee on Science and Astronautics) and Texas Congressmen Olin

TABLE 1. Johnson Space Center Buildings — Construction Costs and Size

	Building	Contract Amount	Sq. Ft.	Cost/Sq. Ft.
1	Auditorium/Public Affairs	$1,099,749	51,840	$21.21
2	Office Building	4,420,487	209,610	21.09
3	Cafeteria	463,180	22,330	20.74*
4	Flight Crew Operations Office	1,944,573	111,911	17.38
7	Flight Crew Operations Lab	825,337	39,621	20.83
8	Technical Services Office	918,204	58,023	15.82
10	Technical Services Shop	1,655,906	77,381	21.40
12	Central Data	907,366	65,930	13.76
13	Systems Evaluation Lab	1,320,670	72,579	18.20
15	Instrument and Electronics Lab	1,344,632	74,277	18.10
16	Spacecraft Technical Lab	1,507,612	97,228	15.50
417	Garage	156,052	6,874	22.70
419	Support Office	259,285	19,170	13.53
420	Support Shops and Warehouse	554,962	44,127	12.58
221	Substation Control	13,244	682	19.42
322	Water Treatment	12,666	511	24.78
223	Sewage Treatment	22,122	326	67.86
24	Central Heating and Cooling	340,091	25,006	13.60**
25	Fire Station	143,349	7,220	19.85

* Excluding kitchen equipment.
** Excluding equipment; building shell only.
Note: Buildings 1 and 2 were later renumbered at the instigation of Deputy Director Sigurd A. Sjoberg who wondered why the main administration offices were not "Building 1."

Houston - Texas - U.S.A.

A 1974 aerial view of JSC denotes the marked changes in the countryside south of Houston, Texas, since the inception of NASA's manned spacecraft program.

Teague, Bob Casey, Albert Thomas, and Clarke W. Thompson; the seven Mercury astronauts; NASA Administrator James E. Webb, Dr. Brainerd Holmes (director of the Office of Manned Space Flight), MSC Director Robert R. Gilruth, and Associate Director Walter C. Williams; and others. It began and ended at the Sam Houston Coliseum. An elaborate barbecue of beef, chicken, pork, potato salad, beans, and all the trimmings followed at the Coliseum for *all* employees of the MSC and their families and guests. It was a rather astounding and heartfelt welcome from the City of Houston and the State of Texas.[48]

Soon after, another Houston welcome was extended to President John F. Kennedy who was making a whirlwind tour of the Nation's space facilities. Twenty-five thousand Houstonians met him at the airport on Tuesday, September 11. The following day almost 200,000 people lined the parade route to cheer the President, Vice President Johnson, Congressman Albert Thomas, James Webb, the astronauts, and MSC officials. Kennedy addressed a full house in Rice Stadium. "The exploration of space," he said, "will go ahead, whether we join it or not . . . no nation which expects to be a leader of other nations can expect to stay behind in this race for space." The Manned Spacecraft Center is evidence, he said, of "how far and how fast we have come, and how far and how fast we must go." Houston, Kennedy said, which was once the "furthest outpost in the old frontier of the West will be the furthest outpost on the new frontier of science and space."[49]

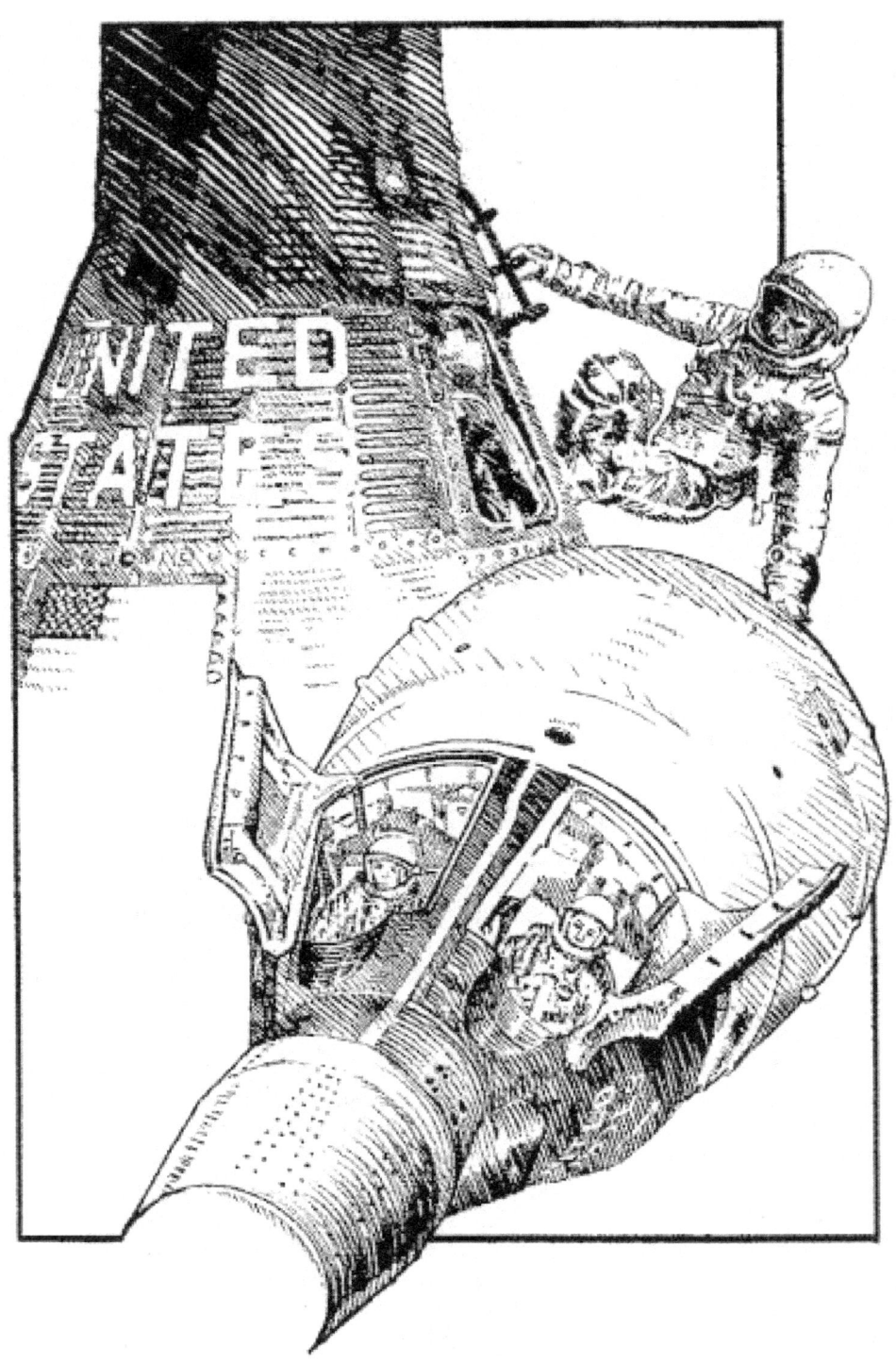

CHAPTER 4: *Human Dimensions*

Few years have been so critical to the American space program as those between roughly March 1962, when old STG and new center employees began relocating to temporary quarters in Houston, and June 1964 when the new MSC formally opened for business. Few years have been so demanding of human energy, effort, and simple endurance. During these years, the Mercury, Apollo and Gemini programs ran concurrently while the MSC was being designed and built. Few years have been so productive. Not only did things get done, but a very important management system or style that became referred to later as the "Gilruth system" became implanted in the organization and culture of the developing space center.

During the spring of 1962, 751 STG/MSC employees moved to Houston from Langley, Virginia, and by July administrators had hired another 689 people who joined the staff in Houston. Personnel worked throughout a dozen buildings in disparate locations in Houston, while construction contracts were being let and buildings built on the site of the new center.[1] But the real business at hand had to do with putting Americans in space, not buildings on Earth.

Not only, said Bob Gilruth, is our mission "to develop here in Texas the free world's largest and most advanced research and development center devoted to manned spaceflight," but the real business at hand is "to manage the development of manned spacecraft and to conduct flight missions." In our work on these missions, he said:

> . . .during the past few months the Manned Spacecraft Center has doubled in size; accomplished a major relocation of facilities and personnel; pushed ahead in two new major programs; and accomplished Project Mercury's design goal of manned orbital flights twice with highly gratifying results.[2]

That was July 1962. By May of 1963, with six more successful manned Mercury flights completed, Mercury ended—and within the year, the first unmanned Gemini vehicle sped into orbit.

Mercury began unpromisingly on August 21, 1959, when the first Little Joe Mercury capsule prototype launch was canceled due to faulty wiring that sent the capsule, without the launch vehicle, on a premature trajectory a short distance out in the ocean from its launch point on Wallops Island. In 1960 there was talk of "slippage" in the space program. Rod Rose remembered that while awaiting delivery of the Mercury capsule, he urged Gilruth to "beat the Russians" by sending an astronaut aloft in a Little Joe module, but Gilruth declined saying that "we're running a research program, not a PR stunt team."[3] That attitude helped provide stability and direction during the high-pressure days of the early sixties.

In 1961 and 1962, amidst the suitcase environment of the move to Houston, Project Mercury enjoyed its greatest successes and the first Apollo systems began flight tests. On May 5, 1961, Alan B. Shepard, launched from Cape Canaveral and directed by the Mission Control Center at Canaveral as were all of the Mercury flights, completed America's first manned space mission. "When Ham (the chimpanzee which had flown

Suddenly, Tomorrow Came . . .

Mercury-Redstone (MR) 3, the United States' first manned spaceflight, was launched from Cape Canaveral May 5, 1961 (left). Astronaut Alan B. Shepard, Jr., piloted the "Freedom 7" to an altitude of 116.5 statute miles, attained a maximum speed of 5150 miles per hour, and landed 302 miles downrange from the launch site. The MR-4 (right), launched July 21, 1961, had an enlarged window hatch, improving the pilot's ability to see. The loss of the escape hatch at splashdown caused the craft to sink, but Astronaut Virgil Grissom was safely retrieved.

earlier test flights) refused to board the capsule, I had to make the flight," Shepard told a large audience at the Johnson Space Center years later (in 1989 during the 20th anniversary of the first lunar landing). Virgil Grissom followed Shepard into space in July. In September an unmanned Mercury capsule made a complete Earth orbit. While public attention focused on the Mercury program, a flawless launch of the first Apollo-type vehicle (a Saturn SA-1) was completed from Cape Canaveral on October 27, 1961. Enos made the first "chimpanzeed" orbital flight aboard a Mercury capsule in November, and finally, Mercury astronaut John Glenn completed the first American orbital mission (4 hours and 56 minutes) on February 20, 1962.[4]

Following Glenn's harrowing return within his capsule-turned-fireball through Earth's atmosphere, the entire flight being one of America's most closely followed news events of modern times, President John F. Kennedy expressed "great happiness and thanksgiving of all of us on the completion of Colonel Glenn's trip." But we have a long way to go in the "space race." ". . . this is the new ocean," Kennedy said, "and I believe the United States must sail on it and be in a position second to none."[5] Scott Carpenter made another significant step across the threshold of space soon thereafter.

The MSC's weekly journal, the *Roundup* described Carpenter's launch aboard "Aurora 7" on May 24:

> ... a massive black silhouette poised on the skyline a mile and a half from the press site where hundreds of watchers held their breaths. Mercury-Atlas 7 hung for agonizing seconds, poised on a column of fire, then rose. She lifted into the low clouds, appeared again above them, flashed into the sunlight and out of sight, her heavy thunder rolling back over the Earth she had left behind.[6]

The flight marked "a major milestone in man's pioneering venture into space," but it almost ended in disaster when fuel and temperature problems aborted the flight earlier than planned, and Carpenter's landing was 250 miles off target. He, as the chimpanzee Ham had been years earlier, was finally located and retrieved.[7]

G. Merritt Preston managed launch operations for Mercury from Cape Canaveral, and the Mission Control Center at Canaveral directed flight operations. To be sure, Mercury flight operations were rather minimal because the capsule was not navigable. As Christopher Kraft explained later, Mercury flight control basically occurred before launch; because once you launched, the main function was to try to maintain contact and wait until it came down. Control center operations changed markedly with Gemini and Apollo. The

Astronaut John Glenn, Jr., enters the "Friendship" spacecraft during rehearsal exercises. Glenn made the first American orbital spaceflight in the Mercury-Atlas 6 craft on February 20, 1962.

Mercury Project Office as well as the home base for Mercury astronauts remained at Langley, Virginia, until November 1963, when the Mercury Project Office closed and Kenneth S. Kleinknecht and most of his staff moved to Houston.[8]

The center in Houston concentrated on the "new" projects mentioned by Gilruth—Apollo and Gemini—and much more so on the former than the latter. As Mercury neared completion, most Mercury project people moved directly to the Apollo program rather than into or through the Gemini program. This ultimately created some special problems for the manned space program.

In December 1961, Project Gemini (originally designated Mercury Mark II), a two-person manned spaceflight program, was initiated to provide experience in flight endurance, rendezvous, and extravehicular activity until Apollo became operational. Thus, for several years before being finally relocated at the Clear Lake site in late June 1964, the work of the space center included the operation of Project Mercury, design and contracting for projects Apollo and Gemini, the design and construction of the Manned Spacecraft Center, the recruitment and training of employees and astronauts, the testing of both Gemini and Apollo hardware and initial flights of both Gemini (Gemini I, April 8, 1964) and Apollo (SA-6, May 28, 1964) systems.[9]

". . . a massive black silhouette poised on the skyline . . . hung for agonizing seconds, poised on a column of fire, then rose. She lifted into the low clouds, appeared again above them, flashed into the sunlight and out of sight, her heavy thunder rolling back over the Earth she had left behind." Thus began Scott Carpenter's venture into space.

The technical challenges of achieving manned spaceflight sometimes seemed less imposing than the human dimensions. Although the space programs seemed to bring America to the leading edge of science and technology, the technology of space may actually have been more in place than the social engineering required to integrate such diverse fields as bioengineering, astrophysics, metallurgy, ceramics, and computer electronics. The management of these large scale endeavors went beyond such experiences as the construction of the intercontinental railroads in the late 19th century or building the Panama Canal in the early 20th century. Even the more recent Manhattan project of World War II and the Polaris missile program differed sharply in their costs, scale, and the extent to which they integrated diverse bodies of knowledge and technologies still in the research and developmental stage. There was little precedent for the most mundane business of determining costs, allocating contracts, and reviewing progress on such a large scale and in such a defined time period.

Moreover, NASA was enjoined to design, build, and operate machines never previously built, and to help create the knowledge and technology necessary to build and fly these machines. The new generation spaceflight vehicles had to be man-rated, that is, be certified as a safe environment for humans and be responsive to human operators. Despite the initial successes of Mercury, whether humans could long survive and function effectively in space had not been resolved. Unlike conventional aircraft, a space vehicle's maiden voyage was its first flight mission. There were no test flights into space. Spaceflight required innovations and inventions in technology, the accumulation of enormous human and material resources, and the development of new management structures and practices. Putting Americans in space was a most difficult assignment by every conceivable measure. The frontiers of space alluded to by President Kennedy were less beyond Earth and more at the site of MSC and its associated NASA installations, and in the workshops and laboratories of the developing American aerospace industries.

Spaceflight involved the Nation's best engineering and scientific talents and energies, and a considerable amount of the public's money. During its first 5 years of operation, total NASA annual expenditures jumped from about $300 million to $5.1 billion. By comparison, federal expenditures on defense (1959-1964) rose from approximately $45 to $55 billion. The administrative budget of the MSC went from $9.2 to $88.5 million between 1961 and 1965, and the direct Apollo budget soared to $2.7 billion.[10]

Despite frequent changes in formal assignments and on organizational charts, the management personnel working on the manned spaceflight programs remained remarkably stable. Most of those who began the program with the STG were there two or even three decades later. Most of those who came into the program at the Houston center remained there throughout their professional careers. Robert Rowe Gilruth was one of those who cast an indelible stamp on America's space program from its conception in 1957 until his retirement in 1973.

George Low, a NACA engineer from the Lewis Flight Propulsion Laboratory, went with Abe Silverstein and others from Lewis to NASA Headquarters in 1958, and became "Gilruth's representative in Washington" before he joined the center staff in Houston in 1964. Low referred to the MSC as:

> . . . Bob Gilruth's center. He built it in terms of what he felt was needed to run a manned spaceflight program. . . . it is clear to all who have been associated with him that he has been the leader of all that is manned spaceflight in this country. There is no question that without Bob Gilruth there would not have been a Mercury, a Gemini, or an Apollo program. Everything we've done, our approach, has grown out of the Bob Gilruth formula for running Project Mercury.[11]

Although the organization changed, Low said, and people came and went, the people who run the center and make the decisions have had primary management roles all the way through from the beginning, and they are people who shared Bob Gilruth's vision of what the center should be. Gilruth did not necessarily initiate ideas or projects; he rarely did so, but freely gave credit to those who did. His great strength was in sorting out the wheat from the chaff, and in inspiring others to accept his decisions.[12]

Thus, Low said, Gemini was Jim Chamberlin's idea, but it was Gilruth who "latched" onto the idea and pushed it into NASA circles, insisting that "we needed to learn how to fly in space in applications more sophisticated than Mercury before attempting to land on the Moon."[13] "Gemini 7/6," involving the orbit and rendezvous of two spacecraft, was another person's idea but adopted by Gilruth as a necessary step in spaceflight. "Bob," said Low, "is more of a leader than a manager. He has ideas; he inspires confidence and knows what's right and what's wrong; but he also expects the rest of us to originate ideas and carry them through to completion." It is Bob and his people who made things go, Low concluded, and added ominously, that "it's when someone comes along who hasn't been brought up under Bob and hasn't learned from him that we have problems."[14]

Low was both right and wrong about Gilruth. He was right that Gilruth inspired confidence and seemed to know instinctively what was right and wrong. He was wrong in attributing the entire space program so singly to Gilruth. Gilruth's success was due in good measure to the fact that he "truly represented" the people working with him. His management, according to Paul Purser, David Lang and others, could best be defined as "management by respect," and although they did not say so specifically, that respect derived largely from the technical expertise which Gilruth shared with his associates. To Gilruth, the STG and those who worked with him were "associates"—just that—not employees or underlings.[15] Thus, the MSC at its best represented a collegial association of engineers gathered together almost fortuitously to complete a task, to build a bigger, better, faster, and more complex machine than ever before had been built. To be sure, the collegiality did have a raw edge. MSC personnel also comprised a pool of talented, young and highly competitive engineers and astronauts who thought that collectively they were very good at what they were doing, and that individually each was better than the other.

Manned spaceflight required not only a regrouping of engineering and scientific knowledge, but a reorientation of the mind-set and culture of the engineering community. Although the engineering expertise of the NASA/MSC community was similar to that of the old NACA, there were distinct differences between the two which tended to be accentuated in the MSC culture. The NACA had been primarily a research, service-oriented organization. NASA, but especially the Houston spacecraft center, became a development-applications-operations organization. Thus, when decision time came at the Langley Research Center for an engineer to join the STG or not to join, or to go to Houston, Texas, or not to go, the underlying incentive had much to do with a personal preference for research or for development.

The cultural delineation between one group of engineers and the other, and indeed between the old NACA and the new NASA, is reflected in part in the careers of Langley engineers such as W. Hewitt Phillips and Robert G. Chilton. Phillips participated in early studies of Earth versus Moon orbital missions and space rendezvous feasibility, but he chose to remain at Langley and concentrate on research. Chris Kraft, who worked under Phillips at Langley for many years before coming to Houston, described him as his mentor and one of the most knowledgeable and ingenious aerospace engineers. Phillips taught him most of what he knew about engineering, Kraft said. After more than four decades observing the growth and programs of NACA and NASA and the Johnson Space Center from the perspective of his laboratories at Langley, Phillips objected to what he called "research by decree." Ideas, he said, cannot be superimposed from the top. And buildings or centers, he said, must be filled with people who can generate ideas.[16]

Robert G. Chilton, who worked with Phillips at Langley in 1959 and 1960, easily chose the STG and Houston. Chilton, as head of the Flight Dynamics Branch under Maxime Fagets' Flight Systems Office became a key ingredient in the "development" aspects of the manned spaceflight ventures.[17] Whereas Hewitt Phillips' work might end with the conceptual and theoretical framework for a space rendezvous, the developmental engineer wanted to make it happen. But the demarcation between the research engineer and the operations or developmental engineer was not nearly so marked as the delineation between the scientist and the engineer.

The basic reaction of the scientific community to Sputnik was to avoid the heroics and concentrate on upgrading the status of science along a broad front in American society. Sputnik and space offered an opportunity for American scientists, but of a different cast than that for the engineer. Most advocated federal support for expanded educational programs, more scientific input in (and control of) weapons development and better working conditions for scientists in federal agencies and projects. Months after Sputnik, the American Association for the Advancement of Science presented a major discussion and document on "science and public policy," but made no mention of a manned space mission. Many scientists and their organizations actively sought to dissuade others from participating in a crash program in space. Only later did the scientific community join in support of an independent civilian space agency—as something preferable to the spectre of a military space agency.[18] But the dichotomy or tension between space as a subject of research and space as the arena for manned flight continued throughout the manned spaceflight programs.

A similar cultural "stress" pervaded the engineering community, where the research engineer who provided the theoretical design for a space capsule stood at some distance from the operations engineer who wanted to fly it. Development became the bridge between the research and the applications or operations engineer. Development, which might be equated to the refinement stage of invention (in which the invention becomes functional and marketable), provided a framework for the spin-off or creation of new ideas and for identifying new applications for old ideas. Development also provided a key element in the unique management style that characterized not just the director, Robert Gilruth, but the entire MSC engineering community. As Max Faget observed, in the early days of the manned flight programs, "there was not a lot of substance to spacecraft technology." Much of what was learned came through experimentation. "We had our own hobby shops," Faget said.[19] In these shops, MSC engineers helped create the new technology of space, which in fact, often was the application of old technology in new ways.

Managing engineers at the space center coupled the older NACA "do-it-yourself" in-house tradition with the newer NASA system of contracted work as an effective management and quality control tool that was really an intrinsic part of the so-called "Gilruth system." Thus, engineering divisions and laboratories at the center became miniature developmental and manufacturing centers where prototypes of flight systems, such as Mercury (Little Joe) capsules, heat exchange devices, or computer hardware and programs were devised or perfected. MSC engineers knew what the contractor was producing and how to manage and direct that production because they participated in the design and had hands-on experience in the fabrication and testing of the product.

Chris Kraft, in fact, explained the Gilruth management system as a "make it work, and if it doesn't work find something that will" attitude. Because of their hands-on experience, NASA engineers could more effectively manage the work of the NASA contractors. It also meant that center engineers became cooperators and collaborators with the contractors, rather than simply purchasers of hardware produced by a manufacturer from a given set of specifications. Managing engineers wanted their contractors to succeed and assisted them in that effort. In the design, manufacture, or operation of components, NASA engineers were usually as knowledgeable and experienced, or more so, than their counterparts in industry. This "nuts and bolts understanding" more than anything else defined the relationship between the engineers and staffers within MSC, and between the center managers and their contractors who produced the final product. Organizational flow charts and diagrams meant little compared to the fact that a group of engineers sat down and tried to do a job they understood a little better than those outside their community.[20] As time passed, the MSC management/contractor relationships became institutionalized in such roles as the subsystem manager, contract representatives, program management offices, and contract change boards.

Under the collegial style of management, the pattern of authority relationships became very suffused. Relationships between programs, divisions, and individuals tended to "float." Communications occurred on both horizontal and vertical levels on a selective, as-needed basis. Program or division heads under such a system operated with considerable authority and responsibility and could assume somewhat more or less of either as they required. They were answerable, not so much to a superior, but to their peers. The collegial system of management worked in part because routine administration was divorced from project management.[21]

Gilruth and his managing engineers concentrated on engineering and left the more routine fiscal and personnel management to carefully selected associates. One of Gilruth's great strengths was his ability to find the right people to do the job and then give those persons full authority and responsibility. Bob Piland described Gilruth as less an administrator and more a "genius in handling people."[22] As time passed and programs developed, management style and structures began to change, but the basic system or concept remained much the same.

Until the move to Houston, Gilruth directed Project Mercury. Charles J. Donlan served as Assistant Director for Mercury until September 1959 when he became Associate Director for Project Mercury (Development) and Walter C. Williams was appointed Associate Director for Project Mercury (Operations). Donlan, who completed work in aeronautical engineering at Massachusetts Institute of Technology in 1938, worked in the Langley Spin Tunnel and the Stability Tunnel before heading the High-Speed 7- by 10-foot Tunnel, and later worked closely with Gilruth before becoming his technical assistant. Williams, who had supervised tests for NACA of the Bell XS-1 in which Charles E. "Chuck" Yeager flew the first manned supersonic flight, appropriately remained the "operating" director of Project Mercury during the move to Houston until he was replaced in a general reorganization that brought Kenneth S. Kleinknecht to head the Mercury Project Office in October 1962.[23]

At that time three project offices (Mercury, Gemini, and Apollo), three functional or line offices (Engineering and Development, Operations, Information and Control Systems), and a variety of support offices reported to the Director. Offices, designations, and work assignments tended to be very fluid and amorphous during the earlier years, and organizational charts at best only reflect a moment in time and imply a rigidity that did not exist.

Gilruth delegated matters having to do with personnel hiring and pay (not recruitment), business affairs, contracting and purchasing to an extremely able team organized by Wesley L. Hjornevik. Gilruth gave him a "wide area of responsibility . . . perhaps wider than most administrative people in other centers in NASA," Hjornevik recalled. When he joined the STG at Langley in March 1961, Hjornevik said, of the 700 people assigned to the STG only 30 or 40 were in support positions while the remainder were all technical people. Administrative support came from the Langley Research Center staff. Thus Hjornevik had to build an administrative support staff from scratch.[24]

Hjornevik recruited Dave Lang from the Air Force to handle contracts and procurement. Lang had been the contracting officer for the B-70 bomber program before joining the MSC group. In having responsibility for negotiating, awarding and administering all contracts for procurement by the center, including the contracts for the research, development and manufacture of manned spacecraft and related equipment, Lang (assisted by the source evaluation boards chaired by an engineer) spent literally billions of dollars during his long tenure, and he did so in a way that would complement rather than lead or impose upon the program and technical work.[25]

What we were trying to do, Hjornevik explained many years later, was to help these people succeed, both the program offices and the contractors. Hjornevik and Lang assigned contract and procurement office representatives to project offices and, when appropriate, to the contractors. These representatives considered themselves staff people for the work to which they were assigned. Thus, when a project office or a contractor needed to purchase a certain piece of equipment or develop a contract or subcontract, the "business" aspects of getting it done would be dispatched as quickly and efficiently as possible. It eliminated much of the hostility, lethargy, and bureaucracy characteristic of large-scale enterprises. It was not, Hjornevik admitted, always successful.[26]

Hjornevik picked Rex Ray from the Atomic Energy Commission to be Chief of Finance because of his work there with private contractors and in auditing contracts. Stuart Clark, a deputy director of personnel with the Army Ballistic Missile Agency in Huntsville, had greatly impressed Hjornevik as a man with "a lot of ideas, who was very personable, and who had experience recruiting the kind of people that we would need in the R&D business." As with contracting and procurement, the personnel officer assigned representatives to program offices and major divisions at MSC so that they could respond immediately to the needs of those offices. This approach eliminated much of the

Suddenly, Tomorrow Came . . .

Dr. Max A. Faget (December 1964), Assistant Director for Engineering Development, Manned Spacecraft Center. Dr. Faget strongly influenced the design and development of every American spacecraft from Mercury through Shuttle.

traditional stress between the administrative and operating divisions.[27]

Hjornevik brought with him from NASA Headquarters Charles (Chuck) Bingman and Phil Whitbeck, who later became Deputy Director for Administration. For the construction of buildings and facilities, Hjornevik recruited a team from widely diverse backgrounds including Leo T. Zbanek, I. Edward Campagna, and James M. Bayne. Zbanek had experience in managing the logistics of heavy construction in the Taconite iron ore system in Minnesota; Campagna had just completed the management of a major research construction project for the Department of Agriculture at Ames, Iowa; and Jim Bayne, an architect, came from an architectural firm in Detroit, Michigan, and "was extremely capable in original design and control" of major projects.[28] The emphasis on cooperation and contractor support complemented the collegial style of management at MSC. A spirit of cooperation and commitment also contributed significantly to the completion of the missions of the MSC.

Those missions, including Mercury, Gemini, and Apollo, were contemporaneous and interdependent programs and were all under way prior to the relocation of MSC to Houston on March 12, 1962. Preceding that move, on January 15, Gilruth organized the center. He created independent project offices for Mercury, Gemini, and Apollo, and the Office of Research and Development (redesignated within the year as the Office of Engineering and Development), under the authority of MSC Assistant Director Max Faget. Faget's office had responsibility for creating and implementing programs for research and development in the areas of space research, space physics, life systems, and tests and evaluations to support and advance manned spacecraft development. Four divisions under him included a Spacecraft Research Division headed by Charles W. Mathews, the Life Systems Division under Stanley C. White, and a Systems Evaluation and Development Division directed by Aleck C. Bond. Each division then established various "branches" which were in turn subdivided into such "sections" as might be required. These could be and were reformatted as often as required, thus the idea again was not to establish a static organizational structure, but to organize to complete the jobs required. Certain elements necessarily retained some permanency. For example, the Structures Branch operating under Mathews' Spacecraft Research Division managed a Heat Transfer Section, a Loads Section, and a Structural Analysis Section, which provided basic engineering analysis and design for any of the program units or contractor projects.[29]

When Kenneth Kleinknecht became Manager of Project Mercury (October 1962), Walt Williams assumed broader functions as Associate Director of the center. The Mercury Project Office had full responsibility for technical direction of the McDonnell Aircraft Corporation and other industrial contractors assigned Mercury projects, and for coordinating Mercury activities and flights with other centers and agencies.[30]

Notably, those who had "done Mercury" that is, participated in the conceptualization and design of the project, such as Max Faget, Paul Purser, Robert Gilruth, Chris Kraft, and Aleck Bond "flowed" into new programs and problems which even now began to supercede Mercury in terms of engineering and management effort. But for the most part, with the notable exception of Chris Kraft, the STG-development types could be found in the "line" or functional offices, such as Faget's Research and Development Office, rather than in the operations or projects offices. It had to do in part with personal preferences or predilections of the engineering character, but it also had to do with keeping a core of managing engineers in the mainstream of all MSC/JSC programs so they might provide the necessary coordination and interfacing with project offices.

"As the organization grew, everybody recruited [was] from outside," Joseph P. Loftus, Jr., Assistant Director for the Johnson Space Center explained. The more senior recruits tended to go to the program offices, the younger recruits to flight operations or the engineering and development type offices where they could be trained in the "disciplines and modes of operations that they were endeavoring to establish," or the "Gilruth system." In addition, as people came off Project Mercury, they often found themselves assigned to technical studies in the development office where their knowledge and experiences could begin to be applied to the new projects.[31] By 1962, Gemini, but especially Apollo, demanded more and more of the center's energies, while the operations elements under Mercury carried that program to its conclusion and developed the expertise to be used when flight operations commenced with Gemini and Apollo.

James A. Chamberlin, formerly Chief of the Engineering Division, became manager of the new Gemini Project Office. Chamberlin had been a key instigator of the Gemini project urging the need to develop operational and flight competencies in preparation for Apollo. Faget suggested the craft contain two astronauts, rather than one, to provide a wider range of flight options. As true with the Mercury office, the Gemini Project Office had authority for technical direction of the industrial contractors, such as McDonnell Aircraft, and had full authority to deal directly with the contractors and with related government agencies.[32] The Gemini program, in part because of its more compressed time frame, sandwiched between Mercury and Apollo, and because of the AVRO contractor-oriented experiences of Chamberlin as well as the Mercury background of the Gemini principal contractor, McDonnell Aircraft, functioned somewhat more autonomously than either Mercury or the following Apollo project.

Mercury engineers moved with Max Faget from Mercury to Apollo-related work, leaving Gemini more in the hands of Chamberlin and the contractors. This transition, or lack of it, was more the product of necessity than of intent. Having three concurrent projects on-line strained the limited personnel of the program. Since Gemini was built upon the previous experiences of Mercury and its contractors, most personnel began to concentrate on the Apollo program, while leaving Gemini to the management of the

Gemini Project Office. But the apparent apartness of the Gemini program resulted in the failure to fully transfer the learning experiences of the Gemini program to Apollo, and later created stress within MSC.

Concurrent with the Mercury and Gemini appointments, Gilruth named Charles W. Frick to manage the Apollo Spacecraft Project Office, with Bob Piland as Deputy Manager. Piland probably had more hands-on experience with the Apollo project than any other space center engineer. Piland was given the job in November 1959 to head an STG study (including H. Kurt Strass, John D. Hodge and Caldwell Johnson) of circumlunar manned spacecraft design and flight, presumably in response to an ongoing study by the Goett Committee on the feasibility of a lunar landing, and by the "New Projects Panel" under Strass which recommended work on a three-person second generation (lunar?) spacecraft.[33]

Many Langley and STG personnel, including Gilruth at this early stage, tended to favor Earth orbital missions, such as a manned space laboratory with a possible lunar landing 10 or 15 years beyond. Others, including Max Faget, supported the idea of a large Earth-launched "Nova" rocket ship that could orbit the Moon and return. Other ideas that quickly began to compete included a lunar orbit from an Earth orbit "sling-shot" launched on a Saturn rocket, a lunar landing using a rendezvous vehicle with a mother ship in lunar orbit (which could have been launched variously by a Saturn or Nova class rocket), and a lunar landing from a rendezvous vehicle in Earth orbit. As interest grew and options became more clear, Gilruth appointed Piland "Chief of Advanced Projects" in September 1960 under Max Faget in what was then the Flight Systems Division. But Gilruth did not acknowledge the existence of an "Apollo" type program and personally preferred Earth orbital missions. He did, however, approve the formulation of "guidelines" for "advanced manned space vehicles" which were addressed to all NASA centers for research and recommendations. These guidelines, or Ground Rules for Manned Lunar Reconnaissance as they were originally styled, gave rather sharp definition to what would become the Apollo program as early as March 1960. Subsequently, in early 1961, a Headquarters committee chaired by George Low concluded that lunar landings could be made either by direct-ascent or by Earth-orbital-rendezvous modes. And of course, Yuri Gagarin's flight and President Kennedy's May 25, 1961, call for a lunar landing redirected energies to consider how a landing should be achieved, rather than whether it should be attempted at all. At that point, in May 1961, Gilruth and Faget created the Apollo Project Office under Piland's direction.[34]

The Project Office approved three concurrent study contracts of $250,000 each to General Electric, General Dynamics, and Martin Marietta. Contacts were also made with Massachusetts Institute of Technology (MIT) engineers for recommendations on possible lunar flight projects. Feedback began to come in from other NASA centers as well. At the conclusion of the contractor and in-house studies, a 3-day meeting attended by more than 1000 persons, including industry and government representatives, was held in Washington where, Piland said, a "huge data dumping" occurred. Out of this convocation came the specifications and work statements for a command and service module suitable for lunar or Earth orbit. A request for contractor proposals (RFP) was released in September 1961, and an evaluation team including Piland, Walter Williams, Max Faget, Wesley Hjornevik,

Dave Lang and others met at the Chamberlin Hotel in Old Point Comfort, Virginia, to review the proposals. In a very close decision, the award for design and construction of a lunar command and service module went to Rockwell International over Martin Marietta. Meanwhile, NASA awarded a separate contract to MIT for the development of a guidance and navigation system. The Rockwell contract was let on December 15, 1961, following which the "old" organization, including Piland's Apollo Project Office which functioned under Faget's Flight Systems Division, was disbanded; and effective on January 15, 1962, the new organization with the autonomous Mercury, Gemini and Apollo Project Offices "reconvened" in Houston, Texas.[35]

At that time, the primary management activities for the center, including research and development activities, also moved "on site," but 2 more years would pass before the center became fully operational and the last large contingent of personnel moved to the site. Paul Purser, Special Assistant to Director Gilruth, and Wesley Hjornevik, Assistant Director for Administration, moved with Gilruth to the new center. Walter C. Williams, the Associate Director, became the Mercury project officer at Langley, with Flight Operations and Flight Crew Operations under his direction. Kleinknecht's Mercury Project Office in Houston maintained liaison with Williams' operations activities and had responsibility for Mercury planning and coordination with contractors and other centers. When Williams moved to Houston in the fall to assume his duties as Associate Director for Operations, the three flight operations divisions included a Preflight Division headed by G. Merritt Preston, a Flight Operations Division headed by Christopher C. Kraft, Jr., and a Flight Crew Division under W.J. North.[36] By the end of 1962, the management personnel of the space center were organized as indicated on figure 2.

The year 1962 had been an incredibly busy but productive one for the spaceflight program. It all took an enormous amount of energy and hard work by the space center personnel. Newcomers (and most were) had to be assimilated and learn their jobs. Dennis Fielder remembers that "everything was in motion" when he arrived on the scene. He described it as a "Brownian motion," the rapid oscillation of small particles suspended in fluids. "What you were supposed to do was not easy to find out," he said. "You had to reach out and capture people."[37]

Meetings were interminable. The MSC Senior Staff met every Wednesday at 9 a.m. and every branch, division, section, and project had meetings—and then representatives of each met with the others. There were so many meetings that Bob Piland in the Apollo Project Office issued a memorandum urging a reduction in the number of meetings and in the conflicting and excessive requirements for participation in those meetings. Practically the entire staff worked long days, 6 and 7 days a week, and took no vacations. Hjornevik issued a memorandum insisting that for reasons of health and morale, every staff member receive (and take) their vacation for a minimum of 2, and preferably for 3, consecutive weeks.[38] Perhaps understandably, family and marital difficulties of NASA/MSC families rose.

Virginia McKenzie remembered "those nerve-racking times" when her husband Joe, with the Apollo Project Office, was spending most of his time traveling—so much so that their children believed that the airport was where their father worked. "Invariably," she said, "whenever Joe was out of town, something would break at home." After taking care

Suddenly, Tomorrow Came . . .

Manned Spacecraft Center, 1962

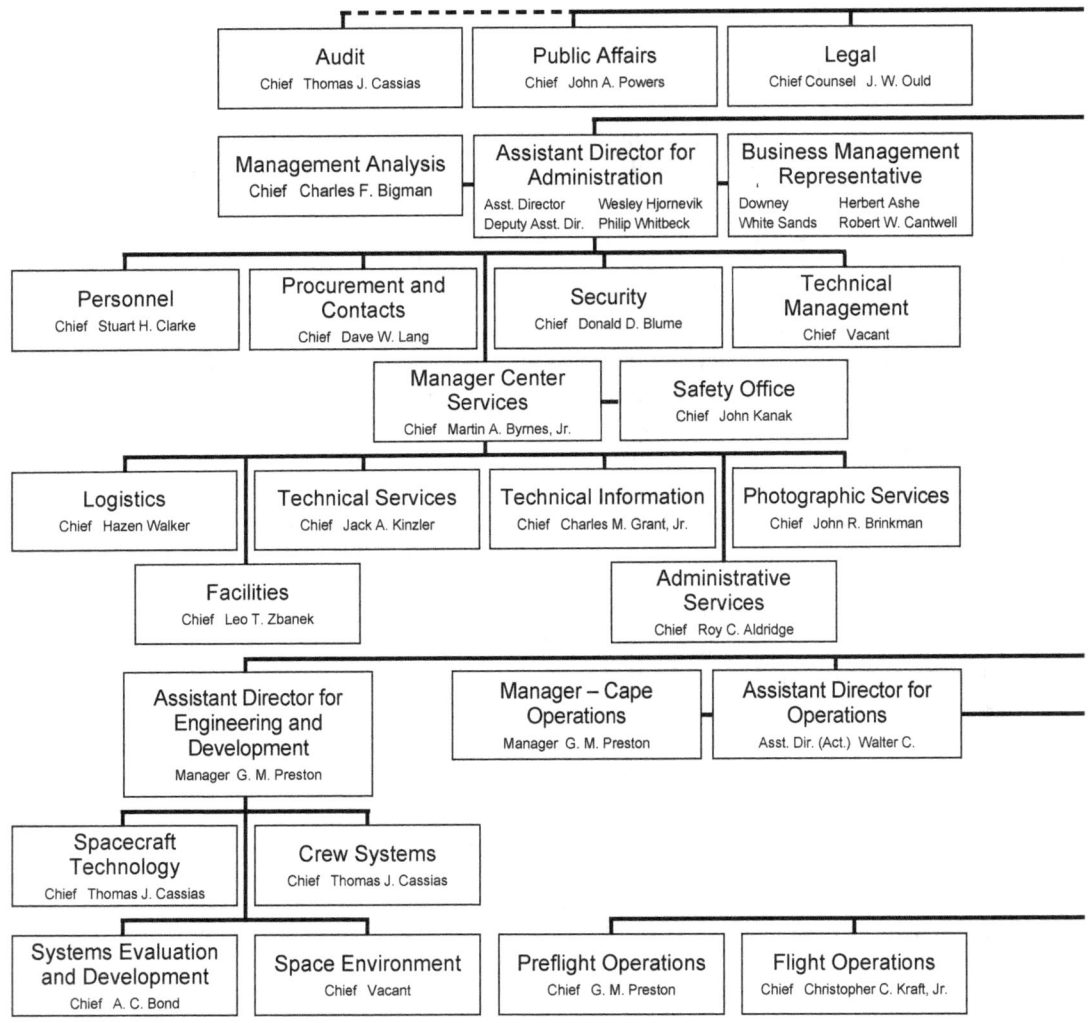

FIGURE 2. Organization as of 1962

Human Dimensions

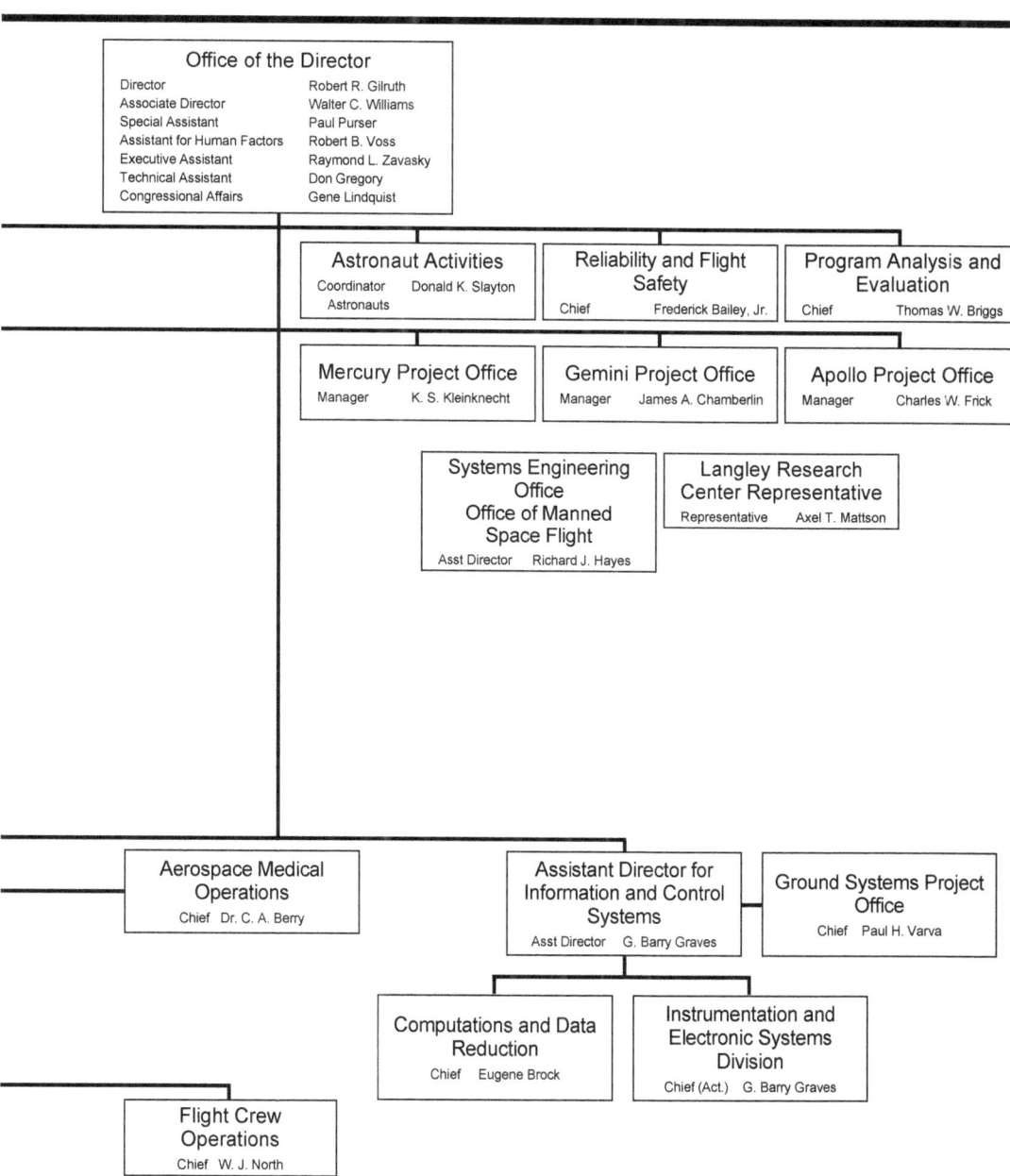

Suddenly, Tomorrow Came . . .

of Kent Slayton while astronaut Donald "Deke" Slayton's wife looked for a home in Houston, Grace Winn felt constrained to tell Slayton later that "he had better stay home more so Kent would know who his daddy is."[39] As it turned out, the manned spacecraft work consumed the entire family and not just the employed spouse.

Following the very successful and celebrated flights by John Glenn and Scott Carpenter (who had been called in to replace Deke Slayton in whom doctors detected an irregular heartbeat), NASA readied the Mercury Atlas-8 flight for astronaut Walter M. Schirra. Schirra was scheduled for six orbits, instead of the three flown by Glenn and Carpenter, and a water-based landing somewhere in the Pacific. The mission sought to check oxygen and fuel consumption, telemetry, and heat control characteristics for extended periods in space.[40]

Schirra flew a virtually trouble-free flight on October 3. The craft reached a speed of 17,500 mph with an estimated perigee of 100 miles and an apogee of 176 miles:

> The return of Schirra and his spacecraft to Earth with almost pinpoint accuracy was an extraordinary tribute to the engineering skills attained by Project

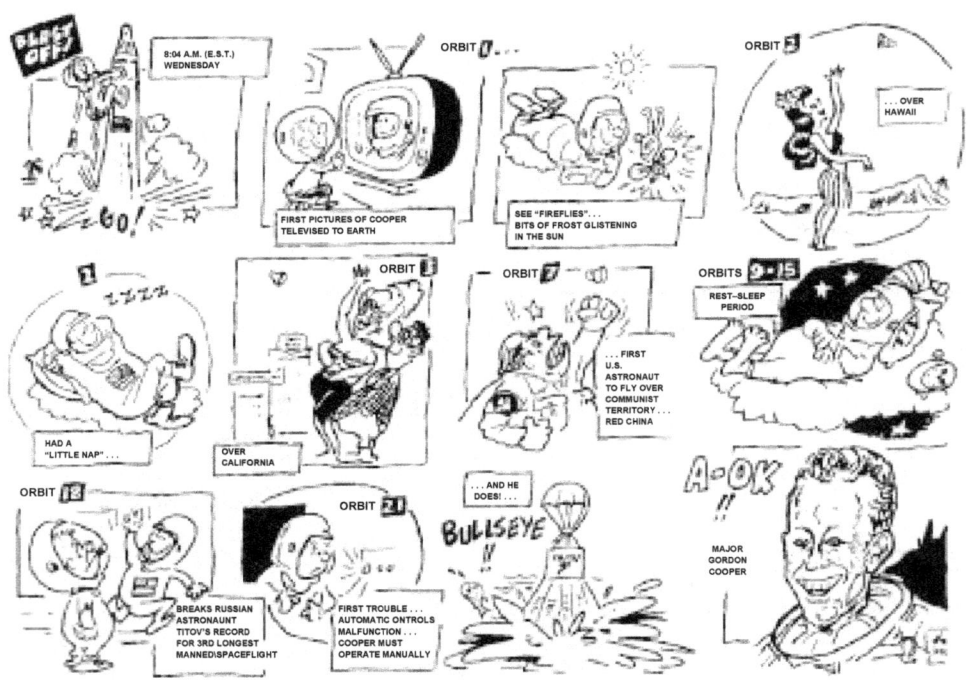

Source: *Roundup* (May 29, 1963)

FIGURE 3. A Chronicle of the Last Mercury Flight

Human Dimensions

TABLE 2. Project Mercury Flight Data Summary

Flight	Launch Date	Maximum Altitude			Maximum Range		Maximum Velocity			Flight Duration: Lift-off to Impact hr:min:sec
		Feet	Statute Miles	Nautical Miles	Statute Miles	Nautical Miles	Ft/sec Earth-fixed	Ft/sec Space-fixed	Mph Space-fixed	
Big Joe 1	9-9-59	501,600	95.00	82.55	1,496.00	1,300.00	20,442	21,790	14,856.8	13:00
LJ-6	10-4-59	196,000	37.12	32.26	79.40	69.00	3,600	4,510	3,075.0	5:10
LJ-1A	11-4-59	47,520	9.00	7.82	11.50	10.00	2,040	2,965	2,021.6	8:11
LJ-2	12-4-59	280,000	53.03	46.08	194.40	169.00	5,720	6,550	4,465.9	11:06
LJ-1B	1-21-60	49,104	9.30	8.08	11.70	10.20	2,040	2,965	2,021.6	8:35
Beach abort	5-9-60	2,465	0.47	0.41	0.60	0.50	475	1,431	976.2	1:16
MA-1	7-29-60	42,768	8.10	7.04	5.59	4.85	1,560	2,495	1,701.1	3:18
LJ-5	11-8-60	53,328	10.10	8.78	13.60	11.80	1,690	2,618	1,785.0	2:22
MR-1A	12-19-60	690,000	130.68	113.56	234.80	204.00	6,350	7,200	4,909.1	15:45
MR-2	1-31-61	828,960	157.00	136.43	418.00	363.00	7,540	8,590	5,856.8	16:39
MA-2	2-21-61	602,140	114.04	99.10	1,431.60	1,244.00	18,100	19,400	13,227.3	17:56
LJ-5A	3-18-61	40,800	7.73	6.72	19.80	17.20	1,680	2,615	1,783.0	23:48
MR-BD	3-24-61	599,280	113.50	98.63	307.40	267.10	6,560	7,514	5,123.2	8:23
MA-3	4-25-61	23,760	4.50	3.91	0.29	0.25	1,135	1,726	1,176.8	7:18
LJ-5B	4-28-61	14,600	2.77	2.40	9.00	7.80	1,675	2,611	1,780.2	5:25
MR-3*	5-5-61	615,120	116.50	101.24	302.80	263.10	6,550	7,530	5,134.1	15:22
MR-4*	7-21-61	624,400	118.26	102.76	302.10	262.50	6,618	7,580	5,168.2	15:37
MA-4	9-13-61	750,300	142.10	123.49	26,047.00	22,630.00	24,389	25,705	17,526.0	1:49:20
MA-5	11-29-61	778,272	147.40	128.09	50,892.00	44,104.00	24,393	25,710	17,529.6	3:20:59
MA-6*	2-20-62	856,279	162.17	140.92	75,679.00	65,763.00	24,415	25,732	17,544.1	4:55:23
MA-7*	5-24-62	880,792	166.82	144.96	76,021.00	66,061.00	24,422	25,738	17,548.6	4:56:05
MA-8*	10-3-62	928,429	175.84	152.80	143,983.00	125,118.00	24,435	25,751	17,557.5	9:13:11
MA-9*	5-15-63	876,174	165.90	144.20	546,167.00	474,607.00	24,419	25,735	17,546.6	34:19:49

Listed range is earth track
Big Joe - MA Development Flight
MR-BD = Booster Development Flight

LJ = Little Joe
MR = Mercury-Redstone
MA = Mercury-Atlas

*Manned Flight

Source: *Roundup* (June 26, 1963)

Mercury-Atlas 9, a Mercury spacecraft boosted into orbit by an Air Force Atlas rocket, carried Gordon Cooper on a 34-hour orbital mission beginning May 15, 1963. It was the last Mercury flight.

Mercury personnel. The spacecraft was spotted from the deck of the carrier as it dived toward Earth at a speed of about 270 miles per hour, leaving behind a vapor trail like a high-flying jet aircraft. At about 21,000 feet the drogue parachute could be seen fluttering behind Sigma 7, and the main chute billowed visibly at 10,000 feet to abruptly slow the plunge.[41]

The flight provided all of the checks and assurances believed necessary for a full 1-day orbital mission.

That mission, the last of the Mercury flights, left the pad at Cape Canaveral in Florida on May 15, 1963. The launch suffered odd delays. After a delay for the weather, reported the *Roundup*, all systems were finally "go": the "Atlas launch vehicle was go; the miniature but no less complicated spacecraft was go; the weather was go; Cooper was go." But the simple diesel engine which must move the gantry away from the firing line would not go. It wouldn't even start. Another try was delayed for a radar problem in Bermuda. Cooper left the capsule and went fishing, but returned the next morning for a perfect lift-off which brought him into 22 orbits during a day and a half in space. On the return, his electrical system failed, and Cooper piloted his craft back to within 4 miles of his target ship. Cooper went on to be the guest of honor at numerous parades and dinners. A parade in Honolulu turned out 250,000 spectators with equal numbers in Washington, D.C. and Houston; Cocoa Beach, Florida, with a population fewer than that number mustered 80,000, and 4.5 million people lined New York City parade routes. And with that grand finale, Mercury went out of business. On June 12, 1963, Administrator James E. Webb announced that there would be no more Mercury shots, and that NASA would concentrate on Gemini and Apollo—as indeed the MSC was already doing.[42]

Despite the successes with Mercury, Apollo was encountering problems—largely organizational "people" type problems, both on the center level and at Headquarters. The Apollo Spacecraft Program Office under Charles W. Frick, who had duty with the NACA Ames Research Center before moving to Convair as the designer and chief engineer for the Convair 880 and 990, attempted to make the Apollo program a "center within the Center." He absorbed more and more of the responsibilities of the supporting functional branches. Cooperation between center directorates and the Apollo Project Office waned. Competition and rivalry developed. Part of the problem derived from Frick's preference for the "industrial" boss style of management rather than the collegial "cooperative" style traditional with the center and virtually required by the contractual programs. Technically, NASA/MSC could not "boss" its contractors, but it could cooperate with, assist, and "manage" them. Finesse and tact were required. Frick finally informed NASA Headquarters that he was being forced to resign, and wanted to know what Headquarters was going to do about it. George Low replied that the matter was a center affair and Frick left. It may have been, in part, that Frick's heart had never been fully in his new job. He never gave up his home in La Jolla, California, and lived "out of a suitcase" in Houston.[43]

Bob Piland assumed the duties of acting Apollo Project Manager, but within months, asked that a permanent manager be appointed to replace him. Gilruth asked that Joseph F. Shea be sent from NASA Headquarters to manage the project, with Piland to resume the deputy role. Shea, according to Piland, had fought the battle at "higher levels" for a Moon-orbit-rendezvous flight—and prevailed. Shea accepted the MSC Apollo position, but only with the understanding that Walter C. Williams would assume duties in Washington as Mission Director—not that Shea wanted Williams in that particular position, but that he preferred *not* to work directly under the authority of Williams at MSC. Williams did go to Washington, and Shea came to Houston, and the exchange eased, but did not resolve, the difficulty of meshing the program offices with the functional divisions of the center. Contractors, for example, might get a favorable evaluation in a project office, only to have it vetoed by the functional engineering and development office. This problem was later resolved when George Low came to the Apollo Project Office and gave the functional offices (and hence the center rather than the project office) authority over contract progress.[44]

Similarly, on the Headquarters level, the growing assumption or "usurpation" of the presumed center autonomy by the missions or project offices led to many difficulties and a general reorganization of NASA management. In September 1961, in response to the developing Apollo lunar program, Administrator Webb abolished the existing program offices and created four new ones including Advanced Research and Technology, Space Sciences, Applications, and Manned Space Flight—the latter to have authority over all spaceflight activity. D. Brainerd Holmes, whom Webb appointed Director, came to NASA from the Radio Corporation of America (RCA) Defense System Division. An electrical engineer, Holmes was project engineer on the Alaskan-Arctic early warning system. Webb and Holmes planned to give Headquarters greater authority over spaceflight programs in the future than it had exercised over Mercury.[45]

Mercury, for all practical purposes, through the cooperation of Silverstein and Low, operated as a NASA center cooperative effort with the STG/MSC assuming the leadership role. The "federalist" style of center association, under which each center enjoyed considerable autonomy but cooperated (usually) in the completion of tasks, conflicted with the centralist or industrial management system which Holmes began to impose, and with the rather fierce spirit of independence which each center and especially the new MSC seemed to be developing.

Holmes had appointed Joe Shea from Space Technology Laboratories as his deputy (before Shea transferred to the Space Center) to concentrate on systems engineering. George Low worked under Holmes on programs. The systems organization, Low said, "tried to run the show technically from Washington; while on the program side we tried to function as we had in Mercury and Gemini, i.e., letting the centers do the work" and Headquarters stepping in to help when needed. Holmes helped in resolving the Apollo lunar versus Earth-orbit-rendezvous issue, and most of the basic Apollo decisions were made during his tenure. He also created a Management Council which brought center directors and associate directors and NASA administrators together in a policymaking body; but by appearing to assume full program authority at the Washington level, he created tremendous conflicts with the MSC managers and with the Administrator and Deputy Administrator at Headquarters. Finally, Holmes was removed from his position, the old Manned Space Flight (program) Office was abolished in September 1963, and George E. Mueller became Associate Administrator for Manned Space Flight, with George Low as Deputy Associate Administrator.[46] Changes in the higher administrative echelons had little impact on progress in the Gemini and Apollo programs, which were already largely in the hands of the contractors.

Suddenly, Tomorrow Came . . .

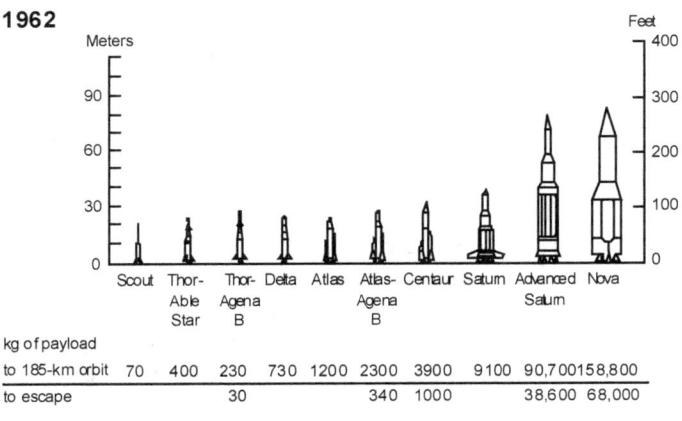

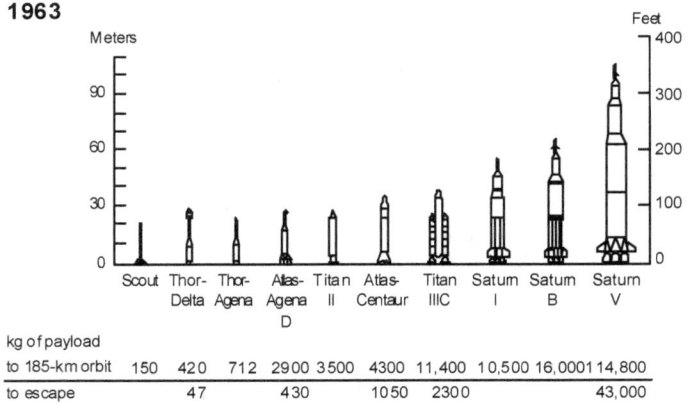

Characteristics

The launch vehicles used by NASA during the agency's first 10 years are illustrated above. Two proposed vehicles, Vega and Nova, are also shown. Two boosters borrowed from the military, Atlas and Thor, were used with several different upper stages. Atlas was paired with Able to create a vehicle for orbital missions. Able, Agena, and Delta were added to Thor to increase that missile's range and versatility. Juno and Vanguard vehicles contributed to NASA's early space science program. Redstone missiles were man-rated to boost the first Mercury astronauts onto ballistic trajectories, and Gemini astronauts rode modified Titan IIs into orbit. Two distinct vehicles, Little Joe I and Little Joe II, were used to test and qualify launch techniques and hardware for the Mercury and Apollo programs. The Saturn family of launch vehicles was developed specifically to support the Apollo lunar exploration venture. And Scout, which changed over time as its engines were upgraded and its reliability improved, was NASA's first contribution to the launch vehicle stable.

Source: Linda Neuman Ezell, *NASA Historical Data Book*, II, *Programs and Projects, 1958 - 1968*, 3, 24.

FIGURE 4. Launch Vehicles

North American Aviation received the prime contract for the three-man Apollo spacecraft vehicle in 1962. Its design required rendezvous capability, accommodations for a 14-day mission for the three-man crew, and the option of accommodating larger crews for shorter missions. An expendable service module and lunar landing module would be components of the Apollo craft. Grumman Aircraft, the prime contractor for the Lunar Excursion Module (LEM), proposed a preliminary design for such a vehicle as early as May 1961 while doing one of the "feasibility" studies for Piland's "Advanced Projects Office." Grumman, in turn, by July 1963 selected six major subcontractors, including RCA, which received the $40 million contract for LEM electronic subsystems and engineering support.[47] The Apollo spacecraft would be the product of a large assortment of industrial contractors and subcontractors working under the relatively close guidance and supervision of an equally diverse NASA management contingent representing numerous branches, divisions, and project offices.

The proposed launch vehicle for the Apollo spacecraft, now designated as a Saturn I, was the responsibility of Marshall Space Flight Center. The Saturn emerged from a considerable history of experimental and design work dating back to initial studies in 1957 by Von Braun's team at the Redstone Arsenal of a rocket booster that could launch heavy loads into orbit (9,000 to 18,000 kilograms or 20,000 to 40,000 pounds). Work by the Army Ballistic Missile Agency, development contracts with Rocketdyne for the Thor-Jupiter engine, and a 1959 contract with Rocketdyne for the Saturn preceded NASA's assumption of responsibility. Douglas Aircraft received the contract for the Saturn second stage, the new Marshall Space Flight Center (Von Braun's group) received program responsibility, and contracts with Pratt & Whitney, Convair, Chrysler and other contractors led to the first Saturn test launch (first stage) in October 1961. The next year, in late April, the second of 10 Saturn (C-1) test and development flights, preparatory to a planned 1964 orbital mission, made a fully successful lift-off. It was powered by eight H-1 engines which developed 1.3 million pounds of thrust and climbed to an altitude of 135 miles in 115 seconds.[48]

The decision to develop a more powerful Saturn booster, even while development of the C-1 continued, led to work on the Saturn 1B by Chrysler and Douglas and work on uprating the H-1 engine by Rocketdyne. Still not satisfied, on November 10, 1961, NASA accepted proposals from five contractors for the development and production of yet more advanced Saturn boosters than the 1B, using Rocketdyne F-1 and J-2 engines. Contracts for three of these Saturn V booster stages were let to North American, and Boeing and Douglas received first stage and third stage contracts, respectively.[49] By 1963, NASA, and especially the MSC and the Marshall Space Flight Center, focused its energy and attention on Apollo. Even while the Mercury project was peaking, Gemini was coming "on line" and the new MSC was under construction.

Finally, in November 1963, Gilruth abolished the Mercury Project Office and completed the reassignment of Mercury personnel to Gemini and Apollo projects. He appointed Christopher C. Kraft Director of Flight Operations and Deke Slayton Assistant Director for Flight Crew Operations and head of the Astronaut Office. Near the end of that month, President John F. Kennedy and Vice President Lyndon B. Johnson arrived in Houston, and the center encouraged employees to see the motorcade. The next day,

November 22, Kennedy arrived at that fatal day in Dallas.[50] The Nation was shocked and deeply grieved by the untimely death of the President—one who was so instrumental in the expansion of the Nation's manned space program. Administrator Webb publicly and the NASA community privately vowed to meet his challenge of May 1961.

Personnel and contractors redoubled their efforts on the Apollo and Gemini projects. Gemini, having been announced in 1962, was "reconfigured" in January 1964, when MSC managers working with North American found it necessary to abort the paraglider landing system in favor of a Mercury-type parachute water landing. But progress was being made on many fronts. Several groups of astronauts were in training, successful unmanned suborbital tests of the Gemini-Titan I were made, Titan IIs were test-fired, a new mission control office was being established in Houston rather than on the Cape as under the Mercury project, new astronaut pressure suits and greatly enhanced ground-based computer control systems for Gemini and Apollo were being developed, and perhaps most importantly, the support and the confidence of the Nation, and certainly of the new President, Lyndon B. Johnson, remained with the NASA missions despite domestic problems relating to race, education, segregation, and the growing involvement of American combat forces in southeast Asia.

And, almost 4 years after its selection, the manned spacecraft people in Houston, Texas, really had their new home. Between February 20 and April 6, 1964, some 2100 MSC personnel moved into their new Clear Lake quarters, and the final move from all leased facilities in Houston to new on-site quarters began on June 24, 1964. No Americans flew in space in 1964. It had been a year, as the *Roundup* summarized it, "of filling the pipeline with hardware," and it could have added, "of filling the MSC with buildings."[51]

George Mueller, the Associate Administrator for Manned Space Flight, summarized the very brief history of the Manned Spacecraft Center for the MSC Senior Staff in October. "We can congratulate ourselves," he said, "on a particularly impressive set of accomplishments while this rapid growth was taking place." He noted that the Marshall Space Flight Center had completed the development of the Saturn I, the Kennedy Space Flight Center (Cape Canaveral) now had tuned its launch operations, and here at the Manned Spacecraft Center, he said, "you have special reasons for pride."[52]

"You began in 1961 with a budget of $200 million; in 1964 the center operated with a budget of about $1.475 billion—50 percent greater than the entire NASA budget for 1961. The center had 4277 employees by the end of 1964, a fivefold increase. You constructed a brand new plant from the ground up," he said. "All of this was accomplished while you were flying six manned Mercury missions, conducting the Gemini program at top speed, and building up to a full head of steam on Apollo." And he added, "The creation of this center—the people, the physical plant, and the associated industry team—in the pressurized environment to which you all have been subjected over the last few years is a remarkable achievement, probably without parallel. The country owes a great deal to Bob Gilruth and all of you for the tremendous job you have done."[53]

Then Mueller went on to review the recent history of human entry into space beginning with Sputnik, the flight of Yuri Gagarin, and the great leap forward in the American manned spaceflight programs through the Mercury, Gemini and Apollo projects.

He talked about the things NASA had promised to do, the things it had failed to do, and the things that must be done, for paradoxically, he explained, "now we are in a sense paying the price for our success." There was a sense of rising expectations on the one hand, but a growing aura of complacency on the other. He also talked at great length about the changing mood in Congress and the high dollar costs of manned spaceflight.[54] Within the 7 years of the flight of Sputnik in October 1957, Americans had arrived confidently and competently on the threshold of space, but the technology, the ground rules and, perhaps necessarily, the management systems were in a constant state of change. Space was no easy business.

CHAPTER 5: Gemini: On Managing Spaceflight

"The first phase of the Nation's second manned space program began like a storybook success" on Saturday, April 8, 1964, when an unmanned, partly instrumented Gemini capsule entered orbit from its launch site at Cape Canaveral.[1] The 12 Gemini flights completed by mid-1966 brought America from the edge of space to outer space, from the pioneering days of Mercury to the lunar landings of Apollo, and into new management techniques including processes like systems and subsystems management, configuration control, and incentive contracting. A major building block in the operations components of spaceflight, Gemini provided an invaluable learning experience in flight control, rendezvous, docking, endurance, extravehicular activity, controlled reentry, and worldwide communications. But the acceleration of Gemini and Apollo programs strained the human resources of the Houston center and created stress and management crises. Although critical in the manned space effort, Gemini was much more and much less than a "storybook" success.

The April 1964 launch of the first unmanned Gemini spacecraft on the shoulders of an Air Force Titan II rocket was followed in May with the launch of the first Apollo vehicle aboard a Saturn I. Both coincided nicely with the final relocation of Manned Spacecraft Center personnel to their new permanent site at Clear Lake. Director Bob Gilruth declared an "open house" for the weekend of June 6 and 7, and took great "personal and professional satisfaction" in welcoming the public to the NASA MSC.[2]

It was an open house that has been extended throughout the days of the Johnson Space Center, helping establish the important precedent that the center and NASA flight missions are for participation in and viewing and use by the public. Almost 80,000 visitors attended the grand opening. They viewed a film about the Nation's space program in the auditorium (later named for Congressman Olin E. Teague) and toured exhibits in the lobby and on the grounds. Exhibits included hardware from Mercury flights, scale models of Mercury, Gemini and Apollo systems, pressure suits, survival gear, and photographs. Outside displays included full-scale mockups of the Gemini and Apollo modules, a Mercury spacecraft, a boilerplate test module and escape tower, and a Redstone launch vehicle.[3]

The open house contradicted the more traditional practice of government agencies and especially the World War II tradition which stated, that the public's "need to know" was rather limited, as well as the older NACA (and academic) attitude that the workplace should be protected from external influences. NACA had generally limited its news releases and public relations activities to an annual report to Congress and a week-long open house at the NACA facilities. Although he held a very understanding view of public relations, John A. "Shorty" Powers, who first served as the public affairs officer for the STG, hailed from the Air Force where news releases were confined largely to rather concise handouts, and he tended to follow this custom. With the very limited staff at his disposal he could do little more. But the Mercury launches gave NASA and the manned spacecraft program a visibility that could not be avoided—despite the preferences of some to do just that. A major shift in public relations came just prior to Alan Shepard's flight in May 1961 when, at an informal gathering over martinis at the Hay-Adams Hotel in Washington, D.C., Roy Neil, with the National Broadcasting Company, discussed the NASA news problem in the company of Walter Williams and Paul Haney from headquarters and Shorty Powers with the STG at Langley. Powers and Haney both supported more public access to information, and Williams finally suggested that the STG put an information officer at the console in the control center to disperse instant and accurate information to the press. Haney and Powers looked at each other and said, "Gee—I wish I had thought of that!"[4] And it was done.

Paul Haney replaced Shorty Powers in September 1963 as director of public affairs at the MSC in Houston. Powers had spent most of his personal time in travel, leaving much of the administrative detail to others and predominantly to Paul Purser, Gilruth's special assistant. Independent branches of the MSC could and did issue their own news releases and information. Moreover, the Houston center and its public information staff seemed inclined to function independently of NASA Headquarters. Administrator James Webb sent Haney to Houston to head the Public Affairs Office, and Powers returned to Washington on special assignment for a year before taking his retirement from the Air Force.[5]

Haney centralized and reorganized the Public Affairs Office of the MSC bringing in John Peterson as his key administrative officer and Roy Alford for public relations. Alford had previous experience as an assistant city manager in Texas and as a military governor of five different states in Japan. Haney and his staff immediately began to encourage an "open door" policy toward the public. He recognized that the managing engineers at the center were people "who rarely ever had to talk to reporters about anything." And he strongly believed that "we in government and particularly in NASA, by accident or design, constantly erect information barriers around our work."[6]

Haney thus encouraged the MSC's open house as a philosophical statement. James C. Elms, who replaced Walter Williams as deputy director of the MSC in November 1963, approved the plan and Robert Gilruth concurred. Haney and his staff organized a VIP welcoming ceremony for the Friday preceding the open house weekend, and recruited some 20 Gulf Coast Chambers of Commerce as hosts for a gigantic, Texas-style cocktail party at a neighboring inn in the evening. About 25,000 guests attended.[7] That and the open house weekend were great successes and clearly helped bind the goodwill of the Houston community (then the sixth largest city in the country) to the MSC.

Although Haney believed that "MSC from the outset has taken a much more understanding view of the public information role than have any of the other centers of NASA," different views of public information continued to create problems within the center and in the center's relations to other centers and with Headquarters. Bob Gilruth discovered that the public's intense interest not only necessitated an open door, but also offered an excellent opportunity for explaining the NASA and MSC mission to the public and Congress. Gilruth explained in the special edition of the *Roundup* prepared for the occasion that in the offices and laboratories of the center "a concerted effort is made by management personnel, engineers, scientists, and many support personnel to assure that the national goal of the United States achieving preeminence in all aspects of space research and exploration is attained." MSC engineers, he said, "conceive the design and specify the systems to be used in these very complex spacecraft; American industry provides the detail design and fabrication."[8] The hand of the private sector, in fact, loomed larger in the Gemini program than it had in the Mercury program.

Work on the Gemini program began as an outgrowth of Mercury. As early as 1959, NASA began considering post-Mercury flights, and McDonnell Aircraft Corporation, which manufactured the Mercury spacecraft, independently began redesigning for an "improved" Mercury. James A. Chamberlin, Chief of the MSC Engineering Division, collaborated closely with McDonnell engineers and with Max Faget's Flight Systems Division in the redesign of Mercury. Faget recalls that NASA Headquarters, having just endorsed the Apollo lunar program, was reluctant to initiate another new venture. Bob Gilruth, Faget said, convinced Abe Silverstein that Gemini was an absolute necessity and Headquarters soon endorsed the program.[9]

NASA awarded a design study contract to McDonnell in April 1961, while center engineers studied design modifications and new flight requirements. The profile for the new craft included that it support a two-person crew for extended flights of up to 14 days and that it have the capability for rendezvous, docking maneuvers, and a land landing. The landing problem led to competitive design studies for a paraglider system by Goodyear Aircraft Corporation, North American Aviation, and Ryan Aeronautical Company. First styled the "Mercury Mark II," plans developed to launch the vehicle aboard a new Titan II rocket being manufactured by the Martin Company of Baltimore, Maryland, for the Air Force. The Air Force would serve as NASA's supplier.[10]

Redesignated "Gemini" in January 1962 by an ad hoc committee at NASA Headquarters, the name was suggested by Alex P. Nagy who identified the two-man space crew with the "twins," one of the 12 constellations of the zodiac. Historian Barton Hacker and James Grimwood believed that Gemini was "a remarkably apt name," because in astrology, "Its spheres of influence include adaptability and mobility—two features the spacecraft designers had explicitly pursued—and, through its link with the third house of the zodiac, all means of communication and transportation as well."[11]

Contracts for the spacecraft, propulsion systems, and landing devices were awarded in November and December of 1961 following approval of the project at Headquarters by NASA Associate Administrator Robert C. Seamans. McDonnell Aircraft, also the prime manufacturer for the Mercury vehicle, received a contract for the manufacture of 12 spacecraft, while North American Aviation received a contract for the development of a paraglider (land) landing system, and the Air Force, acting on behalf of NASA's request, ordered 15 Titan II rockets from the Martin Company.[12] The MSC, which officially came into being almost concurrently with the Gemini program, had responsibility for management of the spacecraft and paraglider systems, while the Air Force managed the propulsion work.

Gilruth created a Gemini Project Office on January 15, 1962, and appointed James Chamberlin project manager (see figure 4, 72). Chamberlin, born in Kamloops, Canada, in 1915, graduated from the University of Toronto in 1936 and received a master's degree from the Imperial College of Science and Technology in London, England, in 1939. After a brief stint with Martin Baker, Ltd., he moved to Montreal, Canada, to work for Federal Aircraft, Ltd. on design modifications of Canada's version of the British Avro-Anson aircraft. During the war he worked on projects for Clark Ruse Aircraft in Nova Scotia, Noorduyn Aviation at Montreal, and the Royal Canadian Air Force. He joined AVRO Aircraft, Ltd. in 1946, where he became Chief of Design and was primarily responsible for the aerodynamic design of a jet fighter and a jet transport and for the overall design of the AVRO advanced interceptor, the CF-105 Arrow.[13]

When Canada canceled the AVRO-Arrow fighter program, Chamberlin was instrumental in moving the contingent of AVRO engineers to the STG. He became Chief of the Engineering Division in 1959 and, in that capacity, was responsible for managing the development and production of the Mercury spacecraft by McDonnell Aircraft. He contributed heavily to the resolution of Mercury designs and to the engineering interface between the capsule and the Atlas launch vehicle. Chamberlin began work on the improved Mercury capsule that became Gemini in February 1961, and in cooperation with other center and McDonnell engineers was the "creative genius" of the Gemini spacecraft.[14]

Chamberlin's work on the Mercury systems in collaboration with McDonnell Aircraft made the Gemini alliance of McDonnell and Chamberlin's Gemini Project Office a natural, comfortable and very productive arrangement. The initial letter contract between NASA and McDonnell, dated December 15, 1961, provided that the company would immediately begin a research and development program which would result in the development to completion of a two-man spacecraft, with a launch vehicle adapter and a target vehicle docking adapter. The company was to manufacture 12 spacecraft, 15 launch vehicle adapters, and 11 target vehicle docking adapters with the "test articles and ancillary hardware," and provide field services, training and liaison with "NASA, other government agencies, NASA associate contractors, and subcontractors." In terms of the agreement, NASA would provide astronaut pressure suits, the spacecraft paraglider, survival equipment, launch vehicles and facilities, and the target vehicle "in orbit." The first spacecraft, with the launch adapter, was to be delivered in 15 months.[15]

The brief technical guidelines supporting the agreement stated that the vehicle should be able to orbit the Earth for 14 days, make "land" landings, rendezvous and dock with a target vehicle in orbit, have pilot-guided reentry capabilities, require simplified countdown techniques and procedures, and test "man's performance capabilities in a space environment during extended missions." The original letter contract would fund McDonnell development at cost-plus-a-fixed-fee with a ceiling of $25 million, to be followed by the final cost-plus-fixed-fee contract at a later date.[16]

The Gemini program soon became beset by a "pattern of rising costs." Cost estimates rose quickly from the original estimate of $350 million made for Mark II in August 1961, to $529 million in October, to $744 million in May 1962, to a final cost of $1.283 billion. By comparison, Mercury program costs totaled $392.6 million while Apollo would cost $29.5 billion.[17] Within a few years, the MSC began an innovative contracting program for cost containment. At the moment, however, given the objectives, the time frame, and the generally robust health of the national economy, money was no object.

There were, in fact, Gemini problems that money could not solve. Jim Chamberlin, considered by many a genius at detailed design work, assumed almost sole authority for the development of the Gemini spacecraft and worked closely with McDonnell in its production. "There was not," according to Paul Purser, "a single change made in the basic design of the Gemini spacecraft or its systems once they were established by Jim. The basic design and systems that he established were workable ones and it turned out to be a highly successful program."[18]

Chamberlin, however, did not seek outside advice, internalized the work of the Gemini Project Office, and tended to ignore the functional divisions of the MSC. Max Faget observed that Chamberlin tended to isolate himself and his project office from other divisions. Rodney G. Rose, working in the Gemini Project Office with Chamberlin, acknowledged that Chamberlin tended to "play his cards close to the vest," but recalled that Faget and the Engineering Division, with its hands full of Apollo, had little time for Gemini. Similarly, Chris Kraft, from the perspective of the Flight Operations Division, thought that the relative independence and isolation of the Gemini Project Office was a "comfortable" arrangement because everyone had other things to do. Nevertheless, as Bob Chilton (then acting head of the Guidance and Control Division) observed, some of Apollo's problems derived from the fact that organizationally Gemini had been too independent of both the Mercury and Apollo programs. Gemini failed to benefit fully from the Mercury engineering and flight experiences and, subsequently, Apollo failed to learn fully from the Gemini experiences.[19]

Joe Loftus concurred that the embryonic engineering directorate at MSC "did not have as extensive a participation in Gemini as they had in Mercury and were to have in Apollo." Many of the MSC engineers quickly came to the conclusion that the Gemini Project Office was too exclusive, and that overreliance on the contractor preempted the role of the center engineer. It was in part, Loftus believed, an extension of the old NACA versus Air Force syndrome where one group of NACA-style engineers was accustomed to in-house, hands-on development and the other group saw its role as purely consulting or advisory to the contracting engineers. It was a problem, Loftus admitted, which was both "knotty" and "classical" and it created severe stress within the center.[20]

The "Gemini management crisis" came to a head in the spring of 1963, but it was by then only one facet of a broad-based organizational and personnel imbroglio. To be sure, the crisis was attributable not only to Chamberlin's predilection for personal control over the Gemini program, but also to the reality that MSC human resources were stretched very thin and already heavily committed to the Apollo program, which had seniority and was considered more urgent. Operations people, as well, were still busily involved with Mercury. Henry Pohl (then chief of the Auxiliary Propulsion and Pyrotechnics Branch in the Engineering and Development Directorate) believed that the management crisis was in part more presumed than real, simply because Jim Chamberlin was not a "good communicator."[21] The organization at every level was young, growing, and inexperienced. But Chamberlin knew what he was doing and so did McDonnell Aircraft; and by temperament and circumstance, Chamberlin and Gemini did not interface with other elements of the center.

Gilruth resolved the dilemma in March by assigning Chamberlin as the Senior Engineering Advisor to the Director and placing Charles W. Mathews at the head of the Gemini office. Mathews, from Duluth, Minnesota, joined NACA in 1943, and headed the STG Operations Division and then the Spacecraft Technology Division during the Mercury years. He brought Kenneth E. Kleinknecht to the Gemini office from Mercury and helped fuse Mercury experiences, and especially its flight experiences, into the Gemini program. Chris Kraft, who worked under him, described Mathews as "an intimate part of the flight operation management team and a key individual in the development of technical policy." Many believed, as did Paul Purser, that Mathews was not only a "smart engineer" but also a good manager. He was "more willing to realize that there are several, in most cases, right ways to do something," Purser observed.[22] In the early fall, Gilruth completed a general reorganization of the MSC and there were related changes in management at NASA Headquarters.

The Gemini program actually made tremendous strides during the administration of Chamberlin, despite the real focus of MSC engineering resources on Apollo and the preoccupation of operations with Mercury. In October, McDonnell delivered the first of 12 Gemini spacecraft to Cape Canaveral, Florida, for preflight checkout procedures. Astronauts were recruited and trained and flight missions planned while life both in the center and without tended to go on largely oblivious to the Gemini program. In March, the same month that Mathews came to the Gemini office, NASA signed a contract with Grumman Aircraft Corporation ($397.9 million) for the development of the Lunar Excursion Module (LEM) for the Apollo spacecraft and Secretary of Defense McNamara visited Houston for a briefing on the Gemini program.[23]

A fourth successful firing of the Apollo Saturn rocket came in April; and in May 1963, Vice President Lyndon B. Johnson told the American Institute of Aeronautics meeting in Dallas that the United States "must forge ahead in space or become a second-rate nation." Astronaut L. Gordon Cooper flew Mercury (MA-9) in May, and following his 22 orbits, he told both houses of Congress and the packed galleries that the public's response to the flight "Shows that Americans want to express their feelings and their confidence that we . . . can conduct peaceful research programs; that we can conduct them openly, and under the surveillance of every man, woman and child in the world."[24] The MSC's and NASA's "public" had become worldwide and NASA intended to maintain an open door to that very large audience.

Cooper's flight ended project Mercury and helped salve the wounded pride inflicted by Sputnik and subsequent Russian space achievements. It also brought the sharp attention of the Operations Division of the MSC to the Gemini program. With a capsule already on location at Cape Canaveral, astronauts in intensive training, and Mercury behind them, Dennis Fielder remembered the sharp shift in attention when Chris Kraft looked at the Gemini spacecraft and said, "we're going to have to fly this."[25]

Flying Gemini would be considerably different from flying Mercury. Gemini was maneuverable. Mercury was not. Mercury was mostly a matter of trajectory and tracking. Gemini provided flight control options. Astronauts became pilots rather than passengers. Gemini operations built upon the Mercury experiences, but it became a substantially different program. The mission control and communications network which existed at the close of the Gemini flights was considerably more sophisticated than that which existed at the time of the last Mercury flight. Operations and mission control really "cut their teeth" on Gemini, designing and building the control center, creating a worldwide communications network, and flying Gemini, while looking ahead to Apollo.[26]

When Mercury flights began there was no around-the-world communications network; long distance voice connections were even more rare. Flight control for Mercury and Gemini required real-time voice contact with the spacecraft; and since the spacecraft would be accessible by a direct radio contact for only about 7 minutes, continuous contact required numerous stations and significant real-time communications between those stations. The rapid and instantaneous handling of data, which became increasingly critical as missions became orbital under Mercury and then maneuverable with Gemini, required computer and program sophistication which simply did not exist at the start of the space program. Worldwide networking required the cooperation of foreign countries with the United States and the cooperation and interfacing of American communications companies with each other. Both were relatively rare phenomena. The most serious hemispheric communications network then being developed by the Department of Defense would have been the North American Air Defense Command or DEW Line radar defense systems. Because most of this work was classified, little of that experience seems to have reached NASA. Human spaceflight "drove" a reformation and near revolution in the civilian sector of communications and computer technology.[27]

John D. Hodge, who headed flight control operations under Kraft, suggests the enormity of the communications problems surrounding the creation of the Mercury control center at Cape Canaveral, from which the Air Force launched its missiles, including the Redstone, and maintained downrange telemetry monitoring stations through the West Indies to the coast of South America. "There always seemed to be some kind of controversy," Hodge recalled, "as to who had the communications responsibility at the Cape site." The Department of Defense controlled cabling inside the fences, Radio Corporation of America (RCA) carried the cable to the fence, and NASA interposed with Western Electric, BellComm and the Bell Telephone system. "Some very strange interface problems occurred," Hodge said. Joining the cables of different commercial carriers violated legislation establishing carrier tariffs and franchises. AT&T's George Vogel, for example, said over and over that AT&T would not knowingly interconnect its services with another carrier. Subsequently, AT&T laid cables to the Cape control center while "oblivious" to the fact that NASA was going to tie everything together.[28] Thus, AT&T could compromise in fact, if not in principle.

Simply resolving disputes with the Air Force over space allocations and who should own and who should control which buildings and what equipment required some considerable diplomacy and expertise. The flight command center at Cape Canaveral operated for a time under a complex arrangement. RCA, as an Air Force contractor under special assignment to NASA, operated the command center on behalf of the Air Force, while the management authority rested in the Tracking and Ground Instrumentation Unit of the STG (located at Langley) which was administered by the Goddard Space Flight Center. Despite the apparent confusion "that was probably one of the best things that ever happened to us," Chris Kraft said, "because the people we got were all experienced and highly competent."[29]

G. Merritt Preston headed Cape Operations under the general direction of Charles W. "Chuck" Mathews who remained at Langley and supervised for a time all of the activities related to flight control, crew systems, and medical operations. In late 1962, Gilruth appointed Walter Williams Assistant Director for Operations to supervise a Preflight Division under Preston, a Flight Operations Division headed by Chris Kraft, and a Flight Crew Operations Division under W.J. North. Chuck Mathews became chief of the Spacecraft and Technology Branch under Faget's new Engineering and Development Division.[30] The organizational changes tended to be more paper than real, because the real work of the various managing engineers did not change significantly.

Chris Kraft and Tecwyn Roberts worked on the design of the Mercury control center, and Fred Mathews with Kraft and Gene Kranz concentrated on flight operations. Porter Brown was the key liaison and support person between Cape Canaveral and STG (or MSC). John Hodge and Dennis Fielder created and refined the communications network between stations. When a network contract went to Western Electric with another to IBM as an associate contractor for work on the computer integration systems, Fielder became the interface between the MSC and the contractor. Howard Kyle focused on the design of the communications systems. Bill Boyer worked on network and station contracts and permissions, and Barry Graves "and his merry men" designed and developed the instrumentation and electronic systems and managed the IBM and Philco contracts. Graves eventually got "crossways" with IBM, Philco and Chris Kraft, contributing to another substantive administrative reorganization in 1963. The problem had to do with the fact that at the close of the Mercury flights, Graves accepted a Sloan Fellowship, and when he returned to MSC in 1963, he was appointed Assistant Director for Information and Control Systems with Paul Vavra in the Ground Systems Project Office reporting to him. The major responsibility of the project office was to build the new Gemini Mission Control Center, but that design work had largely been completed and the contracts awarded. For the most part, however, everyone, including Kraft and Graves, cooperated and worked together and did what needed doing.[31] The work was intensive and exhausting—but also exhilarating. Most recognized that they were doing things no engineers had done before.

The human spaceflight program and those flight control operations which first centered at Cape Canaveral before the Mission Control Center was built at Houston became a critical catalyst in changing the world of communications. The world really did not communicate much before the advent of human spaceflight, certainly not in terms of real-time and voice communications. Mercury, Gemini and Apollo programs inspired cable and voice interfacing, stimulated the development and placement of communications satellites, and helped link the world together as never before. Philco, which had networking experience with the Discoverer satellite program, received the NASA contract to manage and develop the manned spacecraft network.[32] Networking, even assuming the existence of the technical expertise, became a problem of considerable diplomacy and statesmanship. In fact, the State Department was very much involved.

Department of State diplomats negotiated a "government-to-government" agreement with Australia to develop a number of communications and flight control stations, including stations at Muchea, north of Perth, and one in the Woomera Mountain Range. These were under the technical management of Australians and "except for a few company reps and an occasional NASA advisor" the Australians "ran the show." Spain had to be convinced that the Americans were not building a missile site on Spanish soil before agreeing to locate a NASA communications station there. Although Mexico approved a site at Guaymas, Communist and anti-American groups constantly threatened the security of the station. Nigeria approved a site after NASA representatives visited the country, and NASA brought Nigerians to America for briefings and informational tours. Although Zanzibar admitted a mobile relay station, political unrest finally forced NASA to remove the facility to Tennarive, Madagascar. The Bermuda station, already in service by the Air Force, was a primary flight control station.[33]

Two ships, the *Coastal Sentry Quebec* located near Okinawa and the *Rose Knot Victor* off the west coast of South America filled slots that could not be monitored by land-based stations. Life aboard those ships and in the remote stations could be very tedious. Before John Glenn flew in February 1962, Alan Shepard had been sent to the control room of the *Quebec* off Okinawa. Innumerable launch delays created a terrible morale problem and restlessness aboard the ship, which had been allowed to drift hundreds of miles from Okinawa toward a convenient liberty port in Japan, while awaiting the launch. Finally, Shepard called Chris Kraft, explained the problem, and asked for permission to give the sailors liberty in Japan. But Kraft hustled them back to their duty station with the word that indeed the launch was now imminent.[34] Spaceflight had many dimensions.

As Mercury began orbital missions with John Glenn's flight, equipment and procedures needed to be constantly updated. When Gemini came on line, the system established for Mercury had to be completely retooled. The Mercury control center at Cape Canaveral used "off-the-shelf" electronics gear, but more was needed. It became obvious that existing equipment and techniques were inadequate. In 1961, Chris Kraft, Dennis Fielder, Tec Roberts and John Hodge began a serious study of needs and options for an improved control center and directed a study contract awarded to Philco's Western Development Laboratory. While this study developed, IBM received a contract on competitive bids to design and build a complex digital command system which could control the Gemini spacecraft, its target vehicle Agena, and the Apollo craft. This would become part of the Mission Control Center. After Philco and NASA completed the basic control center study, contractors were given the opportunity to bid on construction of the control center. Philco received the award, and in 1963 began work on the new Mission Control Center, which, it had been decided, would be located at the MSC in Houston, rather than at the Kennedy Space Center where Mercury controls were located, or at the Goddard Space Flight Center, where the Mercury flight computer systems were located.[35] Location of the spacecraft control center proved to be a somewhat thorny problem.

Were flight operations to be part of the design center, that is, MSC in Houston or part of the operations center at Kennedy Space Center where Mercury controls were housed? Goddard Space Flight Center had the attraction of being conveniently located near the National Capital and NASA Headquarters. When the Kennedy Space Flight Center was organized, G. Merritt Preston and some of his preflight operations personnel became a part of the new center and strengthened the idea that the control center should remain with the launch operations crews in Florida. Travel also was a factor. Would travel requirements be greater or less if the control center were located in Maryland, Texas or Florida? Could communications be enhanced by locating at the launch area? By the latter stages of Mercury, some of the remote flight control stations, as at Bermuda, were being shut down. It had become more and more apparent that through networking a control center could be effectively established at any number of points in the network. But both John Hodge and Chris Kraft initially suggested that the Florida location might be most reasonable. Further discussions with Walter Williams, Bob Gilruth, and the Gemini and Apollo Program Offices, among others, resulted in the decision to locate in Houston.[36] It was, all agreed, a happy and fortuitous decision which strengthened the relationship between the engineering design and flight operations programs. Mission control and operations soon became a major component of MSC responsibilities.

Had this responsibility been assigned to Goddard Space Flight Center or to Kennedy Space Center, the MSC would likely have been more design/development oriented in the NACA/Langley tradition. As it was, construction on the new advanced control center began in late 1962, and when Gemini 2 flew in January 1965, Mission Control Center in Houston monitored the flights. Houston's mission control directed the Gemini 4 flight in June and all subsequent Gemini and Apollo flights.[37] Although the Mission Control Center at first lacked the flight simulation systems, they soon were added and became important in flight planning, training, and real-time operations. Houston's Mission Control Center did represent the state-of-the-art in modern communications.

Chris Kraft, incidentally, credited IBM and the Philco contractor teams for the design and fabrication of the Mission Control Center. IBM's technical manager, Jim Hamlin, who had worked on the Mercury control center, "was the man responsible for the development of the 7094 system which we used to support Gemini. I can't praise him too highly," Kraft said. And Philco's manager, Walter LaBerge, and the Philco team who worked closely with IBM did an outstanding job, Kraft said. But as an aside, Kraft also mentioned that despite the excellence of the Gemini control system, it, and especially the 7094 computer capability, was inadequate to meet the developing needs of planned Apollo flights. For that, yet a new generation of real-time computers would be needed.[38] Gemini, then, provided a transition in the technology of communications and control from Mercury to Apollo.

Gemini also provided a management transition from the more simple structures used in Mercury to the more elaborate systems-management structures created for Apollo. Bob Gilruth gave James C. Elms, who joined MSC on February 1, 1963, primary responsibility for developing the reorganization that would strengthen the Gemini and Apollo management systems and alleviate some of the confusion that derived in part from the more informal, collegial style of management associated with Mercury. That system could not cope with the multifaceted management responsibilities of Gemini and Apollo. Elms came to MSC (somewhat reluctantly) from his position as Director of Space and Electronics at the Aeronutronic Division of Ford Motor Company. He had previous management experience with North American Aviation and Martin Company. He said his reluctance to join MSC was because "I had a growing family and many expenses and my industrial salary was twice that of the salary I was offered in the government." But Brainerd Holmes (who would soon leave NASA Headquarters and return to industry) was a persistent recruiter and Elms accepted the job.[39]

Elms viewed the MSC organizational problem in very broad terms. "I felt that it would be possible to handle the Gemini program in somewhat the same manner as the Mercury program, but that it would be extremely wise to start taking large steps in the direction required for Apollo using the interim step known as Gemini," Elms explained in an interview some years later. "The methods used to manage the Mercury program could not possibly have been applied to the Apollo Program," he said.[40]

The problems encountered with Charles Frick's industrial-style, Headquarters-oriented control of the Apollo Program Office and Jim Chamberlin's independence in the Gemini office were symptomatic of more than a conflict with the collegial tradition of management at MSC. They reflected very real difficulties in delineating between developmental functions and operations. Thus, disputes between Barry Graves, who was developing spacecraft information and control systems, and Chris Kraft, in flight operations, involved more than personalities. Differences between the Engineering and Development Directorate under Max Faget, and the Operations Directorate under Walter Williams involved in part "deciding where development stopped and operations began." When he arrived at MSC, Elms said, he assumed one of Walter William's hats as Deputy Director for Programs and Development. Williams became Deputy Director for Missions and Operations. Some of the assistant directors, such as Barry Graves, held two different titles and reported to two different bosses. Graves reported to both Williams and Elms. "The situation was a little confusing from the theoretical organization chart point of view," Elms explained.[41]

Although there was a need to clarify roles and functions, the management problem was not truly a matter of "either organization or of personality . . . but of the philosophy of managing a difficult program involving requirements for ultra reliability, schedule, safety, and yet operational flexibility and perfection of equipment without exorbitant costs," Elms said.[42]

Quality control needed to be maintained at the manufacturer's site, not at the delivery point as had been true with Mercury spacecraft where each capsule had been virtually disassembled and reassembled prior to flight. The Gemini program stressed inspection and checkout at the factory. And it was natural and appropriate, Elms believed, that the engineering development divisions would want to maintain control over a product from its design to its completion, while the users in operations would want to participate in the development and manufacturing cycle to help assure the quality of the product. "It seemed to those involved at the time to be a very difficult and almost insoluble organizational problem," he observed.[43]

Viewed in this broader perspective, the appointments and organizational changes at MSC that occurred during the fall of 1963 and the spring of 1964 represented a broad-based and rational attempt to resolve some very difficult management issues. Gilruth's reorganization became effective November 1 and elevated Jim Elms to Deputy Director. Four functional assistant directors reported to Elms: the Engineering and Development Directorate headed by Max Faget, the Flight Crew Operations Directorate under Donald K. Slayton, a Flight Operations Directorate with Chris Kraft the assistant director, and the Administration Directorate under Wesley L. Hjornevik. The Flight Crew and Flight Operations Directorates were newly created, and facilitated the input from operations into spacecraft planning and development. Two program offices, the Apollo office under Joseph F. Shea and the Gemini office under Charles Mathews, also reported to the Deputy Director, as did the Manager of Florida Operations, G. Merritt Preston.[44]

The reorganization involved substantive changes within the various directorates, and especially within the Engineering and Development Directorate which combined with it the Information and Control Systems Offices which had been headed by Barry Graves. The basic elements of the reorganization had to do with creating small, management-oriented program offices with access to all levels of engineering and operations. Instead of program managers having their own engineering staff and in effect creating a minicenter, they would rely on the engineering expertise of the functional (or line) divisions. With this arrangement, a Gemini or Apollo program office could not isolate itself from the center management as had happened in the past, and the managing engineers at the center could provide the interface or flow of experience and expertise which seemed to be lacking to some extent between the Gemini and Apollo programs.[45] As a practical matter, the reorganization required fewer people in fewer meetings, created less confusion, and more clearly, but certainly not perfectly, defined responsibilities and lines of authority.

With these organizational changes in place, James Elms elected to return to private industry on February 1, 1964. "Our center, our agency, and our Nation owe him a debt for his accomplishment. I cannot adequately express to him my own deep and personal appreciation," Gilruth said. Much to the satisfaction of MSC personnel, George M. Low returned to the fold from the Manned Space Flight office in Washington, D.C., as the replacement for Elms.[46]

These organizational and personnel changes occurred while the pace of activity for both Gemini and Apollo programs increased. Upon completion of the successful unmanned suborbital flight of a launch vehicle and Gemini spacecraft in April, MSC announced that Virgil I. (Gus) Grissom and John W. Young would be the prime crew for the first manned Gemini flight, tentatively scheduled for November or December. The backup crew would be Walter M. Schirra, Jr., and Thomas P. Stafford. One more test suborbital flight of the Gemini-Titan system (GT-2) was scheduled December 9 (prior to the manned flight), but it was delayed until January 19 because of a cracked servo valve flange. That fully successful flight on the 19th preceded the launch on March 23 of Grissom and Young, America's first two-man team in a three-orbit mission which successfully tested the maneuverability of the craft. The "Molly Brown," as the astronauts named their craft, landed near Turks Island in the British West Indies.[47] Five more manned Gemini flights were launched and returned safely within the year, and each pushed the frontiers of spaceflight into new dimensions.

The reorganizations of 1963 and 1964 continued to unfold in terms of management efficiencies and cost savings. The growing involvement of American military forces in southeast Asia, the approval by Congress of President Johnson's War on Poverty and other social welfare programs, the rising costs of "cold war" hardware, not to mention the billions being spent on NASA, began to turn America's post-war world of budget surpluses and constant economic growth into slowdowns and deficits. As a result, President Johnson declared a campaign for economy in government and, concurrently, Gilruth announced a cost-reduction program for the space center which was expected to result in saving $2 million in operating costs for fiscal year 1964.[48] The center also anticipated greater savings, as well as

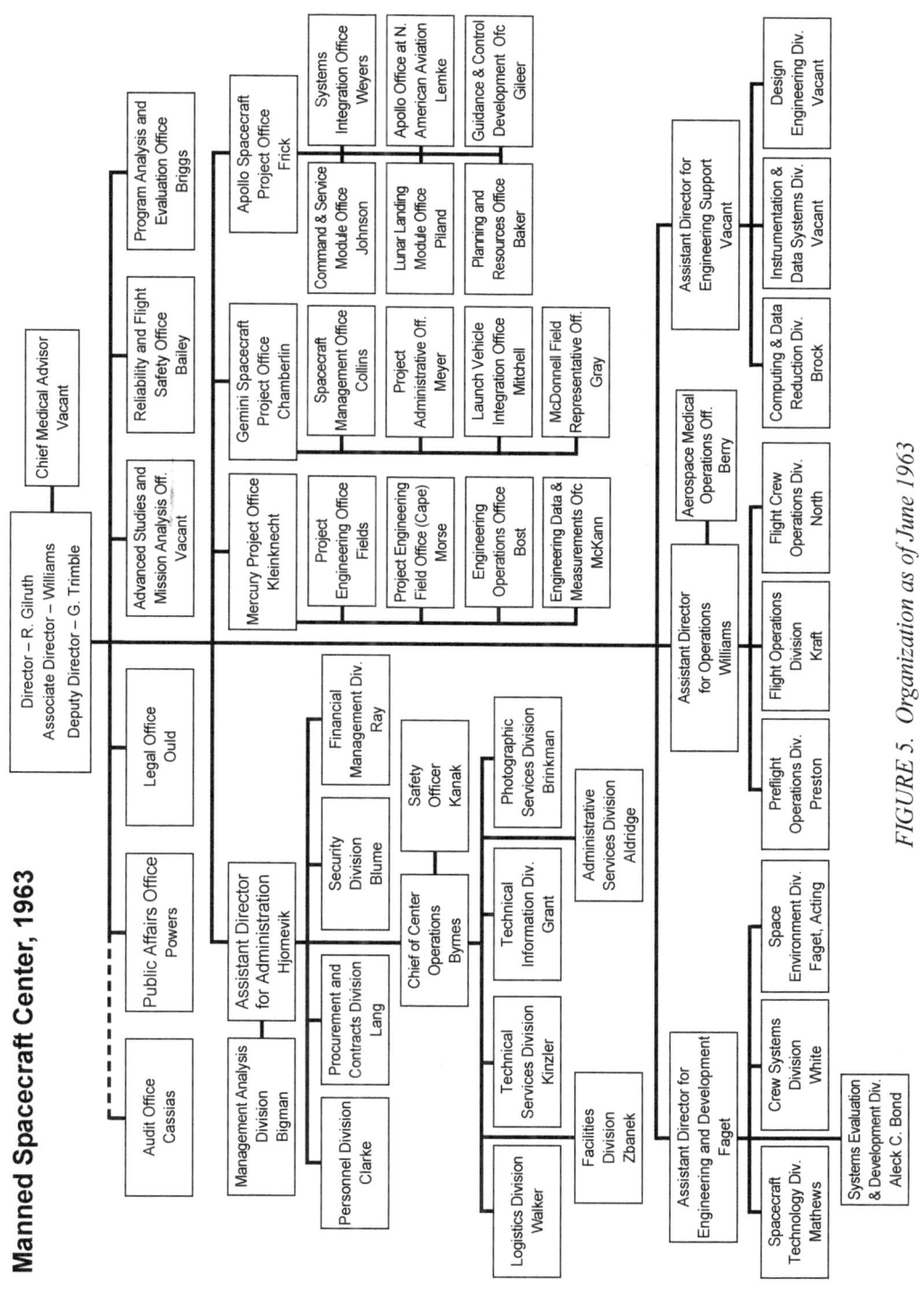

FIGURE 5. Organization as of June 1963

Suddenly, Tomorrow Came . . .

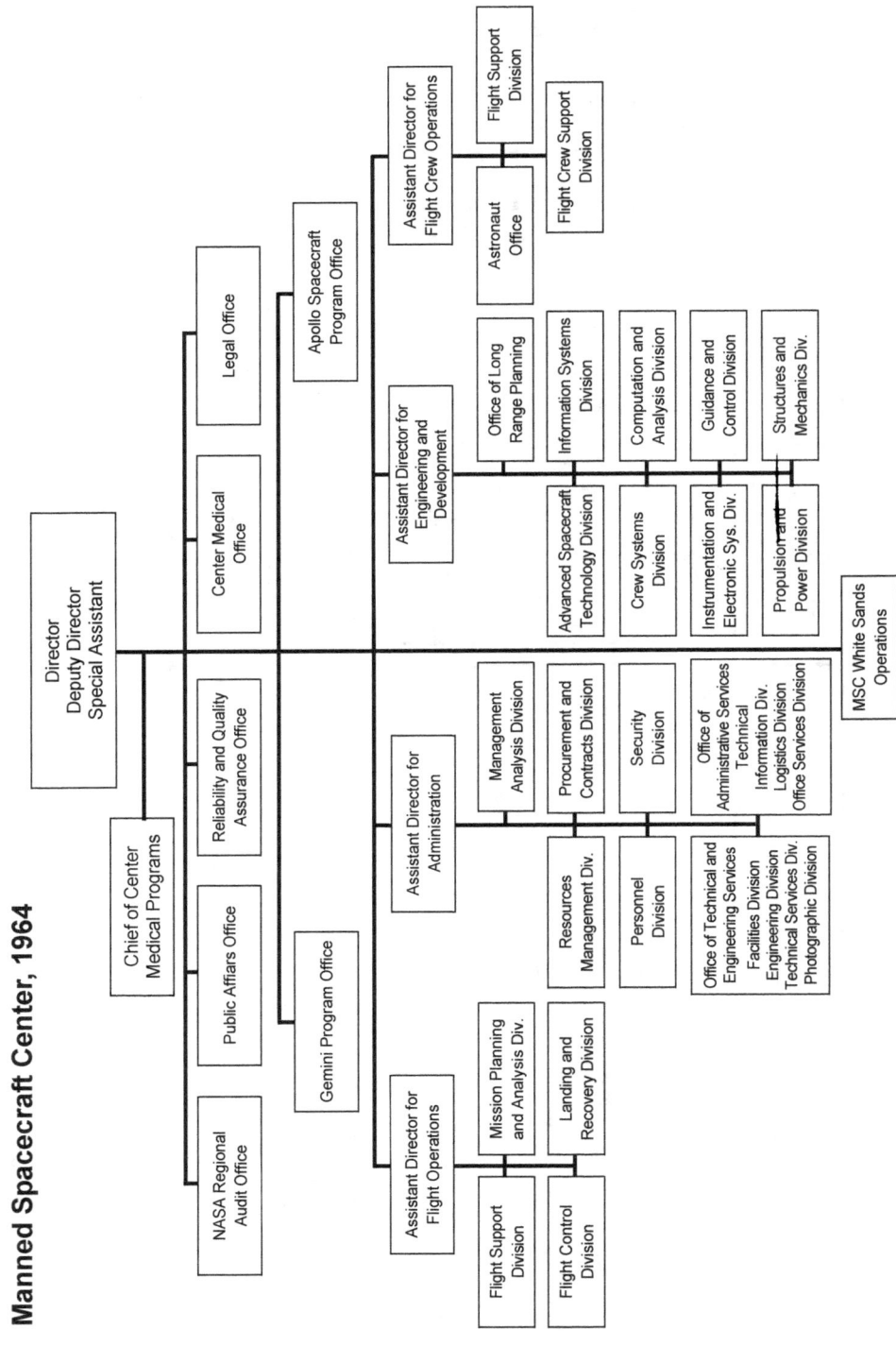

FIGURE 6. *Organization as of December 1964*

technically improved products, through more effective contracting procedures and better controls over design change procedures.

Cost-plus-incentive-fee contracts, the establishment of configuration management panels and boards, and the formalization of subsystem management structures promised to improve the cost efficiency and the engineering effectiveness of the center. MSC and NASA contracts had been awarded usually on the basis of cost plus a fixed fee (CPFF) to ensure the delivery of the product and a profit to the contractor. But, "the fatal flaw of the CPFF contract," Wesley Hjornevik pointed out at a meeting of the Harvard Business School Club of Houston, "is that profit is a function of estimated cost; and being established at the outset of a program it is not affected by how well or how poorly the contractor actually performed the work." Despite competitive bidding, contractors tended to overprice because the fixed-fee or profit was usually a percentage of the cost package. Once the contract was awarded, the only real incentive to simplify design, improve deliveries, or control costs would be the hope of future contracts. "This is not enough," Hjornevik said, and he explained that NASA was designing an incentive fee contract.[49]

Later that year NASA converted the McDonnell Aircraft Company's cost-plus-fixed-fee contract for the delivery of the remaining Gemini spacecraft to a cost-plus-incentive-fee basis. The change was effected after extensive study by both MSC and McDonnell study groups. Kenneth Kleinknecht headed a "Gemini Incentive Task Group" to study and renegotiate the spacecraft contracts. The initial phase involved the establishment of basic incentive criteria and the preparation of the Request for Proposal (RFP) to the contractor. McDonnell conducted its own study and costing analysis and, after negotiations, agreed to convert its $712 million contract to the new formula. The contract provided profit incentives for outstanding performance, cost controls, and timely delivery, and profit reductions for failure to do so. No performance incentive fee would be paid a contractor "on any mission involving the loss of life of a crewman." A weighted evaluation system based on performance, costs, and schedules determined performance levels. Such contracts, it was hoped, would provide an emphasis on reliability, ensure successful mission performance, control costs, and encourage timely delivery.[50]

The McDonnell incentive contract included a provision for regularizing and improving procedures for contract changes and for minimizing the costs of such changes. In developmental-type contracts, which characterized most NASA contracts, changes were part of the business. Changes could be suggested at any level by the contractor, by managing engineers in design and development work, and at the operations level. All of these elements were interfaced through the subsystem managers and by both formal and informal testing and evaluation committees.[51] But the system for approving changes had not been regularized and changes usually added significantly to production costs.

As a result, while the McDonnell incentive contract study proceeded, NASA began to devise a better system for implementing contract changes. This became known as configuration management, which created a flow of contract change information and decision making from the subsystem manager to top-level management where a final decision was approved. As the system developed, a Request for Engineering Change Proposal (RECP) had to first be approved by the individual responsible for that specific subsystem. It

Gemini 3, launched March 23, 1965, carried the first two-man crew (Virgil Grissom and John Young) on a 4-hour and 53-minute orbital mission. The astronauts completed the first piloted spacecraft maneuvers.

then went to the project officer for approval where a Configuration Control Panel (for example, a command space module panel or lunar module panel) reviewed and approved the changes. The panels ordinarily met on a regular weekly basis. From there the change went to the Configuration Control Board at the program level (e.g., for Apollo or Gemini) where representatives from each of the directorates reviewed the changes. At each level, evaluators maintained a register of changes in a system, schedules for reviews, agendas, and pertinent review information, and distributed changes proposed and approved to affected organizations.[52] First implemented in the McDonnell/Gemini contract, the configuration control system became an integral part of the Apollo management structure.

There was a natural reluctance on the part of the government and contractors to accept configuration management because it restricted somewhat their own management independence, but overall it forced discipline into design and development. Interestingly, configuration management derived from the "minimum essential" philosophy of the industrial revolution, which in layman's terms argued "if it ain't broke don't fix it." Or, in industrial language, "if it works as well rough as it does ground and polished—leave it alone." NASA's configuration management philosophy was to build for a high degree of reliability and create redundant systems when in doubt.[53] The success of the Gemini missions reflected in part the development of more mature and disciplined management systems within the MSC in particular and throughout NASA.

Each Gemini flight seemed to produce some quantum "leap" in the mastery of space. Now that the new Mission Control Center was on line, Gemini 4 became the first flight controlled from Houston and the longest duration mission to that date. The Cape Kennedy control center provided backup services for the initial launch and trajectory, and Goddard's computer center provided support for the entire 4-day mission. Three mission controllers, Chris Kraft, John Hodge, and Eugene F. Kranz, directed the flight from Mission Control. Paul Haney sat in the control room and described the flight for the press. Haney had hoped to obtain television cameras aboard Gemini flights but concerns about using additional electronic equipment in the cabins and what might be considered the frivolous nature of such expenditures overrode Haney's appeal. During the mission, an attempt by Command Pilot James A. McDivitt to rendezvous with the orbiting booster rocket failed because of excess fuel consumption during the maneuvers, but Edward H. White made history with NASA's first spacewalk—20 minutes in EVA tethered to the reentry module. "This is fun," he said, and returning to the capsule was "the saddest moment of my life." In that one flight, McDivitt and White logged more time in space than all previous United States astronauts combined.[54] Not long after that flight an African Episcopal Bishop Josiah Mtekateka of Malawi was visiting the Houston area and attended the Friendswood Episcopal Church. He asked the pastor, William Sterling, "Is it true what the drums are saying, that a man has walked in space?"[55]

Two months later, on August 21, 1966, Gordon Cooper and Charles Conrad sped into orbit aboard Gemini 5 and spent twice as long in space as the Gemini 4 astronauts. They completed 17 assigned scientific experiments and a rendezvous with a "phantom" vehicle. The Gemini 6 flight was canceled on October 25, when the Agena-D target vehicle's engine failed after separation from the booster rocket. A very bold decision followed, which Chris Kraft said was initiated by McDonnell Douglas, to launch Gemini 6 (GT-6A) while Gemini 7 was in orbit in order to accomplish the much desired rendezvous. Thus, the critical Gemini 7/6 mission began on December 4 with Frank Borman and James A. Lovell, Jr. at the controls of Gemini 7. They were joined, almost literally, by Wally Schirra and Tom Stafford, who launched aboard GT-6A on December 15 and 6 hours later came to within one foot of Borman and Lovell's module.[56]

Attention shifted for a moment from Gemini to the first suborbital launch of a Saturn IB rocket carrying an unmanned Apollo module (February 26, 1966). The previous month, based largely on the Gemini-McDonnell experience, Grumman Corporation's contract for a lunar landing module and North American's contract for the Apollo spacecraft were renegotiated as cost-plus-incentive-fee contracts. Then Gemini 8, with David R. Scott and Neil Armstrong aboard, completed a successful rendezvous and docking with an Agena target vehicle, but broke off after 30 minutes when the combined craft began to yaw and roll wildly. The astronauts made an early but safe return. The following 3-day mission of Tom Stafford and Eugene A. Cernan in Gemini 9 was beset with problems: rendezvous with the target vehicle failed because a protecting shroud over the adapter had not fallen off, and there were visibility problems during Cernan's EVA. But the mission successfully tested rendezvous maneuvers, including

The interior of Gemini 6 has a striking similarity to the Apollo spacecraft still under development. Gemini 6, launched on December 15, 1965, completed a piloted rendezvous with Gemini 7 launched December 4. Gemini provided important technical and training missions in preparation for the Apollo lunar missions.

Suddenly, Tomorrow Came . . .

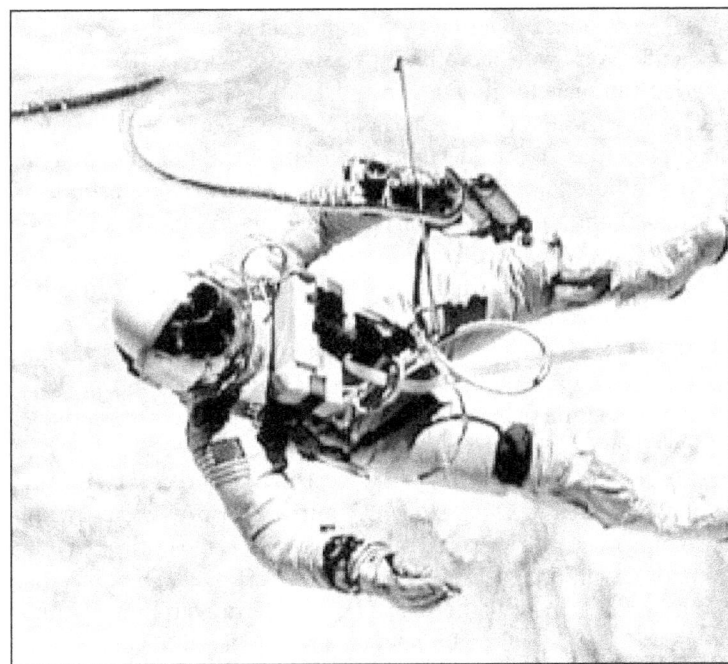

A critical interlude between Mercury and Apollo, Gemini proved the necessary capabilities for lunar flight including rendezvous and extravehicular activity (EVA)—more popularly called a "spacewalk." The photograph shows astronaut Edward H. White floating in the microgravity of space secured to the spacecraft by a 25-foot umbilical cord and a 25-foot tether line. He moved about using a hand-held maneuvering unit.

Gemini 10, launched July 18, 1966, carried astronauts John Young (left) and Michael Collins on a 70-hour plus mission that included rendezvous, docking, and EVA activities.

a simulated rendezvous with a lunar module.[57]

Each of the last three Gemini flights, including Gemini 10 on July 18 carrying John Young and Michael Collins, Gemini 11 on September 12 with Charles Conrad and Richard F. Gordon, and Gemini 12 on November 11 flying James A. Lovell and Edwin E. Aldrin, completed sophisticated rendezvous and docking maneuvers and EVA activities. Michael Collins on Gemini 10, for example, retrieved an experimental package from a target vehicle that had been in orbit since March. Gordon, aboard GT-11, tethered the module and target vehicles together; and on the last Gemini flight, Aldrin and Lovell completed three separate activities outside the reentry module.[58] Gemini had become an invaluable Apollo lunar landing learning experience.

What had been learned as a result of the Gemini program? Many of the answers to that question were not immediately apparent to those who planned and flew the Gemini missions. Dr. Charles Berry, who was the MSC flight physician for the Mercury, Gemini and Apollo programs, concluded many years later that Gemini diminished or repudiated some of the old "straw men" arguments which claimed, in spite of the Mercury missions, that long-term spaceflight would adversely affect humans physiologically and psychologically. The predicted effects of long-term weightlessness included hypertension, hypotension, reduced plasma volume, reduced blood volume, and variously that a person would urinate all the time or not be able to urinate at all, or sleep all the time or not be able to sleep at all. And there were many more bogeymen Dr. Berry

"President Johnson wants to say 'Howdy.'"
Drawing by Whitney Darrow, Jr.; ©1966
The New Yorker Magazine, Inc.

said at a 1989 conference celebrating the 20th anniversary of the 1969 Apollo lunar landing. Gemini proved that during spaceflight vital functions remained normal, stress was tolerable, there was no psychomotor impairment, and cardiovascular deconditioning could occur. Gemini flights confirmed that humans could survive relatively long periods in space, but they did not wholly resolve issues for long-duration flight.[59]

Glynn Lunney, who served as flight director on Gemini and Apollo missions, some two decades later called Gemini a "stroke of genius. The operations team came out of Gemini. Flight crews and ground crews trained on Gemini and came on [Apollo] like gangbusters," he said. On the same occasion celebrating the 20th anniversary of the Apollo lunar landing, Cliff Charlesworth pointed out that Gemini flights had hardware problems, while Apollo had few. Stephen Bales said the Gemini 10 rendezvous was a particularly "historic experience" that prepared the way for Apollo, and Gerry Griffin referred to Gemini as "a whole series of little things—it was to shake out the system."[60]

Gemini recorded a series of "firsts" including the first pilot-controlled maneuvering in space, the first rendezvous, the first docking with another vehicle, the first extended flight of more than a week in duration and extended stays by astronauts outside the spacecraft, and the first controlled reentry and precision landings (albeit not on land as originally planned).[61]

But perhaps the deeper meaning of Gemini had to do with the enhancement of worldwide communications and the management reorganizations and reorientations accomplished at MSC and tested on the Gemini program. The manned spacecraft effort matured greatly during Gemini, and hardly had the program closed than Apollo did, as Glynn Lunney noted, "came on like gangbusters." But the path to the Moon would yet be strewn with many unforeseen obstacles.

CHAPTER 6: The NASA Family

The melding of all of the NASA centers, contractors, universities, and often strong personalities associated with each of them into the productive and efficient organization necessary to complete NASA's space missions became both more critical and more difficult as NASA turned its attention from Gemini to Apollo. The approach and style and, indeed, the personality of each NASA center differed sharply. The Manned Spacecraft Center was distinctive among all the rest.

Fortune magazine suggested in 1967 that the scale of NASA's operation required a whole new approach and style of management: "To master such massively complex and expensive problems, the agency has mobilized some 20,000 individual firms, more than 400,000 workers, and 200 colleges and universities in a combine of the most advanced resources of American civilization." The author referred to some of the eight NASA centers and assorted field installations as "pockets of sovereignty" which exercised an enormous degree of independence and autonomy.[1] An enduring part of the management problem throughout the Mercury and Gemini programs that became compounded under Apollo, because of its greater technical challenges, was the diversity and distinctiveness of each of the NASA centers. The diverse cultures and capabilities represented by each of the centers were at once the space program's greatest resource and its Achilles' heel.

NASA was a hybrid organization. At its heart was Langley Memorial Aeronautical Laboratory established by Congress in 1917 near Hampton, Virginia, and formally dedicated in 1920. It became the Langley Research Center. Langley created the Ames Aeronautical Laboratory at Moffett Field, California, in 1939. After the formation of NASA, Ames expanded its capabilities in research and experimentation in the life sciences and aerodynamics. Under congressional authority, Langley established the Lewis Flight Propulsion Laboratory adjoining the Cleveland Municipal Airport in 1940. As NASA's Lewis Research Center, the facility continued its work on propulsion systems. Its research on hydrogen fuel rockets contributed to the development of the upper stages of Saturn (Apollo) and Centaur rockets, and Lewis scientists and engineers made significant discoveries in solar power, reentry aerodynamics, lifting body concepts, and thermal protection systems. A High Speed Flight Station at Edwards, California, which had been formed in 1946, continued under the same name until it was renamed the Dryden Flight Research Center for Hugh L. Dryden. The Pilotless Aircraft Research Station at Wallops Island, Virginia, which provided hypersonic flight test support for Langley and was the point of origin of many MSC engineers, became NASA's Wallops Station which reported to Goddard Space Flight Center (earlier Beltsville Space Center) in Greenbelt, Maryland.[2]

Three of the NASA centers which were central to the Apollo program had non-NACA origins and very different personalities from those with a Langley lineage. These included the Marshall Space Flight Center in Huntsville, Alabama, the Kennedy Space Center at Cape Canaveral, Florida, and the Jet Propulsion Laboratory (JPL) in California. The JPL, founded in 1944 for work with the Army Air Forces, was operated under contract for NASA by the

California Institute of Technology. JPL had more real identity as a "pocket of sovereignty" because of its independent role in supporting the Army and then NASA, and its unique academic affiliation.

The JPL reported to the NASA Headquarters Office of Space Sciences. It had major responsibilities for lunar and planetary exploration and in that role provided data that helped validate engineering models used for Apollo lunar module development. Through the Ranger and Surveyor programs, which it supervised, JPL provided information on Apollo lunar flight approach patterns and landing sites.[3]

In addition to JPL and an Electronics Research Facility established in Cambridge, Massachusetts, NASA established four post-Sputnik spaceflight centers. These included MSC, Goddard Space Flight Center, Marshall Space Flight Center, and Kennedy Space Center. These four centers were similar in that they tended to operate as development or operations centers while the older NACA centers, including Langley Research Center, continued their traditional concentration on research and technology studies. Goddard Space Flight Center and MSC retained a closer filial relationship with the centers of NACA extraction because of their Langley lineage. They, and especially the MSC scientists and engineers, revered the NACA laboratory-research heritage of autonomy and independence. Marshall Space Flight Center and, to a lesser extent, Kennedy Space Center came out of the military Department of Defense culture. They were more accustomed to working under a central authority and to "systems" approaches to management.[4]

Managing NASA and achieving program objectives not only involved problems of managing a large scale and physically scattered institution that rather suddenly sprang into being, but NASA's component parts were very unlike one another. The changing relations between MSC, Headquarters, and other NASA centers and the tensions which existed within the NASA organization reflected not only the diversity and culture of NASA, but the changing complexity of programs. Spaceflight was an intricate and highly interdependent business and became more so as programs developed through the Mercury, Gemini, and Apollo phases.

Manned spaceflight, initially almost solely the responsibility of the Space Task Group, became increasingly the collective responsibility of all NASA Centers with MSC, Goddard Space Flight Center, Kennedy Space Center, and Marshall Space Flight Center having lead roles. MSC managed the development of the spacecraft, Marshall had responsibility for launch vehicles, Goddard developed the tracking and monitoring networks and emphasized scientific instrumentation and operations for manned and unmanned programs, and Kennedy conducted launches and provided ground support for both manned and unmanned missions. Although center responsibilities became reasonably clear and well-defined by the mid-1960's, spaceflight programs required very careful interfacing and cooperation by the essentially autonomous NASA centers and their equally independent contractors.[5]

Each NASA center had a distinctly different style, personality, and approach to management and operations. They were staffed by civil service employees largely trained in the NACA concept of in-house design, development and testing or, in the case of Marshall and Kennedy personnel, they were accustomed to the arsenal-procurement style of management. The newer manned spaceflight centers had to redirect their efforts into the developmental and operations spheres, as well as to accept their primary role as managers of independent contractors who did the actual construction and fabrication—in contrast to the

Langley in-house research and testing experience. But MSC personnel in particular sought to preserve the "hands-on" engineering associated with Langley, as well as the autonomy and independence consistent with NACA tradition. They, as did Langley engineers and scientists, tended to view themselves as part of a collegial association or federation. This perception contributed to stress between MSC, NASA Headquarters, and other NASA centers.

George Mueller, the Associate Administrator for Manned Space Flight, felt that MSC exhibited an unusually independent attitude, and indeed that the world view of each NASA center was startlingly different from that in NASA Headquarters. Headquarters constantly sought to bring the NASA centers under tighter central control. One such effort was the appointment of Edgar M. Cortright as Director of Langley Research Center in 1968. Some believed that Cortright's experience on the road to the directorship—specifically his project management work at NASA Headquarters—would bring about dramatic changes in the NACA style of "independent" management at Langley.[6] MSC, as a matter of perceived professional integrity and heritage, rather fiercely resisted Headquarters control—not because it was any less committed to the policies and programs established by NASA, but largely because MSC engineers believed that project management could not be separated from center-based technical capability.

There were, to be sure, other reasons for conflicts and stress. NASA engineers and managers, particularly those at the director and administrator levels, were people of great experience and considerable expertise, and by nature independent and competitive. Moreover, MSC attitudes of independence were bolstered in part by the perception that it emerged from the Space Task Group originally charged with the task of putting Americans in space. MSC regarded itself as the lead center in manned spaceflight activities, and looked upon other centers as suppliers and upon Headquarters as the funding agency. Thus, in the following government-industry functional matrix wheel representing the space consortium involved in the Apollo program (figure 7), one might substitute MSC at the hub in place of the NASA Headquarters Office of Manned Space Flight to properly see the program as it was seen in Houston.[7] Very likely the other centers had the same viewpoint.

NASA Headquarters established program goals and objectives, allocated resources (including budget procurement and distribution), and maintained critical interface with Congress, the executive offices, other government agencies, the scientific community and the public. NASA, as true of most large-scale businesses and multinational corporations, operated on the basis of delegation of authority and decentralized management. Administrator James Webb, in 1968, described the NASA management system as one of planned disequilibrium. For 10 years, he said, "we have been constantly seeking to prepare for and organize to meet substantive and administrative conditions which could not be foreseen. We have sought to avoid those concepts and practices which would result in so much organizational stability that maneuverability would be lost."[8]

Although there were pressures for greater central control, through its first several decades of existence a decentralized management style prevailed which seemed to best fit the need for the specific independent tasks being performed by the NASA centers. But this planned disequilibrium also meant fluid organizational dynamics and instability, themselves causes of stress. It also, perhaps, maintained an appropriate environment for the highly motivated, bright, aggressive, and competitive personalities of the NASA community.

Suddenly, Tomorrow Came . . .

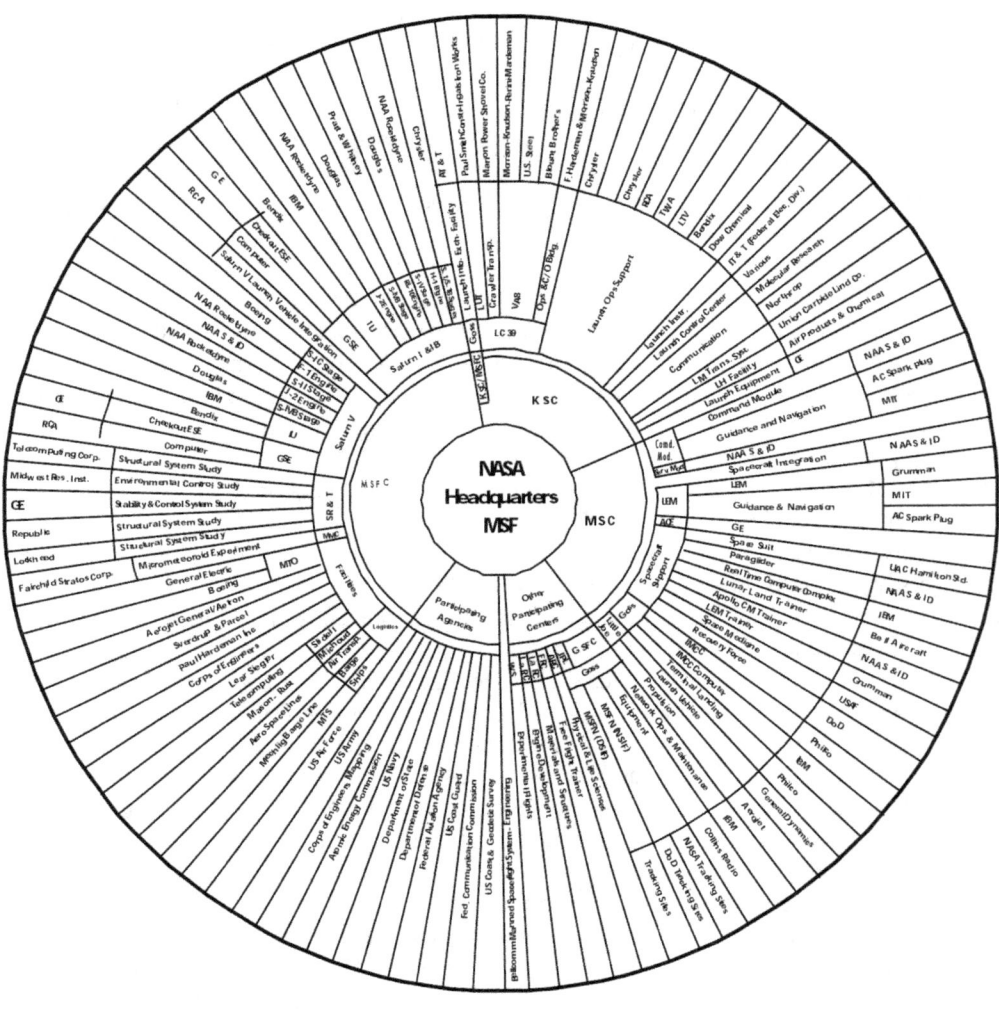

Source: The Apollo Program Development Plan, Office of Manned Space
Flight, January 15, 1965, Apollo Series, JSC History Office.

Figure 7. Apollo Program Government-Industry Functional Matrix

Although organizational structures (or the lack of them), programs, and Congress gave form to the NASA administrative system, individuals within the organization at every level made the system work. Thus, from Headquarters, Abe Silverstein and George Low interfaced with the Space Task Group and MSC during the Mercury years, and during the Gemini and Apollo programs, with George E. Mueller (who served both as Associate Administrator for

Manned Space Flight and Acting Director of the Gemini Program Office) and Samuel Phillips, whom Mueller made Apollo Program Manager with responsibility for planning schedules, budgets, and systems. Bob Gilruth, Director of MSC, with Kurt H. Debus, Director of the Kennedy Space Center, and Wernher von Braun, who directed the Marshall Space Flight Center, served on Mueller's Executive Council which met monthly. These people, with Administrator James Webb, and Deputy Administrator Robert C. Seamans, Jr., set the tone of relations between NASA Headquarters and the centers for the years between 1964 and 1968 when Gemini closed and Apollo made its debut.

Arnold S. Levine (*Managing NASA in the Apollo Era*) credits George Mueller with the administrative changes at Headquarters and within the manned spaceflight centers that resolved the management crises precipitated by Brainerd Holmes' efforts to centralize Apollo program management. Tensions between Headquarters and MSC in particular affected NASA management well into the 1970's. Mueller, however, "restructured the Apollo program so that every functional element at the Headquarters program office had a corresponding element in the center project office." This facilitated a liaison and promoted cooperation without imposing hierarchical direction and control by NASA Headquarters over the centers. For each of the major systems, Mueller made one person singly responsible for performance, costs, and schedules. That person "defended his programs before top management and Congress, set and interpreted policy with his program managers and center directors, and set the terms on which long-range planning would proceed."[9]

Mueller joined NASA in November 1963, upon the departure of Brainerd Holmes, and by the end of 1964 effectually completed the reorganization of the Manned Space Flight Office. A native of St. Louis, Missouri, Mueller received a bachelor of science degree in electrical engineering from the Missouri School of Mines and earned a master of science degree at Purdue University before joining the Bell Telephone Laboratories where he continued research on video amplifiers, television links, and microwave research. He pioneered in work on the measurement of radio energy from the sun, microwave propagation through gases, and the design of low-field magnetrons. In 1946, Mueller joined the faculty of Ohio State University as assistant professor of engineering, where he also continued graduate studies and completed his Ph.D. in physics in 1951. Prior to joining NASA, Mueller spent 5 years with Space Technology Laboratories, Inc. of Redondo Beach, California, serving as Vice President for Space Systems Management and Vice President for Research and Development.[10]

When he arrived at NASA, Mueller's experience in the laboratory and in the commercial side of the space business gave him valuable insights into the management problem. He set about, he said, trying to convince Wernher von Braun at the Marshall Space Flight Center to implement a systems engineering approach. The design and construction of booster engines, manned spacecraft, and electronic guidance systems involved distinctive tasks and products that had to be fitted after they were built. Systems engineering required a strict interface control system. What he wanted, Mueller said years later, was to meld the traditional research strengths of NACA, with the technical know-how of Marshall and the MSC STG experiences. He inserted a "program management system in parallel with the functional systems" and set up what he called a "5-box" management structure which provided for direct communications between like disciplines.[11] Thus many parallel lines of communication existed between

Headquarters and the centers, and between the centers. As Aleck Bond and Jerry Hammack put it, a division head or anyone else could (and did) pick up a telephone and call their counterpart about a problem anywhere in the NASA organization.[12] Spaceflight was a team operation. The team was far-flung and disparate, and communications between them was essential.

Mueller also provided critical liaison between NASA and Congress and between NASA and its contractors. He organized the Apollo Executives Committee, comprising corporate Chief Executive Officers (CEOs) or their representatives who were Apollo contractors. The committee met periodically to keep the contractors apprised of overall program progress, to review problems, and to better develop systems engineering approaches. We worked "quite openly" with our contractors, Mueller said. The committee provided an invaluable and more informal link with the contractors than existed at the center level where relationships were largely defined by the contract. Mueller said in 1989 he did not think it would be any longer possible to create such a body. Legal constraints in the contracting process and changing relationships between NASA and its contractors preclude the close, personal cooperation of earlier days.[13]

It was particularly imperative, Mueller said, that close relations be maintained with Congress. He personally met frequently with the House Science and Astronautics Committee, and monthly with Olin E. Teague and the Manned Space Flight Subcommittee. Teague, as mentioned earlier, took a strong and personal interest in the space program and became an invaluable congressional ally for NASA. Teague and other members of Congress relied heavily on the work of William E. Lilly, who worked under Mueller as the Manned Space Flight Program Control Officer. Bill Lilly supervised program planning, costs, and schedules, and had responsibility for the management of resources and facilities. He was, in effect, comptroller for the Apollo program and highly respected on Capitol Hill. When Lilly gave a figure, it was reliable. Although first impressions suggested that he was somewhat rough or coarse, he made highly polished presentations. Moreover, he was a strategic conduit between lower-level managers at a center and Headquarters who responded to calls for help from individuals stuck with a cost that their own institution could not readily absorb. Oran Nicks, who worked at NASA Headquarters, called Bill Lilly the "unsung Godfather" in Washington of MSC. Nicks also described Mueller as an indefatigable manager who dressed like a math professor and often carried the day in meetings by his perseverance.[14]

As the human spaceflight program shifted from the Gemini to the Apollo program, Sam Phillips became a major conduit between Washington and MSC. His counterpart in Houston was George Low, who had long experience in Washington with Abe Silverstein. The Phillips/Washington—Low/Houston connection proved exceptionally providential. Phillips and Low were enormously respected at every level. Phillips had been manager of the Air Force Minuteman program and Vice Commander of the Air Force Ballistic Missile Division before being detailed to NASA in 1963. Mueller assigned Phillips responsibility for Apollo planning, budgets, systems engineering and "other functions needed to carry out the program." Center Apollo program offices, prime contractors, and special intercenter coordination panels reported to Phillips. Phillips traveled extensively to the centers. He provided strong technical direction, was very conscientious ("dropping in on every detailee" at the Houston center for example) and, according to some MSC engineers who worked with him, "kept George Mueller (who was inclined to go off in every direction) straight."[15]

Kenneth Kleinknecht said that Phillips had a "tremendous understanding of the way to manage and direct a program from the Headquarters level," but he thought that as time passed, the Washington office became too involved in too much detail. For example, MSC's Mission Operations Director, Chris Kraft, had to specifically forbid the Headquarter's Mission Director from intervening in Mission Control Center flight operations during Gemini flights. That individual had the nominal authority but not the experience, practice, and training with the Mission Control team to direct flight operations. Headquarters' job, Kleinknecht said (probably reflecting the view of most of the centers), should be to "sell the program, get the money, and let us do it."[16] With only a few exceptions that generally reflected Headquarters' management philosophy during the Mercury and Gemini programs.

By the end of 1964, Sam Phillips, working with Mueller, the three spaceflight center directors, and other staff officers at Headquarters, developed a comprehensive "Apollo Program Development Plan," which established basic organizational guidelines for the program throughout its existence. The Mercury program, according to these guidelines, "established man's ability to perform effectively in the environment of orbital flight" and developed the foundation for manned spaceflight technology. Through Gemini, they stated, "we would gain operation proficiency and develop new techniques, including rendezvous." Apollo seeks to achieve "preeminence in space and to develop the ability to explore the Moon and return safely to Earth before the end of this decade."[17]

Apollo mission planning envisioned three flight phases including unmanned suborbital and Earth-orbital flights, manned Earth-orbital and long-duration and Earth-orbital-rendezvous flights, and manned lunar flights. The first Saturn IB flight was scheduled for 1966, with manned IB flights in 1967 and unmanned Saturn flights the same year. The next year, 1968, the Saturn V was to be used for manned Earth-orbital flights, followed in 1969 by manned lunar orbit and lunar landing flights. The plan specified that the Marshall Space Flight Center held responsibility for developing the Saturn I, Saturn IB, and Saturn V launch vehicles and engines and providing associated ground support equipment and flight operations support. MSC had responsibility for the Apollo spacecraft with ground and mission support, and Kennedy Space Center was responsible for launch and facilities.[18]

"A large segment of the United States industrial base is required to support NASA in accomplishing these responsibilities," the plan acknowledged. The government-industry functional matrix, mentioned earlier (figure 7) provides a visual representation of the magnitude of the Apollo program. The plan specified that "Whenever possible, matters of mutual concern are resolved by direct communication between participating organizations." When those agreements or concerns affected other centers, they had to be informed. Phillips created 8 standing Intercenter Coordination Panels and 15 subpanels reporting to a Panel Review Board chaired by Phillips. An Executive Secretariat composed of the chairman from the Office of Manned Space Flight and representatives of each of the three field centers set the agenda and meetings of the Panel Review Board and implemented decisions of the board. The Apollo plan also attempted to relate other unmanned space programs, such as the Ranger lunar survey, Surveyor lunar landing surveys, and Lunar Orbiter, to the completion of the Apollo missions.[19] Overall, Phillips' 1965 document offered a clear, comprehensible, and feasible action plan for the Apollo program. Phillips worked very hard to implement those plans.

So did the center directorates and program managers, such as George Low in Houston, who (after the Apollo 204 fire) was the primary interface with Phillips in the Headquarters' Apollo Program Office. Low went to Headquarters with Abe Silverstein in 1958 with a number of other Lewis engineers, including Edgar M. Cortright, William (Bill) Fleming, John Sloop, John H. Disher, DeMarquis D. Wyatt and Warren J. North. There were, in fact, so many Lewis engineers who served on the NASA Headquarters staff that it is appropriate to suggest that one of the great and most direct contributions of the Lewis Research Center to the manned spaceflight program was its pool of managing engineers who staffed the Headquarters program offices.[20] NASA engineers in Houston counted George Low, who came to their center as Deputy Center Director and later as Apollo Spacecraft Manager, among their most esteemed colleagues.

Aleck C. Bond, who managed Systems Test and Evaluation at MSC, worked hand in glove with Low. Low was a "human dynamo," he said, who got up at 5:30 in the morning and jogged, was in his office by 6:30 or 7:00, and kept three secretaries busy all the time. Jerry Hammack, Deputy Manager of Vehicles and Missions in the Gemini Project Office, who regularly put in 12-hour days at the center, remembered seeing Low's little white Ford Mustang in the parking lot when he arrived and there when he left in the evening. They, and most who worked with him, remember George Low as the man who could cut through red tape, maintain good rapport, and get things done. Self-effacing, he always had time to commend others for their work and provided inspiration to all who worked with him.[21] In November 1969, when James Webb turned over the Administrator's job to Thomas O. Paine, Low returned to Headquarters as NASA's Deputy Administrator, and became Acting Administrator upon Paine's resignation.

Low's technical skills related largely to aerodynamic laminar flow and boundary layers, but his management skills were "people" skills. He, with Phillips, helped maintain a generally cordial and cooperative mode with Washington. But MSC managers strongly resisted technical control of projects by Headquarters, and were perhaps even more jealous of their functional offices such as Public Affairs. For example, when the first Apollo orbital missions (Apollo 7 and 8) began to attract tremendous public attention, Julian Scheer (Assistant Administrator for Public Affairs at Headquarters) instructed Paul Haney, the public affairs officer at MSC, that NASA Headquarters would produce the Apollo 8 film rather than it being done in Houston as had been true on all previous flights. The MSC response was: "Your arrangement is unacceptable to this center. We intend to handle film as we have in the past, and have issued instructions to this effect. Your office is receiving a copy of the instructions."[22] George Mueller responded directly to Bob Gilruth agreeing to MSC film management and requesting that the center deliver copies of processed film to the Public Information Officer and the Office of Manned Space Flight in Washington 24 hours after processing, "with whatever release restrictions you may desire to impose."[23]

Although relations between Headquarters and MSC could sometimes be strained, they could be downright difficult between Marshall Space Flight Center and MSC. No two NASA centers were at once so interdependent in terms of their technical work and so independent in terms of their spirit as were MSC and the Marshall Space Flight Center. One built the spacecraft, the other built the engines that made it fly. The interface between MSC and Marshall became much more critical and complex as NASA's programs expanded from

Gemini to Apollo. The MSC was the lead center for Mercury and Gemini and operated under a relatively small and close-knit Headquarters organization, George Low said later. Until Apollo, MSC, Low commented, "had been clearly in charge not only of the spacecraft but also the launch vehicle and the flight operations." Marshall, in other words, first related to MSC more as a supplier than a partner. Moreover, the Redstone-Agena rocket, Kenneth Kleinknecht said, involved much simpler functional interface and required less contact and cooperation between the centers. Apollo, however, changed that because the Saturn rockets and the space vehicle were of an integrated design. Thus the changing nature of NASA space programs helps explain the changing relationship between MSC, Marshall, Headquarters, and the space community.[24]

Marshall and MSC worked on the same team and aspired to the same goals. Both accepted their roles as members of the NASA family, but as Kleinknecht explained, being "brothers" in the same family created special kinds of problems:

> . . . you start working with your brother—sometimes it's harder than working with a neighbor, and that's kind of like what I think we've been through with Marshall. Even the fact that everybody became so dedicated to this program as a national goal maybe made it a little difficult. Everybody was trying harder—worked long hours and always thinking of what we can do to make it better, regardless of whose hardware it was.[25]

It was an institutional form of sibling rivalry—basically healthy and often productive, but frequently annoying.

Although the two centers might be considered "brothers" in the NASA family, they had somewhat different parentage which contributed a bit to internecine strife. The MSC culture came through its NACA/Langley origins. The Marshall Space Flight Center evolved from the Army Ballistic Missile Agency (ABMA) which in a very real sense was uprooted and transferred from Germany's World War II Peenemünde rocket group headed by the irrepressible Wernher von Braun. In the minds of most MSC engineers, Von Braun defined the personality of Marshall and its relations to MSC. The two centers held something of the traditional "brotherly" love-hate relationship. Marshall seemed to demand both caution and a defensive position by MSC engineers on the one hand, and respect and admiration on the other.

MSC's perception of itself as NASA's "lead center" irritated Marshall engineers who prided themselves on being the real pioneers in spaceflight. Von Braun and his colleagues regarded their rocket developments for the German military as an expediency by which they could "indulge in spaceflight operations." Marshall engineers resented their initial role with the ABMA as a supplier or subcontractor to NASA. They regarded NACA and NASA as "an old stodgy short-sighted research organization that kind of got into the spaceflight game politically." There had always been "this background of resentment between ABMA and the Space Task Group, and then between Marshall and MSC," Paul Purser, Gilruth's special assistant observed.[26]

In some respects the modern space age began not with the launch of Sputnik, but rather with the launch of the German V-2 rockets by Von Braun's group at Peenemünde. I.B. Holley, Jr., then an Air Force officer stationed at Wright Field, recalled many years later having attended a meeting shortly after V-E day for a report on the status of German research and development. "Among other things," he said, "the speaker told us about uncovering German plans for establishing stations in space from which to bomb the United States. The idea seemed so farfetched, so impossible, that a roar of laughter swept through the hall." But it was the Germans, he said, who conceptualized the reality of space; it was we who, with the critical assistance of the Von Braun group, "picked up the ball and ran with it." Holley closed his remarks with the story of a Russian cosmonaut and an American astronaut who on passing each other in space, spoke to each other only in their native languages. Finally one blurted out, "Why don't we cut out this nonsense and speak German?"[27]

The philosophical or cultural differences between the two centers were aggravated by the contrast in the style of management and operation. Von Braun, Purser said, "ran his organization at Marshall with an iron hand and nothing was ever decided there without holding a big committee meeting over which Wernher presided and made the final decision.... Gilruth, on the other hand, worked closely with his people and tended to delegate more authority and responsibility to individuals . . ."[28] Purser, who helped establish the initial relationship between the Space Task Group and the ABMA, believed that he and Wernher von Braun developed a mutual respect and friendship.

Purser worked hard, but without considerable effect, to improve the personal relationships between Gilruth and Von Braun. But they were two markedly different personalities. Von Braun, Purser said, had a tendency to "run off at the mouth," while Gilruth always waited until there was a break in the conversation. With Von Braun around, there was never a break in the conversation. And Von Braun inadvertently offended Gilruth on a number of occasions. For example, on one occasion, Purser recalled, Von Braun wrote Gilruth a very condescending letter noting that it was the duty of a teammate to tell a fellow teammate when one of his shoelaces was untied. He warned Bob Gilruth that one of his shoelaces was untied—that being a poor job of wiring done by one of his contractors. On another occasion, Von Braun gave Gilruth a 4-hour harangue about MSC planning to use the Agena rocket in the Gemini program without first consulting Marshall. Later, Purser protested to Jack Keuttner that Von Braun's raving coupled with Marshall's independent proposal to Headquarters for Marshall to head a program for an orbiting laboratory—without consulting MSC—did not help intercenter relationships "one damn bit." Von Braun, Purser said, was unaware of the Marshall proposal and had "lost control of his troops," and when he found out he was at fault he apologized profusely to Gilruth.[29]

Although personal relationships remained cool, the two centers did cooperate and direct intercenter contacts were maintained by the engineers of each center. And most of the MSC engineers retained a genuine respect for Wernher von Braun and Marshall personnel, mixed with a proper dose of caution. Ken Kleinknecht said that Von Braun was a supersalesman. "Wernher," he said, "could sell refrigerators to the Eskimos and even after they had them for 6 months they still wouldn't be mad at him, when they found out they didn't need them." He credited Von Braun with being better known in the space business than anyone else other than perhaps the astronauts, and with having been a significant contributor to the American manned space effort. Before Sputnik, Max Faget said, Von Braun proposed to put an American as a payload on a Redstone rocket for a 5-minute experience of weightlessness. He concurred that Von Braun's spaceflight planning preceded Sputnik and NASA.[30]

The ABMA, which became the core of Marshall, played a largely peripheral role in the Mercury program. "We had a minimum amount of intercourse with Marshall," Bob Gilruth commented. "They did produce the Redstone rocket for us in connection with the suborbital flights of Mercury." But, he added, "we had more than our share of difficulty in working out arrangements with them." Marshall, he said, wanted MSC to send its capsules to Marshall for integration with the launch systems, and Gilruth would not agree to that. He added that "we flew four Mercury spacecraft on the Redstone."[31]

Titan II rockets, used as the Gemini booster, were being developed by the Air Force and its contractors to deliver warheads. Even while vigorously continuing its own missile program, the Air Force reconfigured and man-rated the Titan II rockets for use by NASA's Gemini program. Marshall's role in the Gemini program largely related to intermittent consideration of the use of Agena or even Saturn rockets in the Gemini stack, but Marshall did play a peripheral role in Gemini, rather than having "no part" as Bob Gilruth said.[32]

The Apollo spacecraft, managed by MSC, however, required close cooperation and integration with the Saturn systems being developed by Marshall. Apollo employed multistage Saturn launch vehicles built by different contractors under Marshall supervision, interfacing with the command modules and lunar modules developed under MSC direction. Marshall had a major part in the Apollo program. Marshall accomplished a technical tour de force in the development of the Saturn rocket used to boost the Apollo spacecraft. Unlike for Mercury and Gemini programs, Headquarters provided the interface between Marshall and MSC. As Gilruth noted, during Apollo "the relationships aren't so much between centers now as they are between centers and Headquarters. We now have good relations with MSFC."[33]

What happened is, as George Low indicated, the role of MSC in the Apollo program changed considerably from its role in Mercury and Gemini. "In Apollo, MSC was to be a third and equal partner (with Kennedy and Marshall) under an overall Headquarters Program Office, whereas for Mercury and Gemini, MSC had been a lead center with a relatively weak Headquarters organization." Thus, the initial reorganization of NASA administrative systems under Brainerd Holmes and the establishment of the Office of Manned Space Flight was an attempt to provide centralized direction for the Apollo program with each "lead" center, including Marshall, MSC, and Kennedy Space Center, having its own assigned portion of the program. Holmes' problem, Low believed, was simply that he tried to manage too much of the technical detail from Headquarters. When Joe Shea, who headed the Apollo Program Office, with the technical support of BellComm (Headquarters' contractor management team), began to assume responsibility for the technical decisions in spacecraft development, design, systems engineering and mission operations, "in fact, all the things for which MSC had prime responsibility," it quickly became clear that this kind of effort from Headquarters, directed by people who did not have the experience that the people in MSC had and who were unaware of MSC's independent spirit and rather unique culture "would not and could not work."[34]

Thus, as mentioned earlier, Brainerd Holmes left the Office of Manned Space Flight in 1963, a casualty, in a sense, of the friction generated by efforts to centralize program management. At this point, George Mueller and Sam Phillips, working through such experienced program managers as George Low in Houston, reestablished a more balanced management system that reinstated the basic integrity and autonomy of each Apollo lead center while imposing greater control and surveillance by Headquarters.

Intercenter difficulties and rivalries, continued, however, particularly those between Marshall and MSC. For 2 years before 1967, Faget said, the Marshall center had tried to "get a piece of the spacecraft" and was at work on manned orbital workstations. George Mueller, he said, was giving Marshall "more and more license" in the spaceflight business.[35] In 1965 Houstonians became concerned that Marshall was attempting to usurp the programs and responsibilities of MSC and move programs and personnel to Huntsville, Alabama.

In October 1965, a *Houston Post* story mentioned that Marshall might assume control of the forthcoming Apollo Applications Programs that would extend Apollo work into areas other than the lunar flights. One year later, the *Houston Post* front-paged an article under the ominous title: "Von Braun a Persuasive Voice: Some MSC Tasks Being Moved," with the lead sentence reading, "Some of the work that should be done at the MSC is being steadily transferred, with as little publicity as possible, to the Marshall Space Flight Center in Huntsville, Alabama." With the last flight of Gemini scheduled for November 9, 1966, and the first manned Apollo flight scheduled for December 5, Jim Maloney, the journalist in the story, commented, now "MSC's responsibilities are being diluted." The Marshall Center, Maloney suggested, had run out of things to do just when the acceleration of the war in Vietnam made money for new projects more difficult to come by; so Marshall "officials" had sold NASA the idea that the basic Marshall scientific and engineering organization needed to be maintained as a group. As a result, Apollo Applications Program, that is the use of Apollo hardware and systems for other than Moon trips, was to be assigned to Marshall.[36]

Maloney argued that the completion of Saturn V, scheduled for launch in 1966, marked the end of the road for Marshall, until NASA decided that Marshall should help out with Apollo spacecraft work. And, he said, MSC officials made no fuss of this decision. "None at all. MSC will have plenty of work, MSC officials said." This was a major MSC responsibility, the *Post* reported, that was slipping away to the Marshall Space Flight Center.[37]

There followed some frenetic activity after the *Post*'s revelations of a transfer of programs to Huntsville. A NASA release, dated October 16, 1966, stated that contrary to the information contained in the *Post* article of October 10, "no work has been transferred from the MSC, Houston, Texas. In fact, 200 positions were transferred during this last year from the Marshall Space Flight Center, Huntsville, Alabama, to the MSC in Houston to provide for the buildup of personnel necessary for the Apollo launch control facilities." The article in the *Post*, according to the unsigned NASA memorandum, "does not deal in substantive fact and attempts to establish a case for movement of work from the MSC on the basis of unfounded opinion."[38]

The Houston public and Texas Congressmen remained unconvinced and concerned. Olin "Tiger" Teague wrote William P. Hobby, Jr., President and Executive Editor of the *Houston Post* and Teague's friend, on October 17, suggesting that the *Post* might be "crying wolf." On October 19, George Mueller wrote Teague, who chaired the Subcommittee on Manned Space Flight, to the effect that no MSC projects were being transferred to Marshall, but on the contrary 200 civil service personnel were transferred from Marshall to MSC during the past year. The project relating to the Apollo Telescope Mount, he said, dealt with experiments and not with spacecraft development, and MSFC would develop "Experiment modules designed primarily for astronomical experiments." The mission of MSC continued to include vehicle development, life support systems, astronaut activities, flight operations, medical research and operations, and lunar surface scientific activities, he added. The Kennedy Space Center, he said, will continue to be responsible for launch operations and support.[39] Although it continued to be debated in Congress and within NASA, overlapping program responsibilities, like system redundancies, provided a degree of quality control and engineering alternatives. There were different ways to solve the same problem.

Hobby responded to Teague in early November that the NASA Memorandum sent by Teague tended to substantiate rather than refute the *Post*'s concerns that "responsibility for the development of spacecraft for post-Apollo uses is being shifted to Marshall." Teague took Bill Hobby to task a few days later, saying:

> Bill, every person with whom I talk and who are connected with NASA are glad and happy they moved to Houston. As an example, at Cape Kennedy, Astronaut Cernan came over to me and said, "I just want to tell you how much we enjoy Houston, Texas." On a plane from Ellington Field to Cape Kennedy, Bob Gilruth, George Low and Chris Kraft started a discussion of how pleased they were to be in Houston. I know that we can trust these people and I know that we can trust Dr. George Mueller.

And he added, "I don't believe there is any more of a chance of downgrading the Houston Center than there is of my being one of those going to the Moon."[40] The incident was not the first time that a Texas delegation or constituency rushed to defend MSC (and local interests) from a threatened diminution of programs, funding, or personnel, nor certainly would it be the last.

Although "Tiger" Teague might never make it to the Moon, with the successful completion of the Gemini flights in November 1966 and the launch of two unmanned Apollo craft earlier in the year, the Moon now seemed appreciably more accessible than it had been since the beginning of the manned space program. The first Apollo-Saturn launch was made from the Kennedy Space Center on Cape Kennedy on February 5, 1966. The "stack" began with a Saturn IB first stage, having eight H-1 engines built by Rocketdyne that produced 1.6 million pounds of thrust. The second (S-IVB) booster stage built by Douglas Aircraft featured a single Rocketdyne J-2 engine to which was attached the launch vehicle adapter, service module and command module, headed by the pylon-shaped launch escape tower constructed by North American. Bad weather forced a halt in the launch countdown, but after a 5-day delay, the countdown was resumed on February 25. Only 3 seconds before ignition, falling pressure in two helium spheres on the Saturn forced another delay until, finally, on February 26, 1966, the first successful launch of the assembled Apollo-Saturn system sent the unmanned command module on a 37-minute downrange flight. There were some minor malfunctions, but the system worked.[41] AS-201 marked a significant step forward for the manned lunar landing mission.

The launch of AS-201 was organizationally a much more complex thing than the launches of previous Mercury or Gemini missions. In 1960, when NASA's Space Task Group representatives, G. Merritt Preston and Scott Simpkinson arrived at "Hangar S" at Cape Canaveral, they were given work stations in a janitor's closet. Gilruth recalled how "shocked and disgusted Scott Simpkinson was at the time." Within 2 years, however, the group occupied the entire hangar and a newly constructed engineering building that adjoined the hangar.

Suddenly, Tomorrow Came . . .

Throughout the Mercury flights, MSC had its own launch directors and personnel at the Cape. Relations with the Florida center, Gilruth said, "were quite good."[42]

The launch facilities at Cape Canaveral included the Air Force Missile Test Center, the Space Task Group's launch team, and the Army's Missile Firing Laboratory, originally established in 1952 and transferred in 1956 to the command of the ABMA at Redstone Arsenal, Alabama. The laboratory operated the launch facilities used for the Redstone and Jupiter rockets. Wernher von Braun directed the technical work of the Army's agency, when General J.B. Medaris was in command. Dr. Kurt Debus, one of Von Braun's engineers who fled with him to the west after Germany's collapse, reported to Von Braun for the work at the launch facility. Debus received degrees from Darmstadt University in mechanical and electrical engineering, a dueling scar on his left cheek, a doctorate in 1939, and an appointment as assistant professor at the university the same year.[43]

When NASA acquired most of the personnel and properties of the ABMA and its Missile Firing Laboratory on Cape Canaveral, the launch facility became the Launch Operations Directorate under Marshall. Debus continued to direct the manned flight portion of Cape operations, while unmanned launches were handled by a Goddard team.

On March 7, 1962, NASA separated the launch facility from Marshall and organized it as a Launch Operations Directorate under Debus. The launch facility became a separate Launch Operations Center in July 1962. For the continuation of Mercury flights and through the Gemini program, the Launch Operations Center at Cape Canaveral remained directly responsive to MSC and interfaced with MSC through such individuals as Merritt Preston and Walter J. Kapryan, who became launch director in 1969. Preston became launch operations director for the Gemini program and his STG/ MSC group, permanently assigned

AS-201 liftoff, Cape Kennedy, Florida, on February 26, 1966. This unmanned flight marked the first flight of the Saturn IB first stage and Saturn IVB second stage, and the first flight of an Apollo production command and service module. The Apollo 009 spacecraft was retrieved 5000 miles downrange in the Atlantic Ocean near Ascension Island.

110

to Kennedy Space Center (as it was redesignated in 1964 after President John F. Kennedy's assassination) became the center's Operations Directorate.[44]

During all Gemini launches, MSC retained a tangible presence at Kennedy in the form of old STG personnel who had been reattached to Kennedy. Despite the overriding presence of Debus and the Army/Von Braun legacy and the earlier "janitor closet" confrontation, relations between Kennedy Space Center and MSC were generally cordial. During Mercury and Gemini flights, business tended to be conducted directly between the centers, rather than through Headquarters, but the Apollo program invoked more formal relations through the appropriate office at Headquarters. For whatever reasons, but likely because of the early infusion of MSC/STG personnel into the Cape Canaveral launch center, harmony and cooperation generally prevailed between MSC and Kennedy Space Center.

Donald K. "Deke" Slayton, Director of Flight Crew Operations, and George M. Low, MSC Deputy Director, relax during the AS-202 unmanned flight in August 1966. On April 10, 1967, George M. Low became Apollo Spacecraft Program Manager.

The year 1966, when Apollo-Saturn 201 made its maiden flight, was packed with activity at the Cape. In March after the AS-201 launch, Gemini 8 carrying Neil Armstrong and David Scott was lofted. Gemini 9 followed in June. On July 5, the launch team fired AS-203, an Apollo-Saturn launch without a payload. The flight was intended to study liquid-hydrogen fuel behavior in a weightless environment, and to determine if the third S-IVB rocket stage would retain enough fuel to boost the command module and lunar module into a lunar obit. Engineers decided that it could indeed. Within 2 weeks, Kennedy launched Gemini 10 into a 72-hour Earth-orbital mission; and a month later, on August 25, fired another unmanned Apollo-Saturn system into orbit. This, the AS-202 (originally scheduled to precede AS-203), tested engine firing sequences and the reentry performance of the capsule and heat shield. The final two Gemini craft flew respectively on September 12 and November 11.[45]

NASA now planned to launch its first manned Apollo craft (AS-204) before the end of 1966. But the intensive training of astronauts Gus Grissom, Edward White, and Roger Chaffee, under the supervision of Deke Slayton, was hampered by constant modifications to the command module, which meant that the mission simulator and training procedures constantly required revisions. Moreover, North American (which merged with Rockwell in

Suddenly, Tomorrow Came . . .

AS-204 astronauts Edward Higgins White II, Virgil Ivan "Gus" Grissom, and Roger Bruce Chaffee died when an oxygen-enriched fire swept the interior of the spacecraft during preflight tests at Cape Kennedy on January 27, 1967.

1967) was experiencing production problems with the command module, which was finally shipped to Kennedy Space Center in August but in a state that required considerable engineering work to make it flight-ready. The 012 service module associated with the capsule for the flight was held up for inspection when a similar unit (017) exploded at the factory. By the time these problems were resolved, the AS-204 flight was rescheduled for February 1967.[46]

A launch simulation preparatory to the actual launch was scheduled for January 27. Shortly after noon Grissom, White and Chaffee were in the module on top of the Saturn IB, some 25 or 30 engineers and technicians were in the launch tower adjoining the capsule, and another 1000 technicians, engineers and ground crew were assisting in the launch simulation. The astronauts began removing all gases except oxygen from their space suits and the cabin, as was the standard procedure for all previous Gemini and scheduled Apollo flights. Finally, the cabin pressure stood at 16.7 pounds per square inch of pure oxygen, and the long tests of equipment and procedures continued, with interruptions, long into the afternoon. At 6:30 p.m. someone in the command module cried over the radio circuit, "There is a fire in here!" Within moments the cabin was engulfed in a flash fire of pure oxygen, and the three astronauts were dead of asphyxiation.[47] It was the worst moment up to that time in the history of the manned space program.

The AS-204 fire and the death of the astronauts was a great tragedy and felt personally throughout MSC, Kennedy, Marshall and NASA. "It shouldn't have happened," George Low said later, "it could have happened in Mercury or Gemini, but it didn't." Administrator Webb appointed a Review Board chaired by Floyd L. Thompson and including Frank Borman and Max Faget of MSC and representatives of other centers, the President's Science Advisory Council, and others outside of government. NASA asked Congress to delay a full-scale congressional investigation until the Review Board submitted a report, which Congress agreed to do. The press insisted on public hearings, wanted more direct access to information, and suspected a "cover-up." The Review Board literally presided over the dismantling and review of every component in the cabin and each procedure relating to launch. Information was released to the public in what the press regarded as "small doses" but which NASA declared to be all that was really available—which could have been the case. Investigations were slow but thorough. By April a summary report concluded that conditions leading to the fire included having a sealed cabin with a pressurized oxygen atmosphere, extensive combustible materials within the cabin, vulnerable electrical wiring, plumbing containing a combustible and corrosive coolant, a hatch that could not be opened quickly for escape, and inadequate provisions on the launch site for rescue or medical assistance. The final report was compiled in 3000 pages and 14 booklets.[48]

An independent report by North American employee Thomas R. Baron, who had been fired by the company on January 5 before the fire, implied gross negligence on the part of the contractors and others, but in hearings before Olin Teague's subcommittee, none of the allegations could be supported. Baron and his family died in a car-train crash only a week after the congressional hearings. It did become clear, however, that a General Electric official had warned Joseph F. Shea, MSC's Apollo Program Manager, about the possibilities of fire in the spacecraft before launch, and MSC Medical Director Charles Berry had expressed concern about flammable materials in the pure oxygen environment of the spacecraft.[49] Many Americans, within and outside the government, wondered if the disaster might have a long-term adverse effect on spaceflight and even bring the program to an end.

Already the growing preoccupation with the war in Vietnam and rising government deficits occasioned by that war and by President Johnson's expensive Social Security, Medicare, and War on Poverty programs were contributing to purse tightening by Congress and to a rising disaffection or at least a disinterest in space by the American public. Olin Teague, for example, as a Congressman closely involved with the space program and a vigorous supporter of MSC in Houston, was extremely interested in the repercussions of the Apollo tragedy. One measure of the public pulse was given him the day after the fire by a radio talk show commentator, Lou Martin with station WTOP in Washington, D.C., who took a quick poll of his listeners as to whether the space program should be continued. Of the 59 people who got through on the telephone in his 15-minute time allotment, 70 percent advised either continuing the program at its present level or accelerating the program. Only nine callers suggested curtailing the space effort, and another nine thought it should be abandoned completely.[50]

Despite considerable public contention (and rather remarkably when compared to the aftermath of the *Challenger* accident in 1986), the tragic fire created a new resolve within the NASA establishment and concurrent support from Congress, the Executive and the public.

George Low, among many others at MSC, regarded the fire as a turning point in the entire space program. In January 1969, Low said "the reexamination of Apollo that came as an aftermath of the fire required us to build a different Apollo spacecraft," and most importantly, he added, "it created an entirely different atmosphere among ourselves, our contractors, and within MSC."[51]

Among the immediate repercussions of the Apollo fire was the resignation of Joseph Shea, who was personally devastated by the accident, as Apollo Program Manager at MSC. George Low stepped from the Deputy Director's seat into that chair. Gilruth appointed a "tiger team," including Frank Borman, Douglas Broome, Aaron Cohen, Jerry W. Craig, Richard E. Lindeman, and Scott H. Simpkinson to visit the North American plant in Downey, California, and review production systems and techniques. North American, in turn, replaced its president of the Space and Information Systems Division with William D. Bergen, formerly of Martin Marietta. Bergen's role, with his managers Bastian Hello stationed at Kennedy Space Center and John P. Healy who was to supervise the Block II module production at Downey, was to improve quality, safety, and production review procedures, and to eliminate the problems existing or anticipated by the Review Board. Grumman Aircraft, responsible for the lunar excursion module, intensified its review and quality control processes with the assistance of Richard S. Johnston, an MSC materials expert. All levels in the spacecraft production chain conducted careful reviews of materials being used in the modules.[52]

Max Faget's Engineering and Development Directorate launched a multifaceted testing and evaluation program, headed by Aleck Bond, directed at understanding in detail the characteristics of the Apollo 204 fire and toward the development and evaluation of an array of new and improved fireproof or flame-retardant materials. Joseph Kotanchik's Structures and Mechanics Division conducted in situ fire tests employing Apollo boilerplate command modules, using first the old and then the new materials. Richard S. Johnston, chief of the Crew Systems Division, tested and helped develop nonmetallic materials such as Beta cloth, flame-retardant velcro and other materials that were upgraded and improved for fire safety.[53]

Bond and his team directed tests in MSC stress laboratories, vibration acoustic facilities, space environmental simulation laboratory, and in the thermochemical and structures laboratories on every material that might be associated with spaceflight. The work stressed duplicating the real environment in which the materials would exist in space, and the combinations in which they might be used. "The only way you can understand materials," he said, is to test them in their real environment. "The tests," he said, "contributed to redesigning the space cabin environment and its atmosphere. In the longer run, the tests contributed to a better understanding of terrestrial uses of materials, flight and fire safety, and energy efficient modular design," he said. Bond, who had earlier worked on "man-rating" materials for human use in the environment of space, found these principles applicable for both terrestrial and nonterrestrial environments.[54]

The trauma of the AS-204 fire precipitated a vital new learning experience and a renewed dedication and sense of cooperation among the NASA centers and contractors. Managerially, NASA began to move from a state of planned disequilibrium to one of greater stability. Headquarters began to exert more influence and control. The older NACA traditions of informality, collegiality, and center independence waned under the pressures of an enforced technical collectivism. MSC retained a strong sense of independence, a product in part of its Langley legacy, and, perhaps, its Texas environment. It retained its self-image of being the lead center for manned space programs, a mantle which it assumed in its origins as the Space Task Group and earned in the Mercury and Gemini programs. As was the entire NASA organization, MSC personnel were shaken by failure and the loss of the crew of AS-204, but even more determined to succeed. By the end of 1967, the new Apollo-Saturn 501, renamed Apollo 4, stood atop the new Saturn V rocket ready for launching from the pad at Kennedy Space Center. Apollo soon would be ready to deliver its precious human cargo to the Moon.

CHAPTER 7: Precious Human Cargo

"Spaceflight, like airplane flight," Michael Collins wrote in *Carrying the Fire*, "did kill." But it had never happened in the American space program before AS-204.

> Now this Apollo had destroyed three without even flying one; what was the pattern? Would one disaster follow another . . . ? How could NASA get going again? How many astronauts would decide they hadn't signed up to be incinerated and quit? How many wives would quit if hubby didn't?[1]

The answer, Collins pointed out, was that no one quit, "not husband or wife." Neither did NASA or the American public. Eighteen months later Collins, in orbit about the Moon aboard the Apollo 11 command module, heard Neil Armstrong's message from the lunar module to the Manned Spacecraft Center Mission Control: "Houston, Tranquility Base here, the *Eagle* has landed."[2]

From the very beginning, NASA's astronauts had been some of the most acclaimed and visible heroes America ever celebrated. Being an intensively and highly trained astronaut was difficult enough; being a celebrity compounded the work and the responsibility. Being an astronaut was no mundane or easy business. In the public's mind, the astronaut was what NASA and space were all about. Rarely did engineers, managers, centers, Headquarters, or even Congress and presidents intrude upon the public's space consciousness. That those astronauts were a part of MSC in Houston, and that their words and deeds became public through the center's Mission Control and its Public Affairs Office, gave the center a certain centrality in NASA's space programs in the mind of the public and in the estimation of those who worked there. It was a centrality, to be sure, not accepted by NASA Headquarters or by other NASA centers, but it was an element which helped define the character of MSC and its personnel.

MSC managed the engineering design, development, and construction of the spacecraft; supervised the selection and training of the astronauts who flew the craft; and directed spaceflight operations. In 1969, when Neil Armstrong and Edwin Aldrin stepped out of the *Eagle* onto the surface of the Moon, they were, with Mike Collins, 3 of the 73 individuals selected by NASA since April 1959 as astronauts. The 3 and 70 astronauts were the highly visible and in a sense final elements of that massive effort which thus far involved some $35 billion in public monies appropriated by Congress to NASA, and almost 250,000 employees (some 35,000 NASA civil service employees and the remainder employees of the firms providing contracted services to the space program).[3] Although one-half of NASA funds were involved in other than manned space programs, the astronaut came to represent in the public mind what the total NASA effort was all about. And, to be sure, the manned and unmanned programs were inextricable parts of the total NASA mission.

Although media attention and publications about space have understandably focused on the flights of the astronauts, the processes of their selection and training and the astronauts' relationships to MSC, to NASA, and to their public—the people on Earth—is a more meaningful story albeit a less dramatic one than is the story of the relatively brief flights through space. It is useful to reflect upon the selection and training of the astronauts through the Apollo era, to consider the organization of the Astronaut Office at MSC, and to contemplate some of the human dimensions of being a part of the astronaut corps. Those early years helped set the tone and style, not only of the astronaut corps but also of MSC and of NASA itself. Insight into the selection of the astronauts also contributes to an understanding of the "way NASA works."

There were no real answers in 1957 and 1958 to what an astronaut must do or be. No American flew in space until 1961. But some had come to the very edge of space, and that experience, and considerable speculation and extrapolation, suggested what might be reasonable criteria for being an astronaut. In February 1957, Dr. D.H. Beyer and S.B. Sells, Ph.D., with the School of Aviation Medicine at Randolph Air Force Base in Texas, published the results of their deliberations regarding the "Selection and Training of Personnel for Space Flight." The authors presumed that the return of a spacecraft into the atmosphere would require an extended glide and a conventional landing of a winged craft with a tricycle landing gear—a premise that perhaps fit the Shuttle but not the intervening Mercury, Gemini and Apollo spacecraft. They reasoned that training and experience in piloting jet and rocket aircraft, such as the X-15 then being developed, would be "most useful for transition to spacecraft."[4]

Current hypotheses for space launches, they said, citing Wernher von Braun among others, indicated an acceleration force of nine times the Earth's gravity on the passengers and a configuration approximating contemporary jet aircraft, but a much more complex instrumentation and control system. Given these parameters, the spacecraft pilot fit the mold of "experienced pilots of high performance aircraft." But the critical elements in the selection, they believed, related more to the psychological than to the physical aspects of spaceflight, for "by far the greatest problem involves the implications of a seemingly complete break from the Earth and the protective societal matrix in a small, isolated, closely confined container with few companions."[5]

An astronaut candidate, they believed, must "manifest intense motivation for the project," have a strong ability to cooperate to the point that they could place trust and confidence in associates and win the trust and confidence of those associates. They should have "positive interpersonal attitudes, mature character integration, and emotional stability involving an inner sense of duty, responsibility, self-control and restraint." And they had to be adventurous but not foolhardy.[6]

Although astronautical flight would not be drastically different from aeronautical flight, they admitted that the first space crews would be pioneers who would have to be "their own instructors," but they believed that astronauts would require academic training in "applied and theoretical mathematics, electronics, engineering, navigation, astronomy and astronavigation," as well as intensive courses on the design and construction of the spaceship, instruction in "basic spatial medicine," and training in simulators and near-space conditions. Years later Henry Cooper, Jr., author of *Before Lift Off,* which describes the training of a latter-day space shuttle crew, defined the astronaut as a "highly trained generalist," which seems to fit the early astronaut specifications.[7] The Beyer and Sells report intimated that the physical, psychological and mental demands on an astronaut would be very great indeed.

These theories quickly became confronted with realities. With the organization of NASA in 1958, selecting men for spaceflight became a pressing matter.

In November of that year, Administrator T. Keith Glennan appointed a team of "aeromedical consultants" from the military services for temporary assignment to NASA's Space Task Group. After a brief study requested by Glennan, Wesley Hjornevik urged that a biomedical office be permanently established. Six months after being organized as "consultants," the biomedical team became a permanent component of the Space Task Group. The initial group of medical advisors included Dr. Stanley C. White, described by John A. Pitts in his study of biomedicine in the manned space program as the nominal head of the team and a "specialist in human factors engineering and biotechnology." He worked closely with Robert Voas, Ph.D., a psychologist and a "Buck Rogers" devotee, whose first duty with the Navy involved pilot selection at its Pensacola, Florida, training base. Voas left there for an assignment to the medical laboratory at Bethesda, Maryland, and was attached by the Navy to NASA on its first day of existence, October 1, 1958. The third medical consultant, William S. Augerson, was an Army major, physician, and specialist in human physiology.[8] A fourth physician, Dr. Charles A. Berry, assisted the biomedical team for the Mercury astronaut selections from his position as Chief of Flight Medicine for the Air Force. He later became chief medical officer for NASA.

Dr. Berry, assistant and then chief (1958) of the School of Aviation Medicine at Randolph Air Force Base, provided a direct link between early Air Force space pilot criteria and NASA's astronaut selection. In 1959 he accepted the assignment with the Surgeon General of the Air Force as Chief of Flight Medicine, and in July 1962 joined MSC as Chief of Medical Operations Office. Berry was medical director for both the Gemini and Apollo programs.[9] In addition to NASA's biomedical team and the later appointment of Berry as chief medical officer for MSC, on October 27, 1958, Glennan appointed a non-NASA, independent Life Sciences Advisory Committee headed by Dr. W. Randolph Lovelace II to recommend programs and to assist in defining the qualifications and selection processes for the first astronauts to fly the Mercury vehicles.

Lovelace, who directed the Lovelace Clinic and Foundation in Albuquerque, New Mexico, had for some time been involved in special Air Force crew selection projects and conducted medical examinations for personnel involved in sensitive national security programs. A pioneer in high-altitude, near-space flight studies, Lovelace in 1943 investigated the effects of an extremely high-altitude parachute jump by personally bailing out at 36,000 feet. He speculated that the fall into denser atmosphere might create severe shocks on the human system. It did—and almost killed him—but the special equipment he had designed saved his life.[10]

The Life Sciences Committee produced a set of very broad specifications. The prospective astronaut must pass rigorous physical and psychological tests, have a degree in physical science or engineering, be under 40 years of age, and 5'11" in height. Using these criteria, NASA prepared a draft of a civil service notice for astronaut applicants at the GS-12 to GS-15 level (in 1958 scheduled at $8,330 to $12,770). At this point, President Eisenhower personally intervened to specify that astronauts must be selected from the rolls of current military test pilots. Bob Gilruth later remarked that this decision greatly improved the selection process and ruled out the "matadors, mountain climbers, scuba divers, and race drivers and gave us stable guys who had already been screened for security."[11]

In January 1959, Gilruth met with Charles J. Donlan, Dr. Stanley C. White, George Low, Dr. Lovelace, and Brigadier General Donald D. Flickinger (a member of the Life Sciences Committee and Surgeon and Assistant Deputy Commander for Research with the Air Research and Development Command in Washington) and condensed the rather elaborate specifications to a simplified list of seven. Astronaut candidates must:[12]

1) Have a degree or the equivalent in physical science or engineering
2) Be a graduate of a military test pilot school
3) Have at least 1500 hours flying time including a substantial amount in high-performance jets
4) Be younger than 40
5) Be no taller than 5'11"
6) Be in superb physical condition
7) Possess psychological attributes specified by the Life Sciences Committee

By these specifications the pool of candidates for the astronaut corps would be largely male by virtue of the heavily male-dominated fields of engineering and physical sciences, and by virtue of the prerequisite for test pilot experience. Leaving the pool would be an even smaller segment of men who not only had the academic background but as Tom Wolfe explained it, "The Right Stuff" as hypersonic, daring, high-speed addicts. While they may not have been "matadors, mountain climbers and scuba divers," they were birds of a feather. Although 13 women applied for the astronaut corps in 1960, and passed the grueling physical and psychological tests, none were admitted into the astronaut corps until 1978 when 5 women became astronauts.

Bob Gilruth, Donlan recalled, had been very uneasy about the broad specifications for an astronaut and was clearly relieved to have the President narrow the qualifications. Once the qualifications were established, Gilruth asked Donlan, whom he had recruited as his deputy for the Space Task Group in October 1958, to "drop everything" and give his full time to the selection of the Mercury astronauts. Donlan then recruited a Space Task Group team headed by Bob Voas to review the records of 473 military test pilots. Voas's group selected 110 pilots as potential candidates and divided those into 3 groups. Each group was then invited to the Pentagon in Washington, D.C., under orders marked "secret," for a preliminary briefing and personal interviews. The first two groups met separately in Washington on February 2 and 9. Abe Silverstein and George Low explained the nature of the program and invited those who were still interested to report to NASA Headquarters in civilian clothes (to heighten the "peaceful" intent of NASA programs) for further briefing and tests. Of the 63 interviewed in the first 2 groups, 80 percent indicated they were interested and would be available for more rigorous testing. After personal consultations and more interviews, the list of candidates was narrowed to 32 individuals, and the third interview group was canceled.[13]

Next, in five groups of six and one group of two, the astronaut candidates reported to the Lovelace Clinic, beginning on February 7, for a week of what Voas described in an understatement as an exhaustive series of examinations. Deke Slayton described the tests as medical experimentation rather than a physical examination. The astronauts were the rats being tested. Michael Collins, although he came through the tests several years later when the examination procedures had been somewhat moderated, luridly described the procedure and NASA later incorporated his description in the official record as an example of a "humanistic perspective of what all those tests were like":[14]

> Inconvenience is piled on top of uncertainty on top of indignity, as you are poked, prodded, pummeled, and pierced. No orifice is inviolate, no privacy respected. . . . Cold water is poured into one of your ears, causing your eyeballs to gyrate wildly as conflicting messages are relayed to your brain from one warm and one cold semicircular canal. Your body is taped with electrocardiogram sensors and you are ordered onto a treadmill, which maintains its inexorable pace up an imaginary mountain road. As the tilt becomes steeper, the heart rate increases, until it finally reaches 180 beats per minute . . . Your fanny is violated by the "steel eel," a painful and undignified process by which one foot of lower bowel can be examined . . .[15]

And the "shrinks," Collins said, take over where their compatriots leave off.

A psychological and stress evaluation of the astronaut candidate was conducted by the Air Force, with the assistance of Army and Navy specialists, at the Wright Air Development Center Aeromedical Laboratories at Wright-Patterson Air Force Base in Dayton, Ohio. One of the tests, Deke Slayton recalled, was to lock the candidate alone in a totally dark room for an extended time. "What are you supposed to do in a dark room?" he exclaimed many years later with some residual disgust. "Go to sleep!" And that's what he did. Collins admitted that when he came to the Rorschach (inkblot) tests, instead of describing a scene as he had the previous year as "nineteen polar bears fornicating on a snowbank," and thus incurring the displeasure of his examiner, because "I want to fly to the Moon, badly I want it, . . . I will describe that white card in any way that will please them." But he admitted that "second-guessing shrinks is not easy."[16] The first group of candidates began their 6 days of psychological evaluations on February 15.

The final step in the selection process occurred at Langley Research Center where a group representing both the medical and technical fields evaluated the data from the Lovelace Clinic and Wright-Patterson Laboratories. Donlan, who presided, announced that each of the final candidates would be reviewed in alphabetical order. Those candidates, he

Astronaut Group I, selected in April 1959, included: **Front row:** *(left to right) Walter M. Schirra, Jr.; Donald K. (Deke) Slayton; John H. Glenn, Jr.; and Scott M. Carpenter.* **Back row:** *(left to right) Alan B. Shepard, Jr.; Virgil I. Grissom; and L. Gordon Cooper, Jr.*

said, met the basic physical and psychological requirements. The final decision, however, rested largely on the nontechnical evaluation of the person's resourcefulness, interest in the program, and "survivor" instincts. A good number of the prospective astronauts, however, earlier withdrew from consideration because they believed that the space program might be a very short-lived program, and because it did not seem to contribute to their promotion and career enhancement as military officers. Moreover, many pilots and the Experimental Test Pilots Association, he said, believed that the Mercury program required a "salmon in a can" rather than a real test pilot. In a 2-hour meeting, the review committee selected seven finalists.[17] Those seven were invited to NASA Headquarters where their names were publicly announced at a press conference on April 9. They were:

Lieutenant Malcolm S. Carpenter, U.S. Navy
Captain Leroy G. Cooper, Jr., U.S. Air Force
Lieutenant Colonel John H. Glenn, Jr., U.S. Marine Corps
Captain Virgil I. Grissom, U.S. Air Force
Lieutenant Commander Walter M. Schirra, U.S. Navy
Lieutenant Commander Alan B. Shepard, U.S. Navy
Captain Donald K. Slayton, U.S. Air Force

John Glenn described the press conference as "wild and woolly" and Slayton said it was a "shocker." The astronauts had moved from a very closed, protected environment onto center stage.[18] Few with NASA, unless it might have been Walter T. Bonney, the public information officer, anticipated the extent of the public reception of the astronauts or the continuing "media event" that became a part of the astronauts' lives.

Lieutenant Colonel John A. Powers, who came from the Air Force Ballistic Missile Division's lunar probe program to NASA, said he had some sense of the public's interest, but Bonney had a better feel for the situation. Walt Bonney, he said, came to the conclusion "there was going to be a scramble to get exclusives, inside personal stories, etc., out of these guys." Bonney wanted to allow the astronauts to sell their personal stories. After NASA approved the idea, Bonney contacted Leo D'Orsey, a prominent Washington, D.C., attorney, who agreed to represent the astronauts as their agent at no charge. The astronauts and their families agreed to combine, sell the rights to their personal stories on a single contract, and let D'Orsey handle the negotiations. After negotiations with *Saturday Evening Post*, *Look*, AP, UP, and several syndicates, D'Orsey accepted an offer from *Life* Magazine for $500,000, to be distributed equally among the astronauts.[19]

John Glenn described D'Orsey as "one of the best friends the astronauts had" who "gave us sage and wise counseling." He took care of contracting problems, helped with public relations, and served as a very good elder statesman. One problem that D'Orsey helped with was insurance, Glenn recalled. Although each had some coverage as test pilots, additional insurance coverage for the space program seemed unobtainable. D'Orsey finally got one company to agree to insure John Glenn for $100,000 for his (and America's) first orbital flight in space for a premium of $16,000 for the 5- or 6-hour flight. "Leo," Glenn said, "worried about this," and decided that he would not bet against Glenn and pay $16,000 from the astronaut's fund to the insurance company, but instead would personally write a check to Glenn's wife Annie for $100,000. D'Orsey did so, Glenn said, and gave the check to a third party to hold. When Glenn returned, Leo told Glenn "how glad he was to see him back down safely," because he could tear up the check.[20] Some insurance executives may still be regretting that they failed to issue Glenn a policy. It would have been the lowest priced, highest return on advertising ever. But few, inside or outside NASA foresaw the public's interest in astronauts and space.

Glenn reasoned that if the astronauts were to permit people to come into their homes and interview families and children and be part of their life, there should be some compensation for their loss of privacy. Although other members of the press criticized the *Life* contract arrangements as a use of the space program for private gain, the end result was, as Powers suggested, that the astronauts' privacy was protected more than it might have been by precluding "free" access to the astronauts. And most importantly, the astronauts were not thrown into competition with each other (as modern athletes might be) for media contracts and profit.[21]

The astronaut public relations problem continued to reappear in various forms. Tension between the astronauts and the Public Affairs Office at MSC was a continuous problem, Powers recalled. All seven astronauts, he said, really enjoyed the exposure, but as test pilots they instinctively rebelled at having to spend time talking to the media. Glenn, who was most adept at handling the media, commented that "life in the gold fish bowl did cause some problems. Everywhere we went, it seemed there had to be the press conference, the extensive press coverage and photography session, and while this helped support the program," he admitted, "sometimes it was carried to extremes." Eventually media pressures leveled off.[22]

Glenn recalled the enormous volume of mail directed to him and to the other astronauts. It came from heads of state and from ordinary people all over the world. Following his initial flight into space, Glenn received more than 350,000 pieces of mail. He didn't know what to do with it. Finally, NASA took it over, or more accurately, Steve Grillo in Administrative Services at NASA Headquarters assumed responsibility and established procedures to make sure that every letter was answered.[23] Every astronaut received dozens of invitations a week to speak, some from Congressmen and high officials who were difficult to ignore.

During the first 10 months of 1963, John Glenn received through official channels 1400 requests for appearances, while the other 6 astronauts, 4 of whom had not yet flown, had 700 requests. When George Mueller proposed that MSC release each astronaut for 2 weeks each year for public relations work, Gilruth responded that to do so would not only disrupt the training program, but once it became known that requests for astronaut appearances might be honored through political channels "most of the 300-plus per month requests will arrive at the Administrator's office and create quite a workload there."[24]

Tensions existed at every level between the desire on the one hand to accommodate the public's interests and the growing level of extra program activities required by the astronauts and NASA administrators. The astronauts' private lives did require protection. Publicity also flew in contradiction to the traditional government and Department of Defense security consciousness. The old NACA had little experience with public relations. Moreover, television was rapidly changing the management of public affairs. Government agencies, ranging from the White House to NASA to the specific NASA centers, such as MSC, held ambivalent and often contradictory ideas about dealing with the public and the media.

The White House, beginning with the Kennedy administration, became involved in the astronauts' public affairs problem on several levels. President Kennedy advised Administrator Webb to minimize the number of commitments by astronauts of a nonoperational nature. Jerome Weisner, President Kennedy's science advisor, prepared a brief memorandum in March 1961, prior to Shepard's historic flight in Freedom 7, for McGeorge Bundy, President Kennedy's national security advisor, expressing alarm about "pressures from the press" and that "press and TV for on-the-spot coverage of the first manned launch" could lead to the launch becoming "a Hollywood production," and might jeopardize the mission and have "catastrophic effect." Weisner advised that the press pressures must be met "with firmness" in order to promote the safety of the astronaut and a successful mission. But then he added, "It is my personal opinion that in the imagination of many, it will be viewed in the same category as Columbus' discovery of the new world. Thus, it is an extremely important venture and should be exploited properly by the administration."[25]

President Kennedy developed a rather close relationship with John Glenn and that too created problems. On one occasion, NASA's chief counsel Paul Dembling said, while water skiing with President and Mrs. Kennedy, John Glenn (on behalf of the astronauts) personally urged the President to fly yet one additional Mercury flight. Kennedy called Administrator Webb about the matter and Webb is supposed to have responded, "Who's running the Agency? If you want to run the Agency, appoint yourself a new administrator," or words to that effect. A series of White House meetings followed that conversation, Dembling said, and as a result the White House took a different approach to astronaut affairs.[26]

The line between wanting to help the astronauts and wanting to help oneself was often very thin and indeterminate. Frank Sharp, an enterprising real estate developer in Houston, for example, offered each of the first seven Mercury astronauts a new house in "Sharpstown" when the center moved to Houston in 1962. Powers got approval of NASA's general counsel, the astronauts, and D'Orsey, but when word was leaked to the press "a large unpleasant flap" followed. Powers admitted he probably exercised poor judgment, because "nobody gives you anything for nothing, and it was obvious Mr. Sharp certainly had plans for exploiting the fact that the seven astronauts lived in his development."[27] Although it was declined, the gesture by any standards was rather munificent.

NASA Headquarters began to feel that MSC tended to be excessively permissive in matters relating to the astronauts' outside activities and their business and financial arrangements. Robert C. Seamans finally sent a memorandum drafted by Paul Dembling and Walter Sohier in the General Counsel's Office requiring Headquarters' concurrence on any outside astronaut activities, including business arrangements and public engagements.[28]

Life in the fish bowl worked both ways. In answer to questions about the toughest part of their flights, some of the astronauts responded that it was the press conference. Gus Grissom built a house near Houston with no windows on the side facing the street. "He simply did not want people peering in his windows." After his first flight, John Glenn's home in Arlington, Virginia, had to be guarded by county and state police to ward off the curious and literally to protect his property.[29] For some, such as John Glenn, life in the gold fish bowl proved very rewarding. Glenn won election to the U.S. Senate in 1974. Others, however, found their postflight experiences difficult.

In part, for their own protection and to exercise greater control over both their professional life and their home life, the astronauts organized their own office or division within MSC. Their first several years with MSC were organizationally unstructured. Bob Voas, involved in the initial selection process as a Navy psychologist, was assigned to develop training programs and to a lesser extent look after the astronauts' administrative needs. Eugene Horton, who worked with "Shorty" Powers in the Public Affairs Office, served variously as the astronauts' "Executive Officer" and press secretary. Voas, with the assistance of Joseph Loftus, Raymond Zedekar and others, developed a training regimen, but Slayton said that basically "we were doing training on our own."[30] That soon changed.

During their first years with NASA, the astronauts reported individually to Bob Gilruth. Gilruth told them, when they first reported for duty, John Glenn recalled, that they were chosen because they were experienced engineers and test pilots, and they could apply that experience to the new area of testing spacecraft. "If there was anything at anytime in the program that we didn't like," Gilruth told them, "we had free access to him with our complaints." Gilruth promised that they would be happy with the spacecraft before it flew and that no one would push them into anything. Moreover, Gilruth told them, anytime anyone became dissatisfied or wanted out they were free to go back to their parent services.[31] Gilruth maintained personal contact and a personal interest in those whom he referred to as his "precious human cargo."

As the astronauts grew closer they began to develop their own informal structures and associations, and as time passed these often became institutionalized. Each of the first seven astronauts, for example, assumed responsibility for reviewing specific design aspects of the mission. Slayton, for example, became responsible for escape systems; when Gemini came on line, Gus Grissom became the astronauts' liaison with Gemini. Astronauts visited all of the contractor facilities, and worked closely with MSC managing engineers. In doing so they strengthened the cooperation between the contractor and MSC. Slayton became an unappointed group leader who attended the weekly staff meetings with Gilruth. As new astronauts joined and the programs developed, first the Astronaut Office and then the Flight Crew Operations Division became formal parts of MSC. That is not to say that organizational systems and the training programs were ad hoc arrangements, rather they emerged and developed as the understanding of the requirements of spaceflight grew and as training systems and equipment caught up with the needs. One aspect of astronaut training throughout the Mercury, Gemini, and Apollo programs was that changes in the equipment constantly necessitated changes in training procedures. Moreover, the production of training equipment quite often lagged behind the basic equipment design changes. Thus, despite every effort, there were aspects of "make-do" in the astronaut training regimen.

Bob Voas completed an initial outline of an astronaut training program about the time the first seven astronauts came on board in late April 1959. The program essentially involved a "ground school" phase and a flight test phase. Conveniently, the ground school occupied the remainder of 1959, while Mercury capsules and components were being built. For the first 3 months, the astronauts met with various engineers involved in the design and construction of equipment, including the capsule, booster, range, tracking, and recovery systems; onboard equipment; computers; environmental control systems; and navigational systems. They attended seminars and courses relating to basic sciences such as astronomy, meteorology, and aviation physiology. They visited contractor assembly plants and other NASA installations. Voas devised a rather tightly constructed training calendar which included 7 hours per week of leisure time to be devoted to flight training and physical exercise.[32]

Voas advised establishing a training committee including himself as the training advisor, Douglas to serve as flight surgeon and direct life support training, Harold I. Johnson to handle simulators, George Guthrie to produce a pilot's handbook and monitor program arrangements, and Raymond Zedekar to provide flight and overall program coordination. The committee would meet each Friday morning and complete and approve the training schedule for the following week, with a report to the Chief of the Operations Division (Charles W. Mathews). Every Friday afternoon the committee reviewed and discussed the agenda for the following week with the astronauts.[33]

Training hardware to be developed included a Missions Procedures Simulator, a Mercury capsule mockup with instruments and displays linked to an instructor-trainer console. An Environmental Controls Trainer would be a pressure capsule used to train in life support and emergency systems. Another Escape and Recovery Trainer would be a non-pressurized boilerplate mockup used to train for landing and recovery operations. An Air-Lubricated Free-Attitude Trainer (ALFA, designed and developed at MSC) trained astronauts in manual control skills while undergoing extreme roll, pitch and yaw changes. A Multi-Axis Spin-Test Inertia Facility Trainer (MASTIF), developed by engineers at Lewis Research Center, came on line in February 1960 and gave astronauts experience at tumbling at 30 rpm along three possible axes. Couches, flight instruments, computers and lesser components completed the list of training devices.[34]

One of the great difficulties in training for spaceflight was that nothing on Earth could quite simulate a space environment; moreover, no one really knew what that environment might be like. Developments in training clearly had to await new information anticipated from flight experiences. Acceleration forces and weightlessness were known factors. Training at the Aviation Medical Acceleration Laboratory centrifuge in Johnsville, Pennsylvania, and flights aboard the Air Force's C131, the Navy's F9F-2 or later in an Air Force KC-135 aircraft provided limited experience (60 seconds) with weightlessness.[35]

Voas prepared a Mercury project training summary in 1963 which concluded that overall the training program appeared to have been successful, but that it had been a learning experience for everyone. The training devices were simple and rudimentary, simulation for spaceflight was in its infancy, and the training program was on an accelerated schedule.[36] Over the next few years, new and improved laboratories and training facilities came on line at MSC and the training regimen became more intense and sophisticated.

With the conception of the Apollo Moon-landing program, training began to shift in emphasis and purpose. Almost concurrently, insofar as the astronauts were concerned, Gemini came on line as a training program for Apollo. Slayton said they were trying to make a "fighter plane" out of the Gemini craft, and pilot training became more critical. Longer duration flights required more emphasis on celestial navigation, science applications, environmental adaptations, and survival training. Astronauts spent extensive time in pressure suits underwater as a simulation for EVA in a weightless environment. Communications, computer, and control systems changed markedly with Gemini and even more as Apollo systems were designed and produced, all requiring new training programs and apparatus.[37] As they progressed from Mercury through Gemini and the Apollo vehicles to the Shuttle, American spacecraft, by design, became eminently more flyable.

Because of the rapid pace of the design, construction, and launch of Gemini systems, training for each Gemini flight depended to a major degree on the preceding flight. In its latter phases, NASA launched a Gemini mission about every 2 months. As missions evolved, training plans were formulated in concert with the crews. For Gemini, astronaut Edwin Aldrin said, "there was a lot more crew participation" in setting mission profiles.[38] Apollo mission planning, on the other hand, and its training regimen were more carefully structured before they reached the Astronaut Office. Both Gemini and Apollo training differed sharply from Mercury training in that flight involved crews of two and three persons, each having far more flight-related and nonflight-related tasks to complete.

Training activities for the Apollo missions were structured so that the training for one was a building block for the following mission. Apollo crews did, however, exercise some influence and independence in doing things their own way, and this caused some conflicts with mission specifications. On the more subjective flight decisions, such as how to go about performing specific tasks (such as undocking), when or how to perform inflight inspections, or whether to fly "heads up or heads down," the crew and its commander generally made the decisions.[39]

More frequent flights and accelerated programs meant that additional astronauts were needed. Congressman Olin Teague and others began to talk about organizing an "astronaut academy" similar to the Nation's military academies. MSC administrators, more realistically, began to discuss recruiting more astronauts and organizing the astronauts into a regular branch or division.

According to Deke Slayton, recently grounded from Mercury flights because of a suspected heart condition, Wally Shirra, Gus Grissom, and Alan Shepard decided that, "hell, if we're going to have a boss, why bring somebody in from the outside and superimpose him on us?" They decided they wanted Slayton to be their boss, so Slayton got Gilruth's approval and "we organized the astronaut office," he said. In addition, Gilruth appointed Warren J. North, then with NASA Headquarters staff and formerly a test pilot for the Lewis Research Center, as head of a new Flight Crew Support Division in the spring of 1962, which became the Flight Crew Directorate a year later. Slayton thereafter recommended flight crew assignments to the center director. Gilruth independently made those assignments through Mercury 7. Slayton selected crews for Gemini, Apollo, and Skylab. He, Warren North, and Alan Shepard comprised the selection board for the second group of astronaut candidates.[40]

Specifications for the second field of astronaut candidates changed slightly, but significantly. The age limit was lowered from 40 to 35, educational qualifications were broadened to include degrees in biological sciences, and while flight experience required "experience as a jet test pilot," that experience could be achieved through the aircraft industry or NASA or by having graduated from a military test pilot school. Thus, the second astronaut draft opened the door to civilians and to persons with scientific as well as engineering credentials (table 3).[41]

In 1963, for the third recruiting effort, flight requirements were lowered to 1000 hours, non-test pilots were qualified, and the age limit was lowered to 34. Instead of prospective candidates being prescreened by NASA or by the military services, the call extended to

TABLE 3. NASA Astronaut Selections, 1959 to 1969

Group I/April 9, 1959 (7 Selected)

Scott Carpenter (USN)
Gordon Cooper, Jr. (USAF)
John Glenn, Jr. (USMC)
Virgil "Gus" Grissom (USAF)

Walter Schirra, Jr. (USN)
Alan Shepard (USN)
Donald "Deke" Slayton (USAF)

Group II/September 17, 1962 (9 Selected)

Neil Armstrong (civilian)
Elliot See (civilian)
Frank Borman (USAF)
James McDivitt (USAF)
Thomas Stafford (USAF)

Edward White, II (USAF)
Charles "Pete" Conrad (USN)
James Lovell (USN)
John Young (USN)

Group III/October 8, 1963 (14 Selected)

Edwin "Buzz" Aldrin, Jr. (USAF, Ph.D. astronautics)
William Anders (USAF, M.S. engineering)
Charles Bassett, II (USAF, B.S. engineering)
Alan Bean (USN, B.S. engineering)
Eugene Cernan (USN, M.S. engineering)
Roger Chaffee (USN, B.S. Engineering)
Michael Collins (USAF, B.S., U.S. Military Academy)

R. Walter Cunningham (USMC, M.S. physics)
Donn Eisele (USAF, M.S. astronautics)
Theodore Freeman (USAF, M.S. engineering)
Richard Gordon (USN, B.S. chemistry)
Russell Schweickart, (civilian, M.S. astronautics)
Clifton Williams, Jr. (USMC, B.S. engineering)

Group IV/June 28, 1965 (6 Selected)

Owen Garriott (Ph.D., engineering)
Edward Gibson (Ph.D., engineering)
Dr. Duane Graveline (M.D., medicine)

Dr. Joseph Kerwin (USN, M.D., medicine)
F. Curtis Michel (Ph.D., physics)
Harrison "Jack" Schmitt (Ph.D., geology)

Group V/April 4, 1966 (19 Selected)

Vance Brand, (civilian, B.S. engineering)
John Bull (USN, B.S. engineering)
Gerald Carr (USMC, M.S. engineering)
Charles Duke (USAF, B.S. engineering)
Joe Engle (USAF, B.S. engineering)
Ronald Evans (USN, M.S. engineering)
Edward Givens, Jr. (USAF, B.S. Naval Academy)
Fred Haise, Jr. (civilian, B.S. engineering)
James Irwin (USAF, M.S. engineering)
Don Lind (civilian, Ph.D. physics)

Jack Lousma (USMC, M.S. engineering)
Thomas Mattingly, II (USN, B.S. engineering)
Bruce McCandless, II (USN, M.S. engineering)
Edgar Mitchell (USN, Ph.D. aeronautics
  and astronautics)
William Pogue (USAF, M.S. mathematics)
Stuart Roosa (USAF, B.S. engineering)
John Swigert, Jr. (civilian, M.S. aerospace science)
Paul Weitz (USN, M.S. engineering)
Alfred Worden (USAF, M.S. engineering)

Group VI/August 4, 1967 (11 Selected)

Joseph Allen, (Ph.D., physics)
Philip Chapman (Ph.D., instrumentation)
Anthony England (M.S. physics)
Karl Henize (Ph.D. astronomy)
Donald Holmquest (M.D.)
William Lenoir (Ph.D. engineering)

John Llewellyn (Ph.D. chemistry)
F. Story Musgrave (M.D.)
Brian O'Leary (Ph.D. astronomy)
Robert Parker (Ph.D. astronomy)
William Thornton (M.D.)

Group VII/August 14, 1969 (7 Selected)

Karol Bobko (USAF, B.S. Air Force Academy)
Robert Crippen (USN, B.S. engineering)
Charles Fullerton (USAF, M.S. engineering)
Henry Hartsfield (USAF, B.S. physics)

Robert Overmyer (USMC, M.S. astronautics)
Donald Peterson (USAF, M.S. engineering)
Richard Truly (USN, B.S. engineering)

volunteers from industry, professional groups, and other organizations. There was more emphasis on academic credentials, but most of the astronauts for the first three groups (24 of 30) still came from the military.[42] The fourth group was different. Its selection followed several years of discussion and some controversy relating to the perceived need for astronauts with strong scientific training for the lunar missions.

The Space Sciences Board of the National Academy of Sciences conducted a preliminary study of space research needs in 1962, and in 1963 a special ad hoc committee, retained by NASA and chaired by Dr. C.P. Sonett, submitted a report on "Apollo Experiments and Training on the Scientific Aspects of the Apollo Program." As a result of this work, NASA decided that astronaut selection should be based on both scientific and operational criteria, but that "because of the complex and difficult operational requirements and crew safety, whenever conflict exists between operational and scientific requirements, flight safety considerations demand that the scientific requirements be subordinate to the operational requirements."[43]

Subsequently, the NASA Office of Space Science and Applications cooperated with the National Academy of Sciences in defining specific scientific qualifications desired for the scientist-astronauts to be trained for lunar expeditions. The Office of Manned Space Flight defined the other-than-scientific requirements, notices for applicants were published and distributed in October, November, and December of 1964, and applications were due by January 1, 1965, to MSC. Flight experience or training was not required of these candidates. The applicants were then reviewed by the National Academy of Sciences which ranked the top 50 applicants on the basis of their scientific qualifications. Finally, on June 28, 1965, NASA selected six finalists from the Academy list. The finalists all had M.D. or Ph.D. degrees, two were medical doctors, two were engineers, one a physicist, and one a geologist. All but one, Lieutenant Commander Joseph Kerwin, USN (medicine), were civilians.[44]

The fifth group, selected in 1966, met the requirements established for Group III, but the age limit was raised from 34 to 36. NASA selected 19 astronauts, 4 of whom were civilians, in this round. The next year, 1967, the National Academy of Sciences again screened candidates as in 1965, and NASA selected 11 finalists, all of whom had Ph.D. or M.D. degrees, all of whom were civilians, and all of whom were required to attend jet pilot school for a year before beginning their regular training program at MSC.[45]

A seventh group of astronauts joined MSC in 1969, as transfers from the Manned Orbiting Laboratory Program being canceled by the Department of Defense. Seven transfers were accepted on the basis of their Air Force program qualifications, and by virtue of the fact that they were under 36 years of age.[46] One of these transfers, Lieutenant Commander Richard Truly, became NASA's Administrator in 1988. Nine years passed before NASA recruited any additional astronauts for spaceflight programs. Therein lies another story.

During the 10-year time frame in which the Mercury, Gemini, and Apollo astronauts came on board, "life systems" engineering and astronaut training changed to reflect the experiences and growing body of knowledge about spaceflight. Over time NASA engineers, such as Aleck C. Bond who assisted in the design and planning of the training and testing laboratory facilities at the Houston center and managed the Systems Test and Evaluation Program of the MSC's Engineering and Development Directorate from 1963 through 1967, heightened their senses and their skills in "man-rating" the design and operation of equipment used by astronauts. This process relates, Bond says, to designing equipment to accommodate human use and making it safe.[47] Before NASA, man-rating for conventional aircraft components was largely intuitive and unstructured. NASA engineers refined the concept as a technical engineering tool and a reliability and quality assurance measure.

Bond explained that the Mercury and Gemini programs used military boosters (Redstone, Atlas and Titan II) as launch vehicles. These missiles were designed to provide moderate reliability for a reasonable cost. "This," he said, "was not acceptable for manned spaceflight and thus very aggressive and definitive man-rating programs had to be undertaken to provide the desired safety and reliability of the launch vehicles and also of the spacecraft." General design criteria for launch vehicles and for manned spacecraft required conservative design approaches, redundancy in all critical systems, the use of "off-the-shelf" proven components to the fullest extent possible, and the use of standard design practices. A general design philosophy included the guidelines that no single mechanical failure would cause a mission to abort, and no single failure would result in the loss of life of the crew.[48]

Man-rated design criteria were supported by both standard engineering design reviews and by formal flight safety review panels which included representatives from engineering, operations, flight safety, and the astronauts. Prior to launch, a final mission review included key management personnel who certified the "total vehicle's readiness for launch."[49]

Man-rating criteria also changed between programs. The greater understanding of man's capabilities in space derived from Mercury resulted in Gemini design giving greater reliance on the astronaut for redundancy or backup systems. Piloting successes with Gemini resulted in the Apollo astronaut having more control. The debate over human versus automated control systems waned as the human became more of a system's manager integrated into the electronic controls. The Apollo program was the first to use launch vehicles specifically designed for manned flight; and Apollo, unlike Mercury and Gemini, relied heavily on alternative design approaches and extensive testing of subsystems.[50]

The test facilities of MSC in Houston sought both to test equipment and to train astronauts for living and working in the unearthly environment of space. While the Flight Crew Operations Division and the Mission Operations Division of MSC generally supervised astronaut training, every division of MSC participated, to some extent, in training and in testing materials and equipment. Astronauts, for example, spent many of their thousands of training hours in the laboratories and facilities operated by the Crew Systems Division, Structures and Mechanics Division, and other units of the Engineering Directorate.

The Crew Systems Division validated the "physiological design parameters for manned spaceflight"; that is, it had responsibility to design and test life support systems including space suits, atmospheric instrumentation, and food, water, and waste systems, and to train astronauts in their use. Testing and training were conducted in the laboratories specially designed and built at MSC to simulate space conditions on Earth, albeit in piecemeal portions. Thus the two altitude chambers (a 20-foot chamber and an 8-foot chamber) could replicate air pressures at 225,000 feet and 150,000 feet, respectively. A liquid nitrogen cold-trap associated with the 8-foot chamber could test the characteristics of solids and liquids and heat exchange characteristics. The envirotron chamber associated with the 8-foot altitude chamber could subject an equipped astronaut to a near-vacuum and temperature ranges of –100 degrees F to +400 degrees F.[51]

A separate crew performance laboratory allowed physiological and biomedical tests of astronauts in pressure suits. A flight acceleration facility or centrifuge was a primary training and testing device used to evaluate astronaut tolerance to acceleration and spaceflight stresses. Weightlessness, first reproduced for brief moments by zooming an aircraft, was better simulated by placing the suited astronaut underwater. Later a special tank, called the Weightless Environment Training Facility (WETF), was used to simulate spacewalks or EVA.[52]

Supporting laboratories included a chemistry laboratory used to evaluate "expendables" (carbon dioxide, water, etc.) produced during simulations and spaceflight. The waste management laboratory designed and tested spacecraft waste and water management systems. The crew performance laboratory examined the performance of astronauts within pressure suits. A microbiology laboratory checked bacterial contents of food, water, wastes and blood specimens. A clinical biochemistry laboratory analyzed urine samples, performed hormone analyses, and examined the effects of space-like conditions on the human body. Life support systems, crew provisions and equipment, space suit, and nutrition laboratories studied and tested the performance of the astronaut in space and provided input into the training regimen of the astronauts.[53]

One of the most important devices used in the training of Apollo astronauts was "SESL," the Space Environment Simulation Laboratory operated by the Structures and Mechanics Division of the Engineering Directorate. The offices and corridors of the laboratory, located in Building 32, are lined with photographs of the astronauts who spent so much of their training time in Chambers A and B.

Those chambers can simulate the vacuum of space, the wide ranges in temperature bearing on objects in space, the light and darkness, and varying degrees of radiation intensity. Astronauts, before entering a chamber, are given a full preflight physical examination, enter into a bioinstrumentation area where body sensors are applied, then are outfitted in pressure suits, spend several hours in a special denitrogenation area—and only then enter SESL to begin their tests or training regimen—after which they must go through an equally elaborate and time-consuming exit procedure. During an exercise emergency, repressurization systems can restore chamber pressure from 0 to 6.0 psia in 30 seconds (with oxygen at 4 psia) and can achieve a normal atmosphere within 90 seconds. Each component of SESL, such as the compressed air systems, emergency power systems, solar simulation (carbon-arc lights), vacuum panels, cryogenic panels, control room, electronic equipment, measuring devices, liquid nitrogen, and gaseous helium systems, requires teams of technicians and engineers working in tandem for prolonged and critical simulation tests.[54] SESL was only one of the elaborate test and training facilities intrinsic to the mission.

Mission operations training sought to replicate as closely as possible the flight plan and required the coordination and training of all elements that would be involved in an actual flight. Mockups or boilerplate models of spacecraft, linked to Mission Control, gave the astronaut a hands-on simulation and absorbed countless hours of training. Each program and each mission required a unique flight plan and specialized training. Very special training equipment sometimes had to be designed for each mission. It was very difficult to simulate equipment that was itself still being designed. Over time the astronaut graduated, as technology improved, from a mechanical Link-type pilot training device to a highly sophisticated Shuttle Mission Trainer which could simulate most of the flight possibilities that

Suddenly, Tomorrow Came . . .

The Space Environment Simulation Laboratory

Exhaustive tests were run on manned and unmanned Apollo spacecraft. This is a view of Apollo 8 in Chamber A of the SESL at MSC.

With the advent of the space age came the need for new testing and development facilities. In 1961 an ad hoc group of Space Task Group engineers began designing and drawing the specifications for test facilities to be built at MSC.

Space could not be duplicated on Earth, but many of its characteristics including weightlessness, audio and radio wave qualities, temperatures, and vacuum could be replicated. The Space Environment Simulation Laboratory [SESL] proved as essential to the design and testing of space vehicles as was the wind tunnel for aircraft.

The two chambers were completed in 1965. The external measurements of chamber A, 65 feet in diameter and 120 feet in height, made it unique. Chamber B was 45 by 43 feet. The two chambers were also unique because they were man-rated and had high vacuum performance and solar simulation fidelity.

The first tests in SESL took place in January 1966 in chamber B for the qualification of Gemini suits and associated EVA life support systems—as used by Edward H. White during his EVA during Gemini 4. The initial tests in Chamber A occurred later in 1966, on the Apollo Block I spacecraft for the purpose of demonstrating its adequacy for manned Earth-orbital missions. The tests revealed several design flaws and procedural and process errors that were corrected before the first Apollo flight. Later tests in 1968 to certify equipment for lunar flight conditions revealed several design anomalies and procedural errors that were corrected before subsequent lunar flight.

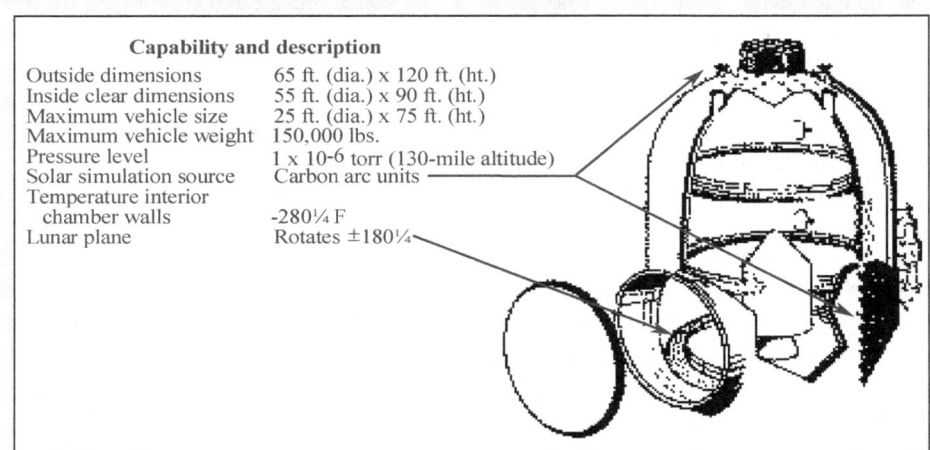

Capability and description	
Outside dimensions	65 ft. (dia.) x 120 ft. (ht.)
Inside clear dimensions	55 ft. (dia.) x 90 ft. (ht.)
Maximum vehicle size	25 ft. (dia.) x 75 ft. (ht.)
Maximum vehicle weight	150,000 lbs.
Pressure level	1×10^{-6} torr (130-mile altitude)
Solar simulation source	Carbon arc units
Temperature interior chamber walls	$-280°$ F
Lunar plane	Rotates $\pm 180°$

a pilot might encounter during a mission.[55] Training and testing were constants in the spaceflight business and consumed a major part of the energies of MSC personnel, but the public and the press rarely ventured into these deeper recesses. What the public saw and heard were the astronauts who, willingly or not, became celebrities.

The growing numbers of astronauts, accelerated training and missions coupled with the Gemini and Apollo programs, and the enormous increase in public interest created more and more stress in the area of public relations. Through the Mercury program and in the early stages of the Gemini flights, public affairs people at NASA Headquarters had an almost incidental relationship to the operating NASA centers. Public information for the manned spaceflights was derived from MSC and the Cape (Kennedy Space Center). There was, according to Julian Scheer who joined the Headquarters information staff in 1962, "little coordination, little cooperation" and a lot of frustration between public affairs people at Headquarters and those in the centers. Field people were not getting proper direction or supervision from Headquarters, he said; rather, the field centers were the "tail wagging the Headquarters dog." When George Mueller became Director of Manned Space Flight, Scheer was named Deputy Assistant Administrator for Public Affairs (circa March 1963). Subsequently, to develop a more coordinated and centralized public affairs policy, Scheer sent Paul Haney to MSC to relieve John "Shorty" Powers, who had functioned effectively but very independently.[56]

Alfred Alibrando became the public affairs officer at NASA Headquarters, working under Julian Scheer, and Headquarters strengthened its functional supervision over public relations, but never wholly displaced the independence of the center offices, particularly the independence of MSC which generated 80 percent of NASA's media releases. Haney, Scheer said, worked well with Bob Gilruth and George Low at MSC, and a coordinated and effective public affairs program developed over time. Houston established public affairs personnel assignments for the missions and submitted them to Headquarters for approval. Press releases were initiated in Houston and sent to Headquarters for approval and production. Live commentary for flights was a Houston responsibility.[57]

Scheer credited the Houston center for developing very strong public affairs programs including exhibits, public tours of space equipment and paraphernalia (such as Gordon Cooper's Mercury capsule), public programs (as in the Teague theater), and astronaut public speaking engagements. Some of the astronauts accepted foreign speaking engagements. Gordon Cooper and Charles Conrad, for example, went to Europe and Africa, Richard Gordon and Neil Armstrong spoke in South American engagements, and Walter Schirra and Frank Borman went to the Far East. Edwin Aldrin, who flew Gemini XII and the Apollo 11 lunar landing mission said, "Being in the public eye continually without any particular isolation was not a situation I relished in Gemini, and I certainly didn't look forward to an intensification after the lunar landing." Aldrin felt that public relations activities of MSC and Headquarters were not well coordinated and that each astronaut often became his own public relations resource.[58] The astronauts, despite tremendous workloads and little private time, generated widespread public goodwill and support for NASA programs.

Because of the affiliation of the astronauts with MSC and to some extent because of the linkage provided by the Mission Control Center to the astronauts in flight, MSC, and to a lesser extent the Kennedy Space Center from which the missions were launched, achieved a greater identification in the media and the public mind as being the essence of NASA and the American space program. This undoubtedly contributed to some friction between the various elements of the NASA community, but it also resulted in MSC, in cooperation with the Headquarters Public Affairs Office, developing unusual proficiency in the area of public relations.

The open door, visitor-oriented public exhibit policies adopted by MSC produced some operating inconveniences, but large dividends in goodwill. The Educational Programs and Services Branch of the center's Public Affairs Office, established by Paul Haney and directed by Eugene Horton, brought legislators, teachers, students, and the general public to the center's "campus" for information, orientation, and a sense of public participation in the space adventures of NASA. Horton stressed the benefits of spaceflights to Americans. "Where else," he told audiences, "could one buy a decade of technological and economic growth, national pride, and wholesome family entertainment for the price of four cinema tickets a year per family?" America's investment in space, he explained in 1970, was less than one-half of one percent of the gross national product. Space technology, he emphasized, not only resulted in sharper X-ray pictures, longer lasting paint, faster dentist drills, smaller TV cameras, weather detection and tracking satellites, communication satellites, new medical instruments and fire protective materials and devices, but it has been most important as a successful management approach to solving overwhelmingly complex problems. And he stressed, as NASA has stressed, that this has been done "within full view of the whole world."[59]

Despite the competition of worries at home and wars abroad, the American public and people throughout the world became drawn with fascination to the flights of Apollo following the enormously successful mission of Apollo 4, the first flight test of the Saturn three-stage launch vehicle, launched on November 9, 1967. As the authors of (NASA's) *Chariots for Apollo* (1979) explained:

> Technically, managerially, and psychologically, Apollo 4 was an important and successful mission. . . . The fact that everything worked so well and with so little trouble gave NASA a confident feeling, as [Sam] Phillips phrased it, that "Apollo [was] on the way to the Moon."[60]

Bob Gilruth congratulated center personnel for their achievement. The successful launch of the Saturn V and the perfect performance of the Apollo spacecraft in flight and during reentry at lunar return speeds," he said, "make Apollo 4 a major milestone for the entire program." Despite the problems of the past year, he said, "Our goal continues to be a lunar landing in this decade. With the continued dedication and personal commitment of our entire staff, this goal can be met."[61] Rejuvenated, the entire NASA organization, with the cooperation of the contractors, redoubled efforts to put a man on the Moon within the decade.

On January 22, 1968, a second Saturn IB test flight, this carrying the first unmanned lunar module with an unmanned command and service module, lifted off from the pad at Cape Kennedy. Production delays on the lunar module (LM-1) resulted in its delivery to Kennedy Space Center 7 months after originally scheduled, and the investigations and systems reevaluations that followed the AS-204 fire further delayed a test of the module. Now it flew. The LM separated successfully from the S-IVB stage, and after two independent revolutions, ground control fired the descent engines for a programmed 38-second burst. Four seconds later the engines stopped, under an automatic impulse which signaled that the vehicle was not accelerating fast enough. Mission control evaluated the situation (as would happen under a manned flight situation), and an alternate flight program was implemented. The descent engine was fired twice successfully under the new program, and the mission was determined to be a success. In February, the LM reentered the atmosphere and its "fiery remains" plunged into the Pacific southwest of Guam.[62]

Apollo 6 fared less well, but despite countdown delays, engine failures aboard the second stage of the Saturn, and an emergency burn by the third stage which lifted the unmanned command module and LM into a much lower orbit than planned, the command module was retrieved and the mission objectives achieved.[63] Despite the problems with the mission, Apollo 6 was largely ignored by the media and the public. Launching unmanned vehicles into space did not tickle the public fancy nearly so much as manned flights—and there were more pressing concerns.

Americans were deflected from their recent interest in space launches by the growing and more difficult involvement of American military forces in southeast Asia and by the heightened racial confrontations at home. Rising federal expenditures on war and welfare began to affect NASA budgets. President Lyndon B. Johnson announced a few days before the launch of Apollo 6 that he would not seek reelection; and on the day of the launch, Martin Luther King, Jr., a leader in the civil rights movement, was assassinated in Memphis, Tennessee.[64]

Nevertheless, the resumption of manned spaceflights beginning with the launch of Apollo 7 in October 1968, and the subsequent lunar expeditions, rekindled an excitement at MSC, within the NASA community, and throughout the United States that has rarely been duplicated. The "space race" against time, against money, against technology and human frailty, and against the Russians was being won many Americans believed. Americans needed heroes and a victory, and that victory, if not in southeast Asia or along the Iron Curtain, could well be in space. The astronauts became something greater than life, and space became for a time the "opiate of the masses" and the media.

The flights of Apollo, however, were far more important than a media event. NASA and the American space program had already invoked changes in life on Earth, changes still largely imperceptible to the casual observer. Astronauts were flying in machines that a decade earlier had not existed. They were being trained and tested in laboratories designed and constructed to simulate conditions not of this Earth. The men and women who built and operated the machines were as indispensable as the spacecraft that carried them into space. Ten years before there had been no astronauts. Only now, in the mid-1960's did aerospace begin to replace aeronautics in the American lexicon. The technology upon which spaceflight depended ranged the gamut of human knowledge and experience. The contractors and subcontractors who built the machines in which the astronauts trained and which they flew into space covered a broad spectrum of American technology and industry. Almost unwittingly, and sometimes unwillingly, NASA and the space programs were putting Americans and people of the world on a new learning curve. The atomic age with its more defined technology that emerged in war began to yield to the more broadly conceived technology of space dedicated to peace.

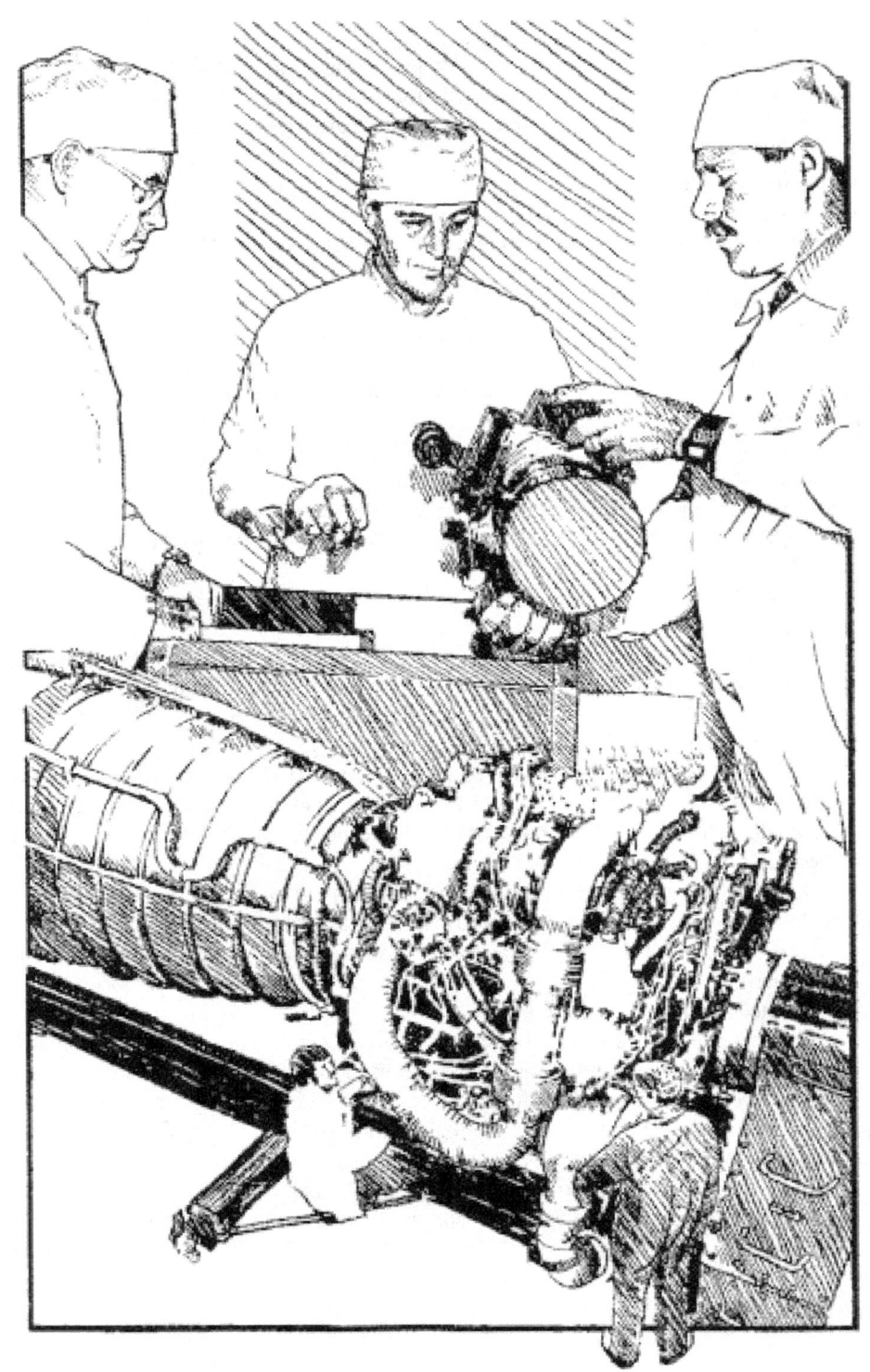

CHAPTER 8: A Contractual Relationship

Through its NASA contracts, the United States Government mobilized a large segment of American industry comparable to that ordinarily achieved only during the exigencies of war. Space, and more specifically the Apollo lunar program, required a massive effort by a fledgling aerospace industry still struggling to find its own identity. NASA grew massively in size and changed markedly in its configuration from a NACA-like and largely passive research organization into a research/development and mission-oriented agency whose primary business became project management and systems engineering. NASA engineers, trained and with experience in the laboratory and test facility, became managers of people. Whereas in war the government generally mobilized the Nation's resources through conscription, regimentation, and regulation, NASA mobilized the aerospace industry through a contractual relationship.

Arnold S. Levine in *Managing NASA in the Apollo Era* attributes the enormous increase in government contracting after World War II to the basic virtues of contracting out to the private sector, the limitations of formal advertising, and the demand for special skills in the management and integration of complex weapons systems. Contracting allowed the government to tap the technical experience and capabilities available in the private sector, which was not bound by civil service hiring and retention regulations—albeit labor legislation and union contracts certainly had an effect. Moreover, contracting allowed a very real flexibility in work force and budget expenditures.[1] While the government required it, the reality was that through contracting, NASA built a strong and diverse supporting political framework for government-funded space-related programs, and concurrently helped develop a broad-based technological strength throughout the national economy.

NASA's contracting drew on both the Air Force's heavy reliance on independent contractors for design and delivery and the Army's traditional arsenal or in-house production and design capability. NASA engineers, at least through the Apollo years, maintained an in-house capability allowing them to keep the design and technical skills to effectively direct, lead, and manage the NASA contractors.[2] Thus, the NASA-contractor relationship could best be defined as a partnership rather than a customer-client relationship.

Willard F. Rockwell, Chief Executive of North American Rockwell, described the contractors as the "unsung heroes of the Apollo program." Space programs altered the character of the private sector of the old aircraft industry and changed the traditional relationships between government and the private sector. North American Aviation, for example, founded in 1928, merged with Rockwell-Standard Corporation, an automotive parts manufacturer, to become North American Rockwell Corporation, a major aerospace firm and the prime contractor for the Apollo space vehicle.[3] North American and NASA's many contractors became more than suppliers or manufacturers of space components; they became "cooperators" with government and other firms in the design, development and assembly of spacecraft.

North American figured prominently in NASA's Apollo lunar effort. An understanding of North American's relationship as a contractor to the MSC offers an illuminating insight into the basic nature of NASA's Apollo program and its management. It also provides a study of the dynamics of American industry and of the relationships of industry to government.

John W. Paup, North American's program manager for Apollo, returned to work for North American's Space and Information Systems Division in 1961 following a stint with Sperry Rand Corporation. His initial job, working under the Division President Harrison A. "Stormy" Storms, was to oversee North American's proposal to NASA to design and construct the Apollo spacecraft. J. Leland (Lee) Atwood, president of the corporation, had reorganized North American's Missile Division as the Space and Information Systems Division the previous year. NASA selected North American's Rocketdyne Division, under Samuel K. Hoffman, to be the primary contractor for each of the Saturn propulsion systems under the management of the Marshall Space Flight Center. In November 1961, North American's Space and Information Systems Division received the Apollo spacecraft contract administered by MSC. North American became the only company with hardware in every part of the Apollo "stack," including the command module, service module, lunar module, instrument unit, launch escape system, and first, second and third stages of the Saturn rocket.[4]

A NASA contract involved some tensions between the Agency and the contractor in that the manufacturer naturally desired to maximize profits. The manufacture of space components required man-rated production quality, redundancies, and to an extent the sublimation of costs to quality and assurance. That plus the cooperational aspects of production were new experiences for industry. Although cost-plus-fixed-fee and special incentive contracts ameliorated the basic conflict of interests between the producer (the contractor) and the consumer (NASA), there would, of course, be no perfect solution.

A more basic communications problem also pervaded the contracting relationship. Contractors changed from program to program and project to project. Their knowledge and experience, as a result, tended to be very limited and defined. John Paup explained that "Government is the customer. The contractor is the producer. The experience that had been gained in building spacecraft in Mercury, and subsequently in Gemini, was known to the customer; it was not as well known to the contractor." But, he said, the contractor had more and better experience in handling "the magnitude of the Apollo program." Thus the government, he said, possessed the technical and operational knowledge, but the "real live program knowledge" was best known by the contractor.[5] Managing the Apollo program required fusing the knowledge and energies of diverse and traditionally competitive firms to produce a product that no one of them could independently produce. NASA management, contractors and subcontractors were indispensable to each other and to the program.

Program management required close communications and coordination between NASA Headquarters and centers, and between the project or program managers at the center level and the contractors. Effective management required overcoming, or at least subordinating, the independent and competitive instincts of cooperating firms. The contractor-subcontractor relationship provided an effective vehicle for doing that. Systems engineering and NASA contracting, however, confronted an inherent cultural tension within the engineering design/manufacturing relationship.

North American Aviation began in 1928 during the height of the 1920's stock market boom as an investment holding company headed by Clement N. Keys. North American bought interests in Curtiss Aeroplane and Motor, Transcontinental Air Transport, Curtiss Flying Service, and Douglas Aircraft. It acquired all of Eastern Air Transport, 27 percent of Transcontinental, and 5 percent of Western Air Express, and invested in Sperry Gyroscope, Ford Instruments, Intercontinental Aviation, and Berliner-Joyce Aircraft. General Motors acquired control of North American in 1933 when Ernest R. Breech became Chairman of the Board. General Motors merged its Fokker Corporation of America with North American's Berliner-Joyce to establish General Aviation Corporation which for the first time brought North American directly into aviation manufacture.

When the Air Mail Act of 1934 required the separation of aircraft manufacture and airline operations, North American was made a separate manufacturing concern headed by J.H. Kindleberger, and airline operations were consolidated as Eastern Airlines under Eddie Rickenbacker. General Motors sold its interest in North American in 1948. John Leland "Lee" Atwood became president, and Kindleberger became North American board chairman. By 1964, North American's $2 billion annual income derived heavily from government contracts associated largely with Apollo, Saturn S-II, and Minuteman production. North American Aviation merged in 1967 with Rockwell-Standard Corporation, a major producer of automotive parts, to become North American Rockwell Corporation. John R. Moore, previously the executive vice president for North American, headed the new corporation called the Aerospace and Systems Group, of which the North American Aircraft Division remained a component. Subsequently, the corporation became Rockwell International Corporation.

The command module mockup under construction by North American Aviation at Downey, California, shows the more spacious three-person interior of the Apollo spacecraft.

Source: Russ Murray, *Lee Atwood... Dean of Aerospace* (Downey, California: Rockwell International Corporation, 1980).

The "manufacturing people build from the details up. Engineers design from the top down." The engineer formulates a basic design first and then proceeds to the detailed design of the pieces. The manufacturer, on the other hand, wants detailed design on the little pieces first so that it can plan and design the tools to produce the many parts of the whole. In some respects, the development of a spacecraft was something like creating a continually growing and changing organism where each part could affect the nature of the whole system. Inasmuch as each part, as well as the whole system, was going through constant design, test, and evaluation, the "problem of making schedule, dollars and performance come out acceptably" was brought sharply into focus on the Apollo program.[6] Communications, or knowledge transfer, was a fundamental necessity of the system. It had to occur within each center between the program and project offices and the line divisions, from center to center, and between the center and headquarters. The contract manager at the center then became the agent for technology and information transfer between NASA and the contractor. But it was also critical that the learning experiences and technology of one space program were transferred to the next.

To help facilitate that transfer, Dr. George E. Mueller (who became NASA's Deputy Associate Administrator for Manned Space Flight in 1963) organized the Gemini-Apollo Executives group in 1964. The idea was to facilitate the transfer of information and experiences from the Gemini program to the Apollo program. The need was very real. Apollo contractors were trying to reinvent the wheel that had already been invented by Gemini contractors, thus the Apollo program in 1964 had fallen 6 months behind schedule and costs were spiraling. John F. Yardley, McDonnell Aircraft Company's manager for Cape operations, said that the immediate result of that Gemini-Apollo Executives meeting was to finish the program 2 months ahead of schedule and save large sums of money for the government and the contractors.[7] Those savings were effected largely by eliminating a duplication of effort among contractors, by sharing the expertise or "how-to" between contractors, and simply by facilitating the transfer of NASA's experience to the private sector. Meetings were held periodically through the following years.

At the conclusion of the Gemini program, Gemini and Apollo executives met on January 27, 1967, for a "final review" of Gemini, with an agenda to review the "lessons learned" from Gemini that would benefit Apollo. That same day President Lyndon B. Johnson signed the world's first space treaty. And at noon of that same day, Virgil Grissom, Edward White, and Roger Chaffee boarded Apollo spacecraft 012 (AS-204) for launch simulation tests. That afternoon, as the tests at Cape Kennedy progressed, in Washington, D.C., President Johnson addressed the Ambassadors of Great Britain and the Soviet Union, high American officials, and representatives from 57 foreign nations on the occasion of the signing of the international treaty committing those nations to the peaceful use of space and prohibiting weapons of war in outer space. Johnson said, "This is an inspiring moment in the history of the human race." Later that afternoon, tragedy struck. News of the Apollo 204 fire, Lyndon Johnson said, "hit me like a physical blow."[8] The news struck the Gemini-Apollo Executives no less forcefully.

In the separate meeting which began that morning and was scheduled to run through the 28th, the business executives whose companies built the Gemini and Apollo systems were discussing: "How do we assure that the maximum transfer of recorded experience from Gemini to Apollo takes place?" Yardley summarized for his counterparts in the Apollo program McDonnell's Gemini experiences so that "our Nation's space program can make the best use of our collective knowledge." But he noted that McDonnell was not closely involved with Apollo and had little knowledge of where the program stood or what practices were being used.[9]

One of the greatest construction difficulties with Gemini, he recalled, was related to the "weight critical" configuration of Gemini. This resulted in launch delays, lengthy retest periods following modifications, and a lower level of reliability than desired. Modular construction, on the other hand, facilitated the testing of each independent module without researching an entire system. And the interface between modules was kept as simple as possible. In contrast to Mercury's "layered" construction, Yardley explained, each Gemini unit had to be individually accessible and individually removable. We learned too that aircraft construction techniques did not meet spacecraft requirements, and as a result Gemini "pioneered in a number of areas such as all brazed propulsion system plumbing, crimped electrical connections, salt-free coldplate brazing, etc." McDonnell learned too that product and design changes were inevitably required as a result of testing and flight operations. It was necessary for McDonnell or any manufacturer to participate heavily in flight operations in order to close the response time for necessary modifications. It was also critical to incorporate changes with test operations, that is, to modify testing and evaluation procedures to accommodate the changes. All of this required close coordination between the manufacturer and test personnel. The management tool used, Yardley noted, was daily meetings at the "most detailed level between all organizational elements."[10] Space manufacture required an unusual degree of integration and cooperation between government and business and between ordinarily independent and often competitive private firms. Changing technology created changing social structures, and those changing structures facilitated yet more advances in technology.

The independent and competitive nature of American corporations, and particularly those in the developing aerospace-related industries, traditionally tended to preclude cooperation and the flow of technical knowledge from one to the other. This is, incidentally, one reason why managing engineers and executives such as John Paup, John Yardley, Robert Seamans, Jim Elms, and George Mueller, among many others, tended to move with some frequency from one corporation to another or from government to corporations. The exchange or transfer of managers allowed a transfer of knowledge otherwise discouraged by the competitive nature of the corporate world. The experiences of John Paup and North American Aviation are representative of those in the industry, and illustrate the growth and maturation of the aerospace industry.

John Paup, who as previously mentioned went to North American's Space and Information Systems Division from Sperry Rand in 1961, joined Milton Sherman, Charlie Feltz, and Norman Ryker (all of whom had X-15 or Jet Propulsion Laboratory contract experience) in developing North American's proposal for an Apollo spacecraft. The Apollo program and the contractor relationship to NASA began well before President John F. Kennedy established a lunar landing as an Apollo goal. NASA held an industry conference on July 29, 1960, to announce and describe the parameters of the Apollo program as it then existed. Potential bidders then met at Langley where they were briefed by Bob Gilruth and members of the Space Task Group. Subsequently, NASA awarded independent feasibility study contracts to General Dynamics, General Electric, and Martin Company; and from these studies and in-house work, NASA developed the Request for Proposal (RFP) which included a statement describing the nature and specifications of the work. North American received the RFP in July. One hundred North American engineers worked on the three-volume proposal which was delivered before the October 9, 1960, deadline. Next, Paup, with Company President Leland Atwood, Division President Harrison Storms, and others conducted an oral briefing on the North American proposal at the Chamberlin Hotel at Point Comfort, Virginia, in competition with teams from Convair, General Dynamics, General Electric, Lockheed, Grumman, and other prospective contractors.[11] The competition was itself a historic occasion. Paup recalled later:

> . . . this was undoubtedly the biggest procurement the government had ever considered and it was the first wide open big competition that NASA had ever run. So, there was a lot of bigness to this proposal activity—the trade journals were full of it, and much publicity was given to such an exploration. The whole idea of going to the Moon, of course, was a big thing. But, moreover, it required a big business operation. Everybody concerned with, or that speculates in big business, really had been waiting to have something like this to talk about for a long time.[12]

The North American delegation from Downey, California, traveled to the briefing in two separate aircraft, Paup remembered, to minimize a "possible catastrophe to our effort." But after reaching Washington, D.C., safely, the combined delegation boarded another plane for the trip to Langley only to be forced to circle for an interminable time because of bad weather. Finally, with about 10 minutes to spare, they made their meeting—only then to be stymied by the fact that the hotel's electrical plugs did not match their projection equipment. Finally, he said, Milt Sherman "took out his pen knife, cut the plug off, peeled back the wires, and sat there all during the briefing" holding the two bare electrical wires into the connections. "Our reliability was assured by man, not by any system or mechanism or planning—maybe that's a message for the Moon program," he added.[13]

The announced major elements of the Apollo-Saturn program included the design and construction of the spacecraft, the launch vehicle, launch facilities, control centers, and tracking network facilities. The spacecraft project itself, for which North American competed against four other bidders, comprised a command module which housed the crew and would be the only part of the vehicle to reenter the atmosphere and land, a service module which would include a propulsion system to return from the Moon but would not itself descend to Earth, and a lunar landing module which included propulsion systems for decelerating and landing on the Moon. A contract for the lunar landing module was to be awarded to a subcontractor at a later date. The contract envisioned the launch of the Apollo spacecraft by a NOVA-type rocket (not yet designed) using a direct lunar approach, or possibly a spacecraft launch and direct return.[14] The options of a lunar orbit rendezvous and the Saturn V propulsion system (with approximately 65 percent of the thrust of the envisioned NOVA rocket) were at first largely ignored, but would later be reinstated as part of the program.

That decision, coming almost 2 years after the initiation of the Apollo work, changed the configuration of the spacecraft.

Once the contractor presentations were completed, a NASA Source Evaluation Board, chaired by Max Faget and including Robert Gilruth, Robert O. Piland, Wesley Hjornevik, Kenneth S. Kleinknecht, Charles W. Mathews, James A. Chamberlin, and Dave W. Lang from MSC, George M. Low, A.A. Clagett, and James T. Koppenhaver from Headquarters, and Oswald H. Lange from the Marshall Space Flight Center, began the meticulous and intensive work of evaluating the proposals. Some 190 persons representing all major elements of NASA and a few representatives from the Department of Defense reviewed the proposals and made independent reports to the Source Evaluation Board. The Board divided itself into subcommittees, assisted by panels of specialists, and submitted evaluations based on a weighted scale of 30 points for the technical qualifications of the proposal, 30 points for the technical approach, and 40 points for business management and cost factors.[15]

Of the five competing contractors, General Electric proposed to collaborate with Douglas Aircraft, Grumman Aircraft, and Space Technology Laboratories. McDonnell proposed to team with Chance-Vought, Lockheed, and Hughes Aircraft. General Dynamics proposed to work with AVCO. The Martin Company and North American proposed to work as prime contractors and subcontract specific components of the work. The Source Evaluation Board then submitted its analysis and very close summary ratings (6.4 to 6.9 for the 5 proposals on a scale of 0 to 10) to the Administrator for a final decision.[16]

On November 28, North American, which the media had not considered a major contender, received word that it had been awarded the NASA Apollo contract. The decision rested in part on the greater attraction of dealing with a single primary contractor as opposed to a consortium of contractors. What followed in those first 6 months of "getting to know you" and getting organized was an intense effort to resolve unanswered questions and issues within North American and between North American and MSC representatives who would manage the $934 million contract. "Problems," Paup said, "came at us wave, after wave, after wave. The days were long, the excitement was intense, and the period was wonderful from the point of view of participating. But nevertheless they were tiring and exhausting days."[17] And then the honeymoon ended, and the hard, unceasing work began.

In December 1960, North American's Space and Information Systems Division signed on its first four major Apollo subcontractors. Collins Radio would manage the spacecraft telecommunications systems. Garrett Corporation's Air Research Division would handle environmental control equipment. Honeywell, Incorporated contracted with North American to develop the stabilization and control system, and Northrop Corporation's Ventura Division was assigned the parachute Earth landing work—later abandoned by NASA. NASA subsequently selected General Electric to oversee the integration of the Apollo space vehicle with the launch vehicle and to assure system reliability. North American then signed up the Marquardt Corporation to build the reaction-control rocket engines for the spacecraft, Aerojet-General to develop the service module propulsion system, Pratt and Whitney to build the Apollo fuel cell, and AVCO Corporation to design and install heat resistant (ablative) material on the spacecraft's outer surface. Over the next few years, numerous subcontractors went to work for North American's Space and Information Systems Division on Apollo hardware and design.[18] The North American prime contract thus became an umbrella contract creating a consortium of firms to accomplish a task that no one of them singly could complete.

No one had previously designed and built a vehicle to carry men to the Moon and back. There were, to be sure, ongoing Mercury (and by 1964) Gemini experiences to draw upon, but Apollo was different. Just as North American and MSC managers began to get their organization and efforts in focus, the work changed rather significantly. On July 11, 1962, now a year and a half into the Apollo spacecraft work, NASA announced that instead of building a spacecraft that would make a descent onto the surface of the Moon from an Earth launch or from Earth orbit (as envisioned in the original design concept), a lunar excursion module (LEM) separated from a command module in a lunar orbit would make the descent. The decision required major design changes in the lunar spacecraft and in the design and construction of the LEM, the inclusion of new rendezvous and docking apparatus, and new tests and procedures for every component. It also resulted in the award of a prime contract for the LEM to Grumman Corporation. Grumman and North American engineers held their first meeting to discuss the design of the LEM and its interface with the Apollo spacecraft on January 14, 1963.[19]

The lunar-orbit-rendezvous (LOR) decision was itself a critical historic moment in the Apollo-Saturn program. Engineers first preferred the direct ascent NOVA rocket technique "pictured in science fiction novels and Hollywood movies." But the technical realities and cost of such a battleship-sized, fuel consuming monstrosity led to the Earth orbit option. Advanced Saturn rockets, already in production, could launch a vehicle into Earth orbit from which a lunar mission vehicle could be launched and docked on its return. Once the NOVA rocket was rejected, NASA engineers, particularly those at MSC and Marshall Space Flight Center, supported the Earth-orbit-rendezvous (EOR) concept, in part because it obviously would be a prototype for a permanent space station.[20] But neither option was selected.

Serious study of LOR possibilities began at Langley Research Center in the Lunar Mission Steering Group, led by Clinton E. Brown who headed Langley's Theoretical Mechanics Division, and in a special Rendezvous Committee, chaired by Dr. John C. Houbolt who was assistant chief of the Dynamic Loads Division. William H. Michael, Jr., a member of Brown's study group, produced a monograph describing the advantage of parking the Earth-return propulsion part of a lunar spacecraft in orbit around the Moon during a lunar landing mission. Chris Kraft commented later that he thought this study became the seminal piece in the LOR decision. In 1960, Houbolt, with Ralph W. Stone, Clinton E. Brown, John D. Bird and Max C. Kurbjun, formally submitted a report to associate administrator Robert Seamans advising the LOR concept for the proposed Apollo program. Strong objections within NASA, however, centered on the proposition that the development of problems in a lunar orbit would be far more unsolvable and hazardous to the astronauts than would be problems in an Earth orbit.[21] Thus the lunar orbit concept remained dormant and discounted.

John Houbolt, however, refused to let the issue lie. In November 1961 in a private nine-page letter to Seamans, he bypassed the NASA hierarchy which was overwhelmingly indisposed to the lunar orbit concept. "I fully realize that contacting you in this manner is somewhat unorthodox," he admitted, "but the issues at stake are crucial enough to us all that an unusual course is warranted." Seamans responded 2 weeks later to the effect that new studies of that option would be initiated. Although Houbolt remained a pariah in the NASA

A Contractual Relationship

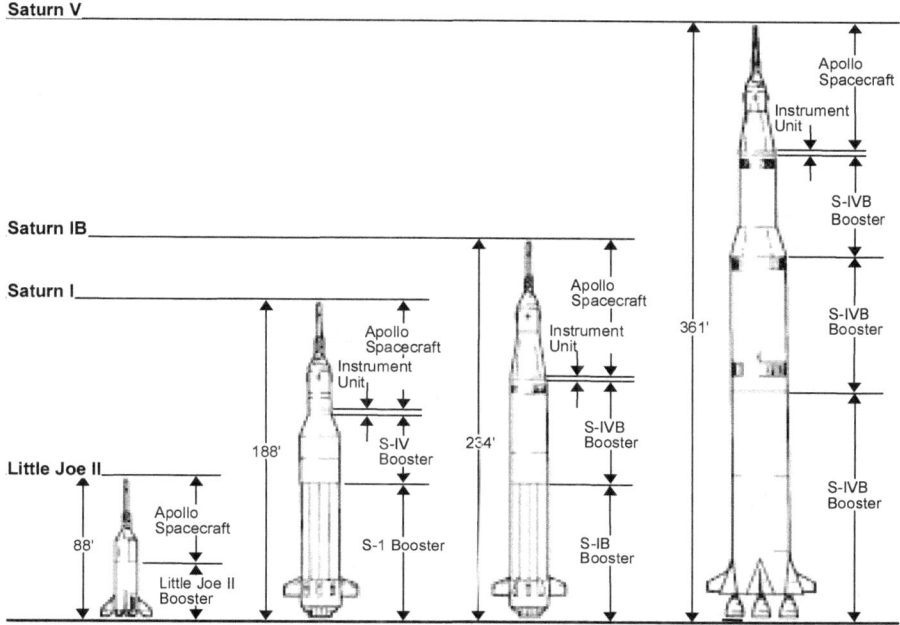

FIGURE 8. *The Apollo Stack*

community for some time, new tests and analyses did support the LOR over the EOR technique. Subsequently, Bob Gilruth and MSC personnel shifted to the LOR approach, and they were soon joined by Wernher von Braun's engineers in Huntsville. Gilruth and Von Braun, with their engineers, then persuaded James Webb and other administrators to support LOR. President Kennedy's science advisor, Jerome Wiesner, however, continued to oppose the lunar rendezvous. Nevertheless, on July 11, 1962, long after work on the Apollo spacecraft had begun, Webb and Seamans announced during a press conference that the projected lunar mission would employ a lunar orbit approach.[22] As previously mentioned, it meant considerable retooling, redirecting, and retesting for Apollo-Saturn contractors.

Instead of lengthening the time frame for the lunar mission, the Houbolt decision probably put NASA and its contractors back on track, and ultimately resulted in saving time and money. A NOVA rocket, many believed, required far greater power, fuel, and money. The Houbolt decision meant that existing technology and equipment could be better applied in the lunar landing program. For example, the decision was reached in August 1962 that the Apollo command module could use Rocketdyne engines being developed for the Gemini spacecraft, rather than requiring newly designed engines as projected originally.[23] Most importantly, the lunar approach championed by Houbolt proved eminently successful.

Engineers made a number of other critical engineering and design decisions in 1963 and 1964. For example the virtual impossibility of accurately measuring fuel masses in a zero-gravity environment was in part solved by measuring the radioactive particle emissions from the propellant. The final resolution involved the use of a capacitants system for the service module propellents. Heat shield problems were solved by using an open-faced fiberglass honeycomb filled with ablative material. In February 1964, NASA decided to reject the planned land recovery of the Apollo spacecraft after exhaustive tests and design and weight problems ruled out a paraglide landing system. NASA elected to use parachutes and a water landing as used with Mercury and ultimately Gemini. The development of long-endurance and restartable fuel cells by Pratt and Whitney engineers resolved many of the weight, electrical and power problems in the spacecraft. Although the production of what became known as the Block I command module continued and would be used in initial Apollo-Saturn test flights, design of the Block II module, which included lunar orbit qualifications and the LEM configuration, was approved in November 1964, and construction on it began immediately under a new North American Apollo program manager, Dale D. Myers. (Later, in 1970, Myers left industry and served with the Department of Energy before replacing George Mueller in NASA's Office of Manned Space Flight to head the space shuttle program.)[24] As the work developed and became more focused, NASA added more contractors, and primary contractors, such as North American, used more subcontractors.

North American recruited subcontractors in much the same fashion as did NASA. It issued RFPs, established source selection boards, adopted a point rating system (similar to a system North American previously used on its F-108 fighter and B-70 bomber contracts), and submitted the recommendations to the North American Source Selection Board which made the final determination for all subcontracts.[25] North American's major Apollo subcontractors are listed in table 4.

NASA contracts and subcontracts attracted enormous public attention and competition. And those who received the contracts worked, as did the astronauts and NASA engineers, under the public eye, or as Gilruth put it, they pursued "life in a goldfish bowl."

Noticeably, and perhaps expectedly, members of Congress also kept a keen eye on the award and placement of NASA contracts. Administrator Webb particularly worked under the gaze of Congress. MSC personnel did so as well—but to a lesser degree. Texas' congressional delegation, and notably Olin E. Teague, who chaired the powerful NASA Oversight Subcommittee of the House Committee on Science and Astronautics, maintained a constant vigil on the award of the greater and lesser NASA contracts and on the subcontracts let by the primary contractors—particularly those in Texas. Teague, as did other Congressmen, received annual status reports on NASA contract awards from NASA's Financial Management Division. Those reports identified the contract recipient by firm, city, county, total number of contracts received, and dollar value of the award. At the close of 1967, for example, Texas contractors and universities received a total of $866,571,000 in NASA contracts and grants, of which the larger portion ($643.5 million) went to Houston firms. Texas-based contractors and those firms with large divisions physically located in Texas with NASA contracts in excess of $100 million included IBM, General Electric, Philco Ford in Houston, and LTV Aerospace in Dallas. Teague also maintained independent contact with many primary and small contractors and frequently requested and received financial data from Administrator Webb as well as from many of the primary contractors. His office, judging by his correspondence, became something of a clearinghouse between many prospective NASA contractors and the Agency.[26]

TABLE 4. Major Apollo Subcontractors
(contracts in excess of $1 million)

Company	System	Approx. Value as of April 1966 (in millions of dollars)
Accessory Products Company Whittier, California	Helium transfer unit, valves, and assemblies	$ 2.2
Aerojet-General Corporation Space Propulsion Division Sacramento, California	Service module propulsion motor	62.6
Aeronca Manufacturing Company Middletown, Ohio	Honeycomb panels	12.5
Amecom Division College Park, Maryland	C-band and S-band antennae	1.4
Applied Electronics Corporation of New Jersey, Metuchen, New Jersey	Pulse code modulation systems	1.0
AVCO Corporation Research & Advanced Development Division, Willington, Massachusetts	Ablative heat shield	30.0
Beech Aircraft Corporation Wichita, Kansas	Supercritical gas storage system	19.2
Bell Aerosystems Company Buffalo, New York	Positive expulsion tanks for reaction control system	10.1
Beckman Instruments, Inc. Fullerton, California	Data acquisition equipment	2.7
Collins Radio Company Cedar Rapids, Iowa	Communications and data	88.2
Control Data Corporation Government Systems Division Minneapolis, Minnesota	Digital test command system	9.4
Cosmodyne Corporation Torrance, California	Liquid hydrogen, liquid oxygen ground support equipment and unique detail spares of liquid hydrogen and liquid oxygen transfer units	4.3
Dalmo Victor Company A Division of Textron Belmont, California	Main communications (deep space) antenna systems	2.1
Electro-Optical Systems, Inc. Micro Systems, Inc. (Subsidiary) Pasadena, California	Temperature and pressure transducer instrumentation	8.7
Garrett Corporation AiResearch Mfg. Division Los Angeles, California	Environmental control system	45.3

TABLE 4. Major Apollo Subcontractors (continued)
(contracts in excess of $1 million)

Company	System	Approx. Value as of April 1966 (in millions of dollars)
Leach Corporation Azusa, California	Apollo flight qualification recorder	$1.0
Ling-Temco-Vaught, Incorporated Dallas, Texas	Selective stagnation radiator system	1.2
Lockheed Propulsion Company Redlands, California	Launch escape and pitch control motors	7.9
The Marquardt Corporation Van Nuys, California	Reaction control motors for service module	29.8
Microdot, Incorporated Instrumentation Division South Pasadena, California	Stress measurement system	1.5
Motorola, Inc. Scottsdale, Arizona	Digital data up-link	9.5
Northrop Corporation Ventura Division Newbury Park, California	Earth landing system	36.6
Radiation Incorporated Melbourne, Florida	Automated telemetry data processing system (during vehicle testing)	3.5
RCA Electronics Princeton, New Jersey	Television cameras	3.8
General Motors Corporation Allison Division Indianapolis, Indiana	Fuel and oxidizer tanks	8.4
General Precision, Inc. Link Division Bingham, New York	Mission simulator trainer	36.6
General Time Corporation ACRONETICS Division Rolling Meadows, Illinois	Central timing system	4.5
Giannini Controls Durate, California	Reaction control gauging system	8.1
Gibbs Manufacturing and Research Corporation (Hammond Organ Co.) Janesville, Wisconsin	Mechanical timers and mechanical clocks	1.8

TABLE 4. Major Apollo Subcontractors (concluded)
(contracts in excess of $1 million)

Company	System	Approx. Value as of April 1966 (in millions of dollars)
B.H. Hadley Company Pomona, California	Pressure helium regulator unit and liquid hydrogen tank vent disconnects	$1.5
Honeywell Minneapolis, Minnesota	Stabilization and control	98.4
Kinetics Corporation Solana Beach, California	Power transfer and motor driven switches	1.8
Rosemount Engineering Company Minneapolis, Minnesota	Transducers and MASS flowmeter	1.1
Sciaky Bros., Incorporated Chicago, Illinois	Tooling, welding and machinery	1.7
Simmonds Precision Products Tarrytown, New York	Propellant gauging mixture ratio control	11.4
Thiokol Chemical Corporation Elkton Division, Elkton, Maryland	Escape system jettison motors	3.0
Transco Products, Inc. Venice, California	Telemetry antenna system (R&D)	1.0
United Aircraft Corporation Pratt & Whitney Aircraft East Hartford, Connecticut	Fuel cell	60.3
Western Instruments, Incorporated Newark, New Jersey	Electrical indicating meters	2.0
Westinghouse Electric Aerospace Electrical Division Lima, Ohio	Static inverter conversion unit	4.9

Source: Ralph B. Oakley, "Historical Summary, S&ID Apollo Program," North American Aviation, January 20, 1966.

Teague became particularly concerned that many small contractors were becoming victims of the intricacies of dealing with the government:

> It is obvious to me that many small business firms are not made aware of the subtleties involved in NASA contracts and tend to deal with NASA procurement and engineering personnel on a basis of personal trust, much as they do in commercial practices. However, many of them soon discover that they cannot rely on the verbal promises and assurances of NASA personnel and hence are heavily and unfairly penalized because they chose to trust the "word" rather than a written document.[27]

On the other hand, he observed, larger NASA contractors seemed to have little trouble in getting their contracts adjusted when financial or other problems developed.

Teague expressed bewilderment about the inconsistency of government which expressed full support for the preservation of small business firms, but permitted government procurement lawyers to protect government agents who made oral assurances to small businesses which resulted in their failure to receive reimbursement for "work honestly and faithfully performed."[28] Although he may for a time have provided some measure of relief for small businessmen by being a "go-between" for them with larger firms and NASA, the financial and contractual arrangements by NASA tended to become more complex and difficult, and less accessible to small businesses. Nevertheless, many small businesses became successful large businesses because of their NASA contracts, and made significant contributions to the space program.

Contracting related not only to engineering and design, but also to technical services and maintenance. MSC contracted out facilities and grounds maintenance, food services, on-site transportation shuttle systems, library services, and archival services. Procurement stressed purchases from existing stocks of private suppliers. Thus the economic impact on the community and the state tended to be very broad. While the romance and adventure of spaceflight certainly attracted admirers and attention, the very great economic impact of MSC on the Houston, Texas, and Gulf Coast communities generated strong self-interested, but no less real, public support for NASA programs.

Magazines such as *U.S. News and World Report* began to talk about a new space frontier in the "Southern Crescent." Texas, Louisiana, Mississippi, Alabama and Florida, stretching along the Gulf Coast, became the home of new NASA centers and installations and their hosts of private contractors and suppliers. Nationally, NASA contracts rose sevenfold between 1960 and 1965, from 44,000 to some 300,000 contracts being managed by NASA engineers and scientists, while the number of NASA employees rose to three times that of 1960 levels. It meant that relatively fewer NASA managers were managing more contracts, and that more of NASA's engineers were becoming contract administrators and project managers.[29]

Public interest in the Apollo-Saturn program heightened as real space hardware began to be shipped to MSC, Marshall Space Flight Center, and Kennedy Space Center. North American and its subcontractors first designed and built mockup or simulated modules and components for tests and evaluations. These were usually followed by the construction of boilerplate modules built on the design and weight specifications of the Apollo spacecraft, but not man-rated. The boilerplate modules were used for a variety of tests (such as water impact, parachute, flotation, launch compatibility) and for training. North American delivered its first boilerplate modules in September 1962, and by mid-1964 some 30 boilerplate modules were being used for various tests, including 5 launches on Little Joe II rockets. In February and again in May 1965, Apollo boilerplate modules were adapted to successfully launch Pegasus satellites into orbit. Finally, in October 1965, and basically on schedule, North American delivered the first actual Apollo spacecraft, SC-009, to Kennedy Space Center.[30]

North American's mobilization of resources for Apollo production peaked in 1965. In that year alone, North American's Space and Information Systems Division added more than 5000 employees, and division employment, almost wholly concentrated on Apollo production, peaked at 35,385 persons. Subcontractors similarly were in full production by 1965. Over 400,000 people were at work on NASA's space programs by mid-year. NASA expenditures from 1964 through 1967 approximated $5 billion each year and declined thereafter. The comparative level of NASA economic activity indicated by table 5 should be viewed in the context of a few broader economic parameters. During the same 4-year period (1964-1967), federal defense expenditures rose from $50 to $70 billion per year, the cost of government health programs with the introduction of medicare rose from $1.7 to $6.6 billion a year, and total federal expenditures increased by about one-third.[31]

Growing federal expenditures and inflation generated efforts at economy in all federal agencies, including NASA. One of the anomalies of the space program is that NASA's

TABLE 5. NASA Budget and Personnel Status

Fiscal Year	Appropriated	NASA	Contractors (estimated)	Total
1959	$330.9 million	--	--	--
1960	523.6 million	10,000	37,000	48,000
1961	966.7 million	17,000	58,000	75,000
1962	1,825.3 billion	22,000	116,000	138,000
1963	3,674.1 billion	28,000	218,000	246,000
1964	5,100.0 billion	32,000	347,000	379,000
1965	5,250.0 billion	33,000	377,000	410,000
1966	5,175.0 billion	34,000	360,000	394,000
1967	4,968.0 billion	34,000	273,000	307,000
1968	4,588.9 billion	33,000	235,000	268,000
1969	3,953.0 billion	32,000	186,000	218,000
1970	3,696.6 billion	31,000	135,000	166,000
1971	3,333.0 billion (request)	30,500	113,000	144,000

Source: FY 1971 Interim Operating Plan News Conference, Apollo Series, JSC History Office.

budget and the number of contractor-employed personnel began a downward slide well before the first Saturn-Apollo launch. At the time the Apollo program enjoyed its greatest successes (1969-1970), NASA operated with 25 percent less money than the peak 1965 budget, and total contractor personnel had declined by almost 50 percent. MSC's civil service employment peaked at 4731 (full-time equivalent employees) in 1967, and 2 years later support contractor personnel peaked at 14,276. Under the duress of tightening budgets, MSC began in 1965 and 1966 to convert more of its cost-plus-fixed-fee contracts to cost-plus-incentive-fee contracts as had been done initially with McDonnell Douglas for the Gemini program. The motivation was both economy and efficiency. Negotiations for the conversion of the North American Apollo contract were completed in December 1965 and approved by NASA in January 1966.[32]

Major Apollo contractors now included Aerojet-General Corporation which produced the service module engine. AVCO Space developed the command module heat shield. Bell Aerospace produced the lunar module ascent stage engine. BellComm provided systems engineering and support to Headquarters. Bendix produced the lunar surface experiments package and command module instrumentation systems. Boeing monitored the integration of Saturn-Apollo components. Collins Radio provided communications and data subsystems. Eagle-Picher Company produced batteries for Mercury, Gemini, and Apollo systems. Garrett Corporation specialized in environmental control systems which allowed the astronaut to survive in a hostile environment. General Electric reviewed all programs for reliability and quality assurance. The AC Electronics Division of General Motors manufactured the guidance and navigation system. Grumman (discussed more fully in a following chapter) was the prime contractor for the lunar module and used many subcontractors as did each of the other prime contractors.[33]

Minneapolis-Honeywell worked on stabilization and control systems for all of the spaceflight programs. IBM helped design and build the Mission Control Center (unit 6) and the Saturn V instrument control unit, and International Latex Corporation built space suits. Lockheed manufactured the launch escape and pitch control engines, and Marquardt Corporation produced the reaction control system for the lunar module and the service module. Motorola provided the digital command system; Northrop the landing system; RCA the television, guidance and communications equipment; Raytheon the command module computers; and Space Technology Laboratories (TRW) the lunar module descent stage engine.[34]

Finally, Thiokol built the launch escape tower motor and TRW Systems, Inc. provided trajectory analysis support. United Aircraft's Hamilton Standard Division developed the environmental system for the lunar module, and as previously mentioned, Pratt and Whitney designed and built the fuel cell power units. Philco-Ford, also reviewed in a following chapter, was the primary contractor in the design and development of the Mission Control Center. The most involved of these contractors, of course, was North American, which was the prime contractor for the Apollo.[35] Not only did American business profit financially from the space program, but American industry achieved significant technological growth by doing things that had never been done before.

Fuel cells, metal and ceramic manufacturing, computer design, environmental systems, and human health benefited greatly. Manufacturing processes experienced critical breakthroughs in specific technology such as wraparound tooling, cold bonding, and multilayer circuit board soldering; and even precision hole-drilling through steel, titanium, and composite honeycomb materials had broad applications throughout American industry.[36] Space was business, but more importantly it was a learning business.

Clearly, part of the product of that learning would be the development of new technology, that is, how to create and build new machines, how to operate those machines, and how to do things better or differently than had been done before. A great part of that learning experience, it was anticipated, would have to do with the physical properties of objects and materials found in space—and more specifically on the Moon. Preoccupied with building rockets and spacecraft, NASA engineers gave little thought to what those astronauts might find if and when they landed on the Moon, or the consequences such an event might have on the human learning experience.

In 1964, Elbert A. King, Jr., and Donald A. Flory, who joined the MSC Space Environment Division the previous year, submitted to Max Faget, the Director of Engineering and Development, plans for a laboratory to receive lunar materials where they would be repackaged for distribution to scientists for study. After several refinements, plans for the laboratory went to Headquarters for approval and funding. Headquarters responded "cautiously," David Compton explained in his history of the Apollo lunar exploration missions, and noted that such a laboratory and distribution of lunar samples would be the responsibility of Headquarters. Willis Foster, who headed the Manned Space Science Division at Headquarters, tentatively approved $100,000 for a laboratory design study rather than the $300,000 requested, and created an ad hoc committee of Headquarters personnel and MSC scientists to study the problem.[37] But the Lunar Receiving Laboratory almost got lost in the delays resulting from Headquarters' and MSC's attempts to resolve their differences over the proposed laboratory, in the disputes that began to arise between the science community and NASA, and in Congress' growing desperation to cut government expenditures.

While the NASA ad hoc group studied and "ruminated on the need for a receiving laboratory," at the instigation of Homer Newell, Associate Administrator of the Office of Space Sciences and Applications, a special advisory committee including three persons from the Space Science Board and two academic scientists representing the broader scientific community, met to discuss management and distribution of lunar materials. The committee advised that studies of lunar materials should be conducted by the scientific community at large, that a receiving laboratory need not be located at MSC (unless it could be properly staffed), and that strict quarantine procedures should be imposed to prevent possible biological "back-contamination" by extraterrestrial materials. Headquarters and MSC jousted for a time about who should manage the receiving laboratory, how it would be designed, and how much it would cost. Just about the time NASA concluded its deliberations, Congress decided that a Lunar Receiving Laboratory would not be needed and struck funding for it from the authorization bill.[38]

Rehearings resulted in the restoration of the $9.1 million request for the laboratory, but Congress meanwhile reduced NASA's overall budget for facilities construction by $18.2 million and its administrative budget by $23.9 million. These cuts, Compton suggests, did affect the construction and operation of the Lunar Receiving Laboratory. Nevertheless, with designs developed by the Oak Ridge National Laboratory of the Atomic Energy Commission and Headquarters' approval of the contract award by MSC, construction began on the

$7.8 million Lunar Receiving Laboratory in July and August 1966 with contracts let to Warrior Constructors, Inc. of Houston for preliminary work, and to National Electronics Corporation of Houston and Notkin and Company of Kansas City, Missouri, for completion.[39]

As did design and construction, staffing became mired in seemingly interminable committee studies, in an apparent disinterest by scientists to work in Houston, cost problems, interagency discussions (as between NASA and the Public Health Service), and general bureaucratic procedures and irresolution. Finally, in 1967, MSC Director Bob Gilruth created a Science and Applications Directorate with authority over the Lunar Receiving Laboratory being completed. While a search for a decision on staffing the laboratory and for a permanent laboratory manager slowly unwound, Gilruth appointed Joseph V. Piland, who managed the laboratory construction, as the laboratory's acting manager.[40] The problems developing with the Lunar Receiving Laboratory illustrate the growing complexity of the space business and of doing business with the government. Decisions became increasingly difficult as broader elements of society were affected by those decisions.

Government contracts and firms doing business with the government required more and more supervision, not only by NASA managers, but by a growing host of "outside" government agencies. Beginning in the 1960's the creation of new regulatory agencies and bodies accelerated, with 20 new agencies being added in the decade of the seventies alone. The Environmental Protection Agency created by the National Environmental Policy Act (1969), the Occupational Safety and Health Act of 1970 which established the Occupational Safety and Health Administration (OSHA), plus affirmative action, small business, disadvantaged business laws, and changes in worker compensation, tax codes, and reporting procedures were but a few of the growing administrative burdens encountered by government contractors and, to be sure, by firms doing business anywhere. These regulations, as will be seen in a later chapter, created problems but also opportunities.

Given the inherent difficulties of building a machine that was being designed as it was being built, plus the growing complexities of the business environment, not to mention the weakening economy and signs of weakening congressional and public support for space expenditures, and a real decline (after 1966) in the number of NASA contracts and the rising costs of doing business (including higher interest rates and inflation), aerospace firms (despite their apparent successes) were growing more and more financially exposed and vulnerable. Just as Apollo began to fly and great achievements and expectations in space emerged, the technical infrastructure for space ventures outside of NASA, that is, the private contracting community upon which the entire space program ultimately depended and which received roughly 80 percent of NASA expenditures—as well as NASA itself—faced declining budgets, employment reductions, and rising complexities and costs in the manner of their doing business.

In 1967 these problems seemed somehow remote. There was pressing business at hand. NASA engineers and North American and its subcontractors concentrated on the corrective actions, design improvements, modifications, and production of the Saturn-Apollo spacecraft. Grumman pushed development of the LEM. Construction finally began on the Lunar Receiving Laboratory. The first launch of a flight-ready spacecraft (017) aboard the Saturn V occurred, in November 1967 (Apollo 4). NASA flight-tested a lunar module (LM-1) aboard a Saturn IB in January 1968 (Apollo 5), and in April the launch of a Saturn V stack (Apollo 6) proved that the mission support systems could respond to emergencies caused by serious malfunctions in the propulsion systems. The spacecraft and its components lurched into totally unplanned orbits (Apollo 6), but flight directors were able to save and control the flight.[41] NASA decided it was time to return to flight.

On October 11, 1968, astronauts Walter Schirra, Donn Eisele, and Walt Cunningham awaited lift-off in the redesigned Apollo 7 spacecraft secured to the Saturn IB booster. Eleven years earlier, the Soviet Sputnik began orbiting Earth and created that massive response that brought NASA into being and the astronauts to their place at this appointed hour. Only 15 months before, with the destruction of AS-204, the entire Apollo program was forced into a reevaluation and redesign. The future then seemed dark. Now, a lunar landing within the decade finally seemed truly possible.

Apollo 7 made a smooth lift-off and returned on October 22, following an 11-day mission which proved the space-worthiness of the vehicle and the astronauts. The flight also featured the first live television broadcasts from a manned spacecraft. If it had ever lagged, the public's interest in spaceflight rekindled.[42] Interestingly, within a few years of that launch, each of the three astronauts aboard Apollo 7 became business managers in the private sector, perhaps lending credence to the close interrelationship between the public and private sectors of the space-related economy.

Cunningham, a Marine aviator with an undergraduate and graduate degree in physics, resigned from NASA in 1971 to organize and become president of a Houston-based company called HydroTech Development. In 1976 he became a senior vice-president and Director of Engineering for 3D International in Houston. Donn Eisele, the command pilot for Apollo 7, left the astronaut corps in 1970 and became the Technical Assistant for Manned Flight at the Langley Research Center before retiring from NASA in 1972 to become the Peace Corps Director in Thailand. Later he joined the Oppenheimer investment firm in Ft. Lauderdale, Florida. Walter (Wally) Schirra, known in the astronaut corps for his good humor and practical jokes (once, when asked for a specimen, he delivered to the nurse a 5-gallon jug of water discolored with iodine), flew Mercury, Gemini, and the Apollo 7 flight. He retired from NASA and the Navy in 1967 and became Chairman and Chief Executive Officer of Environmental Control Company in Colorado, before becoming Director of Marketing-Powerplant and Aerospace Systems for Johns Manville Corporation.[43]

In a sense, the flight of Apollo 7 marked the apogee, that is, the high point, of the Apollo program insofar as the contractors were concerned. By now all Apollo systems were in full production. The design and manufacturing problems had been seemingly resolved. Employment levels among NASA contractors were declining, as were the dollars being spent on Apollo and other space-related contracts. How had NASA and its contractors gotten from the point where manned spaceflight had been at best an idle dream to the moment when man's first step on the Moon seemed both imminent and practical? What would come after Apollo? And what would flights to the Moon mean to people on Earth?

Dr. Edward C. Welsh, Executive Secretary of the Space Council, addressed these questions in a talk to the Science Industry Committee of the Metropolitan Washington Board of Trade. Putting federal money into space, he assured everyone, is not taking dollars away from anybody—"every bit of that money [spent on space] is spent right here on Earth, rather than out on the Moon or some other heavenly body." And this financial investment, he stressed, "is bringing in substantial returns to people in every state of the union." Space activity, he said, "is both productive and creative. It puts to work—producing, creating, and doing—some of our most valuable resources such as skilled manpower and modern facilities." NASA's contractual relationships "brings together into a constructive team all of the major elements in our country devoted to technical progress" and technological leadership.[44]

The space program, Welsh believed, developed methods, techniques and procedures which increased the efficiency and profit of a broad spectrum of American enterprise, within and without the aerospace industries. Economic benefits included worldwide communications systems, global weather data and forecasting, and navigational aids. Manufacturers learned new things about heat, metallurgy, alloys, plastics, and ceramics. Computer and electronic technology experienced a veritable revolution, in good measure because of inputs and incentives from the space program.[45] Education, he said, benefited from the space program, not only by direct assistance in the form of scholarships and fellowships and laboratories and research grants funded by NASA, but in the broader dimensions of new knowledge about the heavens and the Earth and of humankind. Medical instrumentation improved markedly as a result of electronic applications from the space program and was beginning "to revolutionize the equipment of clinics, hospitals, and doctor's offices." Concurrently, national security and international relations were greatly enhanced by America's space program. It helped depict and disseminate the Nation's vitality and strength "in ideas, in technology, in freedom, in standards of living, in education, and in objectives for peace."[46] All this while the Moon and a lunar landing still seemed so distant? Welsh's rhetoric and NASA reassurances seemed to fall on a growing number of deaf ears, as Americans at the close of the decade began to weigh the costs of the War on Poverty, the war in southeast Asia, the cold war and rising federal deficits and inflation.

America's mobilization for space peaked in 1965. It had thus far been a unique experience. The mobilization had been peaceful, and with peaceful intent, and was accomplished through conventional free-market mechanisms and notably by contracts. Through the mechanism of the primary contract and subcontracts, with oversight by NASA technical managers, space business became an integrated collectivist enterprise. Almost one-half million Americans, about 35,000 NASA employees and 410,000 contractor employees, were at one time directly involved in the space program. The numbers involved and the dollars committed to space began to decline long before the Apollo program peaked. By 1969 Apollo was a product of a full decade of effort by a broad spectrum of American society. By 1969 the Nation had committed $37 billion of its resources and a considerable portion of its technical expertise and personnel to NASA. Because of NASA and the national space programs the world was changing, but the nature and extent and necessity of those changes was still not at all clear to most Americans.

Soon they, and the other people on Earth, began to see themselves from a different perspective. American astronauts aboard Apollo 8 circled the Moon. Those from Apollo 11 orbited the Moon, landed on the Moon, walked upon its surface and returned safely to Earth.

They gazed upon Earth from another body in the solar system. Through their eyes, the people of the world saw Earth and themselves as they had never been seen before. No one, nor life on Earth, would be quite the same again.

CHAPTER 9: *The Flight of Apollo*

The design and engineering of machines capable of taking humans into space evolved over time, and so too did the philosophy and procedures for operating those machines in a space environment. MSC personnel not only managed the design and construction of spacecraft, but the operation of those craft as well. Through the Mission Control Center, a mission control team with electronic tentacles linked the Apollo spacecraft and its three astronauts with components throughout the MSC, NASA, and the world. Through the flights of Apollo, MSC became a much more visible component of the NASA organization, and operations seemingly became a dominant focus of its energies. Successful flight operations required having instant access to all of the engineering expertise that went into the design and fabrication of the spacecraft and the ability to draw upon a host of supporting groups and activities.

N. Wayne Hale, Jr., who became a flight director for the later Space Transportation System (STS), or Space Shuttle, missions, compared the flights of Apollo and the Shuttle as equivalent to operating a very large and very complex battleship. Apollo had a flight crew of only three while the Shuttle had seven. Instead of the thousands on board being physically involved in operating the battleship, the thousands who helped the astronauts fly Apollo were on the ground and tied to the command and lunar modules by the very sophisticated and advanced electronic and computer apparatus housed in Mission Control.[1] The flights of Apollo for the first time in history brought humans from Earth to walk upon another celestial body.

Apollo is perceived in modern times as the ancients' sun-god, a god of light and of the heavens whose chariot raced across the night skies like a shooting star. Greek mythology ascribes to Apollo much earlier and more simple roles. He appears in Greek writings variously as the god of agriculture, the protector of cattle and herds, the deity of youth and manhood, a warlike god, and a god of prophecy, of healing, and of music (so long as that music came from the lyre). At the height of Greek civilization, as Athens particularly began to colonize throughout Ionia and the Mediterranean world, Apollo became a maritime deity, the "dolphin" god who accompanied emigrants on their voyages. Thus in modern times, fittingly perhaps, another Apollo carried the first voyagers from Earth to a distant heavenly body. "Houston, Tranquility Base here, the Eagle has landed," astronaut Neil Armstrong radioed from the lunar surface to the Mission Control Center as the Apollo 11 mission touched down on July 20, 1969.[2] The journey from here to there had been fraught with peril, difficulties, and bold decisions, and had been made possible by tens of thousands of people who never left Earth.

Other than the astronauts, those most directly involved in the Apollo flights were the personnel at MSC who held and managed those fragile, invisible, extended lifelines to the command service module (CSM) and the lunar excursion module (LEM). (The LEM later became known simply as the lunar module (LM) after NASA's associate administrator for Manned Space Flight, George Mueller, protested that "excursion" in the title sounded a bit frivolous.) In coordination with Goddard Space Flight Center, Mission Control linked the spacecraft to its launch and recovery crews; to a worldwide tracking and communications network; to elements of the technical and scientific personnel at every NASA center; to engineers and specialists at Kennedy Space Center, Marshall Space Flight Center, MSC, and other NASA centers as needed; and to a host of contractor engineers scattered around the United States and the world. The Department of Defense (DoD) supported flight operations in staffing and maintaining the tracking and communications network, in the operation of recovery fleets, and in the deployment of medical and rescue forces. The National Weather Service and the National Oceanic and Atmospheric Administration constantly monitored weather and ocean conditions for launch, flight and recovery operations. The National Laboratories, particularly the Los Alamos Laboratory, provided support for the development and operation of lunar surface experiments. For every astronaut in space, there were many thousands of persons on duty on Earth.

The Apollo program included 11 piloted missions: 9 went to the vicinity of the Moon, and 6 of those landed men on the Moon. The first manned Apollo flight, an Earth-orbital mission lofted by the Saturn IB, flew on October 11, 1968, only 5 days after NASA Administrator James Webb retired and relinquished his duties to Thomas Paine, who became the Acting Administrator. In December 1968, astronauts orbited the Moon; in March 1969, rendezvous and docking procedures were checked in an Earth orbit; in May, Apollo 10 tested equipment and procedures in a lunar orbit and in July, NASA achieved John Kennedy's goal of landing men on the Moon and returning them safely to Earth. There followed in November another, more extended, sojourn on the lunar surface. Then Apollo 13, the only Apollo flight of 1970, failed in its mission but succeeded in returning its passengers safely to Earth.[3] Not only had the design and engineering of machines capable of taking humans into space evolved over time, but so too did the philosophy and procedures for operating those machines in a space environment.

In 1961, when manned lunar flights were being seriously debated, Max Faget recalled that "the basic understanding of the venture was quite primitive." A ship returning from a lunar voyage faced a much more difficult injection into the Earth's atmosphere than did one in Earth orbit. It would be traveling much faster. It had to hit the Earth's atmosphere at the right angle. Too shallow an angle and the vehicle might "skip" off the Earth's atmosphere; too steep an approach would result in certain incineration. Moreover, the human body's adaptation to space might be different from adaptation to Earth orbit. Communications and control over vastly greater distances than Earth orbit were untested. The unknown weighed far more heavily than the known.[4] That, of course, is precisely what made the enterprise so challenging and exciting.

Faget pointed out that the decision to land a vehicle on the Moon from lunar orbit had a major impact on the design and construction of Apollo. Lunar rendezvous meant that Apollo would require two spacecraft: a command and service module for the flight to lunar orbit and back and a separate lunar module for descent to the surface of the Moon and return to lunar orbit rendezvous with the command ship. Moreover the lunar orbit decision markedly affected operational techniques.[5]

Eugene F. Kranz, who served as Chief of the Flight Control Division at MSC throughout the Apollo flights, reconstructed the progression in flight operations from

NASA and DoD — Partners in Recovery
by Jerome B. Hammack, Chief,
Landing and Recovery Division for the Apollo Missions

Early on, NASA decided to have water landings for space capsules (capsules in the early days, then spacecraft as we became more sophisticated) both because water would provide a softer landing and Earth is more water than land. But who was going to recover the capsule? The Navy had most of the ships and the Air Force, and indeed the Army, to assist in this vital part of the mission.

It did not take much persuasion by NASA to get the DoD to become a partner in this vital area of space missions. As things evolved, the DoD set up a single point of contact (the commander of Patrick Air Force Base) through which NASA would levy recovery requirements for each mission. For recovery activities, I was his NASA counterpart. My division—the Landing and Recovery Division (LRD)—was composed of about 100 people, most of whom were engineers. We developed flotation collars and locator beacons, coordinated various recovery hardware on the development of the capsule, and—most important—worked out the mission operations recovery phase of the mission. That phase included training the astronauts in a tank and in open water. The open water part of the training was the most fun. LRD procured its own vessel (an LSD) from the Army, modified it with a handsome bridge, and "sailed" out into the Gulf of Mexico. After putting the astronauts in a capsule in open water, the flotation collar would be deployed, and the helicopters would fly in to recover the astronauts from the side of the capsule and hoist them up into the helicopter. Then a specially designed davit crane would lift the capsule from the water onto the deck of the ship. After several such exercises, the good ship "Retriever," as it was called, would return to port trailing many fishing lines.

The DoD requisitioned ships and aircraft from line units and assembled a recovery task force. In the early days, a typical recovery task force consisted of four ships and several dozen aircraft: helicopter and fixed-wing. The primary recovery ship (usually an aircraft carrier) would be stationed at the primary landing point and three secondary landing points were covered by other type ships (such as destroyers, minesweepers, escort ships). The aircraft would be uprange and downrange of the primary landing point and at contingency landing points throughout the world. The ship requirements were passed to two Navy commanders—one in the Atlantic and one in the Pacific—who each led a Commander Task Force (CTF). The Atlantic unit was CTF-140 and the Pacific unit was CTF-130. The commanders were usually two-star admirals with collateral duties. (For example, the CTF-130 commander was also the commander of Pearl Harbor Naval Station.) Each commander had a staff of officers to plan the support details. Aircraft search requirements were passed to the Air Force Rescue Command where search and rescue aircraft such as the C-130 were assigned.

The ships would embark prior to lift-off in order to be at their assigned stations at the beginning of the mission. Each ship carried its company of officers and crew as well as the LRD recovery engineers and coordinators. The LRD group was responsible for training the ship's crew and briefing them on mission details and characteristics of the capsule—especially any hazards such as the hypergolics and other toxic fluids. We were a good team filled with life and good humor (I was often the object of the Navy pilots' high-spirited schemes) and with the importance of the work we were doing.

One of the big concerns surrounding early Apollo lunar flight recoveries was the fear of contaminating the Earth. Some scientists feared the astronaut crew would bring back pathogens from the lunar surface and pushed for an isolation system. Although the chance for something like that was remote, given the hostile and sterile environment of the Moon, no one came forward to say it could not be. So the plan was to pick up the crew in the capsule, transport it to the carrier deck with the crew inside, and then have them walk through a tunnel into a mobile quarantine facility (MQF). (The MQF was a highly modified Airstream trailer that supposedly would contain any lunar pathogens.) The crew would remain in the MQF until the carrier docked and the capsule was transported (by Air Force cargo airplane) to Houston and placed alongside an elaborate lunar receiving laboratory. The astronauts would continue to live in the MQF for several days to make sure they did not develop any diseases and that no lunar pathogens were present.

However, sometime before the first Apollo lunar flight, the scientists asked that the capsule air vents be closed after landing. We (LRD) objected to sealing up the crew in the moist, hot conditions of the south Pacific. In fact, LRD personnel were already concerned about lifting the capsule with the crew inside—concerned about a possible crane malfunction. For the safety of the crew, we proposed that they emerge from the capsule in the usual manner after splashdown, be scrubbed down with various disinfectant solutions, and put in a quarantine area in the helicopter. After landing on the carrier, the crew would then walk through the tunnel to the MQF. This method was finally approved. After several missions, it became apparent that lunar pathogens were not a problem, and the MQF procedure was removed from the recovery plan.

Ironically, toward the end of the Apollo program—toward the end of water landings—shipboard cranes were so improved that the last recovery operations used the crane to lift the capsule with the astronauts inside. Landing points had also become more precise. Recovery ships were generally in such close proximity to splashdown that little ship maneuvering was necessary.

Mercury to Apollo. He described Mercury operations (where he served as head of the Flight Operations Section in the Flight Control Operations Branch under John Hodge) as a "part-time" business. "The thought processes [for Mercury] were closely attuned to conventional aircraft, that is they were five-mile-a-minute thought processes." Operations people spent perhaps 3 weeks planning a Mercury mission. Mission rules and pilot operating procedures were contained in a 10- to 12-page pilot's handbook similar to that used for a military aircraft mission. The approach to Mercury was simplistic. Spaceflight operations were novel, and operators were novices. First Mercury and then Gemini flight experiences provided critical training for Apollo flights. Operating teams learned particularly that space was a vastly different environment, that part-time operations would not work, and that flight planning, training, preparation, and new organizational structures and greatly broadened support bases must be developed.[6]

John Hodge, Assistant Chief for Flight Control, agreed that the entire concept of flight operations was being constructed out of "whole cloth." But the conceptual design of the Mission Control Center and the basic principles of Apollo operations were completed even before Alan Shepard made the first suborbital flight on a Redstone rocket in May 1961.[7] Flight operations required a great deal of foresight and a lot of learning by doing.

Kranz's association with flight began at a relatively early age and covered the full spectrum of NASA history from Mercury through the Shuttle. During World War II, his mother ran a boardinghouse located close to a USO (United Services Organization) which attracted a continual stream of transient military types. One of these, he remembered, was Billy Huffman, a combat photographer who flew numerous Ruhr bombing missions; and another was Rinehart Brandt who flew in the Battle of the Coral Sea among other engagements. Kranz developed a keen interest in flying and spent his free time around Franklin Field, Ohio. In high school he wrote his thesis on interplanetary flight and then attended Parks College of St. Louis University where he received a degree in aeronautical engineering. After a time as a test pilot with McDonnell Aircraft, he entered the Air Force near the close of the Korean War, spending time at Lackland, Spence, Laughlin, and Williams Air Force Bases, before a 15-month Asian tour with the 13th Air Task Force "showing the flag." When General Curtis LeMay decided that the Air Force did not need anymore fighter pilots and scheduled Kranz for "tanker" school, he opted to return to the more challenging and exciting life as a McDonnell flight test engineer.[8]

Kranz, in Formosa when the Soviets launched Sputnik, was indelibly impressed. The Soviets had it and the United States did not! When the Space Task Group was formed, Chris C. Critzos, who became Christopher C. Kraft's executive assistant in the Flight Operations Division, encouraged Kranz to join them. Gene Kranz said that his wife also encouraged him, thinking that their family life would become more stable and that he could also enroll in school in Virginia for graduate study.[9] So the Kranz family went to Virginia, and in short order moved to Houston.

He became personally involved in every Mercury, Gemini, and Apollo flight. As programs shifted from Mercury to Gemini to Apollo, operations management became complex and deeply layered. "We applied the 'new knowledge' obtained from Mercury on Gemini," he said. The longer duration Gemini flights required far more intensive and sophisticated flight planning and preparation. Operations were now geared to a real-time, one-on-one interface with the astronauts. Flight control teams stood mission "watches." Flight directors began to develop flight "gouge" sheets, which established responses for given conditions and situations. Ed Nieman compiled the information into a formal systems handbook for flight operations. Finally, about the time of Gemini flights 6 and 7, flight controllers began to address the problem of malfunction procedures (that is, the development of conditioned responses to difficulties). The very critical problem-solving function during flight operations began to become systematized.[10] Spaceflight operations largely involved real-time (instant) problem solving.

For example, during the flight of Gemini 8, the vehicle began a rolling motion shortly after a redocking maneuver and as it passed out of contact with the ground stations. Assuming that the Agena rocket rather than the Gemini spacecraft was at fault, flight controllers ordered a shutdown of the attitude control systems which only accelerated the motion. Then, when ground control decided to separate the two vehicles, "everything went to hell in a handbag." The point was we had made a "100 percent wrong call." That taught us, among other things, that problems with the system needed to be fully resolved before flight, that all malfunction procedures needed to be carefully reviewed, and that the flight operations teams and astronauts required intensive training in malfunction procedures. In the Gemini 8 case, close attention to mission rules, reliance on thought processes and reactions ingrained by practice and simulation, plus (John Hodge thought) some heroic piloting by Neil Armstrong resulted in stabilizing the vehicle and a safe return. Overall, although flight remained a continual learning process, Gemini experiences generated confidence in the equipment and in operations procedures.[11]

Max Faget agreed that Gemini was indispensable in developing the flight control techniques and procedures necessary for Apollo orbital rendezvous. Mercury and Gemini flight experiences defined the general philosophy of the interplay between the Mission Control Center in Houston and the astronauts in the spacecraft, and established the flight interrelationship between the NASA operating teams, hardware contractors, and contractor flight controllers.[12] By the time Apollo 8 rolled out on the launch pad, flight operations, while always a learning process, had sharpened and improved in comparison to early Mercury and Gemini operations.

Although the flight operations organization retained its general characteristics typical of Mercury days, that is with a Flight Control Operations Branch, an Operational Facilities Branch, and a Mission Control Center Branch reporting to the Chief (or Director) of the Flight Operations Division, the depth of the organization expanded rapidly during Gemini flights and in anticipation of Apollo, and the function of the branches or sections became more definitive. A brief comparison of the organization charts characteristic respectively of Mercury, Gemini, and Apollo flight operations (figures 9 to 11) depicts better than a lengthy narrative description the changing complexion of the operating systems. The organizational changes were actually much more fluid than the static tables indicate and, as characteristic of MSC, there were many relationships and semiformal structures that simply defy charting.

Notably, throughout most of the operational phases of Mercury, Gemini, and Apollo, irrespective of what the organizational charts suggested, the same lead persons were doing much the same job they did from the beginning. Chris Kraft, who began as assistant to Charles W. Mathews (whom Dennis Fielder described as the "grandfather" of flight operations) in the Flight Operations Division, became chief of the division in 1962, and John Hodge moved to the position of assistant chief. Under a reorganization in 1964, Kraft became Assistant Director of MSC for Flight Operations and Hodge became Chief of the Flight Control Division. When Hodge moved to Assistant Chief of Flight Control and then to head the Flight Control Division, Gene Kranz, who had been Hodge's assistant in the Flight Operations Branch, replaced him. Concurrently, Dennis Fielder headed the Operational Facilities Branch, and Tecwyn Roberts was head of the Mission Control Center Branch. Interestingly, but for Kraft and Kranz, flight operations leadership relied heavily on the Canadian AVRO contingent (Fielder, Roberts and Hodge). During much of this time, of course, the Gemini Program Office was headed by James A. Chamberlin who led the movement of the AVRO engineers from Canada to NASA.

Although the list of "pioneers" in flight operations is too lengthy to fully develop and can be gleaned in part from the various Flight Control Division organizational tables, there was a remarkable continuity in the ranks. Jerry Brewer, for example, described by Fielder as a "dynamic personality" and very management-oriented, helped design long-term ground support systems. Robert F. Thompson, who headed the Shuttle Program Office, contributed significantly to the design of the recovery system. Bill Boyer and Howard Kyle helped develop the worldwide and real-time communications systems. Howard W. Tindall coordinated data from all divisions for the Apollo program. Much of that data came to him from the Mission Planning and Analysis Division where John Mayer, whom Chris Kraft referred to as "Mr. Mission Analysis," presided. Glynn Lunney, Clifford E. Charlesworth, John S. Llewellyn (all in flight dynamics), and Jerry Hammack, who moved from the Gemini Project Office where he served as Deputy Manager of Vehicles and Missions to head the Landing and Recovery Division for the Apollo flights, were among those who "cut their teeth" on Mercury and Gemini before tackling Apollo. By the time Apollo was ready to fly, MSC had become an operations-oriented organization with three directorates (Medical Research and Operations, Flight Crew Operations, and Flight Operations) supporting the Apollo flights (figure 12).

The state of readiness for Apollo operations rested heavily on Mercury and Gemini experiences. Those experiences, however, could not fully prepare anyone for Apollo flight. Apollo would go beyond the Moon and out of sight and sound of any point on Earth. It carried with it a two-stage space vehicle designed to land on the Moon, separate, and return to a rendezvous with the Apollo command module. Those who flew in the LM, unlike those who flew in Mercury, Gemini, or the Apollo command module, could not return directly to Earth in their craft.

Unmanned Apollo test flights resumed in November 1967, about one year following the AS-204 fire, and continued through 1968 when finally, in October, astronauts Walter M. Schirra, Jr., Donn F. Eisele, and R. Walter Cunningham flew the Apollo 7 command module on an Earth-orbital mission following a launch on a Saturn IB rocket.[13]

As 1968 neared its close, there had as yet been no manned flight tests of either the Saturn V rocket scheduled to launch Apollo to the Moon or of the capsule that would bring astronauts from the command module orbiting the Moon to the lunar surface. This was the year when Robert F. Kennedy and Martin Luther King both fell to assassin's bullets and when race riots erupted in every major city. President Lyndon B. Johnson said he would not seek reelection. Production of the LM was seriously behind schedule and NASA faced declining budgets. Remarkably, the major news event of the year had to do with a space voyage of exploration.[14]

Apollo flight plans called for carefully staged flight increments which would first test the Saturn V in Earth orbit; and in a following flight, test the LM in Earth orbit. Similar test flights would be flown in lunar orbit before a lunar landing was attempted. Nevertheless,

Flight Operations Division, 1962

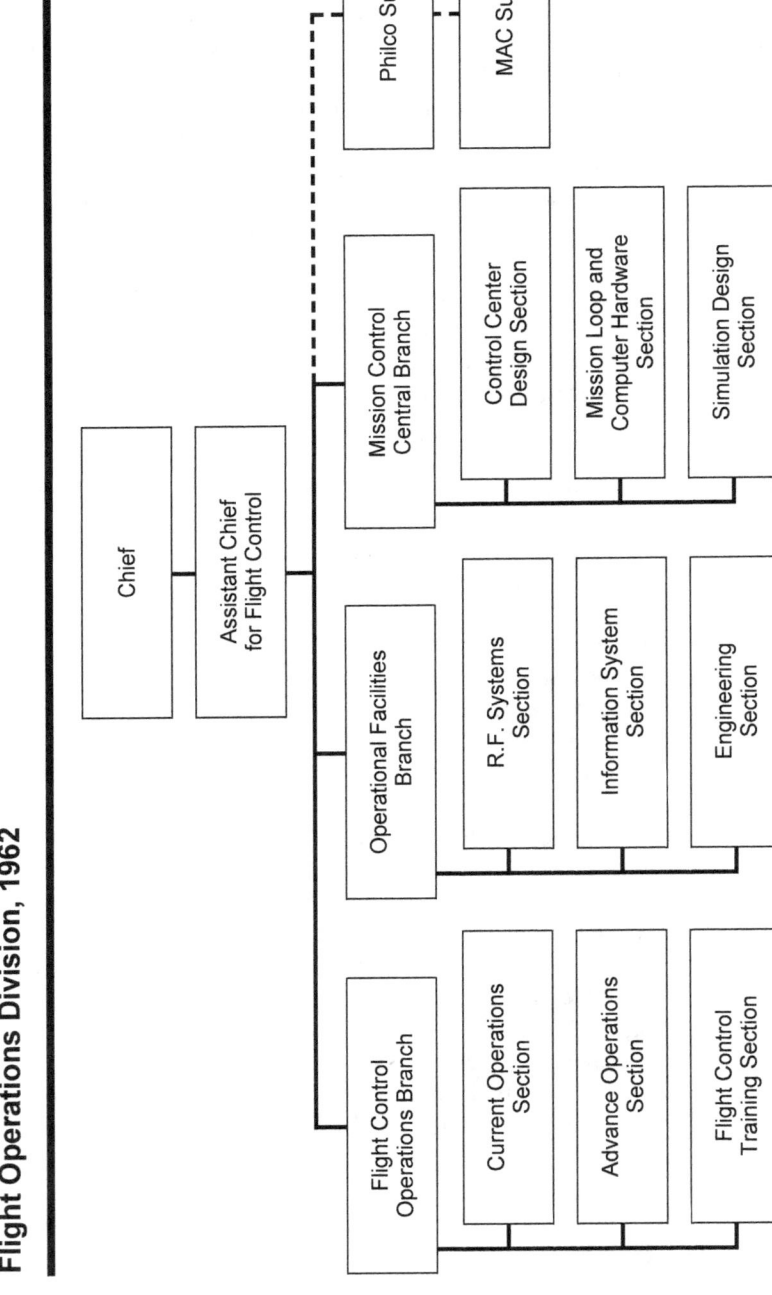

FIGURE 9. *Flight Operations Division Organization as of December 1962*

Flight Operations Directorate, 1964

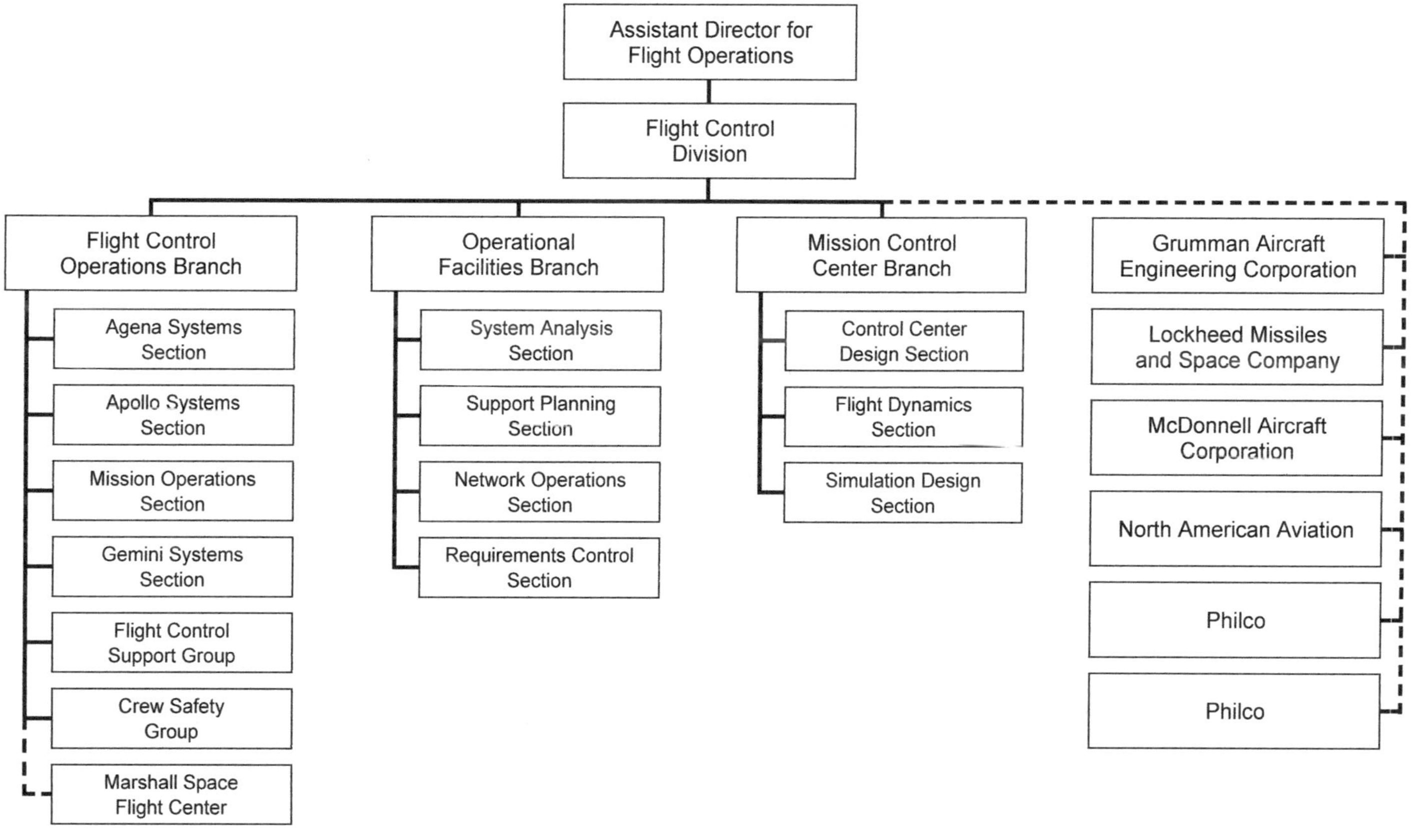

FIGURE 10. *Flight Operations Directorate Organization as of January 1964*

Assistant Director for Flight Operations
- **Flight Control Division**
  - **Flight Control Operations Branch**
    - Agena Systems Section
    - Apollo Systems Section
    - Mission Operations Section
    - Gemini Systems Section
    - Flight Control Support Group
    - Crew Safety Group
    - Marshall Space Flight Center
  - **Operational Facilities Branch**
    - System Analysis Section
    - Support Planning Section
    - Network Operations Section
    - Requirements Control Section
  - **Mission Control Center Branch**
    - Control Center Design Section
    - Flight Dynamics Section
    - Simulation Design Section

Contractors:
- Grumman Aircraft Engineering Corporation
- Lockheed Missiles and Space Company
- McDonnell Aircraft Corporation
- North American Aviation
- Philco
- Philco

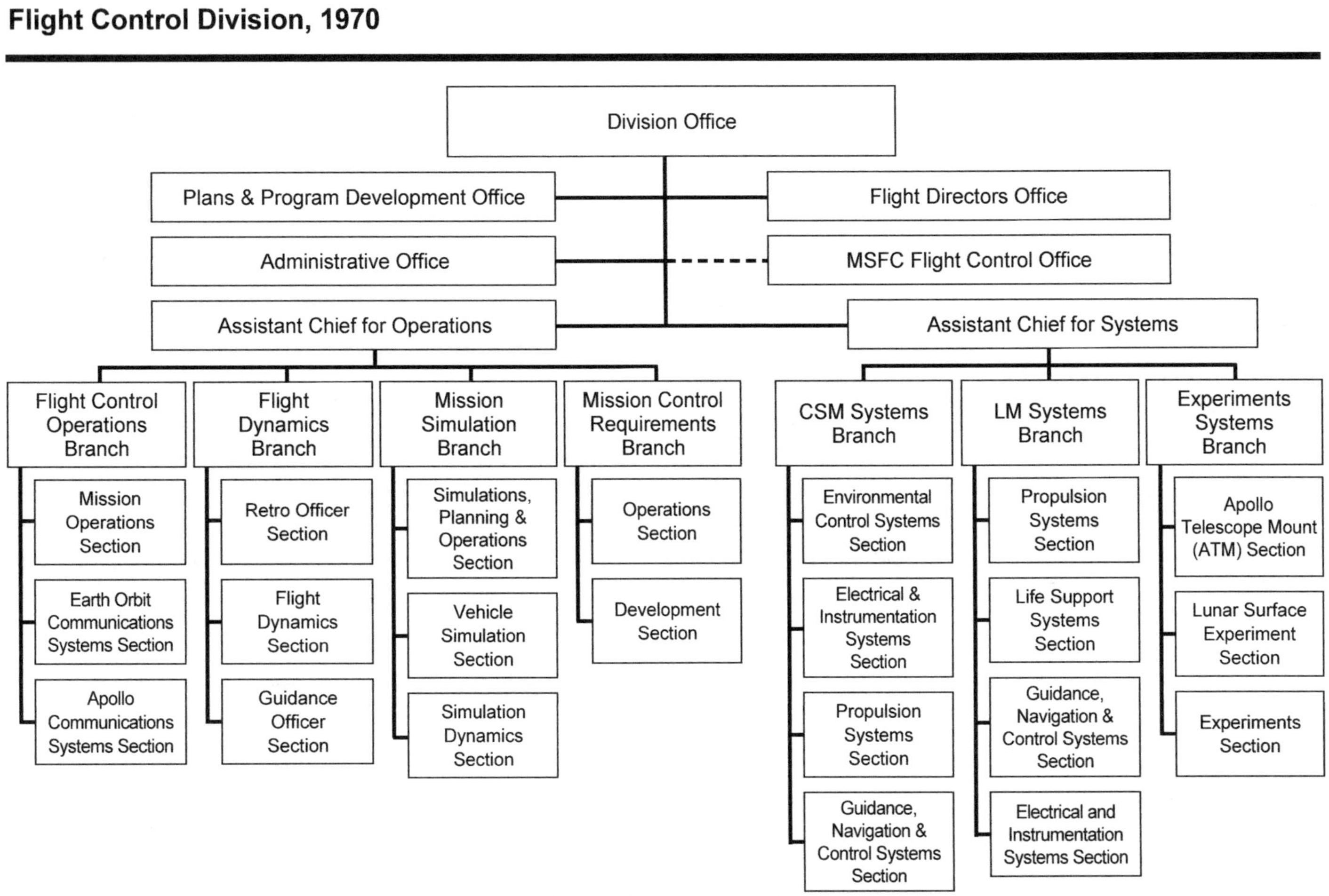

FIGURE 11. Flight Control Division Organization as of March 1970

George Low, manager of the Apollo Program at MSC, set the events in motion which resulted in leap-frogging or consolidating manned Apollo test flights into one bold lunar orbital mission. Low and Bob Gilruth first considered the possibilities in July, and then broadened their discussion to include Chris Kraft and Deke Slayton. On August 7, 1968, Low asked Chris Kraft, Director of Flight Operations, to develop a flight plan for an Apollo lunar mission. Frank Borman, who would fly in Apollo 8 to the Moon, recalled that he, Bill Tindall and Chris Kraft worked out a feasible flight plan in one afternoon.[15]

Low, with Carroll Bolender, Scott Simpkinson, and Owen Morris, then flew to Kennedy Space Center on August 8 to discuss a manned lunar flight with Apollo Program Director Sam Phillips, Kennedy Director Kurt Debus, and others. MSC Director Robert Gilruth endorsed the idea on August 9. That same day he, Low, Kraft, and Deke Slayton flew to Huntsville to meet with Kurt Debus and Rocco A. Petrone from Kennedy Space Center, Sam Phillips and George Hage from Headquarters, and Wernher von Braun, Eberhard Rees, Ludie G. Richard, and Lee James of the Marshall Space Flight Center. That group, representing NASA's manned spaceflight "field" centers, endorsed advancing the schedule for a manned lunar orbital flight. Next, on August 14, a representative group from the manned flight centers and their contractor representatives met with Deputy Administrator Thomas O. Paine. That body, with Paine, ratified the proposal to convert the Apollo 8 mission to a lunar flight.[16]

Apollo 8 was scheduled to be the first manned Apollo launch by a Saturn V, and it was originally scheduled to test the manned lunar module in an Earth orbit. The idea for changing it to a lunar flight originated with Low, was developed and refined cooperatively by managers from Kennedy Space Center, Marshall Space Flight Center, and MSC, and then presented to Headquarters for approval. During the initial discussions, Administrator Webb and George Mueller were both in Vienna. When informed by telephone on August 14, Mueller was distinctly cool to the idea, and Webb was "shocked by the audacity of the proposal" and inclined to say no; but after Paine cabled a detailed explanation, Webb instructed him to proceed with lunar flight plans but not to publicly divulge the plans.[17] The inception of the Apollo 8 lunar flight plan provided an interesting example of Headquarters-center relations and of the essentially cooperative or collegial style of NASA management.

Meanwhile, Apollo 7 (which, incidentally, carried NASA's first manned in-flight television camera providing live coverage to the ground) made an eminently successful flight. MSC and other NASA units continued to study the lunar flight idea. George Mueller met with the Apollo Executives in early November and received their strong endorsement for a manned lunar orbital flight. Thomas Paine, now Acting Administrator following James Webb's retirement, listened to presentations from Sam Phillips, Lee B. James (the Saturn V manager at Huntsville), George Low, Chris Kraft and Rocco Petrone. He then received Gerald Truszynski's affirmation that the tracking network would be ready, and obtained mission support from DoD before approving the Apollo 8 lunar flight for December 21, 1968.[18]

That decision began NASA's "assault on the Moon." Cliff Charlesworth, Flight Director for Apollo 8, recalled that "Apollo 8 was the highlight of the Apollo program. The commitment to do the flight took a lot of courage." A manned lunar-ready (Block II) command module had only flown once (Apollo 7). A Saturn V had never been used to boost a manned vehicle into space. The "deep space" voice communication system had obviously

Suddenly, Tomorrow Came . . .

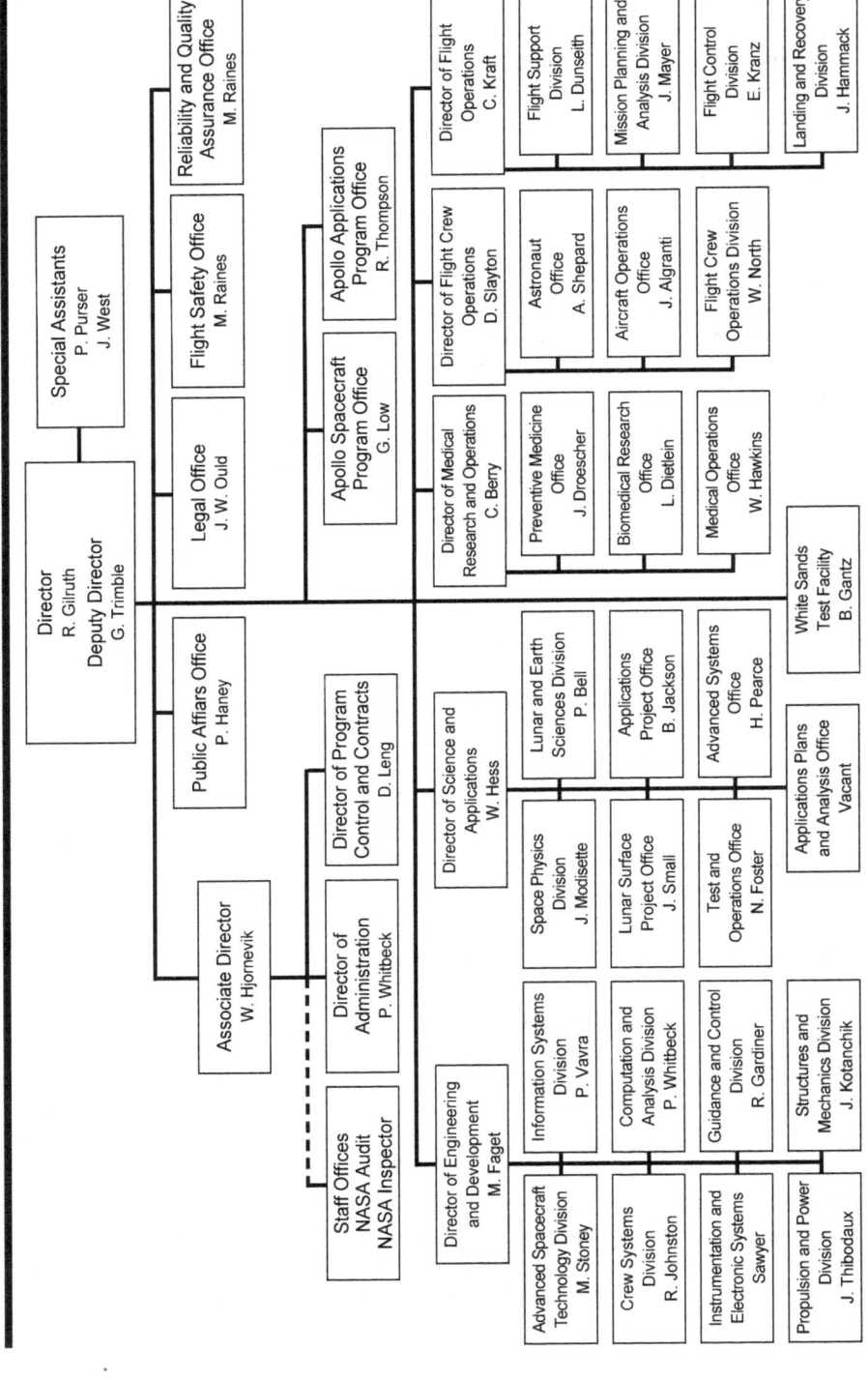

FIGURE 12. *Operations-oriented Divisions as of June 1968*

The Apollo 8 lunar flight marked a giant step toward NASA's lunar landing. Here, Clifford E. Charlesworth, the Apollo 8 "Green Team" Flight Director, is seated at his console in the Mission Control Center.

never been used. Moreover, directing a vehicle from one moving celestial body to another involved what Ron Berry called "a fascinating (but difficult and complex) bit of math, physics and geometry." But Owen Morris and others, were "targeting junkies."[19] And some very helpful data had been secured within the past few years from unmanned probes to the lunar surface.

Although there had been a number of earlier Ranger misadventures, Ranger 7 sent the first high-resolution photographs of the lunar surface back to Earth in 1964. The success of Rangers 7, 8, and 9, from July 1964 through March 1965, provided data on the size and distribution of lunar craters and boulders. In May 1966, Surveyor 1 successfully soft-landed on the Moon, confirmed the bearing strength of the lunar soil, and returned thousands of images. By February 1968, five of seven Surveyors successfully landed and confirmed the suitability of several Apollo landing sites. During 1966 and 1967, five Lunar Orbiter missions carrying high-resolution and wide-angle cameras helped in mapping 97 percent of the Moon's surface. While the Ranger, Surveyor, and Lunar Orbiter unmanned flights provided hard data and encouraged confidence, there were still nagging uncertainties about a lunar voyage. Many speculated that simply leaving the Earth's gravity could cause serious illness. Others worried that returning astronauts could return with contaminants that might endanger human life and Earth's ecology. In spite of these concerns, Leland Atwood's (North American) endorsement of the mission at the Apollo Executives meeting probably reflected the views of most: "This is what we came to the party for."[20]

Apollo 8 flew as scheduled on December 21, 1968, carrying Frank Borman, James A. Lovell, Jr., and William A. Anders on 10 orbits around the Moon. The entire mission was conceived, planned and "trained for" in a 6-month time frame. Gerald Griffin, one of the mission flight control directors, remembers that Apollo 8 "was kind of scary. It pushed the system faster. It showed an ability to take risks." As for Kranz, at first the Apollo 8 lunar flight decision "irked the hell" out of him because it confused and set aside the careful planning and training for the Apollo flight schedule. He complained to Chris Kraft, but he wasn't asked to vote, he said. But the Apollo 8 lunar decision, Kranz added, involved the "management of risk." It meant, in effect, taking a greater risk then in order to reduce risks in later flights.[21] Most of those who returned to Johnson Space Center in July 1989 to celebrate the 20th anniversary of the lunar landing of July 1969 concurred that Apollo 8 made landing in that decade possible.

But Apollo 8 was an uneasy flight. Charlesworth remembers, preparations for the flight were difficult, "but we launched." The astronauts made two orbits of Earth before burning the S-IVB for translunar trajectory, but the trip out "was not uneventful." We decided to test the engine on the way out to be sure it was working right "and it did not work right!" Everyone in Mission Control rushed to deal with the engine. As tension mounted, one of the astronauts reported being ill. A "sick astronaut committee," which included Gerry Griffin, turned to deal with this problem, but feared that indeed the sickness might be caused simply by leaving Earth's gravity, and thus would be incurable. Concentrating wholly on the engine problem, Charlesworth thought the astronaut should "be sick and be quiet about it."[22]

MSC Director Robert R. Gilruth and Flight Operations Director Christopher C. Kraft monitor Apollo 9's Earth-orbital mission of March 1969.

Charlesworth and everyone in Mission Control worried that the engine either might not fire or might not fire correctly for the lunar orbital insertion. If it did not burn correctly, a return to Earth might be impossible. Tension was "thick enough to cut with a knife" Chris Kraft remembers. But the lunar orbit insertion burn worked. Then, Mission Control personnel were "spooked" when Apollo disappeared behind the Moon. But it came back around and made 10 "hard work" orbits around the Moon, before a successful engine burn headed the Apollo home to Earth and a safe landing.[23] Apollo 8 was one of the most significant lunar flights. It resolved many of the unknowns and accelerated the entire lunar landing effort.

It created "an astounding international awakening" commented Owen Morris, who headed the Lunar Module Engineering Division Management team under George Low. Perhaps it was in the understanding that humankind stood closer than ever before to the unknown and the creative processes of the heavens and the Earth, that the astronauts aboard Apollo 8, led by Frank Borman, elected to read from lunar orbit on Christmas eve the first 10 verses of Genesis: "In the beginning God created the heaven and the earth. And the earth was without form, and void . . . " God created from that the land and the sea, and darkness and light. "... and God saw that it was good."[24] The text might have included that one day men would walk upon other heavenly bodies.

While it received less public attention than the lunar flights, Apollo 9, which flew an Earth-orbital mission in March 1969, was one of the most interesting of all Apollo flights. Apollo 9 made the first flight test of a manned LM. Kranz, who assigned flight directors, assigned himself to those flights closely associated with LM operations, including Apollo 5 (the first unmanned LM flight), 7, 9, 11, and 13. Apollo 9 was the first time that flight controllers operated a dual system—that is, one separate flight operation with the command module and concurrently another with the LM. The communications load, workload, and problem trouble-shooting load now doubled.[25]

The nice thing about Apollo 9 was that there was a lot of free time to experiment and get acquainted with the systems. During Apollo 9 operations, the Mission Control teams established procedures for use of the LM as a lifeboat, tested engine burns, and tried lowering every function of the command module and the LM to their lowest possible level—malfunction procedures which later proved invaluable during the Apollo 13 flight. Kranz felt intrigued by the keen sense of competition between North American contractor representatives and flight controllers and Grumman contractor representatives and flight controllers, and by the general level of excitement.[26] With a lunar landing now a tangible reality, the excitement and energy level of all those associated with NASA and the Apollo programs rose precipitously.

The LM was the first vehicle built for humans for nonterrestrial use. There was little engineering and design history to work from, other than that provided by Earth-orbital flights and unmanned lunar vehicles such as Ranger and Surveyor. Surveyor, incidentally, made significant design contributions, especially to the landing gear and Doppler radar systems. Using existing knowledge, managing engineers completed the Statement of Work for LM in June 1962. The Request for Proposals, released in July, produced nine proposals. North American, which held the primary contract for the Apollo spacecraft, was precluded from the competition—over the company's strong objections. McDonnell Aircraft chose not to enter the competition. After some delays and reviews precipitated in part by the President's Science Advisory Committee, NASA awarded the LM contract to Grumman Aircraft Engineering Corporation of Bethpage, New York. The cost-plus-fixed-fee contract for $387.9 million (including the Grumman fee of slightly over $25 million) was signed by NASA and Grumman on January 14, 1963. Constant changes engendered in part by the experimental and innovative nature of the product being manufactured, changes in specifications, production delays, subcontracting problems, and cost overruns resulted in costs reaching $1.42 billion.[27] Perhaps because of the remarkable achievements of the LM, those cost overruns failed to provoke a public or congressional protest.

Engineering guidelines provided that although there would be no provision for in-flight repair, redundancy (or backup systems) would be sufficient to assure that "no single failure can endanger crew safety." In addition, low weight was an "ultimate premium" in design and construction decisions. Each pound of inert weight lowered to the lunar surface and returned to the command ship required an additional 3.25 pounds of propellant. Each pound of LM weight then, added 4.25 pounds to the payload of the Saturn-Apollo system, with commensurate fuel requirements for the Saturn V (resulting in approximately 50 pounds of added weight for each pound of inert weight lowered to the Moon).[28]

Although LM construction encountered many delays, including that caused by the AS-204 fire, by July 1967 Grumman announced that the vehicle would soon be assembled and ready for flight testing. Grumman, as did North American for Apollo, served as the primary contractor for the LM and used many subcontractors. Space Technology Laboratories, Inc. (STL) and Rocketdyne both worked on rocket engines with a throttle control system; and Manned Spacecraft Managers, in a rare reversal of a primary contractor's decision, selected STL for the throttleable LM descent engine. Bell Aerosystems produced the ascent engine, and Hamilton Standard Division of United Aircraft developed the environmental control systems. Although NASA substituted batteries for LM electrical power, Pratt and Whitney (a division of United Aircraft Corporation) developed the electrical power fuel cells (then a very advanced technology). RCA produced the rendezvous and landing radar systems. Other Grumman subcontractors included TRW/STL, the Allison Division of General Motors, Radiation, Inc., Marquardt Corporation, General Precision, Inc., and the Garrett Corporation.[29]

Low strengthened the coordination of the LM and command module projects in 1967 by appointing a resident manager from his Apollo Program Office at MSC to North American. Wilbur H. Gray became the Resident Manager to North American Aviation in Downey, California. Kenneth S. Kleinknecht was made Manager of the Command and Service Module for the Apollo Spacecraft Project Office under Low at MSC, and Dr. William A. Lee, formerly an Assistant Project Manager, received responsibility for the LM, with specific authority over Grumman's design, development, and fabrication of the module.[30]

Another unique "tool" of the lunar missions was the specially designed "extra-vehicular mobility unit," or the astronauts' space suit and battery-powered backpack, which provided a cooled and revitalized atmosphere in which to live. Although obviously similar to and drawing upon EVA experiences from Gemini, lunar suits were much more complex because of the enhanced active cooling system required. Production of a suitable lunar suit proved correspondingly difficult. An initial production agreement between Hamilton Standard and International Latex Corporation failed to produce a suitable lunar suit. While testing continued with the Gemini suits, new competition for an Apollo space suit between Hamilton Standard (with B.F. Goodrich), David Clark Company (which developed the Gemini suit), and International Latex resulted in Hamilton Standard retaining the "backpack" contract while International Latex developed the suit on an independent contract—and the MSC provided systems integration.[31] In a word, a lunar mission involved many untried tools and operational techniques.

Apollo 10, launched May 18, 1969, carried humankind one step closer to the Moon. The command module named *Charlie Brown* and the LM called *Snoopy* completed 31 lunar orbits and successfully demonstrated crew support systems and operational procedures aboard the command and lunar modules. Eugene A. Cernan, who with Thomas P. Stafford and John W. Young flew Apollo 10 and accomplished a separation and rendezvous with the LM while in orbit, reflected that the greatest thing he brought back from his flight was simply the "feeling and the majesty" of it all. Earth, he said, "is overpoweringly beautiful."[32]

On July 16, Apollo 11 left Earth for a mission to land men on the Moon. The two previous lunar orbital flights and the imminent lunar landing awakened among the astronauts and many of those who participated in the programs, and among the general public who

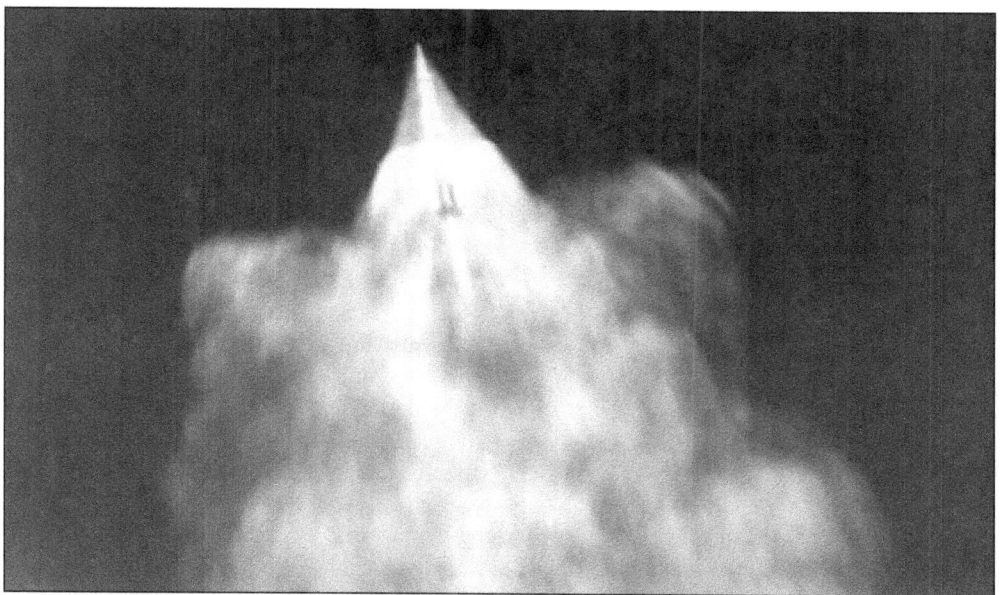

This shot of the Apollo 11 launch taken at 40,000 feet at 9:32 a.m. on July 16, 1969, from Air Force EC-135 captures the surrealistic image of the 7.6 million pound thrust Saturn V first-stage booster just prior to separation of the S-1C stage. Apollo 11 is the first NASA lunar landing mission.

merely observed from a distance, a new perspective of life in the universe. It denoted an awakening perhaps not unlike that triggered by Copernicus's realization that Earth was indeed not the center of the universe, but that Earth was one of many heavenly bodies that revolved about one of perhaps many suns. It was an awakening that Michael Collins shared with the world. Collins, who flew the command module during the Apollo 11 lunar landing, put it rather simply, "A lot of the things we thought were not important—really are!"[33]

The big moment! George M. Low watches a television monitor during the lunar surface EVA of astronauts Neil A. Armstrong and Edwin E. Aldrin, Jr.

For those with Gene Kranz in the Mission Control Center that was an understatement. No thing in lunar flight was unimportant. The training for the Apollo 11 lunar landing mission was particularly rugged. "Nothing we do today (said Kranz referring to the Shuttle missions in 1990) can compare." As crews became more experienced, discipline tightened rather than relaxed. The simulated lunar training missions had almost been clairvoyant, but the mission was fraught with peril and

Suddenly, Tomorrow Came . . .

Tensions remained high throughout the 195-hour Apollo 11 lunar mission. Here Flight Director Clifford E. Charlesworth (center) and Eugene Kranz (to his left) prepare for the change of shifts.

problems and those problems took everyone's mind off the fact that "we were landing on the Moon."[34] Mission Control's job was to solve problems and there were plenty to solve.

Just prior to the lunar landing attempt, Gene Kranz gave a speech to the Mission Control crew that Steve Bales thought "Patton would have been proud of!" Kranz told them that the success of the mission depended on them, and he had every confidence in them. Then he locked the doors to Mission Control and the flight control crew became even more concentrated and intent on their work.[35]

As Apollo made its final orbit of the Moon, the LM, named the "*Eagle*" by its two-man crew Neil Armstrong and Edwin Aldrin, separated nicely from the mother ship on the back side of the Moon. But when the command module cleared the Moon and communications resumed with Mission Control, communications and telemetry between Mission Control and Apollo were bad. At this point Mission Control had about 5 minutes in which to abort a lunar landing. After a careful check of all systems, Kranz decided to continue. Then, with the *Eagle* 4 minutes away from its landing, the crew discovered that the LM's altimeter and velocity gauges were in error. Those problems were corrected. Next, the crew reported a computer alarm. A quick flight control analysis resulted in a "judgment call to continue." Concentration throughout the Mission Control Center was intense. One final abort option existed in the last 27 seconds of the flight. In that brief time, a complete status check through all of the Mission Control desks cleared the *Eagle* for a landing. But even then, as the *Eagle* prepared to set down, Neil Armstrong was forced to

These artist depictions of the lunar surface and a lunar landing reflect a new awareness by Americans of non-terrestrial bodies.

A view of Mission Control Center during Apollo 11 EVA.

override the planned landing program in order to avoid rocky and dangerous terrain at touchdown. But the *Eagle* did land. For a time NASA was not sure just where.[36]

The first real sense Kranz had that the lunar landing had been completed was when he noticed people in the Mission Control viewing room cheering. Suddenly, the flight control team lost concentration. At that moment of stunned realization, Kranz simply stopped thinking.[37] But thought and work resumed quickly as the flight control crew replacement shift came on duty.

President Richard M. Nixon flew to the recovery area in the central Pacific to welcome the Apollo 11 crew upon the completion of the historic mission. President John F. Kennedy's 1961 charge to put an American on the Moon within the decade and return him safely to Earth had been achieved.

Suddenly, Tomorrow Came . . .

Apollo 11's lunar module casts a long shadow upon the Moon and symbolically upon the future of Earth.

Clifford E. Charlesworth replaced Kranz as Flight Control Director for the "lunar surface shift." Already a new problem was on the floor. Mission Control was anxious to take advantage of every spare minute on the Moon's surface. Mission planning included two "activity" options once the *Eagle* landed. One, favored by the medical teams, was for the astronauts to rest and sleep; the other, favored by Mission Control on the premise that the astronauts would be unable to sleep, called for an EVA on the Moon's surface. The debate raged in the control room. Armstrong, on the Moon, put in his two cents worth; "sleep now just doesn't make sense." So the astronauts and Mission Control won. When the EVA did come, Charlesworth was very nervous, and it was a relief to have the astronauts back in their pressurized cabin.[38]

Tension heightened as the *Eagle* prepared for the lift-off from the Moon's surface and the rendezvous with Collins in the Apollo command module. Glynn Lunney took over the lunar orbit return "watch" from Charlesworth. Although he defined the flight control job as one that depended wholly on the kind of decisions that had to be made (and which required instant and programmed responses), Lunney regarded lift-off as something different. There were really no decisions to make. Either the ascent engine worked or it didn't. "We thought we knew how to handle the rendezvous."[39] And of course they did. Apollo 11 brought two men to the surface of the Moon and returned them safely to Earth.

Paradoxically, following that great moment of triumph, some subtle but disturbing crosscurrents developed which would follow NASA for the rest of its days. President John F. Kennedy's 1961 call for America to land a man on the Moon and return him safely to Earth within the decade had been achieved. Narrowly construed, there was nothing more to do. Some Americans, perhaps skeptical, cynical, or disturbed about the concurrent costs of war, welfare, and space, believed that society might benefit more in the short term from war or welfare than from expensive flights in space. Although Lyndon Johnson personally supported NASA from the time of its creation, by 1964 and increasingly through 1968, his administration put pressure on NASA to reduce expenditures lest growing federal expenses for the war in southeast Asia, social programs at home, and space overheat the economy. President Richard M. Nixon continued to cite those budgetary pressures as inflation added to the rising financial crises. Had the commitment to space been too narrowly defined? Was the mission complete? NASA continually tried, but could never satisfactorily explain, "Why go further?" Although projects such as Skylab and the Shuttle were under development, NASA had difficulty justifying those programs and explaining what would be next after Apollo. NASA, to be sure, was almost totally preoccupied with its present mission and successes. Spaceflight, for most NASA personnel and contractors, needed no justification more elaborate or sophisticated than that of Michael Collins: "We are wanderers."[40] But there was more.

No single event in the history of the world has raised the prestige of any country such as the Apollo program has," commented William B. Bergen, President of North American Rockwell. When President Nixon "took off on his trip around the world immediately after Apollo 11 . . . no matter where he went, the main thing in peoples' minds around the world was, this is not President Nixon of the United States. This is President Nixon of that country that put the fellows on the Moon!"[41] The lunar landing of July 1969 gave the American people a shot of adrenaline and a resurgence of faith after a long and bruising succession of real and imagined cold war adversities. NASA, MSC, and personnel in Mission Control were little short of ecstatic. Apollo 11, most thought, was only the beginning.

"What we did after Apollo 11," reflected Gerald Griffin, Flight Control Director for Apollo 12, "was nothing short of fantastic." Problems of a stratospheric dimension literally struck Apollo 12 during launch from the pad at Kennedy Space Center on November 14. Lightning struck the spacecraft during the launch, and although the Saturn V continued flawlessly, the Apollo electrical systems failed. There had been no planning or simulations for such an event. Kraft came to the control room and advised Griffin, "Young man, we don't have to go to the Moon today." When Griffin ran a systems check, John Aaron, then a 25-year-old recent graduate of Northeastern Oklahoma University, asked the astronauts to reset the fuel cell relay switches. When they did, the lights came on again. Aaron then recommended that Apollo make an extra Earth orbit while the regular scheduled lunar orbit check list was reviewed. After the review, all systems were go, and the flight completed its lunar mission in spectacular fashion. As Glynn Lunney said later, "Aaron was the right person at the right time in the right place. Griffin and the entire flight control crew did a great job."[42]

Once the systems check was made Griffin urged the others to "get on with it," and the Saturn-IVB thrusters pushed the vehicles toward the lunar destination. After separation of the lunar craft, Richard F. Gordon piloted the command module *Yankee Clipper* on a total of 34 lunar orbits. The LM, called *Intrepid* by its crew, landed on target in the Ocean of Storms, 182 meters from the Surveyor 3 unmanned spacecraft. Alan L. Bean, the LM pilot, and Charles Conrad, flight commander, spent 31 hours on the Moon's surface, 7 of those in EVA. They walked about collecting samples of rock and photographing the surface—and the Earth as it had never been seen before. They hiked over to the Surveyor, examined it, photographed it, and brought back a camera and smaller parts for study. They collected 34 kilograms (about 75 pounds) of lunar rocks and brought them back to Earth.[43]

Astronaut Joseph P. Allen, who worked the Apollo 12 flight as the Capsule Communicator, had the occasion to scold his son (then only 3 years old) shortly after the Apollo 12 flight: "Why are you dumping dirt on the kitchen floor?" The child replied: "these are Earth rocks, not dirt."[44]

More than any other discovery or phenomenon, the Moon rocks signified the scientific portent of the lunar expeditions. Because of the tremendous emphasis put on performing operations, science appeared to have been relegated somewhat to the back seat of NASA's Apollo program. The stress that developed between the scientific community and the engineering community at the inception of the Mercury program, never fully dissipated through the Gemini and early Apollo flights, despite efforts within and without NASA to obtain a better accord and accommodation. Many scientists thought that spaceflight, with its heavy developmental focus, neglected the sciences and siphoned federal funds from science programs. The apparent imminence of a lunar landing brought into sharp focus within NASA and the scientific community the opportunities and responsibilities commensurate with a lunar expedition. The dialogue (previously mentioned) over how to handle lunar materials and design a Lunar Receiving Laboratory signaled a quickening of scientific involvement in space operations. Until at least 1967 "getting there" reflected the major dynamics of the Apollo program. But what happened once you got there? A flag raised? A salute? Moon rocks? Or more?

The status of science within MSC rose measurably as the understanding of the Apollo program began to go beyond a physical landing on the Moon. In January 1967, Center Director Bob Gilruth removed the old and sometimes ignored Space Science Division and Experiments Program Office from the Engineering and Development Directorate, and upgraded the office to an independent and co-equal Science and Applications Directorate. The new office, Gilruth explained, "reflected the growing significance and responsibilities of the center in these areas" and "will act as a focal point for all MSC elements involved in these programs, and . . . provide the center's point of contact with the scientific community." He appointed Robert O. Piland, who had been managing the Experiment Program Office, as the Deputy Director while searching for a scientist to head the new directorate.[45]

The Science and Applications Directorate comprised a Lunar and Earth Sciences Division which was itself compartmentalized into two segments. John Eggleston, designated as the special assistant to the director, focused on scientific experiments and applications. A mapping sciences branch and a geophysics branch reported to Eggleston. The Lunar Receiving Laboratory, with Joseph V. Piland (Bob Piland's brother) serving as acting manager, was operated by the various branches of the Lunar and Earth Sciences Division. The Geology and Geochemistry Branch had responsibility for the lunar sample laboratory and functions related to astronaut geology training, mission simulation, lunar surface definition, and scientific lunar surface hardware or tools. The Biomedical Branch controlled quarantine, medical and bioscience functions of the Lunar Receiving Laboratory, and the Engineering and Operations Branch provided the detailed operation, planning and program control functions of the laboratory. The Mapping Sciences Branch and the Geophysics Branch reported to Eggleston and retained more of an operations rather than scientific orientation.[46] Although some embittered scientists thought in some respects it was a belated reorganization with only a "scientific flavor," the new directorate facilitated the scientific work which became particularly significant during the flights of Apollo 14 through 17.

On February 17, 1967, Gilruth appointed Dr. Wilmot N. Hess to head the Science and Applications Directorate. Hess came to MSC from Goddard Space Flight Center where he had served as Chief of the Laboratory for Theoretical Studies. Before that he was a nuclear physicist on the teaching and research staff of the University of California and headed the University's Lawrence Radiation Laboratory at Livermore, California, before joining NASA in 1961. His reputation came largely from his work in high-energy nuclear physics, neutron scattering, cosmic ray neutrons, and studies of the Van Allen radiation belts.[47]

Dr. Hess and scientific projects sometimes conflicted with engineering and flight operations objectives. Engineers wanted to be sure their machines could fly to the Moon before they became too concerned about what would happen once they arrived. Operations people wanted their best pilots at the controls of those machines. Scientists began to suspect that the scientist-astronauts recruited earlier for the Apollo missions were being systematically excluded from flights in favor of test pilots. Although Hess and the new Science Directorate helped tilt the last three Apollo flights toward a strong science profile, Apollo continued be an elaborate exercise in flight operations.

There were several significant characteristics of the personnel who comprised the operations teams. As Gerald Griffin observed, "we were a bunch of young people, most of us in our twenties and thirties. We had more responsibility at age 30 than most people will have in a lifetime." Kranz, who headed the Flight Operations Division, was extremely thorough and disciplined thought Griffin, and the Mission Control room always ran in a very businesslike atmosphere.[48]

During the mission, the flight director made all real-time decisions. This unwritten rule seemed threatened, Kranz recalled, when Headquarters began assigning mission directors (such as William C. Sneider, Chester M. Lee, Thomas H. McMullen, and George H. Hage) to the control room during Gemini flights. Some of these on occasion "walked in and tried to take over," but most properly served their role as observers. During one flight, Kraft became extremely upset when the Headquarters Mission Control representative attempted to intervene in a flight director's decision. A "mission directive" from Sam Phillips subsequently made the mission director a broker for broad policy decisions only between Headquarters and the center. Longer-term flight decisions were made by the flight director in consultation with the mission director and other appropriate offices.[49] Mission Control, in actuality, involved a synchronized response by hundreds of operators and managers.

The operating stations in the Mission Control Center physically surrounded the flight director's console. To the right of the flight director, the CapCom (or capsule communicator—always an astronaut) relayed voice information and instructions to the astronauts from Mission Control. A guidance officer (who monitored onboard navigation and computer control systems) collaborated with the flight dynamics officer (FIDO) in planning maneuvers and trajectory. A booster systems engineer; propulsion systems engineer; guidance, navigation and control systems engineer; electrical, environmental and consumables systems engineer; and instrumentation and communications system engineer monitored their respective systems and provided a liaison with contractor and engineer support groups in a myriad of staff support rooms located throughout Building 30 (and beyond, if needed). A ground control officer coordinated tracking and data information with Goddard Space Flight Center, while a computer supervisor had responsibility for Mission Control hardware and software. The flight surgeon monitored the crew's health and provided personal counsel. A space radiation analysis group provided constant readings and recommendations for the surgeon. Mission rules established procedures for solar flares and major radiation phenomena. The public affairs officer provided a continuous commentary and linked the flight to the news media and the public. Brian Duff, who replaced Paul Haney as public affairs officer, over considerable opposition convinced Gilruth that reporters should be admitted into the Mission Control Center during missions for in-flight press conferences. The position reconfirmed NASA's open access policies, but also created a potential public relations problem when things went wrong. A display board, similar to that found in a battleship's "war room" (combat control center) provided constantly updated data and flight positions superimposed on a world chart. All of these information systems and personnel were

Suddenly, Tomorrow Came . . .

An imprint destined to permanently alter human affairs—a simple human footprint on the lunar surface.

immediately responsive to the flight director. Mission rules one through six, Steve Bales said, were that irrespective of what the other rules stated, the flight director "may do whatever is necessary to complete a successful mission."[50]

During the flight of Apollo 13, the flight director and flight control team faced extraordinary crises. Glynn Lunney, who directed the Apollo 11 return flight, recalled Apollo 13, launched April 11, 1970, as the ultimate test in dealing with a problem." The third lunar mission, carrying astronauts James A. Lovell, Jr. (mission commander), John L. Swigert, Jr. (command module pilot), and Fred W. Haise (lunar module pilot) lifted off in something of an already "routine" fashion from the pad at Cape Kennedy. The flight, to be sure, had experienced a preflight problem. John Swigert stepped in to replace Thomas K. Mattingly 24 hours before lift-off, when the flight surgeon determined that Mattingly had been exposed to the measles. Swigert called Mission Control from space to ask someone to mail his tax return for him before the April 15 deadline. And there had been a problem in emptying and refilling one of the oxygen tanks in the service module. Other than that, the trip out was going fine. Things went well until the third day as the craft approached the Moon. Then, Jim Lovell remembered, "I heard a loud hiss-bang." An alarm light came on the control panel. Then two more lights. Then others. "A wave of disappointment swept through the spacecraft. We were in deep trouble."[51]

An oxygen tank in the service module—which affected oxygen, water, and electrical supplies in the command module—exploded. The second of the two oxygen tanks began losing pressure. The command module, which housed the three astronauts, was about to lose

The Flight of Apollo

Millions of people around the world were awakened to a new perspective of life in the universe when astronauts Neil A. Armstrong and Edwin E. Aldrin landed on the Moon and raised the flag.

its oxygen, water, and electrical supply. Water for consumption was important, but water for cooling the electrical equipment was critical. The mission and the astronauts were in very deep trouble. Kranz and the Mission Control team conducted a quick but thorough assessment of the situation. The first problem was to check the instrumentation to be sure that the readings being received were accurate. The second effort was to try to preserve what was left of vital supplies of water, oxygen and electricity. This involved a program of "progressive downmoding"; that is, eliminating all unnecessary consumption, step by step, but analyzing each step to see how it might affect the operation and living environment. It was a process, Kranz said, of "orderly retreat."[52]

Almost fortuitously, during the long "practices" by Mission Control with the Apollo 9 Earth-orbital flight, the flight control team headed by Kranz established procedures for using the LM as a "lifeboat." They had also experimented with "throttling down" the command module. Fortuitously too, Kranz said, he had John Aaron, Arnold D. Aldrich, and Philip C. Shaffer analyzing the data. Mission Control shut down first one, and then the second fuel cell in the command module. The LM's guidance system was then aligned with that of the command module in anticipation that the major guidance system would become inoperative. Then the astronauts were sent to their lifeboat while all systems in the command module were shut down. The LM had a very limited and fragile environment, and it too was held to minimum capacities by Mission Control.[53]

Mission Control decided that course corrections using the service propulsion system could not be risked, in part because of the lack of electrical power, but also because of the risk that the service module had been structurally weakened by the explosion. The descent propulsion engine of the LM would be used to put the crippled spacecraft into a return trajectory and to insert the craft into Earth's atmosphere. These were major problems. Minor, but equally deadly problems, such as the accumulation of carbon dioxide in the LM, constantly confronted the operators. The entire MSC and NASA organization rallied to the crisis:

> When word got out that Apollo 13 was in trouble, off-duty flight controllers and spacecraft systems experts began to gather at MSC, to be available if needed. Others stood by at NASA centers and contractor plants around the country, in touch with Houston by telephone. Flight directors Eugene Kranz, Glynn Lunney, and Gerald Griffin soon had a large pool of talent to help them solve problems as they arose, provide information that might not be at their fingertips, and work on solutions to problems they could anticipate farther along in the mission. Astronauts manned the CM and LM training simulators at Houston and at Kennedy Space Center, testing new procedures as they were devised and modifying them as necessary. MSC Director Robert R. Gilruth, Dale D. Myers, Director of Manned Space Flight, and NASA Administrator Thomas O. Paine were all on hand at Mission Control to provide high-level authority for changes.[54]

Gene Kranz remembers that following the successful burn to put the spacecraft on a free-return trajectory to Earth, three MSC Directors—Deke Slayton (Flight Crew Operations), Chris Kraft (Flight Operations), and Max Faget (Engineering and Development) offered varying procedural advice, which ranged from Slayton's concern that the crew needed sleep and rest, to Kraft's concern about power consumption, to Faget's concern about heat control. Finally, a decision was made to turn the astronauts to work on a program for passive thermal heat control. When those attempts failed, the entire procedure was reset and this time it worked![55]

Words can never wholly recapture the thought processes, analyses, energy, and emotion that went into the return of the endangered Apollo 13 astronauts. On April 17, the astronauts left their sanctuary in the LM and returned to the crippled Apollo CM for the reentry, jettisoning the LM. Operating on battery power alone since the explosion, the Apollo command module reentered the atmosphere and landed the weary and chilled astronauts within a mile of their recovery ship, the *Iwo Jima*. "If I had to explain our success," Kranz reflected years later, "it had to be the confidence in our own management."[56]

Thus, before the close of the decade of the 1960's, NASA had designed and its industrial contractors had built machines that could successfully take humans to the Moon and return them safely to Earth. NASA astronauts and operations personnel learned to fly those machines and to respond to the constant problems and surprises that spaceflight brought. Moreover, they developed the discipline and the systems to respond to problems. Spaceflight put Americans on a new and steeply graded learning curve. Spaceflight tested machines and human mettle and intellect under conditions and in an environment never previously encountered.

Thus far, not one but two machines had landed people on the surface of the Moon and returned them safely to Earth, and three more machines had carried humans in orbit around and behind the Moon and returned them safely as well. During the next 2 years

Apollo made four more successful flights and landed astronauts and their increasingly sophisticated equipment on the lunar surface. And then Apollo never again flew to the Moon.

CHAPTER 10: "After Apollo, What Next?"

At the height of the successes of the Apollo program the Nation entered a period of great malaise about space. Even as the astronauts spent almost 3 days on the lunar surface during each of the last three Apollo missions pursuing scientific objectives and research that might broaden human knowledge, NASA and its space programs were on the defensive. While Americans walked and drove a lunar vehicle on the surface of the dusty, rock-strewn surface of the Moon, the earthly ground from which these operations were conceived, constructed and flown became shifting sands of public opinion.

"What are the causes of this phenomenon?" asked Congressman Olin E. Teague, Chairman of the House Subcommittee on Manned Space Flight and a vigorous proponent of American space programs. That "continuing, strong sense of public pride in our space program," that "exhilaration that culminated magnificently" with the Apollo landing on the Moon has passed, Teague noted in 1971. Apathy had set in, "or worse," the space program came under abuse and attack. Why, in the restructuring of national priorities, a restructuring that began well before the flight of Apollo 11, had space slipped close to last, Teague wondered?[1]

In the time between the flight of Apollo 14 in February 1971 and Apollo 17 in December 1972, NASA, American space programs, and the Manned Spacecraft Center met some of their most formidable challenges. NASA's post-Apollo future became entangled in the web of politics, budget cuts, and Apollo program prerogatives. Apollo had its nemeses from the beginning. Its costs were one. Its seemingly single, goal-oriented lunar landing objective was another. War and welfare, and specifically the cold war and the War on Poverty, were others. The close of the Saturn-Apollo program and the confusion and indecision that finally brought NASA into the post-Apollo world of spaceflight is somewhat complex and convoluted.

Although President John Kennedy's memorable charge to the Nation in 1961 to send a man to the Moon and return him safely to Earth helped galvanize the Nation's energies, it was so singly goal-oriented that having been achieved there was nothing further to do. For example, American political parties with single goals, such as free silver or prohibition, rarely survived the accomplishment of their goal. Even in the early years of Kennedy's administration and certainly during Johnson's administration, America's space programs and the focus on space began to be diluted by many new and growing concerns.

The cold war helped define America's goals in space, but it also erected a host of countervailing social and economic forces or conditions. Cuba and the Bay of Pigs diverted American energies. Construction of the Berlin Wall, begun in August 1961, deflected public attention to Europe for much of the decade. The Cuban missile crisis in October 1962 brought the United States to the very brink of war with Russia. President Kennedy was assassinated in Dallas on November 22, 1963. In the 1960's hundreds of thousands of refugees fled to the United States from Cuba, Hungary, and East Germany. Lyndon Johnson declared war on poverty in 1964, and American military forces began bombing North

Vietnam in an undeclared war in southeast Asia. Racial confrontations and violence erupted in the cities—Los Angeles, Chicago, Atlanta, among others—and Martin Luther King died from an assassin's bullet in Memphis in 1968. Gross national product almost doubled during the decade of the 1960's and so did federal expenditures. The rate of inflation more than doubled. The decade in which the plan to put an American on the Moon and return him safely to Earth was one of America's best of times and one of its worst. In retrospect, it is perhaps remarkable that America's manned spaceflight program occupied so great a part of the Nation's energies and interests. It was one thing, despite problems such as the AS-204 fire, that seemed to be going right.

During the same few years (between 1958 and 1962) that Mercury, Apollo, and Gemini programs were conceived and initiated, NASA and aerospace industries began giving thought to programs that might go beyond the Apollo lunar landing. What might be logical extensions of the Apollo-Saturn effort? How could the technology, expertise, and capital generated from Apollo be applied to other ventures in space or on Earth? While it concentrated its energies and resources on building machines that could carry humans into space, NASA and MSC did consider tangentially what those people might do once they arrived in space, and how Apollo might be harnessed to other tasks. The final Apollo missions and the almost anticlimactic Apollo-Saturn Skylab and Apollo-Soyuz sequels are critical elements of a NASA search for identity that became very intense throughout the decade of the 1970's.

As early as 1959, a NASA committee headed by Harry J. Goett, which included George Low and Max Faget as members, established a general framework for NASA space missions and established a tentative priority for those missions. The outline included Mercury, unmanned probes, a "manned satellite," a manned spaceflight laboratory, and a Mars or Venus landing. The committee suggested the following NASA missions in order of their priority:[2]

1. Man in space soonest—Project Mercury
2. Ballistic probes of the planets
3. Environmental satellites
4. Maneuverable manned satellite
5. Manned spaceflight laboratory
6. Lunar reconnaissance satellite
7. Lunar landing
8. Mars-Venus reconnaissance
9. Mars-Venus landing

The maneuverable manned satellite would have been a vehicle parked permanently in orbit and used for communications, electronic data gathering, and navigation, but without the capacity to return to Earth. The manned spaceflight laboratory became a prototype for Skylab and provided a conceptual beginning for the later space station. Remarkably, this 1959 study established a basic design for future space programs and instigated considerable thought. Ideas and preliminary designs for spacecraft and space station configurations began to appear from a variety of sources within and without NASA.

By 1962, aerospace engineers and managers were seriously deliberating and studying the feasibility of a permanent space station orbiting Earth as a laboratory and staging platform for manned flights to Mars. McDonnell Aircraft proposed a one-person space station based on a Mercury capsule. Rene A. Berglund at Langley proposed an inflatable laboratory extending from a Mercury spacecraft nucleus. A NASA Headquarters staff study headed by Bernard Maggin recommended development of a manned orbital facility. The Space Task Group at Langley (before its designation as MSC) considered using an Apollo spacecraft and a Saturn second stage for an orbiting laboratory, and asked assistance from Ames Research Center. Canada's AVCO Corporation proposed a Gemini-Titan configuration for a space station in 1962.[3]

The Langley Research Center (within Langley's Spacecraft Research Division) created a Space Station Program Office headed by Edward H. Olling, which initiated preliminary studies of structures and configurations, life systems, operations and logistics, docking and rendezvous mechanisms, and associated engineering studies. In July and August 1962, Langley held a formal debriefing, attended by Robert Gilruth, Max Faget, Aleck Bond, Charles W. Mathews, Walter Williams, Paul Purser and others, effectually transferring the information and part of the responsibility for manned space station work to the MSC being established in Houston.[4]

Concurrently, Lewis Research Center, the newly established Marshall Space Flight Center, and other aerospace industries, including North American Aviation, were examining space station configurations. Goodyear Aircraft Corporation developed models and prototypes of an inflatable 150-foot diameter space station which were submitted for review and consideration to teams at Langley Research Center (in 1961) and Lewis Research Center (in 1962). Gene McClard, with Marshall Space Flight Center's Saturn Systems Office, formally presented Marshall's proposal for an inflatable-type space station (which it referred to as an inflatable-structures experiment) to NASA's Management Council in October 1962. Douglas Aircraft also developed, at Wernher von Braun's request, an unsolicited proposal to adapt a Saturn-IVB stage as an Earth-orbiting, manned space laboratory. As early as 1952, before Sputnik and NASA, Von Braun had proposed an "artificial moon" space station concept employing a fixed hub and an inflated, rotating doughnut-shaped outer chamber. Later, MSC and Langley researchers rejected inflatable structures as being unsuitable, undesirable, or not feasible for a man-occupied space station given the emerging likelihood that a space shuttle system could deliver the materials for a large, permanent station into orbit at a reasonable cost. It is interesting, however, that Grumman designed a Mars mission on the hypothesis that a space station would be an intrinsic part of such a mission because not only would a mission require an orbiting station as a staging base, but any interplanetary vehicle would necessarily resemble a space station as a long-duration habitat for humans.[5]

The Air Force canceled its X-20 Dyna-Soar project in 1963 and began work on a manned orbiting laboratory. NASA continued studies of a manned orbital research laboratory at Langley, while MSC pursued designs for an Apollo "X" two-person orbiting laboratory using an Apollo LM. MSC also worked on the preliminary design of an Apollo orbital research laboratory and a large orbital research laboratory. At Headquarters in December 1964, Joseph F. Shea began discussing with Samuel C. Phillips the Apollo missions that might follow a successful Moon landing. Senator Clinton P. Anderson,

Chairman of the Senate Committee on Aeronautical and Space Sciences, recommended to President Johnson that the Air Force manned orbiting laboratory program be merged with NASA's Apollo X program. NASA and the Department of Defense promised to collaborate but to preserve the integrity of each program. During the spring of 1965, the Office of Manned Space Flight and MSC, with contractor support from North American, Boeing Company, and Grumman Aircraft Engineering, conducted an intensive study of Apollo Extension Systems—that is, a study of missions or programs that might logically follow the completion of a lunar landing using Apollo systems and knowledge. Max Faget chaired the briefing before a large NASA audience at MSC in May.[6] Although no firm proposals developed from the conference, serious thought began to be applied to post-Apollo possibilities within and without NASA.

Olin E. Teague's NASA Oversight Subcommittee (of the House Committee on Science and Astronautics) began gathering information from NASA center directors about "where you feel your center should be going into the 1970's." His concern, Teague said in a letter to Bob Gilruth at MSC, had to do with future efforts and future center missions. Gilruth promised a full staff analysis of MSC's organization, functions and personnel requirements given several alternative missions. Teague addressed similar letters to Administrator James E. Webb and Secretary of Defense Robert McNamara. He also requested status reports from major Apollo subcontractors (Grumman and North American). James Webb visited MSC twice in August 1965 to discuss Apollo missions after a lunar landing. Webb stressed that such missions should use "off-the-shelf" hardware and be "cheap." In early September, Gilruth invited "serious consideration" by his staff on the next major mission to be undertaken by the manned spaceflight program.[7]

The Senate Committee on Aeronautical and Space Sciences held hearings on "Space Goals for the Post-Apollo Period" in late August 1965. George Mueller advised the committee members that NASA planned Apollo flights with experimental packages, extended orbital missions, and extended lunar surface missions. The post-Apollo period, he explained, included programs emphasizing Earth-orbital missions that would produce direct economic benefits. Communications satellites would be one such benefit. Other program alternatives included extensive lunar exploration and operations; planetary exploration and scientific missions; a combined "maximum effort" program; and finally, a balanced, cost-effective combination of Earth orbit, lunar and planetary exploration, and science. Only a few weeks prior to his testimony, Mueller established a Saturn-Apollo Applications Program Office, headed by Major General David Jones, detailed to NASA by the Air Force, with John Disher as deputy director. As David Compton and Charles Benson commented in their history of Skylab, by 1965, well before the first successful manned flight of Apollo, NASA had given 6 years to space station study and at least 3 years to post-Apollo planning.[8]

It is significant that the inception and development of the Apollo program occurred within a broad conceptual framework that included space stations, unmanned probes, orbiting laboratories, a shuttle transport system, and manned lunar and planetary missions. The lunar landing objective provided a sharp, definitive focus for space initiatives, but possibly also resulted in narrowing and limiting those objectives. The several billions of federal monies being directed annually to NASA also created criticism and dissent from potential beneficiaries of federal spending. Defense and welfare programs, as well as the traditional local "pork barrel" type funding all had growing appetites. Even those who supported NASA program funding argued over the program distribution of those funds. Scientists became increasingly restive, arguing that unmanned satellite programs would produce greater knowledge and scientific benefits than the far more expensive manned lunar landing program. This dissent, in part, contributed to NASA's decision to use Apollo systems for scientific investigations in space.[9]

By 1965, Congress' funding approach to the space program had changed from "what can we do for you?" to "what can you do without?" It got worse. By 1966, Congress and the administration were no longer asking what NASA could do without, but were deciding for themselves what NASA might do without. The spiraling costs of defense, military commitments in southeast Asia, and Great Society social programs pressed heretofore seemingly unlimited federal resources and fueled fears of inflation. Following a program review by the Bureau of the Budget, in Congress Teague's NASA Oversight Committee began a review of Apollo program costs, progress and program management. While that was going on, Vice President Hubert Humphrey pledged the administration's continued support for space programs in a speech to the Aerospace Industries Association in Williamsburg, Virginia, but he explained that "immediate national security requirements have necessarily limited the funds available" for space. This is regrettable but inevitable, he said. In the future, the space program must accept as its guiding principle "to get the largest possible return on the public funds we have already put into facilities, trained manpower, boosters, spacecraft, and all our other accumulated space assets." It means, he said, that we must exploit to the maximum all that the Apollo program has produced for us, and that we do not start from scratch after the Manned Orbiting Laboratory program, but that we seek out every possible application of that hardware and expertise. Humphrey talked about communications satellites, research in space, environmental benefits that might be derived from space surveillance, international understanding and cooperation in space, and the energizing force derived from space programs that permeates the economy. But he also implied that funding for space programs would be decreasing, not increasing. In direct communication with Administrator Webb, President Lyndon B. Johnson was more explicit. He told Webb, "by God, I have got problems and you fellows are not cooperating with me. You could have reduced your expenditures last year [1965] and helped us out, you didn't do it!"[10]

On the House floor during the appropriations debate in August 1966, Congressman Teague blamed NASA for a lack of advanced planning and warned the Agency that "space does not have the same high priority it once had." As if to illustrate the point, the Senate staved off several attacks on NASA's budget by declining margins. Senator William Proxmire, who led separate moves to reduce space spending by first half a billion dollars and then by $150 million, said that NASA would still be "a fat cat" even if it lost the $150 million. Administrator Webb called for a national debate on where the Nation wanted to go in space, and warned that much of the momentum and investment in space might be lost unless new goals were chosen and funding sustained. Teague supported the call for a national debate, but a comprehensive study by his staff placed the burden for recommending broad objectives on NASA. It asked the Agency to establish specific missions and to identify the costs and benefits associated with each—no later than December 1. Teague's

439-page staff study, headed by James E. Wilson, asked for a reevaluation of space programs in the light of President Johnson's Great Society goals and the budgetary constraints caused by the war in Vietnam.[11]

At a symposium sponsored by the American Institute of Aeronautics and Astronautics (AIAA) in October, panelists responded to the general question, "After Apollo, What Next?" News moderator Peter Hackes said that Congressman Teague had repeatedly posed the question of a post-Apollo program and the need to define it *now*. "Obviously, Vietnam has affected NASA's budget as it has all other governmental agencies," but although budget problems will affect the scheduling of future missions, they do not preclude the establishment of goals. And then he asked Edward C. Welsh, Executive Secretary of the National Aeronautics and Space Council, to describe the administration's position on future space goals. Welsh responded very briefly that there would be future space goals and that the key would be to defining the proper mix. An observer at the hearings later told Teague that everyone present felt that the reply was "completely nonresponsive."[12]

The fact was, according to a private study conducted by Thomas D. Miller (a financial analyst for Arthur D. Little, Inc.), NASA had been too preoccupied with the lunar landing program to give much thought to longer range goals. In truth, although Welsh obviously understated the situation, NASA did have rather well enunciated program goals, but now the problem had become establishing priorities and as Welsh said, "the proper mix." Moreover, although NASA had a coherent plan for the next 20 years of spaceflight, the escalation of the Vietnam War and Great Society programs had reduced the priority of spaceflight in the minds of decision makers. Economies were being forced on NASA and all nondefense government activities. Miller estimated that sustaining NASA flight programs would require a fiscal year 1968 budget level of $5.5 to $5.6 billion, rising to $7 billion by 1976. Although pessimistic on achieving that level of funding, Miller believed that NASA programs would be supported at a lower level because of the continuing competition with the Soviet Union, and because the space program employed a large proportion of the Nation's scientific and technical personnel and contributed significantly to the financial health of the aerospace and electronics industries.[13]

The reality of budgetary constraints and the mismatch of future planning with budgetary projections began to affect and alarm NASA center managers. Gilruth sent George Mueller an analysis of what he believed to be a critical situation, with copies to Wernher von Braun and Kurt Debus. Gilruth said that his concerns stemmed in part from the lack of a definite goal or direction for the future of manned spaceflight. Future program planning, he advised, was employing a launch rate higher than the in-line Apollo launch rate. Apollo Applications Program (AAP) missions proposed to use Apollo hardware for purposes "differing significantly" from those intended. Gilruth feared that economic considerations were driving future plans to use equipment in ways that were inconsistent with its technical capabilities. Moreover, the many changes going on in AAP plans, occasioned by a "steadily shrinking" AAP budget, were causing the diversion of management attention, effort, and funding from the mainline Apollo programs. Gilruth strongly urged support for a large, permanent, manned orbital space station, and strongly advised against using the LM for any operations other than that for which it was specifically designed. He suggested more use of unmanned lunar probes, better and more accelerated mission planning, fewer and more efficient missions, and improvement in the quality of projected scientific experiments. There was, Gilruth believed, a critical mismatch between AAP planning, the opportunities for manned spaceflight, and the resources available.[14] Effective planning required close cooperation between Headquarters and the technical centers and that was not always evident. Headquarters, in fact, tended to reserve planning as its primary duty, while the centers concentrated on development and operations.

The budget and future planning crises became more acute as Congress began deliberating the federal budget for 1968 and the war in Vietnam consumed more money and more American lives. Walter G. Hall, a friend of Olin Teague's who lived in Dickinson, Texas, near MSC, reported to Teague in January 1967, that "morale among many of the folks at NASA seems to be deteriorating." Some of them, he said, "feel that there is nothing in 'the mill' after the Apollo program." Hall asked Teague to send him comparative figures for MSC employment over the past 3 years with projections for 1967. Teague answered:

> The truth of the matter is, Walter, I am surprised that the morale of the space people is not lower than it is. We do not have a follow-on program as we should have—all because of money.

Teague predicted a nationwide decline in the number of space program employees in 1967 from 400,000 to 200,000 people. Grumman employment at Bethpage, New York, had peaked and was going down very rapidly. More would be known after the President's budget message (January 24), Teague wrote, and he promised to have some good information when he came to Houston.[15]

J.P. Rogan, Vice President and General Manager for Douglas Aircraft's Missile and Space Systems Division, said that our Nation *must* answer the post-Apollo question and "very soon! Apollo is only a beginning!" We must convert the beginning technology derived from Apollo into building blocks of future space capability. The blocks include a long-duration orbital experience, reusable spacecraft, reusable launch vehicles, nuclear powered rocket stages, and improved secondary power systems. An orbital laboratory would require a reusable spacecraft (but not necessarily a reusable launch vehicle), and he estimated that the evolution of an operational reusable spacecraft would require at least 9 years of development.[16] What would become the Space Transportation System (STS), or Space Shuttle, began to emerge from the exigencies of a national (and NASA) budget crunch.

Budget concerns also reinforced a more conservative or practical approach to space. The President's Science Advisory Committee estimated that total government expenditures on space in 1967, including NASA ($4.9 billion), the Department of Defense ($1.6 billion), the Atomic Energy Commission ($181 million), and a small allocation to the Environment Sciences Service Administration, totaled $6.7 billion. The Science Advisory Committee recommended a limited extension of Apollo programs for the purposes of lunar exploration, upgrading the unmanned program for the exploration of nearby planets, a program to prepare the technology and personnel for long-duration flight, vigorous exploitation of space applications for "national security and the social and economic well-being of the Nation," and exploitation of near-Earth orbit scientific and astronomical experiments and research. The committee recommended construction of a permanent space station in the mid-1970's, but advised that a decision could await biomedical studies and a decision on the desired pace of effort toward manned planetary travel. The report urged a better mix of manned and unmanned programs and advised that economic benefits from space operations would more likely be derived from *unmanned* rather than manned vehicles.[17] Although industry, scientists, politicians, and NASA administrators all agreed that the United States should clarify the post-Apollo program, by the close of 1967 no firm objectives had developed other than the objectives imposed by budgetary constraints—and those were significant.

NASA's congressional appropriations peaked in 1965 at $5.25 billion and began a steady decline thereafter. Congress, of course, as Arnold Levine explains in his study of NASA management, was not solely to blame for NASA budget reductions. The Bureau of the Budget (later the Office of Management and Budget), responsive primarily to the Executive Office, reviewed NASA's budget requests before they went to Congress. After 1967, the Bureau acted on the premise that the White House would no longer intervene, but would tacitly approve any NASA budget reductions. Thus, prior to congressional action, the Bureau began curtailing budget requests, forcing NASA to choose between programs the agency believed were merely desirable and those deemed essential to the Agency's mission.[18]

Among the very specific results of congressional and Bureau of the Budget fiscal constraints were canceling 2 scheduled Apollo Moon flights (reducing the number to 17), reducing Surveyor unmanned flights from 10 to 7, closing NASA's Electronic Research Center in Boston, freezing the critical NASA "excepted" positions to 425, canceling

TABLE 6. NASA Budget Requests and Appropriations FY 1959 to 1971 (millions of dollars)

Fiscal Year	Administration Request	Amount Appropriated	Percent Cut
1959	$ 280.0	$ 222.8	20.4
1960	508.3	485.1	4.6
1961	964.6	964.0	—
1962	1,940.3	1,825.3	5.9
1963	3,787.3	3,674.1	3.0
1964	5,712.0	5,100.0	10.7
1965	5,445.0	5,250.0	3.6
1966	5,260.0	5,175.0	1.6
1967	5,012.0	4,968.0	0.9
1968	5,100.0	4,588.9	10.0
1969	4,370.4	3,995.3	8.6
1970	3,715.5	3,696.6	0.5
1971	3,333.0	3,268.7	1.9

Source: Thomas P. Murphy, Science, Geopolitics, and Federal Spending (Lexington, Mass.: Health Lexington, 1971), 364.

NASA's NERVA (nuclear) rocket research inherited from the Atomic Energy Commission, and eliminating developmental work on a large solid-fuel rocket engine and a smaller liquid hydrogen engine. Budget office "suggestions" in 1966, reinforced by recommendations of the President's Science Advisory Committee in 1967, led to the closing of the Air Force's manned orbiting laboratory program and the transfer of its hardware and astronauts to MSC in 1969. Future programs, such as the AAP, were particularly vulnerable to budgetary pressures.[19] Thus, even as Congress and the public wondered "After Apollo, What Next?", the answer was: less and less.

AAP funding, that is, the budget for post-Apollo planning and operations, suffered severely. First funded as the Apollo Extension System in 1966, NASA allocated $51.2 million to future planning from its total budget of $5.175 billion. The figure rose to $80 million in 1967. In 1968, under pressure from Congress and the Executive Office, NASA requested $454.7 million for its AAP, received $347.7 million, but then was forced to allocate only $253.2 million in order to cover shortages elsewhere. Webb convened a post-Apollo advisory group, chaired by Dr. Floyd Thompson from Langley Research Center and including center directors, which met variously at Washington, D.C., and each of the three manned spaceflight centers during 1968. That advisory group discovered "unresolved difficulties" with projected post-Apollo programs rather than reaching agreement on future programs.[20]

Paul Purser, special assistant to the director, and others at MSC were of the opinion that, while the center should take the lead in developing future manned spaceflight program options, there would be no real budgetary enabling action until either a successful completion of the lunar landing mission or a sudden and large reduction in Department of Defense spending.[21] Perhaps in part because of this philosophy, and because indeed MSC itself was by no means declining or experiencing current budgetary difficulties, but rather was operating at its peak and maximum effort during the Apollo flights in order to achieve a lunar landing before the end of the decade, MSC focused on its present, real-time, operating program.

The successful lunar landing by Apollo 11 in July 1969, sparked a wave of optimism throughout NASA regarding the viability of future Apollo applications. George Mueller, Apollo program head at Headquarters, summarized NASA's goals in a teletype message to "all stations" in September. Mueller suggested that while the Apollo program had served the Nation well by providing a clear focus for the development of space technology, a balanced program was needed which would focus on a manned planetary landing in the 1980's. The memorandum identified two directions for manned spaceflight programs. The first involved the further exploration of the Moon with possibly the establishment of a lunar surface base. The second was the continued development of manned flight in Earth orbit leading to a permanent manned space station supported by a low-cost shuttle system. The projected schedule included the operation of two Saturn V-launched Earth-orbital workshops (1972 and 1974), a lunar orbiting station in 1976, and an Earth-orbit space base with a possible Mars landing in 1990.[22]

This rather startlingly clear future programs definition was derived in part from intensive studies during the previous 6-month period by a NASA task group headed by Milton Rosen. Rosen's group reported to Homer Newell, who chaired a NASA Planning

Steering Group. Newell's group in turn coordinated its work with a Special President's Space Task Group created by Richard M. Nixon after his election in 1969. The Space Task Group included the Vice President, Secretary of Defense, NASA Administrator, the President's Science Advisor, and observers from the Department of State, Atomic Energy Commission, and Bureau of the Budget. The Space Task Group presented a strong endorsement for continuing space activities using existing and to-be-developed capabilities. It supported the development of a modular space station and a reusable space transportation system to serve the station. The group noted that schedules and budgetary decisions should be subject to presidential choice and determined in the normal annual budget and review process.[23] It was this report that gave Mueller and NASA a clear signal for future programs. It was in the budgetary processes, however, that future programs again clashed with fiscal realities.

Notably, although 1969 AAP budget requests totaled $439.6 million, emergency action by Congress to reduce federal spending at all levels through the Revenue and Expenditure Control Act of 1968 resulted in only $135.5 million of new money going into post-Apollo planning. In 1970, NASA submitted a request for $345 million for AAP, but Congress and the budget office reduced the operating budget to $288.1 million.[24] Thus, by 1970, the stark reality seemed to be that NASA simply could not truly answer the question, "After Apollo, What Next?"

NASA, however, had reached a consensus on its recommendations to Congress. During the House floor debate on the NASA authorization bill in 1970, Olin Teague asked Dale D. Myers (Associate Administrator for Manned Space Flight), Kurt Debus (Director of Kennedy Space Center), Bob Gilruth (Director of MSC), Eberhard Rees (Director of Marshall Space Flight Center) and Wernher von Braun (now Deputy Associate Administrator for NASA) to send him a personal letter explaining why it was important to "move forward" with the manned spaceflight program. Myers explained that post-lunar landing objectives were different from the "clearly defined national goal of the last decade." But the new multiple programs emphasizing economy and direct space technology benefits were no less challenging and important. Debus and Rees stressed the importance of developing a reusable shuttle. Von Braun said that "with the space shuttle and the space station we will have the space age equivalent of the jet liner." Gilruth, referring to the Apollo 13 mission, told Teague that "we cannot expect to push back the space frontier without some difficulty," and he stressed the reusable Earth-to-orbit shuttle as the key to post-Apollo activities.[25] Within NASA the 4- or 5-year search for its future identity had begun to bear fruit. The aura of success and at least temporary resurgence of national support and interest in space helped produce that united front. But the answer to the question, "After Apollo, What Next?" lay outside the NASA community.

NASA historian E.M. Emme, assisted by Acting Administrator Thomas O. Paine, summarized the critical transition between the outgoing administration of Lyndon B. Johnson and the incoming administration of Richard M. Nixon. Management after 1969, Paine explained, was trying to get the "greatest possible space program returns within severe dollar constraints imposed by the FY 1969 austerity budget of President Johnson and Congress." NASA was giving a lot of consideration to long- and short-term planning to help the 91st Congress in deciding NASA's future direction and level of effort. The moment was critical. Given declining public support and lower budgets for the past 3 straight years, even after the achievement of the manned lunar landing, "NASA's future fate . . . remained far from assured." Administrator James E. Webb, who left NASA in 1968, had been replaced by an interim administrator. A Republican administration was entering the White House, but Congress was controlled by the Democratic party. Space vied for public attention in 1968 as the crew of the U.S.S. *Pueblo* was being released from an ignominious North Korean captivity. Thousands were starving in Biafra. Armed combat interrupted a truce in South Vietnam. Arabs and Israelis were on the brink of war, and Communist China exploded a fourth nuclear device.[26] It was an untimely time to decide, "After Apollo, What Next?"

Since at least 1967, NASA's budgetary and political stance had been increasingly defensive. The tone of agency-congressional relations had been to prevent budget cuts rather than to seek new funding tied to new programs. Olin Teague, among others, argued that NASA (and MSC) were remiss by failing to formulate earlier and more definite post-Apollo answers. Now, in 1970, the answer rested in Congress and the White House.

For personnel at MSC it was perhaps just as well. All hands focused intensely on the work at hand, which had to do with successfully completing the flights of Apollo. Gerald D. Griffin, who joined MSC in 1964 as a navigation and control officer on Gemini flight operations teams and began work as a Flight Director on Apollo 7, described the Apollo work as having to do with transportation research. Once we had the capability to fly, only then could we begin to think science or post-Apollo operations. From the time of MSC's creation, its engineers and personnel had their hands full with the present and gave relatively little thought to the past or the future. There was, from the inception of Mercury through the last flight of Apollo and the Skylab and Apollo-Soyuz flights, no repose or pause.[27]

Aleck C. Bond, assistant director for Chemical and Mechanical Systems under Max Faget's Engineering and Development Directorate at MSC, remembered that, at least through 1970, personnel at MSC sensed no problems or cutbacks in space program activities. At that time, personnel layoffs had occurred only among contractor support groups. Center managing engineers were all civil service employees and were secure and generally insulated from NASA budget reductions. Only in later years, Bond recalled, did the impact of the budget reductions of the 1967-1970 era become clear.[28] Center employees, caught up in the intensity and exhilaration of Apollo flights and somewhat protected from the externalities of politics and budgeting, were for the most part oblivious to problems on the horizon.

"I never saw any change in attitude or morale," Griffin commented, "especially in the operations end." Personnel had no time to worry about the world; in fact, MSC people were somewhat insulated from the world by virtue of their intense focus on their work. Houston, Griffin said, might as well have been 100 or 500 miles away. Washington, D.C., in a sense was even farther. The space center comprised a very close community. Discipline was very much like the military, but wholly voluntary. There was a close camaraderie and a sense of purpose and mission. "We had a strong government, industry, academic team. We all worked very closely to get the job done. We dealt one to one, not at arms length with our contractors. Throughout the center, and especially in Mission Control, we had a sense of arrogance—we thought we were pretty good!"[29] Although Griffin did not use the term which later became popularized, center personnel, including the astronauts, had the "right stuff" to get the job done. Center personnel at every level became gripped by a restless energy and excitement. They were one with the astronauts, doing things they had never done before, seeing things they had never seen before.

The successful lunar landings Apollo 11 and 12, and even the harrowing flight of Apollo 13, proved the ability to fly. With the flight of Apollo 14 on January 31, 1971, NASA began to move beyond the goal enunciated by President Kennedy a decade earlier. Griffin first noticed a difference in flight operations beginning with Apollo 14. Subsequently, the installation of scientific experimental packages (the J-capsule) on Apollo 15 through 17 gave the Apollo missions a distinct scientific tilt. Those flights represented a consolidation of efforts after AS-18, 19, and 20 were canceled. Training was correspondingly more diverse and intensified. Griffin, who served as the lead flight director of the Apollo 15 and 17 missions, remembered that mission planning changed from the emphasis on flight operations to scientific experiments. As a flight director, for example, Griffin went on the geologic training missions with the crews. He learned more about geology than he had learned all his past life, he recalled. There was a big change in us, "from the scarf-over-the-neck space era, down to hard rock science."[30]

There was, after Apollo, a rather large change in Griffin's career as well. He went to Headquarters as Assistant Administrator for Legislative Affairs in 1973, then to Dryden Research Center and from there to Kennedy Space Center as Deputy Director, and from Kennedy back to Headquarters to head the Office of External Relations. In 1982 Griffin returned to the (now) Johnson Space Center to replace Chris Kraft as Center Director. Griffin retired from NASA in 1986 to head the Houston Chamber of Commerce and spearhead an effort to reinvigorate the local economy (in part through promoting space technology related industries) which was then suffering severely from the collapse of the oil boom of the seventies. Subsequently, he joined a Houston firm specializing in executive officer searches.[31] Griffin reflected, in part, the restless energy and excitement of so many of the MSC personnel who were determined to continue to do things they had never done before and see things they had never seen before. Few ever truly seemed to retire.

There was simply too much to do and it was exciting work. Changes in top management personnel at MSC after 1969 reflected the center's Apollo operations orientation. Chris Kraft left the Flight Operations Division in 1969 to become Deputy Director, and then Director of the Center in 1972 when Robert Gilruth retired. Sigurd A. Sjoberg succeeded Kraft as Director of Flight Operations, and in January 1972 became Kraft's Deputy Director. When Griffin became Center Director, he named C.E. Charlesworth (also from Flight Operations) as his deputy (1982-1983). In July 1983, Charlesworth moved to become Director of Space Operations, and Griffin appointed Robert C. Goetz, from Langley, as his new Deputy Director. Thus, many of the center's post-Apollo top managers emerged from Apollo flight operations and were in a sense one aspect of Apollo "extensions."

Apollo was a totally preoccupying "present" experience. Following some design changes resulting from the Apollo 13 cryogenic oxygen tank failure, it was time to fly again. Although the lunar objective of Apollo 14 was similar to that of its crippled predecessor, its mission and apparatus were extended to enable the astronauts to gather more information and lunar specimens. The crew of Apollo 14 were rather unique and distinctive human specimens in themselves.

Mission Commander Alan B. Shepard was one of the original seven NASA astronauts and America's first man in space. In 1969 he was grounded with an inner ear problem. Taking a desk job in the Astronauts Office, Shepard used his new "spare time" to organize a bank (which grew rapidly) in Baytown, Texas, and purchase a small bank in Houston. "The banker," as his colleagues at MSC called him, also began raising quarter horses, drilling for oil, and investing in land. But what he really wanted to do was fly. So under an assumed name (Victor Paulis), Shepard entered a hospital in California where a doctor performed a then pioneering type surgery which resolved his ear problem and got him restored to flight status.[32]

Stuart Allen Roosa, the command module pilot, joined the astronaut corps in 1966. Roosa, who was to be the backup command module pilot for Apollo 16 and 17, left NASA in 1976 to enter private business. Edgar Dean Mitchell, a native of Hereford, Texas, with a doctorate in aeronautics and astronautics from Massachusetts Institute of Technology (1964), piloted the LM. Mitchell left NASA in 1972, and founded the Institute of Noetic Sciences (having to do with the working of the mind), organized several private companies, and authored a book entitled *Psychic Exploration: A Challenge for Science*.[33] Apollo 14 left the pad at Kennedy Space Center before dawn on January 31, 1971.

Shepard and Mitchell flew the LM to the Moon's surface in the Fra Mauro highlands (that was to have been the landing site for Apollo 13) and walked, with their two-wheeled cart and equipment transporter, near the rim of Cone Crater where (during two EVAs totaling 9.5 hours) they gathered 43 kilograms (about 95 pounds) of lunar rock samples. The command module returned the crew to a Pacific splashdown almost exactly 9 days after launch. It had been a good mission, with all objectives accomplished, and returned a veritable treasure of geologic knowledge. Apollo 14 also satisfied medical officers and scientists that the elaborate quarantine procedures used to avoid Earth contamination from lunar microorganisms were no longer needed.[34] Lunar soil and space was dead. Scientific inquiry was excited and alive.

Apollo 15, scheduled to fly on July 26, was the first of the "J" missions, containing special scientific experiments aboard

A gathering of Apollo 14 flight directors at Mission Control Center: Glynn S. Lunney (seated on the console), M.P. Frank, Milton L. Windler, and Gerald D. Griffin. Griffin, standing next to Lunney, became Director of JSC in 1982, replacing Dr. Chris Kraft.

Suddenly, Tomorrow Came...

The LM "Falcon" is photographed against the barren landscape during Apollo 15 lunar surface EVA at the Hadley-Apennine landing site. Note tracks of the lunar roving vehicle in the foreground. Astronauts David R. Scott and James B. Irwin were on the lunar surface while Alfred M. Worden piloted the command module in lunar orbit.

David R. Scott photographed James B. Irwin as he worked on the lunar roving vehicle. St. George Crater is 5 kilometers or 3 statute miles in the distance beyond Irwin.

the LM, and in a special compartment of the command module for experimental work during its lunar orbits. Commander David R. Scott and LM pilot James B. Irwin took the LM to the surface of the Moon near the foot of the Apennine Mountains and adjacent to Hadley Rille. One of the most remarkable aspects of Apollo 15 was that the lunar astronauts brought with them an automobile—a battery-operated, four-wheeled lunar roving vehicle. The explorers traversed about 28 kilometers (17 miles) across the lunar surface and collected 77 kilograms (about 169 pounds) of rock and soil samples. Other than for difficulty with core drilling experiments, the scientific and geologic, as well as the photographic, product of the mission exceeded by far anything yet accomplished. Moreover, Earth watched the ascent of the LM from its lunar base on live television. The astronauts returned safely, following the failure of one of the three landing chutes and a somewhat harrowing descent to the recovery point.[35] But public interest, praise, and admiration for the astronauts and the mission accomplishments were short-lived. What Teague had identified first as public apathy had indeed shifted to abuse and attack (that is, if the media provided some measure of the public pulse). Public criticism erupted over what NASA labeled a case of "poor judgment" by the flight crew.

Mission accomplishments were overlooked as the media concentrated on what Congressman Les Aspin (D-Wisconsin) termed a case of immoral, if not illegal, conduct by the astronauts. The astronauts carried with them to the Moon and back 400 specially stamped and canceled first-day covers which were to be sold by a friend who would put the proceeds into a trust fund for their children. About 100 of the covers were sold in Europe for $1,500 each, according to Aspin, but when the story became public they abandoned the scheme and declined to accept any money. Aspin recommended that all should be released from the space program. Although in September the astronauts were welcomed to the Capitol and a visit with the President and selected officials, the controversy continued to brew through the following year.[36]

In October 1972, CBS News with Nelson Benton substituting for newscaster John Hart, posed the proposition that, with the Moon program winding down, ex-spacemen were being touched by the sordid spirit of free enterprise. Understandably so, Commentator Steve Young suggested, inasmuch as the spirit of free enterprise affects most Houstonians. He identified Houston as the "Baghdad of the Bayou, where business is booming everywhere from petrochemical plants to shiny hotels, shimmery office buildings, and complexes." Astronauts were being snapped up for their glamour and name recognition. Scott Carpenter, Buzz Aldrin, Shorty Powers, and Wally Schirra were doing television commercials. Alan Shepard took commemorative medals to the Moon, and the Apollo 15 crew "enhanced astronomically" the value of the covers they took to the Moon, he said.[37] This was, to be sure, a reflection of a distinct mood change in America. These were not, other than perhaps for the remarkable achievements of the Apollo missions, the best of times.

Colonel James Irwin resigned from NASA and from the Air Force in July 1972. He founded a religious organization called High Flight Foundation in Colorado Springs, Colorado, and wrote a book entitled *To Rule the Night* which described his early life, selection into the astronaut corps, and experiences on the flight of Apollo 15. Scott left the astronaut corps in July 1972 to become a special assistant for the Apollo-Soyuz flight, and then served as Director of NASA's Dryden Flight Research Center in Edwards, California, before leaving NASA to organize a private business venture, Scott Science and Technology, Inc., in Los Angeles. Alfred Worden also left MSC in 1972 to serve as Director of Advanced Research and Technology at Ames Research Center, and retired from NASA and the Air Force in 1975 to establish Energy Management Consulting Company

Apollo 15 splashdown.

The lunar surface viewed with a 35 mm stereo close-up camera.

The Lunar Receiving Laboratory. Dr. Elbert King describes procedures for handling lunar materials to Dr. Von Engelhardt, Director of the Mineralogical Institute at the University of Tuebingen, Germany.

and later a helicopter charter and aircraft management company in Florida. He also took time to write a book of space poetry, *Hello Earth: Greetings from Endeavour*, and a children's book, *A Flight to the Moon*.[38]

Apollo 16 flew from Kennedy Space Center on April 16, 1972, carrying commander John W. Young, command module pilot Thomas K. Mattingly, and LM pilot Charles M. Duke, Jr. A number of minor mechanical and computer malfunctions kept the crew and mission control constantly solving problems, but Young and Duke completed a total of 71 hours on the western edge of the Moon's Descartes Mountains. As on the previous mission, the lunar rover enabled the astronauts to cover considerable territory, collect a quantity of diverse geologic specimens, and complete numerous scientific experiments. There were some minor mishaps, such as a broken electronic cable, that prevented completion of some of the experiments, but overall the mission produced new and valuable information. The crew conducted seismic, surface magnetometer, heat flow, cosmic ray, gravimeter, meteorite, atmospheric and ultraviolet experiments among others, in addition to more conventional geologic mapping and sampling.[39]

The last lunar Apollo mission left Earth on Pearl Harbor day, December 7, 1972. The official mission review noted that Apollo 17 "was the longest mission of the program (301 hours 51 minutes 59 seconds) and brought to a close one of the most ambitious and successful endeavors of man. The Apollo 17 mission, the most productive and trouble-free lunar landing mission, represented the culmination of continual advancements in hardware, procedures, and operations."40

America celebrated. Harrison "Jack" Hagan Schmitt, the first civilian-scientist to fly a lunar mission (Schmitt became U.S. Senator from New Mexico, 1976-1982), with command module pilot Ronald E. Evans and Eugene A. Cernan (the Apollo 17 commander and eleventh man to walk on the moon) kicked off an 11-week "postflight tour" at Super Bowl VII in Los Angeles on Sunday, January 14, 1973. They visited 25 states and the Nation's Capitol. They met 13 governors, a considerable number of mayors, and President Nixon. They visited the major NASA contractors, including North American Rockwell, Grumman, Boeing, Bendix, Teledyne-Ryan, Chrysler, and Martin Marietta. They visited all of the NASA centers except Marshall Space Flight Center, which declined the visit due to the press of other programs.41 It was well done. The race was over. America had won. By the close of 1972 the United States had launched 27 manned spacecraft into space and returned them safely to Earth; 34 individuals had traveled in space, 17 of them more than once; and 12 had walked on the Moon. Eleven three-man Apollo flights were launched. Two were Earth orbital, two were lunar orbital, and six orbited the Moon and landed there. Unlucky Apollo 13 used a free-return trajectory, meaning it flew around the Moon without assuming orbit.42

Where do we go from here? Or why should we go anywhere? Skylab, using Saturn rockets originally scheduled for lunar operations and a product of AAP planning and budget constraints, received President Nixon's

Apollo 16, with astronauts John W. Young, Thomas K. Mattingly II, and Charles M. Duke, Jr., accomplished the fifth lunar landing in the Moon's Descartes area. During three EVAs totaling more than 20 hours, astronauts Young and Duke collected 94.3 kilograms (about 207 pounds) of lunar material.

Scientist-Astronaut Harrison H. Schmitt is photographed on December 13, 1972, next to a huge, split boulder during the final Apollo (17) lunar mission. The photograph was taken by Eugene A. Cernan. Ronald E. Evans flew the support system module in orbit about the Moon.

The Apollo 17 splashdown, marking the end of the Apollo lunar programs, occurred very near the U.S.S. Ticonderoga on December 19, 1972, about 350 miles southeast of Samoa in the Pacific Ocean.

support in 1970. It was to be a large orbiting workshop, using systems originally developed for Apollo (Apollos 18 and 19). The orbiting workshop/space station had evolved from that early vision of Von Braun and others to become a "dry" S-IVB capsule with a limited Earth-orbit life that would be manned by three different mission crews: the first for 28 days, and the next two for 59 and 84 days, respectively (originally scheduled for 56 days each). Slippages in scheduling caused largely by budget problems, delayed the Skylab missions to 1973-74. Another project using a Saturn-Apollo system was under consideration, but not yet authorized. For several years, the Nixon administration had been negotiating quietly with the Soviet Union for a combined U.S.S.R.-U.S. spaceflight as yet another means of de-escalating the cold war. That mission would fly in 1975, but it was still being defined and developed in 1972.

In a meeting at the Peaks of Otter Lodge in the Blue Ridge Mountains in September 1972, center directors agreed that the coming decade would be considerably different from the past one. They discussed the future of NASA and the sense of "gloom and doom" that had begun to spread with the end of the lunar program. The consensus was that it was time for those who were weeping to get out so the rest of us could get to work. The work ahead had largely to do with seeing that NASA's future would be bright, innovative, and creative. NASA had to maintain adequate in-house capabilities, provide technical support to other government agencies, support more nonspace projects including local environmental projects, and stress low costs in terms of payloads, satellite design, and launch vehicles.[43] As did the entire NASA organization, MSC began considering its changing role in the post-Apollo era.

Frank A. Bogart, Associate Director of MSC (1969-1972) and a former Comptroller of the Air Force, during his last year before retiring conducted a careful organizational review of the center and queried each directorate about gaps, overlaps, personnel problems, and the role of the center and how the directorate and its divisions might support that role. The responses were diverse, but usually pointed. Some thought the center's organizational structure was simply out of date—and it was. The directorates did not have clearly defined roles and missions for post-Apollo operations. It had an "organizational interface" problem. There were overlaps in responsibility for the development and delivery of hardware. The center could not hire new people, particularly young engineering graduates fresh out of college. Contractor management responsibilities had become fragmented. The future work of the center would be much more diffuse and, in a sense, more difficult. The center would continue to be a busy place, for a decline in flight missions would be offset by multiple and more diverse programs relating to Earth resources and scientific studies. The old "prime" program system followed under Mercury, Gemini and Apollo had been replaced by a multifaceted group of programs requiring considerably more coordination and interface with other centers, Headquarters, and a multiplicity of contractors, few of whom would equate to the prime contractors of times past.[44]

Following the successful landing on the Moon by Apollo 11 in July 1969, space began to slip from the top rank of American priorities. War in southeast Asia and a war on poverty in America attracted growing concern, human energy, and federal money. Although NASA began considering man-in-space programs that might go beyond Apollo as early as 1959, most of NASA's energies, and especially the work at MSC, concentrated on designing, building and flying the Mercury, Gemini and Apollo spacecraft. A more defined Apollo Extension/Apollo Applications Program focusing on a manned orbital laboratory or space station began to develop in the mid-1960's. Concurrently, funding for ongoing programs and especially financial support for future programs such as AAP became constrained. Financial spending on the Apollo program and space peaked in 1965, 5 years before Apollo enjoyed its greatest operational successes with six lunar landings.

The seemingly sudden retrenchment in America's space programs was in fact the product of devolution over a period of years. Budget constraints and the rise of alternative national priorities resulted in efforts to achieve objectives in space with lower cost, more efficient machines and programs. Skylab became a cost-effective interim substitute for a long-duration space laboratory or station. A space shuttle was intended to provide a low-cost, reusable transportation system from Earth to a destination in Earth orbit. The destination, originally conceived as a large, orbiting space station that would be a very long-term space habitat and perhaps a way station to distant planets, would presumably follow the development of a low-cost space transportation system.

Suddenly, Tomorrow Came . . .

This Apollo 17 view of Earth is one of the most published photographs in the world. In explaining the impact of spaceflight, this one photograph has been worth many thousands of words.

By 1972, when President Nixon authorized development of the Space Shuttle, manned space programs by no means had been abandoned, but rather "slippage" in the broad time frame and other economies were intended to preserve the industrial and technological expertise derived from space, provide time for the absorption and distribution of the new knowledge, and preserve national leadership in space while concentrating national energies and wealth on war and the public welfare. MSC now concentrated on flying the Skylab and Apollo-Soyuz mission, designing and constructing a space shuttle, and reorganizing for the more complex and multifaceted post-Apollo era.

CHAPTER 11: Skylab to Shuttle

"Perhaps the most far-reaching event of the year just passed," Bob Gilruth told MSC personnel in his Christmas message at the close of the first year of the new decade, "may be the clear emergence of the Space Shuttle as the keystone of virtually all future efforts in space." But the past year had been difficult and the future was uncertain. NASA and its contractors were having to adjust to "lowered priorities and shrinking resources."[1]

The center began a transitional phase in the 1970's, shifting from its focus on managing Apollo to the role of managing the design and development of the Space Shuttle, or as it was first officially known, the Space Transportation System (STS). MSC became the lead center for the Space Shuttle Program. Apollo folded into the ensuing Skylab and Apollo/Soyuz programs. The center provided support for the Earth Resources Program intended to extend knowledge of the Earth and its resources from the vantage of space and space technology. The center continued a strong operations role through 1975 with three manned Skylab missions of 28, 59, and 84 days, respectively, in 1973-74, and a joint Russian-American space rendezvous and docking maneuver 2 years later.[2]

The new era required new management approaches and new thought processes. Concerned about the center's situation and role within the changing political and budgetary climate for space, Gilruth commissioned a special internal self-study or situation report. The report recommended "a not so subtle change in internal MSC thinking." Now, more than ever, MSC had to cooperate with other centers and with Headquarters. The old autonomy and independence could no longer be sustained. The center's image (within the NASA community) as being intellectually arrogant had to change. Its technical capabilities had to be sublimated to work more closely in concert with the expertise and strengths of other centers. Future spaceflight goals were expected to be less specific than for Apollo, and the level of funding and public support far less assured. There would be lower budgets and fewer personnel. Budget pressures conflicted with the center's and NASA's traditional engineering philosophy that hardware should be of the best design and best production quality possible irrespective of costs.[3] The center had to develop new thought patterns, reform its internal organization, and create new intercenter and contractor relationships, while preserving its old confidence and high technical standards.

The new era also brought new leadership. In January 1972, Chris Kraft replaced Robert Gilruth as Director, and Sigurd A. Sjoberg became Kraft's Deputy Director. Gilruth went on to a 2-year stint (January 1972 to December 1973) as Director of Key Personnel Development reporting to the Deputy Administrator in Washington, D.C. He was responsible for identifying near-term and long-range potential candidates for key Agency positions. Gilruth retired in December 1973 but continued as a consultant to the Administrator for several more years. At Headquarters, James C. Fletcher, with a doctorate in mathematics and physics and a research and teaching background at Harvard, Princeton and the California Institute of Technology, left the presidency of the University of Utah to replace Thomas O. Paine in May 1971 as NASA Administrator. Rocco Petrone, formerly Director of Launch Operations at Kennedy Space

Suddenly, Tomorrow Came . . .

Center, moved from the Apollo Program Office to replace Homer E. Newell as NASA's Associate Administrator. Dale Myers became Associate Administrator for Manned Space Flight in the Headquarters office. Wernher von Braun stepped down as head of the Marshall Space Flight Center in 1970 and was succeeded by Eberhard Rees, while Kurt Debus turned over the reins of the Kennedy Space Center to Lee R. Scherer in 1974.[4] New leadership and new program requirements produced substantive reorganizations at every level.

As the Apollo lunar missions wound down, MSC began to focus on the Apollo program "extension" —Skylab. Gemini personnel were the first to migrate to the Skylab program. Skylab, in truth, was an interim poor-man's concession to space station planning. It had roots in space station planning, Apollo applications, the Air Force manned orbiting laboratory program, and, more specifically, in studies centering in the Marshall Space Flight Center on how to use the Saturn-IV or S-IVB stage. Skylab could make further use of the Apollo systems. Budgetary constraints and an interest in an immediate and recognizable "payoff" from space operations were direct incentives to Skylab program development. Marshall Space Flight Center engineers and Douglas Aircraft engineers contributed to early Skylab designs, which were among the space station configurations being considered throughout the decade of the 1960's. The orbiting laboratory idea evolved, by 1966, to a "cluster" concept linking an Apollo spacecraft to an "orbital workshop." The concept had the attraction of using "surplus" Saturn IBs and Apollo spacecraft (and a Saturn V to launch the heavier workshop). Workshop experiments in earth sciences held the promise of direct and immediate economic benefits for people on planet Earth. Moreover, Earth-orbit missions cost less than lunar or planetary flights. Lower costs and the promise of immediate returns seemed to be a reasonable response to the public's growing disenchantment with outerspace projects. Even so, as David Compton and Charles Benson explained in their history of Skylab, Congress and the Johnson administration were at best lukewarm in their support of NASA's orbiting workshop plans.[5]

Mechanical failures during the May 14, 1973, launch landed Skylab in orbit overheated and starved for electricity. Astronaut EVAs saved the workshop and proved again the worth of humans in space. Skylab crews set new space endurance records and completed more experiments than originally scheduled, clearing the way for space experiments in the Shuttle era.

MSC, which became involved in the orbital workshop planning only in late 1965, at first opposed any configuration that did not provide a minimum of 0.1 gravity (to be created by spinning the workshop or station on an axis). The real hurdles remained budgetary and political. Finally, in July 1969, NASA received approval for an orbiting workshop and three rendezvous missions to begin in 1972.[6]

Skylab moved rapidly from inception to flight status between 1970 and 1973. The workshop, multiple docking adapter, and propulsion systems were managed by Marshall Space Flight Center, and MSC directed the astronaut training, scientific experiments, and Apollo spacecraft modifications and flight operations. NASA contracted with North American Rockwell for modifications to the Apollo spacecraft, with McDonnell-Douglas for two orbiting workshops, and with Martin Marietta for the docking mechanism. The Air Force contributed astronauts from its canceled orbital laboratory program, provisioning by its food and diet contractor (Whirlpool Corporation), and spacesuits from its contract with the Hamilton Standard Division of United Aircraft.[7]

The MSC Skylab Program Office, managed by Kenneth S. Kleinknecht, directed the necessary center resources (including its contractor and university associates) into the Skylab program, and it was the contact point for all other NASA and government agencies (including the Marshall Space Flight Center and the Kennedy Space Center) involved in any way with Skylab. The Mission Office, Management Operations Office, Engineering Office, Manufacturing and Test Office, Orbital Assembly Project Office and Apollo/Skylab Program Support Offices then interacted with their counterparts in the line and functional divisions of MSC or other NASA centers.[8] Although there were substantive organizational changes within MSC in the early 1970's, the essential elements of effective project management continued to be defining responsibilities and retaining a flow of communications between organizations at every level. The traditional elements of collegial association and doing what was required to get the job done remained intact.

By March 1973, just prior to the launch of the Skylab orbital workshop, the center's organization reflected the configuration indicated in figure 13. As usual, the charts and tables do not represent the actual fluidity of the organization and the interlacing lines of communications between the offices, divisions, and personnel.

To retain business management expertise in the director's office, Chris Kraft named William R. Kelly special assistant for management upon the retirement of General Frank Bogart in early 1972. A 1953 graduate of the Georgia Institute of Technology, Kelly went to work for General Electric following a tour in the U.S. Navy testing the J-79 jet engine. He joined the Mercury Program Office of MSC in 1962, transferred to the Apollo office in 1963, and became the Division Chief for the Institutional Resources and Procurement Division under Phil Whitbeck in 1970. These were difficult times, Kelly (who replaced Whitbeck as Director of Administration in 1981) recalled: "An erosion began in 1970. All of a sudden we had to lay people off. Why, when one was succeeding so well, did one have to begin firing employees? It seemed un-American to win and then lose."[9]

"Our new, young, capable employees took the brunt of the reduction in force (RIF). We lost all of the young people in my division," Kelly recalled. "In 1970-1971 we had about 10,000 contractor employees and 4700 civil service employees at the center. The numbers fell to 6000 contractor employees and 3200 civil service employees within a few years."

Suddenly, Tomorrow Came...

Johnson Space Center, 1973

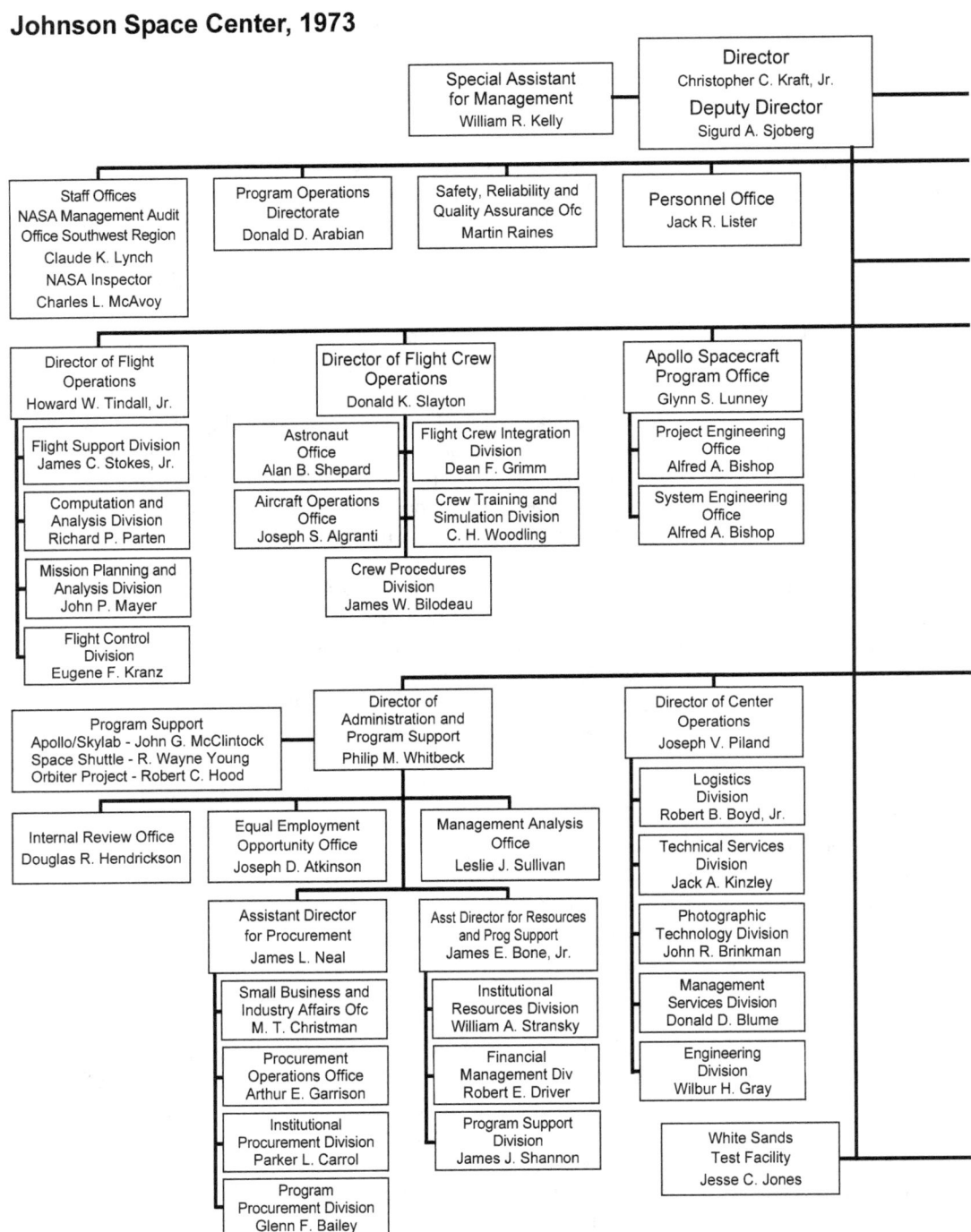

FIGURE 13. *Organization as of March 1973*

Skylab to Shuttle

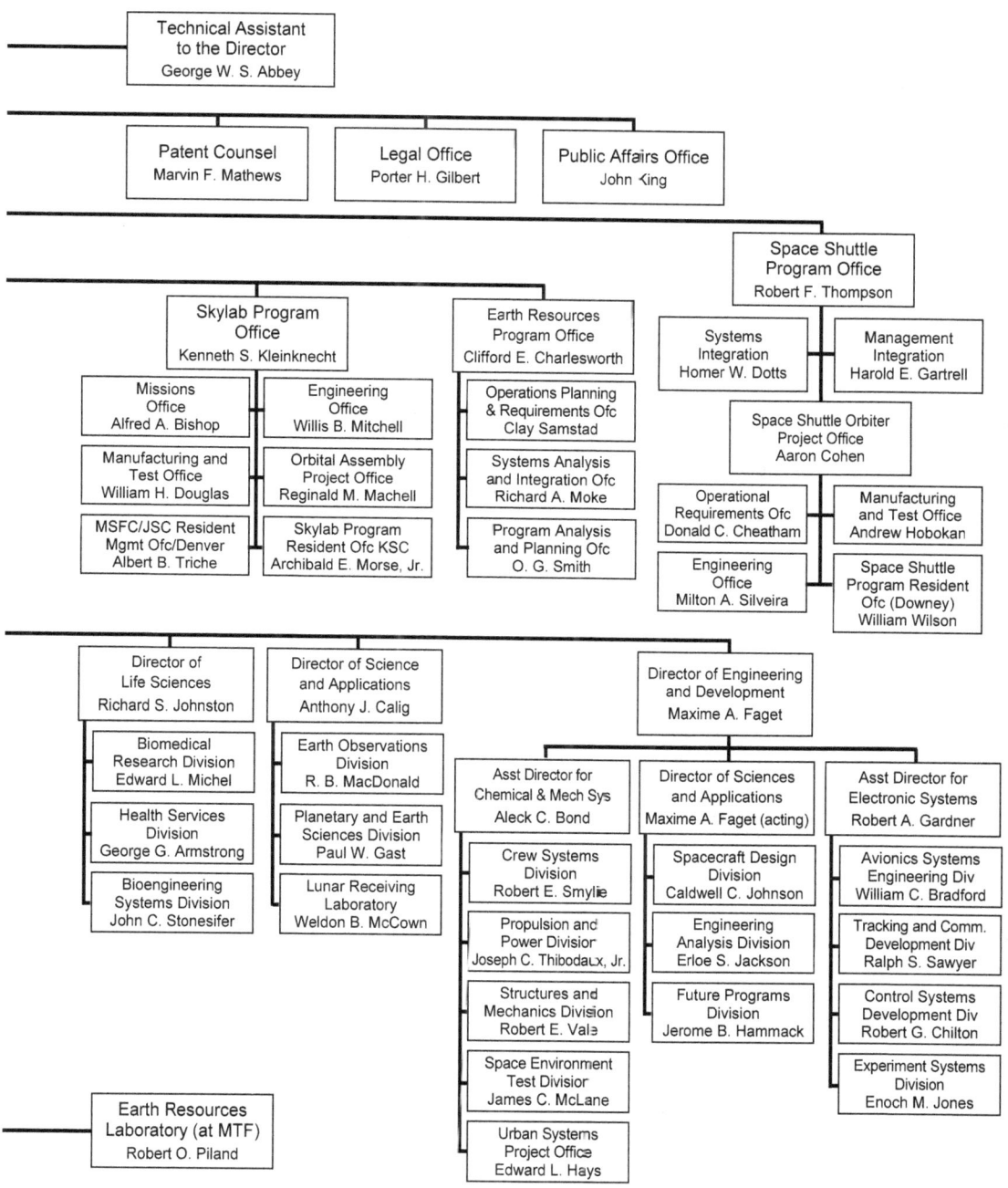

213

Suddenly, Tomorrow Came...

There were several long-range by-products of the reductions. The center lost its young people and part of its future; it lost much of its "hands-on" learning experience because of the in-house personnel reductions; and the center began relying more on its contractors.[10]

While the center reduced its civil service staff by about 600 persons between 1970 and 1972 through retirements and transfers, contractor employment dropped even more precipitously. Budget guidance from Headquarters indicated that MSC might "be required to achieve a lower employment level for the end of FY1972 than had previously been planned." Kraft announced in February 1972 that although the new ceilings had not been established as firm requirements, the center was required to prepare contingency plans to reduce its personnel beyond the levels previously anticipated before the close of the current fiscal year, make deeper cuts for fiscal year 1973 (to a new ceiling of 3727 employees), and reduce its average GS grade by .05—which meant reductions in pay or no promotions for many employees.[11] The center established an "outplacement center" to help the "riffed" employees find new positions. Many employees elected retirement. This was a sad beginning for a year which brought the highly successful Apollo 16 and 17 lunar explorations.

During this difficult period, the activity level and program responsibilities of MSC were being increased rather than decreased. The last two Apollo lunar missions flew in 1972, the orbital workshop and three manned Skylab missions were all launched in 1973. The Mission Control Center and recovery teams were on constant duty throughout the year, and throughout it all training and preparations were in progress for the 1975 Apollo-Soyuz joint mission. Indicative that the center had embarked upon a new era, but one drawing heavily from its past history and resources, in early 1973 MSC became the Lyndon B. Johnson Space Center.

Former President Lyndon B. Johnson, who had retired to his ranch on the banks of the Pedernales River near his boyhood home at Johnson City, Texas, died on January 22, 1973. "Few men in our time," President Richard M. Nixon said, "have better understood the value of space exploration than Lyndon Johnson. . . . Johnson drew America up closer to the stars, and before he died he saw us reach the Moon—the first great plateau along the way." It was he who helped draft, introduce, and enact the legislation creating NASA. Johnson's senatorial colleague, Lloyd Bentsen, introduced a Senate Resolution to rename MSC for Johnson: "Just as the Houston facility is a physical center of the space program," Johnson was perhaps, "the spiritual center." Bentsen called Johnson the "father of the space program" and NASA Administrator James C. Fletcher described him as the "principal architect of this Nation's space program." MSC personnel were pleased to have NASA's Texas center renamed for Johnson. The name change was effective on February 17, and a formal dedication was held at the center on August 27, 1973.[12]

While dignitaries and Johnson Space Center (JSC) personnel memorialized Lyndon Johnson, the second contingent of Skylab astronauts circled the heavens above in a near-Earth orbit. Skylab made a somewhat faltering start with the launch of the unmanned orbital workshop on May 14. A micrometeorite shield was ripped off during its launch. One solar panel was lost and the other failed to properly deploy, cutting electrical power available to one-half that expected. The loss of the shield also raised the temperature levels within the station to dangerous levels. JSC and Marshall Space Flight Center personnel mobilized for intense collaborative studies of the situation and developed what appeared to be workable solutions within a matter of days. Astronauts Charles Conrad, Jr., Joseph P.

Lady Bird Johnson and Center Director Chris Kraft during JSC dedication ceremonies on August 27, 1973. Just as the Houston facility was thought of as the physical center of the space program, Lyndon Johnson was remembered as "the spiritual center."

Kerwin, and Paul J. Weitz went to the Marshall Space Flight Center in Huntsville, Alabama, for a quick study on repair and restoration of the workshop. On the morning of May 25, they launched from Kennedy and docked with the workshop about 6:00 in the evening. During a number of EVAs, the astronauts erected a parasol of ultraviolet resistant materials "conceived, developed and constructed" by JSC engineers and technicians within one week, cleaned up the debris from the damaged meteoroid shield, and freed the undeployed solar power array. They initiated or completed a large number of experiments and returned to Earth after 28 days in space. The mission passed the Soviet's spaceflight endurance record, and turned a near-disaster into an outstanding success. While the Skylab crew was in orbit, 36 JSC employees retired from federal service as part of the NASA funding cutbacks.[13]

The Skylab 3 crew, including Alan L. Bean (commander), Owen K. Garriott (science pilot), and Jack R. Lousma (pilot), accompanied by two spiders, a swarm of vinegar gnats, a half-dozen mice, and an aquarium of fish, docked with the orbital workshop 8½ hours after lift-off on July 28. Bean had been the fourth man to walk on the Moon. Garriott had a doctorate in electrical engineering from Stanford University, and Lousma held a commission in the U.S. Marine Corps and degrees in aeronautical engineering. The crew encountered stability problems caused by faulty thrusters aboard the command module, had difficulty adjusting to weightlessness, and observed and photographed one of the largest solar flares ever recorded. Fifty-nine days later they returned with a new endurance record and a long list of scientific experiments completed—but the experiments that sought to study the circadian rhythm of pocket mice and gnats failed because of broken circuits providing power to the experimental package.[14]

On October 1, 1973, while astronauts Gerald P. Carr, Edward G. Gibson, and William R. Pogue trained for the final Skylab mission, NASA celebrated its 15th birthday. Skylab 4 launched from Canaveral on November 16, and 8 hours later docked with the workshop. The crew spent a record 84 days, 1 hour, and 16 minutes in space—including Christmas Day, 1973. The astronauts observed that the Earth looked terribly small from space; it was a "tiny blue island in the vast sea of space." Moreover, its populations seemed to be crowded into very small hospitable zones amidst vast areas of desolation. They were also the first humans to observe a comet from space. They photographed the passage of Comet Kohoutek, only recently discovered by Professor Lubos Kohoutek from an observatory in Czechoslovakia.[15] The Skylab crew returned on February 8, 1974. The Skylab office at JSC closed the following month.

A new JSC organizational review concluded that the center could no longer use Apollo-type program or project offices for the multiple programs anticipated in the future. Too many project offices and too many divisions and branches created fragmented responsibilities, scheduling difficulties, and operational inefficiencies. Project offices, it was recommended, should plan, coordinate and direct activities, while the regular line organizations such as administration, engineering, science, medicine, flight, and flight crew should provide technical and administrative support for all programs. Lead personnel from the line offices were to be attached to the program office to coordinate project planning with line office support. Personnel from other centers and "user agencies" needed to collocate with their respective activities offices at JSC. Finally, hardware development at JSC should remain the single responsibility of the Engineering and Development Directorate.[16]

As the Apollo era wound down, personnel recruiting and retention became an increasingly critical problem. Unless the center could, in the face of current personnel freezes and budget cuts, devise means to recruit young administrative and technical personnel, it could "run out of gas." Some economies could be obtained by consolidating support contractor work. For example, thermal and structural analysis and computer programming services were being provided by numerous contractors. Consolidation of service contracts could reduce costs and probably improve service.[17] Shrinking resources, as well as program diversification, mandated center reorganization and operating efficiencies.

While engineers worked to launch and operate Skylab, JSC experienced many consolidations and reorganizations. Thus the old Administrative Directorate became two new offices. The new Administration and Program Support Directorate headed by Philip H. Whitbeck focused on business management in support of the program offices and other center organizations, while a Center Operations Directorate headed by Joseph V. Piland had responsibility for facilities and center support services.[18]

The program offices (Apollo, Skylab, and Shuttle) reported to the Center Director through the Deputy Director. The Administration and Program Support Directorate assigned its own program managers to each program office.[19] Reminiscent of the subsystem managers assigned to contractors, the Administration Directorate's program managers provided greater flexibility in serving the needs of the project offices (figure 14).

The center reorganizations sought to address the seemingly interminable problem of effectively linking permanent center line directorates and support services to the program or project offices. Each of the past programs usually required the unreserved energy of the entire center. But those programs, Mercury, Gemini, Apollo, and Skylab, were relatively short-lived. When each ended, organizational structures had to be realigned to the new program. Long-term programs such as the Shuttle, accompanied by many diverse programs and projects, would require a constant shifting of center resources. A necessary but previously nonexistent mechanism for more responsive and efficient allocation of resources would be to institutionalize advanced planning.

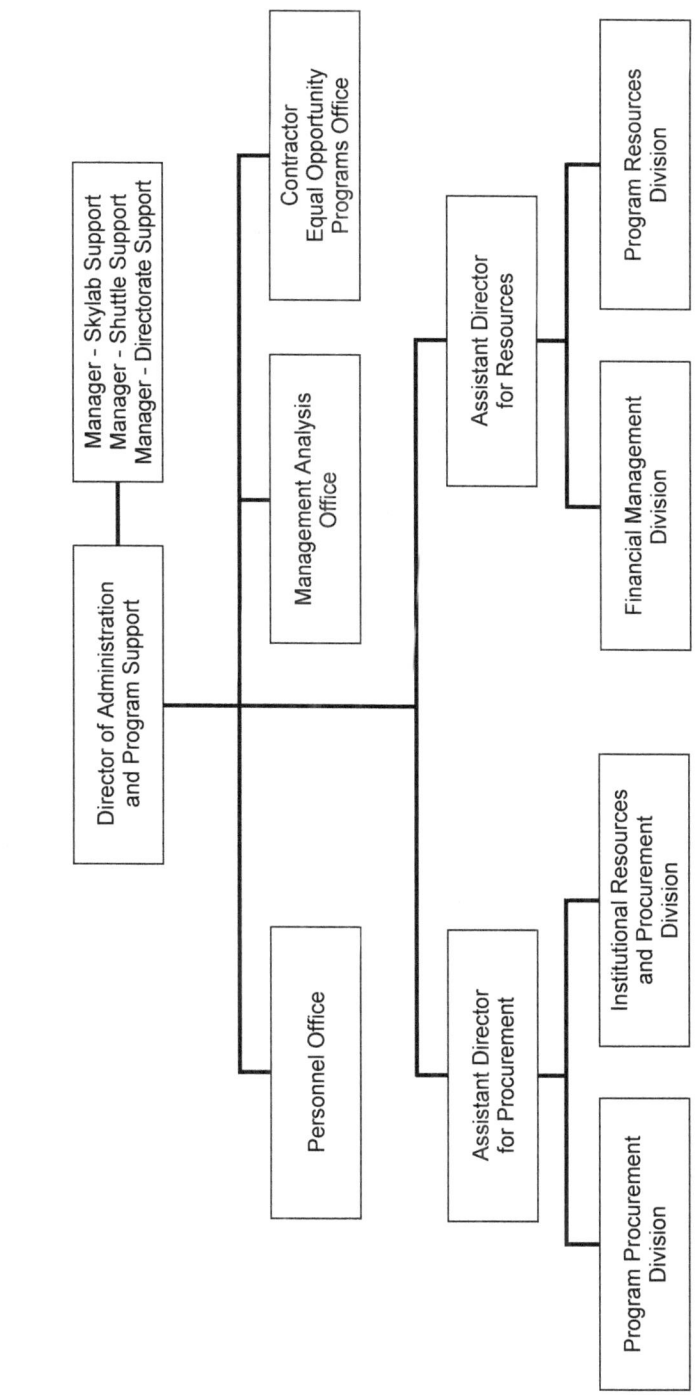

FIGURE 14. Organization as of January 1971

Advanced planning evolved from the ad hoc Apollo Extensions and Apollo Applications study groups to become institutionalized as the Advanced Missions Program Office. That functional office was now transferred to a new Future Programs Division under Jerome B. Hammack (Chief), which reported to Max Faget's Engineering and Development Directorate.[20] The Future Programs Office became the Advanced Programs Office within the Engineering and Development Directorate and much later advanced planning was elevated to the program office level with the creation of a New Initiatives Office reporting to the Center Director.

Other organizational changes came in response to a shift in spaceflight program emphasis. Whereas, at least through Apollo 11, spaceflight programs concentrated on development and operations, later Apollo and then Skylab flights focused on operations, science, and applications. Although a Science and Applications Directorate was established under Wilmot N. Hess in 1967, it held a secondary if not controversial status at the center.[21] A reorganization of the Science and Applications Directorate by the new director, Anthony J. Calio, in 1971 denoted the rising significance of science and applications in spaceflight programs. The Lunar Sample Office supervised the Lunar Receiving Laboratory. A Planetary and Earth Sciences Division included Physics, Geophysics, Geology and Geochemistry Branches, and a separate Earth Observations Aircraft Program Office provided data to correlate with information from space missions (table 7).[22]

A special Earth Resources Experiments Package (EREP) Investigations Office was created in January 1972 to manage the contracting phase of Skylab experiments. The manager of the EREP Investigations Office, O. Glenn Smith, on the staff of the Skylab Program Office, managed the contracting for the Science and Applications Directorate.[23]

As did the Administration Directorate, the Science and Applications Directorate coordinated with the program offices through assigned managers. Thus, Charles K. Williams, assigned to the Engineering Office of the Skylab Program Office, was responsible for the engineering and integration of the EREP being developed by the Science and Applications Directorate into the Skylab systems. He provided a single point of contact for all the program offices, line offices, contractors, and all organizational elements relating to the EREP and its integration into the Skylab program.[24] Through such interfaces, program management became integrated into the line functional divisions of the center consistent with the recommendations of the organizational study completed in 1970.

One of the most fascinating management challenges faced by the JSC and NASA during the post-Apollo initiatives was working with engineering and administrative counterparts in the Soviet Union during the Apollo-Soyuz Test Project. The program began by a direct request of President Nixon that NASA seek to develop significant foreign participation in post-Apollo programs. Nixon was very responsive to a suggestion from Administrator Paine that a joint U.S.-Soviet space project be sought and, with the approbation of the President, in late 1969 Paine sent Mstislav V. Keldysh, President of the Academy of Sciences of the U.S.S.R., copies of reports on long-range American goals in space. He invited the Soviets to join in a cooperative project. Keldysh suggested that a meeting be held by representatives of the two nations to discuss the possibilities. Although a meeting was not scheduled until late the following year, an exchange of correspondence explored the possibilities of a joint docking maneuver by an Apollo and a Soyuz spacecraft.[25]

TABLE 7. Science and Applications Directorate

Office of the Director	A.J. Calio, Director
Lunar Sample Office	J.W. Harris, Manager
Planetary and Earth Sciences	P.W. Gast, Division Chief
Physics Branch	R.J. Kurz, Chief
High Energy Physics	R.J. Kurz, Acting
Astrophysics Section	Y. Kondo, Chief
Geophysics Branch	D.W. Strangway, Chief
Geology Branch	W.C. Phinney, Chief
Geochemistry Branch	P.R. Brett, Chief
Isotope Section	P.W. Gast, Acting Chief
Mineralogy/Petrology	P.R. Brett, Acting Chief
Earth Observations Program Office	A.H. Watkins, Manager
Mission Management Branch	W.A. Eaton, Chief
Engineering and Requirements Branch	B.R. Hand, Acting Chief
Lunar Receiving Laboratory	P.J. Armitage, Manager
Curator	M.B. Duke, Chief
Facilities System Branch	I.E. Campagna, Chief
Laboratory Operations Branch	K.L. Suit, Chief

Source: MSC Announcement 71-54, April 15, 1971, Reference Series, JSC History Office.

In October 1970, Bob Gilruth headed a small NASA delegation for an initial visit to Moscow to discuss a joint venture. Glynn Lunney, who became the center's project manager for the Apollo-Soyuz flight, accompanied Gilruth. The Soviet team was headed by Boris M. Petrov, chairman of the Soviet Intercosmos Council. The representatives agreed to establish "working groups" to study systems for making Soviet and American spacecraft compatible for rendezvous and docking maneuvers. George Low and Mstislav Keldysh then met in January 1971. In April 1972, Low went to Moscow with Glynn Lunney for more intensive discussions. This meeting resulted in a final accord. Subsequently, when President Nixon visited Moscow on May 24, 1972, he and Alexei Kosygin, Chairman of the Soviet Council of Ministers, signed an agreement providing for cooperation and the peaceful use of outer space by the two nations. The leaders specifically approved the Apollo-Soyuz flight being planned and they agreed on a 1975 launch.[26]

While the diplomatic significance of a joint Soviet-American space venture in 1975 was considerable, the engineering aspects of that project faced formidable difficulties. A large delegation of Soviet and American engineers met at JSC in June 1972, and began the very serious work required to make the wholly independently designed and constructed Soyuz and Apollo spacecraft compatible. The two craft would require compatible docking hardware, radio communication modes, rendezvous navigational aids, cabin pressures, and intravehicular communication systems. Moreover, the flights would necessarily require cooperative "live" or real-time management by the launch and flight control teams—one located in the Soviet Union and the other in the United States. Flight teams and ground control teams needed long hours of joint training and simulation. There were marked differences in language, management, and engineering practices that compounded the technical differences. Simple geographical distance made travel and communication difficult. Six technical working groups (including the project directors) with Soviet and American counterparts on each were established to manage the Apollo-Soyuz Test Project.[27]

Glynn S. Lunney took representatives of three of the working groups with him to Moscow in October, accompanied by O.E. Anderson, Jr. from the Headquarters Office of International Affairs, three engineers from North American Rockwell, and secretaries and interpreters. Soviet and American engineers visited and collaborated while cosmonauts and astronauts trained at JSC and the U.S.S.R. Flight control teams in both countries ran simulation exercises. Each control center had direct communications with the other country's spacecraft. The two centers were linked by nine telephone lines, two of them equipped (in 1975) to handle facsimile transfers. Two television links connected the control centers by satellite. During the mission, six Americans and six Soviet mission specialists competent in each other's language and in technical terminology assisted in the other nation's control room.[28]

Three years after its approval, Alexei A. Leonov and Valeriy N. Kubasov were launched aboard Soyuz 19 on July 15, 1975, from Baikonur (Kazakhstan) in the Soviet Union. A few hours after the Soviet launch, Thomas P. Stafford, Vance D. Brand, and Donald K. Slayton were launched from Florida aboard Apollo CSM 111. The two vehicles then "found" each other following a script of very carefully timed and calculated orbital and phasing maneuvers. They began docking maneuvers on July 17 and completed the docking about 12 hours later. When the hatch swung open at 2:17 p.m. Houston time, the television camera showed tangled, spaghetti-like communication cables in the Soviet Soyuz craft. Then the Soyuz commander, Colonel Alexei A. Leonov, a cosmonaut since 1960, and Colonel (soon Brigadier General) Thomas P. Stafford, an astronaut since 1962, greeted each other. "Glad to see you," Leonov said, and the astronauts clasped hands, bridging for a brief time an incredible chasm of space, and the then almost equally insurmountable obstacle of national rivalry and cold war.[29] This is what the world saw and heard. It was an impressive and historic moment. What they did not see was that the moment also reflected the technical as well as political merger of two unlike systems.

The crews, in various combinations, entered each other's compartments, conducted a number of simple scientific experiments, and after 2 days of "joined" operations closed the hatches and undocked. An additional docking maneuver was completed before the final separation on July 19. Soyuz 19 returned to Earth on July 21, and the Apollo crew remained in orbit until splashdown on July 24. During the descent, the astronauts failed to activate two landing switches at the proper time causing nitrogen tetroxide fumes to enter the cabin. The pilot, Brand, lost consciousness and all of the crew became ill. When pure oxygen flooded the cabin, the astronauts recovered and made a safe landing. They were hospitalized for 2 weeks for treatment and observation, before going to Washington D.C., with their families for press conferences.[30]

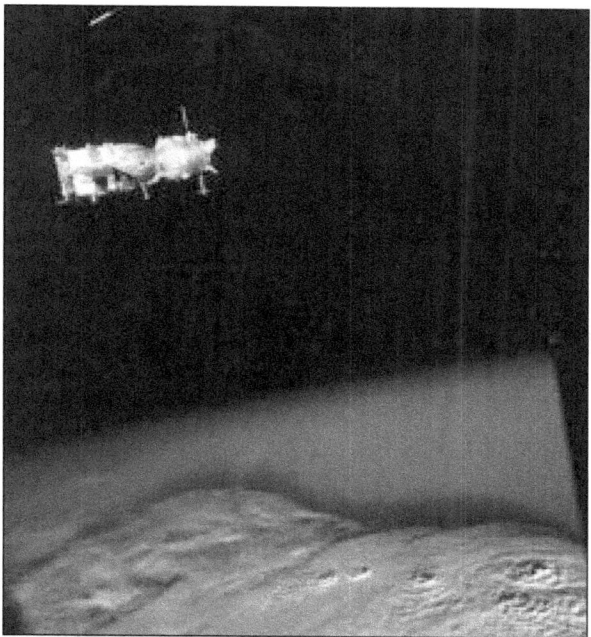

This July 1975 photograph of the Soviet Soyuz spacecraft, taken from the American Apollo spacecraft, is remarkable not only in documenting the technological feat of the U.S.-Soviet space rendezvous mission, but as evidence of a thaw in the omnipresent cold war that continuously threatened to erupt into open hostilities.

President Gerald Ford, who replaced Nixon in 1974 when the latter resigned under "Watergate" pressure, presented the astronauts Distinguished Service Medals, announced Stafford's promotion to Brigadier General, and had a private luncheon for the astronauts and their families. The crew then met briefly with Congressman Olin Teague, attended a reception hosted by NASA Administrator James Fletcher, and finally got a warm welcome back home in Houston. Then the astronauts were joined by cosmonauts Leonov and Kubasov for a 2-week tour of major American cities followed by a tour of Soviet cities.[31] It was at once a triumphant and eloquent finale for the Apollo program, and an uncertain and disquieting epilogue for the Apollo lunar missions. The brief crack in the Iron Curtain soon closed, and the political and technical dividends of the Apollo-Soyuz mission were left to be determined in some indeterminate future.

A hiatus set in. There would be no more manned spaceflights until the beginning of a new decade. But there was work to do and the promise of a bright future. That future, as Bob Gilruth observed, rested on the development of America's new space transportation shuttle system. The Shuttle, as did Skylab, emerged out of early considerations for establishing a permanently manned orbiting laboratory or space station in orbit about the Earth. The establishment of the Saturn-Apollo program, with its emphasis on large payload capabilities and fewer launches, as opposed to smaller payloads and more frequent launches consistent with a reusable aerospace-type vehicle, actually delayed active development of a space shuttle. Plans for a reusable vehicle "moved downstream by one vehicle generation." The primary rationale for the development of a reusable Earth-to-orbit craft, when it finally came to fruition, had to do with cost reductions.[32]

The Marshall Space Flight Center continued work on the reusable vehicle concept, and in January 1963 developed a statement of work for a fully reusable rocket-powered vehicle. The vehicle was to have low acceleration levels to accommodate non-astronaut passengers, easy access to payload bay areas, and multiple reuse of all stages. Marshall Space Flight Center let independent study contracts to Lockheed Aircraft Company ($428,000) and North

Suddenly, Tomorrow Came . . .

By 1977, the Apollo-Saturn hardware had been retired. This S-1C stage of the Saturn V launch vehicle is being transported by barge for display at JSC.

American Aviation ($342,000). There were other in-house and small contractor analyses. Paraglider studies and contracts for land recovery of Apollo systems related both to reusable vehicle concepts and to reducing space vehicle recovery costs. Conceptual ideas for recovery and reuse of space vehicles and Saturn stages included "air snatches," parachute and retro-rocket recovery, hot-air balloons, and expandable wings.[33]

Early concepts for the reusable space vehicle stressed horizontal take-offs and passenger reliability and safety. Air-breathing propulsion, linked with rocket engines, was believed required for initial flight stages. Configurations for reusable craft ranged from a 10-ton orbiter for passenger transportation, to a 50- to 100-ton space truck cargo carrier. Numerous space shuttle studies completed between 1963 and 1967 established, among other things, a preference for a vertical take-off mode and the elimination of air-breathing engines from incorporation in the propulsion system. Studies also proved the feasibility of a fully reusable two-stage launch vehicle, but indicated that at launch rates of four to eight per year, a reusable lifting vehicle coupled with an expendable launch vehicle would be more economical.[34] A decision on adopting the fully reusable or partly reusable system was not reached until 1971.

By 1967, a consensus developed that a new spacecraft, coupled with an updated Saturn I first stage booster, would be useful for several decades and would produce the most savings that could be expected from reusability. The reusable shuttle studies began to tie in with AAP/post-Apollo planning. Max Akridge, in the Program Development Directorate at the Marshall Space Flight Center, attributes the real inception of the

Shuttle to an Apollo applications conference in Houston on October 27, 1966, attended by A. Daniel Schnyer from Headquarters, Harold S. Becker, C.H. Rutland and Akridge from Marshall, and W.E. (Bill) Stoney, Caldwell Johnson, Ed Olling, Carl Peterson and others from MSC. The meeting resulted in an agreement between Marshall and MSC, with the concurrence of Headquarters, that Marshall and MSC should pursue independent studies of a shuttle system reflecting the parameters of a March 1966 statement of work for a Reusable Ground Launch Vehicle Concept and Development Planning Study (which focused on a nine-person reusable vehicle).[35]

But a conference at Headquarters on January 19, 1967, stressed using the present stable of space vehicles without new launch vehicle development in order to reduce the strain on NASA budgets. And despite a sharpening of the definition of a reusable vehicle, prototypes still varied rather widely from a modified Apollo four- to six-person module, to a new nine-man space ferry being studied by McDonnell Douglas, to a rocket plane strapped onto a Saturn launch vehicle. While technical studies continued on the merits of different systems, "policy" deliberations became increasingly pertinent. George Mueller called a meeting with center representatives and contractors at Headquarters on January 6, 1968, to discuss orbital transportation concepts—and *low cost* operations.[36] Mueller outlined the status of the problem at that time:

> Where we stand now is that feasibility generally has been established for reusability. And we have much data on many concepts. We have an uncertain market demand and operational requirements. The R&D costs for fully reusable systems, including incremental development approaches, appear high. Personnel and cargo spacecraft seem to dominate Earth-to-orbit logistics costs. R&D costs for new logistics systems are in competition with dollars to develop payloads/markets (dollars are scarce).

Participants reviewed with Mueller the studies and conclusions reached thus far by the centers and their contractors relating to a reusable space transportation system. They pondered, "What should we do?"[37]

With the encouragement of the Air Force, which was studying orbiting laboratories and aerospace planes, NASA held a number of work sessions in April and May of 1968 to establish a general agreement of what "we should work toward." The sessions produced 10 variations of a statement of work for a logistics space vehicle, but there was a consensus on guidelines that the emphasis for the shuttle would be for space station logistics missions, the payload range would be from 5000 to 50,000 pounds, and it should be operational by the mid-1970's.[38]

In October, Clarence Brown and Max Akridge from Marshall Space Flight Center flew to MSC to talk with Wayne Corbet, Don Hathaway and others about issuing a joint Phase A (concept definition) study request for proposal. MSC issued the request with proposals to be in by November 29. After an evaluation by teams at Marshall and MSC, the two centers sent a joint TWX to Headquarters requesting approval to award a contract. At Headquarters, Robert Voss resolved some problems involving the request with the two centers, and on December 10, a final letter requesting authority to negotiate and award Phase A shuttle contracts, signed by Marshall and MSC representatives and with the concurrence of Voss, went to NASA Headquarters.[39]

Mueller then called Wernher von Braun asking him to support the shuttle rather than continuing to urge an interim vehicle. Von Braun replied, "If Nixon wants to spend $3 billion, who am I to say no?"[40] But Headquarters approval of the contract negotiations remained on hold while Apollo 8 was readied. That was the flight, it should be recalled, that was advanced to a lunar orbital flight through the collaborative efforts of Marshall and MSC. That successful flight, and the subsequent lunar landing in 1969, strongly advanced the prospects for a post-Apollo shuttle program.

On January 23, 1969, George Mueller approved contract negotiations for the initial shuttle design work and specified that the contract negotiation team would include members from MSC, Marshall Space Flight Center, and Langley Research Center, with the negotiation team to be headed by MSC. It should be noted that this initial arrangement very likely established a precedent for the shuttle "lead center" concept by which MSC was designated the lead center for shuttle developments with other centers to provide independent but supporting roles. To be sure, lead center style operations were a strong part of the NACA tradition, wherein each center specialized in a certain expertise. Another precedent for what became a lead center relationship in the shuttle program came from the early days of NASA when, under Langley Research Center "lead," Ames and Lewis Research Centers played strong supporting roles in developing the Mercury and Apollo program concepts.

Now, in the initial Shuttle contracting phase, Langley Research Center was to manage a contract with McDonnell Douglas, two contracts with Lockheed Missiles and Space and with General Dynamics/Convair were to be directed by Marshall Space Flight Center, and MSC was to manage a contract with North American Rockwell involving configurations for an expendable lower stage and a reusable spacecraft. Each contract was budgeted in the $300,000 range. Other feasibility studies of alternate shuttle designs were being conducted privately by Martin Marietta, and the following year new contracts were awarded to Grumman Aerospace/Boeing, Lockheed Aircraft, and Chrysler Corporation. A review team chaired by Headquarters' A. Daniel Schnyer, Director of Transportation Systems, provided a joint review and reported regularly to a space station coordinating group.[41] What this meant was that the "Shuttle era" began for different reasons than did Mercury or Apollo, and it used different management approaches. It involved NASA-wide integrated management systems with a lead center providing direction and coordination but not necessarily control.

During discussions in Houston of the North American Rockwell study on an expendable low-cost launch vehicle and a reusable spacecraft, MSC personnel (and prominently Ray Bradley, John Hodge, and Caldwell Johnson) disagreed rather strongly with ground rules urged by Engineering and Development Director Max Faget. Daniel Schnyer and Robert Voss from Headquarters were also unhappy. Faget urged cost reductions on the launch vehicle configuration through the use of a solid propellant rocket-powered vehicle, eliminating redundant and expensive launch escape systems. He advised working on only one or two launch vehicle/shuttle configurations.[42]

John Hodge believed that during initial discussions of the shuttle, Gilruth and other administrators at MSC generally were "dead against" the shuttle as it had evolved, while

George Mueller at Headquarters was its primary champion and regarded the shuttle as a necessary program to relieve the inevitable hiatus that would follow the close of the Apollo flights. Although the shuttle was keyed initially to a projected space station (for which Marshall Space Flight Center was to be the designated lead center), an internal MSC study completed under Hodge's auspices recommended that NASA's priority be the shuttle. A space station could be attempted only after the completion of a shuttle.[43] But through at least 1970, NASA anticipated working both on the shuttle and the space station as integral components of Earth resources studies and further planetary exploration.

While differences and discussions between the centers, between the centers and Headquarters, and within the centers over the feasibility and design of a space shuttle continued, Headquarters initiated a joint DoD/NASA study on space transportation. Concurrently, it should be recalled, a special task committee under Milton Rosen and a planning steering committee headed by Homer Newell were preparing recommendations for future programs. Mueller next assembled shuttle task teams from Marshall, MSC, Langley, and Kennedy Space Center to help define the tasks and approach to the planned DoD/NASA study. He also appointed a Headquarters task group including representatives from each of the spaceflight centers to work in Washington, D.C., under Leroy Day until the joint DoD/NASA study was completed. Meanwhile, at each center at the end of April, or during the first few days of May, contractors presented their initial findings and data from Phase A studies. On May 5, 1969, Mueller opened a shuttle review conference by noting he expected to have reports from Day's in-house space task group by mid-May and from the joint DoD/NASA study by mid-June. He added that he and Grant Hanson, formerly a General Dynamics vice president now working with the DoD study group, would present the reports to President Nixon's Space Task Group.[44]

Mueller also outlined the developing configuration of the shuttle, and personally urged the selection of a vehicle with a 50,000-pound payload capability (a container 22-feet in diameter by 60-feet long) and a clam-shell bay door or a swing-nose hatch. In describing the work of such a vehicle, Mueller compared it to a pickup truck, a utility vehicle, or a city shuttle. Max Akridge threw in the word "space" and the "space shuttle" nomenclature became the accepted in-house phrasing for NASA's officially designated space transportation system. The total system included launch capabilities, an integrated launch and reentry vehicle (ILRV), a low-cost Earth-orbit transportation system, and a reusable space vehicle.[45]

During the remainder of 1969, study contracts were revised and extended as new data and resolutions developed. Controversy over design, motives and methods was most often intense. Meetings, most of them in Washington but many at the centers, and review sessions with contractors led finally by December 1969 to the production of a report "Summary Report, Recoverable vs. Expendable Booster, Space Shuttle Studies," that was concurred in by MSC, Kennedy Space Center, and Marshall Space Flight Center. The point of the report, according to Max Akridge, was that by the beginning of the new decade, NASA had exhaustingly explored all possibilities for post-Apollo space vehicles and concluded that the best of the possibilities was a fully reusable system.[46] That, of course, is not what came to be.

Significantly, the Shuttle program was to be one element of the broader space goals outlined by President Nixon's Space Task Group in 1969. But as explained in the previous chapter, the definition of goals failed to equate with worsening budget realities. The

Space Shuttle was to have been one component of a multidimensional program including a permanent space station, unmanned probes to the planets, and manned planetary missions. But a year passed with all eyes still on the Apollo lunar missions and the energies of MSC still devoted almost exclusively to current programs and missions.

The Shuttle got put on the shelf during 1970, while Congress, the Executive branch, and the public agonized over Vietnam and double-digit inflation. President Nixon attempted to "wind down" the war in Vietnam, and announced a plan for the phased withdrawal of American troops. However, in April the administration decided to invade Cambodia in order to destroy enemy bases and supplies thus, in theory, improving the ability of South Vietnamese forces to sustain the war without American involvement and shortening American participation in the war. Many Americans, however, saw this simply as an extension of the war in southeast Asia and public protests rose precipitously. The violent deaths of student protesters at Kent State in a confrontation with National Guard troops created deepening public tensions and controversies. While the public was fast losing interest in Apollo lunar landings and was seemingly uninspired by the announced Skylab or projected Apollo-Soyuz programs, plans for a future program such as the Space Shuttle or a space station were even more remote from the public mind.

Only in 1971 did the President approve program development of the Space Shuttle, and although the related programs were not literally abandoned, in reality what remained in place after 1972 was an emasculated Shuttle program without a space station. There were plans for unmanned planetary scientific probes, orbital and suborbital physics and astronomical projects, life and bioscience experiments, basic scientific and aerospace research, and the promise of continuing study on space stations and planetary missions.[47] But for relatively small attention spans during the Skylab and Apollo-Soyuz missions, space for the American public was passing out of sight and out of mind. Gilruth was quite right when he said in December 1970 that the Shuttle would be the keystone of all future efforts in space, but he was unaware that the Shuttle would be for a time the only manned space program under development.

To be sure, the dwindling contractor and civil service personnel at the center had a very full plate from 1970 through 1975, with the conclusion of the Apollo lunar flights, the development and operation of Skylab, and the Apollo-Soyuz mission. As time passed, MSC began devoting more of its resources to shuttle development and management.

Despite the inactivity on the Shuttle front in 1970, there were some important management changes bearing directly on the Shuttle program in that year. At Headquarters, George Mueller resigned as Associate Administrator for Manned Space Flight on December 10, 1969, to accept a position as vice president of General Dynamics Corporation. Controversies within NASA over the Shuttle program, budgetary uncertainties, and personnel changes at higher and lower levels within NASA undoubtedly contributed to Mueller's departure. Dale D. Myers, a vice president and general manager for the Space Shuttle program at North American Rockwell, replaced Mueller. Myers became involved in North American's space-related programs in the 1950's, and was the company's Apollo program manager in 1964, and a vice president and general manager of the command and service module work in 1968. He was tremendously knowledgeable about NASA and its space programs and had a strong record in production and management discipline.[48] He also had a long and effective working relationship with MSC.

Myers assumed his new position keenly aware of the "watershed" situation in which NASA found itself. In May 1969, prior to the lunar landing, Myers expressed concern that a lunar landing might be followed by a tremendous emotional letdown in the aerospace industry. North American employment peaked long before. Company employment was already down by one-half and still falling. He sensed an apparent lack of interest in Congress and a propensity in the public mind to solve budget problems by cutting NASA expenditures. A lot of people, he said, seemed to think that a lunar landing, when it came, was going to be just one big trick. People, he said, have got to develop a much longer view of space.[49] He prepared for what became an inordinately long Shuttle development and production program.

Myers established the Shuttle Program Office under the Office of Manned Space Flight and appointed Charles J. Donlan (an original member of the Space Task Group who had remained at Langley as associate director when the new MSC was located in Houston) program manager. Myers also created a new Shuttle Program Task Group under Donlan's direction to reexamine the technical and management aspects of the Shuttle program. Donlan urged a restructuring of the management system to provide greater coordination and integration among the centers on the Shuttle project, but to establish one center, MSC (Johnson Space Center), as a lead center. He recognized, as did Gilruth, that the Shuttle was intrinsic to future efforts in space and was most closely related to the technical expertise possessed by MSC. Low supported Donlan in the "lead center" concept.[50]

He regarded the lead center management style as a way to preclude some of the conflicts between Headquarters and the field centers that plagued the Apollo program. Low and Myers wanted to hold Headquarters staffing to a minimum. NASA personnel cuts from the 1968 peak averaged one-third by 1973, with heavier cuts being made at Headquarters and Marshall Space Flight Center than at other centers. Thus the lead center concept was appealing as a device to most efficiently use NASA's dwindling resources.[51]

Although he had some reservations, Myers announced in June 1971 that MSC would have "program management responsibility for program control, overall systems engineering and system integration, and overall responsibility and authority for definition of those elements of the total system which interact with other elements . . . (a Level II responsibility)." The center also received primary development responsibility for the orbiter stage of the Shuttle, while Marshall would manage the booster stage and main engine for the Shuttle, and Kennedy Space Center would design and direct launch and recovery facilities. Headquarters (Level I), Myers announced, was to manage the overall program and have primary responsibility for the assignment of duties, basic performance requirements, allocation of funds to the centers, and control of major milestones.[52]

Shuttle project integration, it was felt, could be accomplished with Apollo-type Integration Panels. The panels included members from Kennedy, Marshall and MSC with Marshall and MSC co-chairing the panels. Under the lead center system, integration panels reported to a Systems Integration Office at MSC which in turn reported to a Policy Review Control Board chaired by NASA Headquarters.[53]

Suddenly, Tomorrow Came...

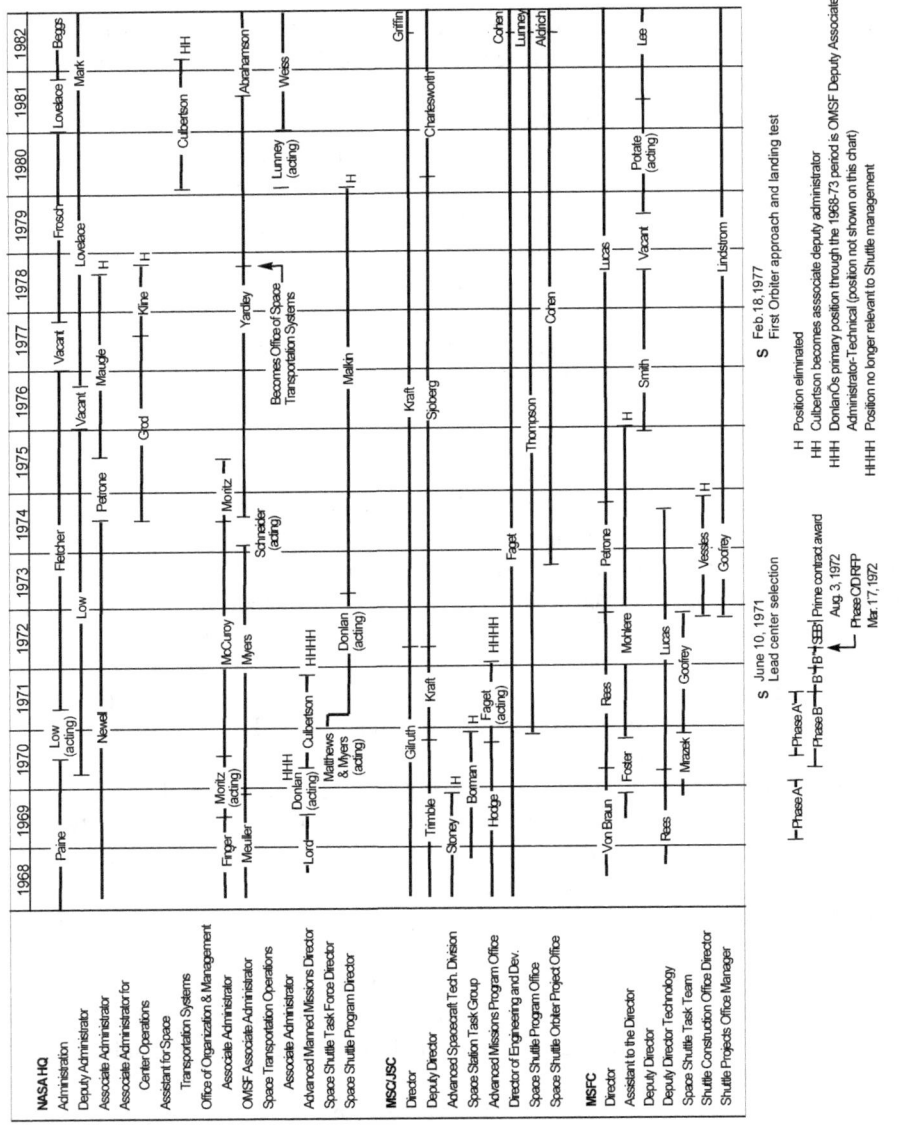

FIGURE 15. *Shuttle Management Timeline, 1968-1982. This chart portrays the development and administrative history of the Space Shuttle from its conceptual stage in 1968 through the first flight in 1981.*

228

Space Shuttle responsibilities at MSC rested on Robert F. Thompson, whom Gilruth appointed Space Shuttle program manager in April 1970. Another of the original Space Task Group personnel from Langley Research Center, Thompson headed the MSC Landing and Recovery Division during the Mercury, Gemini, and early Apollo years; and in 1966 became manager of the Apollo Applications Program Office and then Skylab. As manager of the Apollo Applications Program Office, Thompson "grew up" with the Shuttle. As the lead center program management office, Thompson's office provided a day-by-day integration and overview of all aspects of Shuttle design and development. Thompson said that his office was "concerned about everything happening across the total program."[54] Despite the fact that the Shuttle management system was in place by the end of 1971, the Shuttle remained in a very tenuous and uncertain status.

When NASA Administrator James C. Fletcher replaced Paine in April 1971, he became convinced that a fully reusable Shuttle such as that recommended in the NASA "Summary Report" of December 1969, and which bore a tentative price tag of $10.5 billion, would literally "not fly"—especially with Congress. He initiated a "rigorous restudy and redesign." The redesign study, focusing on the use of an expendable launch stage (which had long been considered a viable alternative to a fully reusable shuttle) lowered costs by about one-half.[55]

Fortified with this new data, on January 5, 1972, George Low and Jim Fletcher met with President Nixon and John Erlichman, his staff assistant, for a review of the Shuttle program. Nixon wanted the Shuttle program to stress civilian applications, but not to exclude military applications. Moreover, he wanted the public informed of any military applications. Nixon was pleased that ordinary people could fly in the proposed Shuttle. He was concerned that the skills of people in the aerospace industry be preserved. He said NASA should stress the fact that the Shuttle was not a "$7 billion toy," but a good investment. Even if it were not a good investment, Nixon continued, "we would have to do it anyway, because spaceflight is here to stay." He did ask that NASA stress international cooperation and participation for all nations. When the meeting ended, Nixon approved the Shuttle program and asked Erlichman to mention to Secretary of State Henry Kissinger the international aspects of the Shuttle and the planned U.S.-Soviet docking maneuver for 1975.[56]

The MSC Shuttle Program Office began "gearing up" for the center's lead role. Thompson added a few personnel to his small program office, and began making arrangements for key personnel from other participating centers to sit on his staff in Houston. Since much of the actual Shuttle design and development would come from the Engineering and Development Directorate, Milton A. Silveira became the representative from that directorate on Thompson's program staff.[57] Under the new organization scheme, program offices would avoid duplicating any technical or functional operations available within the center or at other NASA installations. Duplication and competition between program offices and line offices, such as had occurred under the Gemini and Apollo programs, were to be avoided by preserving the *program* management and administrative integrity of the program office.

For example, traditional administrative tasks not directly program related were performed by the MSC Administrative and Program Support Directorate. R. Wayne Young from the Administrative Directorate was assigned to the Shuttle Program Office to manage contracts and procurement through the directorate. When it became appropriate, the Flight Crew Operations Directorate and the Medical Research and Operations Directorate assigned representatives to the program office.[58] Thus the system sought to preserve the integrity of the line offices, prevent duplication, reduce expenses, and promote overall operating efficiency. The management style was actually not so new. It really reflected a more disciplined subsystem management approach in reverse, that is the subsystem managers (or their representatives) were assigned to the program office. Wesley Hjornevik, it should be remembered, placed contracting and procurement personnel in program and functional offices, and with contractors when it seemed advisable, to expedite and improve program development.

With 6 or 8 years of tumult, confusion and controversy behind it, the Shuttle now began to move from the conceptual stage to the design and production phases. Rocketdyne was selected to design and build Shuttle engines under Marshall Space Flight Center management. MSC awarded separate contracts for developing and testing ceramic insulating materials for Shuttle reentry to McDonnell Douglas, General Electric, and Lockheed. The center also requested proposals for the development of a low-density ablative material and a design study for an orbital maneuvering system (OMS). NASA invited proposals for the development of the Shuttle in March, and in May received proposals from North American Rockwell, McDonnell Douglas, Grumman, and Lockheed. An interim letter contract was approved with North American Rockwell in August 1972, and the final contract was approved in April 1973. North American Rockwell (which became Rockwell International) subcontracted with Fairchild Industries for the vertical tail section of the Shuttle, with Grumman for the delta wings, with Convair Aerospace Division of General Dynamics for the mid-fuselage, and with McDonnell Douglas for the OMS. In early 1974, JSC contracted with IBM for computer software designs for Shuttle support systems. A full-scale center/contractor design review was held in early 1975, and a test model (comparable to the boilerplate Apollo mockups) was completed in 1976. Despite the intensive work on the Shuttle, flight was still 5 years away.[59]

Much of the work on the Shuttle had to do with the design and construction of laboratory and testing equipment at JSC that could handle the intricate electronics and flight simulation tests required for the Shuttle. Most of the existing laboratory and testing facilities, such as the thermal and materials test laboratories, continued in full service. The more sophisticated design of the Shuttle required considerably more advanced test facilities. Computer hardware and software programs, communications, electronics, and even materials used in Mercury, Gemini and Apollo command modules were primitive compared to the degree of sophistication of Shuttle orbiter systems. Space had so advanced the frontiers of aerospace and related technology that earlier systems and the tools used to build and test them were becoming obsolete. The Shuttle Avionics Integration Laboratory (SAIL) coupled with a new Shuttle mission simulator became the primary test and training devices for the development and operation of Shuttle systems.[60]

New technology, or the application of new knowledge in engineering, medicine and the sciences, had much to do with creating tools and laboratories that could manage that knowledge effectively. Much of the post-Apollo work of NASA and of JSC had to do on the one hand with harnessing the existing knowledge in such a way as to advance humankind's explorations in space either by increments or by leaps and bounds, and on the other hand to understand and apply the new knowledge obtained from space both to further the frontiers of space and to improve the welfare of people in the United States and on Earth.

Paradoxically, trauma followed the triumphs of the Apollo program for NASA and MSC. Skylab, the Apollo-Soyuz mission, and even the Space Shuttle with its emphasis on economies were products of the changing political and budgetary environment for space programs, as were the new organizational structures and management approaches.

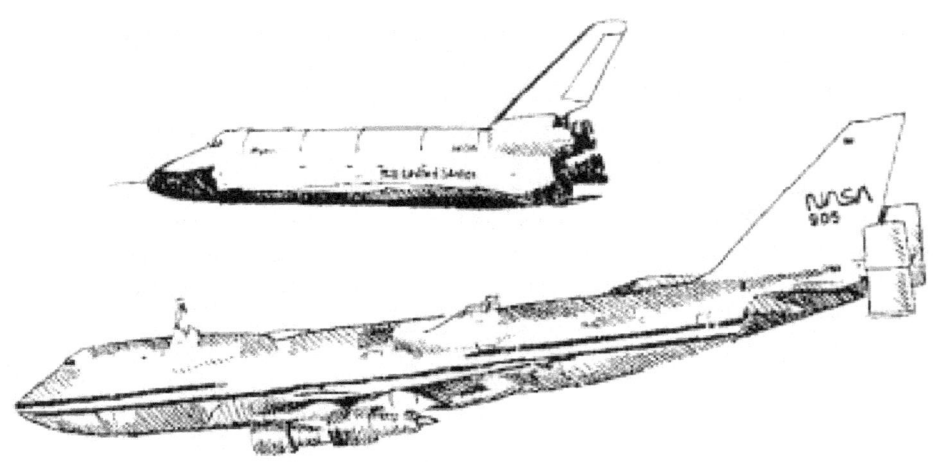

CHAPTER 12: *Lead Center*

Johnson Space Center personnel began work on the Space Shuttle under the cloud of reductions in force, tightening budgets, growing public apathy or outright criticism, and unclear and changing program directives. In the 1970's, JSC evaluated its new status, responsibilities, and capabilities; retooled to provide the engineering expertise demanded by the Shuttle; helped NASA and American society assimilate the lessons learned from Apollo; and recruited and trained astronauts for missions scheduled for the 1970's, but actually flown in the next decade. Between 1970 and 1974, JSC lost one-fourth of its employees, Marshall Space Flight Center lost one-half, and NASA Headquarters accepted the largest proportion of required "RIF" or reductions in force in order to help preserve field operations. Budgetary constraints contributed to the decision to establish a lead center management approach for the Shuttle and affected program development.[1]

The Space Shuttle developed in a markedly different social and technical environment than had the Mercury, Gemini, and Apollo programs. Cost pressures had a direct bearing on its conceptual design and configuration. Costs affected the contract awards, production schedules, and mission planning. Costs affected the style and structure of Shuttle management. Based on budget guidelines for fiscal years 1972 through 1974, NASA Administrator James C. Fletcher described the program as "austere but meaningful."[2] The Sputnik crisis environment of the 1960's became a business-management-cost-effectiveness environment for the 1970's.

A NASA management document called the *Catalog of Center Roles*, issued in April 1976 and revised in December, explained the overall role of NASA as "the conduct of a broad program of research and development aimed at achieving the Nation's goals in aeronautics and space." The NASA field centers possessed distinctive capabilities, technical excellence, and the facilities necessary to accomplish the overall program. JSC's principal role had to do with the development and operation of manned space vehicles and the required support technology and systems.[3]

Headquarters assigned JSC responsibility for the development of the orbiter, that is, the manned shuttle vehicle, and designated it lead center for the management of the entire Shuttle system. This meant that while Headquarters was responsible for planning and policy decisions (Level I), JSC had responsibility for Level II program management relating to systems engineering and integration, configuration, and design and development. Level III project offices such as the orbiter manager at JSC, the booster manager at Marshall Space Flight Center, and the launch and recovery operations manager at Kennedy Space Center had responsibility for their specific projects and reported to and were coordinated and integrated by the Program Manager at JSC. Under the lead center management system, Headquarters effectively delegated engineering and development management to JSC while allocating resources among the centers and among tasks and exercising overall program direction. The Shuttle Program Office exercised cost controls within the funding parameters established by NASA.[4]

The lead center system facilitated the integration of technical capabilities with program management. The Shuttle Program Manager, Robert F. Thompson, came through the NACA ranks into NASA. He earned a degree in aeronautical engineering at Virginia Polytechnic Institute in 1944 and served a 2-year stint as a naval officer before joining NACA in 1947. An original member of the Space Task Group, he later headed the MSC Landing and Recovery Division before becoming manager of the Apollo Applications Program (and Skylab). Now, as Shuttle Program Manager, Thompson had overall responsibility for the development of the Shuttle.[5] Other centers provided support roles and services for the Shuttle program and worked through the JSC Shuttle Program Office.

Thus, Marshall Space Flight Center's project units having responsibility for the design, development, production, and delivery of the Space Shuttle main engine, the rocket boosters, and the external hydrogen-oxygen propellant tanks reported to the JSC Shuttle Program Office. Langley Research Center examined Shuttle payloads and conducted aerodynamic and aerothermal testing. Ames Research Center focused on Shuttle passenger selection criteria, astronomical observation systems, aerothermal dynamic analysis, and materials development. Goddard Space Flight Center provided tracking, data acquisition, and network planning support for Shuttle flights. Dryden Flight Research Center gave direct support to JSC for approach and landing tests of the Shuttle orbiter. Personnel from Marshall Space Flight Center, Kennedy Space Center, and JSC were collocated in certain functional areas at the centers under the authority of the Space Shuttle Program Manager.[6]

The program manager had overall technical responsibility and management authority. He directed, scheduled and planned all elements of design and production, and imposed cost controls on all elements of the program. The office responded directly to Headquarters on matters relating to the Shuttle, and communicated directly with Shuttle project offices at Marshall Space Flight Center, Kennedy Space Center, and JSC.[7]

The Orbiter Project Office at JSC managed the design, development, testing, and production of the orbiter (or manned spacecraft vehicle which is today considered the Shuttle). Aaron Cohen, the Project Office Manager, and a native Texan from Corsicana, completed undergraduate work in mechanical engineering at Texas A&M University, and pursued advanced studies in mathematics and mathematical physics at Stevens Institute of Technology, New York University, and the University of California-Los Angeles before joining MSC in 1962. He became Chief of the System Integration Branch of the Systems Engineering Division, Apollo Command and Service Module Manager (1969-1972), and headed the Shuttle Orbiter Project Office from 1972 to 1982. In 1986, Cohen became Director of JSC.[8]

He organized the orbiter office under a deputy manager with four functional branches each headed by a manager and attached a resident manager to the primary orbiter contractor, North American Rockwell at Downey, California (figure 16). The orbiter project branch managers were in effect special divisions of the center's line directorates collocated in the orbiter office exclusively for work on the Shuttle.[9]

Cohen's orbiter team constantly encountered technical difficulties in the design and construction of the wholly new flying machine. Orbiter management required close and almost constant technical liaison with the line divisions of JSC and other centers. One of the first construction problems on the Shuttle came with the discovery that the contractor used

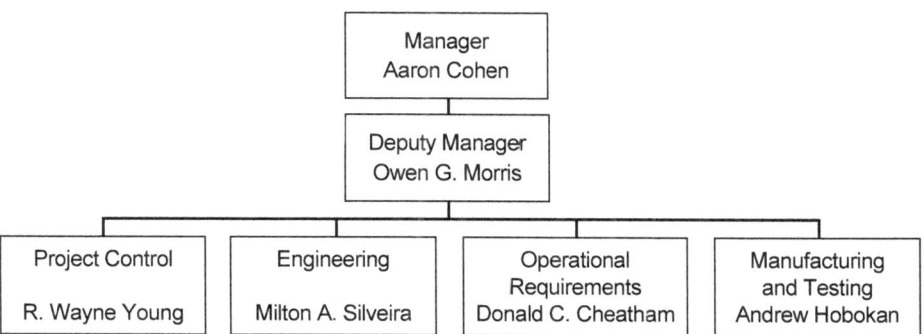

FIGURE 16. *Organization as of 1973*

soft rivets in the fabrication of the forward fuselage. "Why, in this high-technology spacecraft, should we use something so mundane as soft rivets?" Cohen wondered.[10]

The Shuttle and its components required constant testing, adjustment, and frequent redesign or reconstruction. Since the Shuttle was to be the first spacecraft launched on its maiden flight with people aboard, testing, redundancy, and man-rating the systems became more imperative. Testing was a more critical development tool for the Shuttle than had been true in previous programs where unmanned flight testing had been the rule. The Shuttle differed from previous spacecraft, not only in being a more complex flying machine, but in the manner of management and in the greater reliance and emphasis upon ground-testing its components.

Financial resources, always important of course, exercised much greater influence on decisionmaking in the Shuttle program than in previous programs. Thompson spent much of his time as program director attempting to reconcile budgets with elements of risk in the technical development of the Shuttle. Headquarters spent most of its time trying to convince Congress, the Executive branch, and the Office of Management and Budget (OMB) of the essential nature of each technical component of the Shuttle, and of the overall validity of America's space program. Moreover, declining or fixed budgets were being further undermined in the 1970's by the declining buying power of the dollar. Chris Kraft recalls vividly, for example, that when the OMB funded NASA's 1972 budget which had been appropriated with 1971 dollar values in mind, the funding represented a serious budget cut for NASA because inflation had reduced the buying power of the dollar by almost 10 percent.[11]

President Richard M. Nixon justified proceeding with the space transportation system in January 1972, because it would, he said, "take the astronomical costs out of astronautics." But critics in Congress were unconvinced. Senator William Proxmire (D-Wisconsin) said it was a "great mistake and an outrageous distortion of budgetary priorities." Senator Edmund S. Muskie (D-Maine) thought it was extravagant, Senator Vance Hartke (D-Indiana) labeled the Shuttle decision an example of "pork barrel politics," and Senator Walter Mondale (D-Minnesota) called it a ridiculous project.[12]

Shuttle defenders argued that reusable space vehicles could bring operating costs down to one-tenth that of Apollo vehicles. The promise of economies in space operations helped sell the Shuttle program to Congress and to the American people. But the subsequent failure of the Shuttle to achieve those economies became the "original sin" of the Shuttle program. Costs projected on the basis of as many as 50 Shuttle flights each year proved totally unrealistic. Although firm data is elusive because of many different methods for counting costs, Shuttle cost and performance comparisons are indicated in table 8. Generally, the cost of space transportation, as a percentage of the NASA budget, dropped from 56 percent of the budget during the development of the Saturn and Atlas Centaur launch vehicles to about 45 percent of the budget when the Shuttle became the primary launch vehicle (table 9).

Cost considerations affected the configuration of the proposed Shuttle and booster systems. Funding limitations during the start-up years that caused scheduling delays contributed to increasing costs of completion. NASA selected a parallel burn rather than a

TABLE 8. Launch Capability, Launch Vehicle Cost and Performance Comparison

Launch Vehicle	Payload to 160nm due East	Cost per Flight*	Failure Record	Reliability	Cost per Pound**
Delta II	10100	40	6 of 122	0.95	3960
Atlas IIAS	18100	110	6 of 48	0.875	6077
Titan III	27000	130	7 of 115	0.939	4815
Titan IV	44400	180	0 of 2	1.000	4054
Ariane 4	21000	110	5 of 36	0.86	5238
Long March 2e	15200	40	2 of 21	0.905	1645
H-II	22400	70	No flts to date		3125
Proton	38000	75	12 of 120	0.9	1974
Zenit	28000	70	0 of 23	1.000	2500
Shuttle	50000	295	1 of 35	0.971	5900
Saturn V	270000	1145	—	1.000	4241
Energia	250000	634	—	—	2533

Notes:
1. All vehicles' performance normalized to due east launch from the customary launch site.
2. Cost per flight is derived from various sources as a function of the particular vehicle. Cost of launch has a fixed minimum, but may increase significantly depending on negotiated price of interfaces and services.
3. Shuttle costs are the operating budget 1989 through 1994 for all the flights manifested allocated to 10 percent less flights.
4. Shuttle cost per pound is all allocated to payload—none to the orbiter and crew.
5. Saturn V and INT 21 at two per year. Hardware costs were 495 M. The balance is KSC processing.

* Millions of dollars
** Thousands of dollars

Source: Papers of Joseph P. Loftus, Assistant Director (Plans), JSC.

TABLE 9. National Launch Vehicles

	Pegasus	Taurus	Titan II (Castor IVA)	Delta II[a]	Atlas II[a]	Atlas II AS	Commercial Titan III	Titan IV (SRMU)[b]	STS
Responsible agency	DARPA	DARPA	USAF	USAF	USAF	COMSAT	COMSAT	USAF	NASA
Performance, lb. Low Earth polar	600	3000	4200 (10,000)[c]	--	--	16,200	--	32,000 (40,200)[d]	--
Low Earth due east	700	3600	--	11,100	14,300	20,000	31,000	39,000 (49,500)[d]	51,000
Reliability	0.98	0.975	0.96	0.96	0.94	0.94	0.96	0.96 - 0.93[e]	0.98
Cost, $ million	12	25	50	54	62	65	120	150	250

[a] Commercial version in production
[b] Solid rocket motor upgrade
[c] Castor IVA performance
[d] SRMU performance
[e] With upper stage

These vehicles will constitute the national space launch resources for the 1990's. The Air Force Atlas II, Delta II, and Titan II will handle lighter payloads, and the Titan IV with Inertial Upper Stage or Centaur upper stage will handle heavy, high-altitude satellites. The Air Force plans to phase out use of the Space Transportation System (STS) for most primary DoD payloads by 1993 and will use the Shuttle to fly secondary experiments thereafter. The commercial Atlas, Delta, and Titan are used by the commercial satellite (COMSAT) industry. The Pegasus and Taurus are representative of small launch vehicles being developed by the Air Force, the Defense Advanced Research Projects Agency (DARPA), and the commercial launch industry for a potential new class of lightweight satellites and quick-response tactical space support scenarios.

series burn for the solid rocket motor (SRM), in part because of lower cost and less technical risk. NASA selected a single contractor to provide engineering and integration for the Shuttle (Rockwell), and another for the booster (Thiokol) because of cost and management efficiencies. The decision to substitute a partially reusable for a fully reusable Shuttle system resulted in part from the need to reduce initial development costs at the expense of somewhat higher operating costs in the future. The decision to drop the external propellant tanks into the ocean after a suborbital staging, rather than boosting the tanks into orbit and having them deorbit by a solid rocket motor was shown by calculation "cost-effective" and a tradeoff for larger payloads. Although certainly costs always entered into NASA decisions, the Shuttle differed significantly from Gemini and Apollo experiences in that the "fiscal and political environment influenced detail engineering design decisions on a month-to-month, and at times a day-to-day, basis."[13]

Budget reductions often translated into program delays and slippages, which in turn drove total Shuttle development costs higher. In August 1972, Dale D. Myers (Associate Administrator for Manned Space Flight) reported to NASA Administrator James C. Fletcher that Skylab funding was dangerously low and recommended reducing Shuttle funding, if necessary, to preserve Skylab. Caspar Weinberger, Director of the OMB, advised Fletcher that the administration wanted NASA programs to be sustained, but with fewer dollars. Weinberger did anticipate some budgetary improvements in fiscal years 1973 and 1974, but that was an uncertain future. On two occasions, in October and December 1972, Myers recommended canceling the Shuttle program if cost cuts recommended by OMB were approved. The DoD helped rescue the Shuttle program, but recommended payload and reentry and landing configurations that greatly increased development costs. Langley Research Center considered canceling its Shuttle payload studies because of budget cuts.[14] The financial duress continued throughout 1972 and 1973 and beyond.

JSC Director Christopher C. Kraft informed Dale Myers in October, 1973, that proposed budget cuts for 1973 and 1974 would have "negative effects" and constitute "unsound planning."[15] Budget cuts directly affected the planned performance capability of the Shuttle, and resulted in slippages or deferrals in development. In March 1974, Administrator Fletcher explained that the $100 million increase in the NASA budget scheduled for 1975 was less than the amount required simply to sustain existing levels of operation because of inflation and other adjustments. He announced another slip in the flight of the Shuttle from late 1978 to mid-1979.[16]

Fletcher retired on May 1, 1976. Alan M. Lovelace became acting administrator. In the presidential elections in November, Jimmy Carter defeated President Gerald Ford and became President in January 1977. Despite President Carter's favorable statements about America's space programs, money remained short and Congress continued its budget cutting. On May 23, President Carter appointed Dr. Robert A. Frosch, a former assistant secretary general of the United Nations (1973-1975), assistant secretary of the Navy (1966-1973), and associate director of the Woods Hole Oceanographic Institute, to head NASA.[17]

Lovelace and Frosch fought desperately to try to keep the Shuttle from being destroyed by Congress and the administration. NASA's relations with Congress and the administration were either passive or reactive and defensive throughout the 1970's. Chris

Kraft recalled a meeting at Goddard Space Flight Center in early 1977 where he exclaimed, "When are we going to expose the fact that we don't have enough money to do this [the Shuttle]?" Shortly thereafter Kraft became Director of JSC, and Frosh, when briefing President Carter on the Shuttle program alluded to the fact that NASA would likely again slip the Shuttle flight schedule. "What do you mean?" Carter responded. The fact was, Kraft said, Carter's SALT (Strategic Arms Limitation Treaty) talks with the Soviet Union presumed early completion of the Shuttle by the United States. Since the Shuttle comprised a considerable part of the United States' leverage behind the arms limitation talks, President Carter began to move to help get Shuttle budgets and the development schedule back on track.[18] There were still 4 difficult years before the first Space Shuttle flew.

One by-product of the budget difficulties was something of a reconciliation or liaison between NASA and the scientific community, which had been highly critical of Shuttle expenditures on the grounds that they reduced budgets for scientific Earth and unmanned planetary missions. NASA accepted a greater role for the Shuttle as a research vehicle in order to win the support of the scientific community.[19]

At the same time, NASA did receive more support from the DoD, which had become concerned that the Shuttle's demise would adversely affect national security. Deputy Secretary of Defense Kenneth Rush outlined a plan for DoD participation in the Shuttle program, which recognized that "it is essential that DoD continue to support the Shuttle program and that we vigorously plan to utilize the Shuttle's advantages." A special "Air Force User Committee" began studying Shuttle applications and DoD cooperation.[20] With the program under serious fire, NASA welcomed all allies, however disparate. Broadening the Shuttle support base and participation, however, affected configuration, payload planning, and costs. Science and defense received greater priority in mission planning.

Despite the fact that the science, defense, aerospace contractor, and NASA communities became less divisive and more collaborative in their support of the Shuttle, congressional expenditures for the Shuttle and for space, scientific, and nuclear programs remained very lean. Although it is somewhat misleading to compare the first 8 years of Apollo funding with the first 8 years of Shuttle funding, the figures do indicate that Shuttle development costs were considerably less than those for Apollo. It should also be noted that Shuttle dollars, particularly after 1973, had a constantly declining buying power or value as compared to Apollo dollars of the 1960's era.

Not only did the Shuttle emerge in a very hostile budgetary environment, but it offered some new and difficult technological problems. The Shuttle was to be a considerably more complex machine than previous Apollo-Saturn systems. The main rocket engine had to be a high-performance engine capable of being throttled, turned off, and reignited. No space engine had yet been built to do that. No vehicle had yet been built that could be piloted both within and outside of the Earth's atmosphere. The Shuttle was a launch vehicle, spacecraft, and glider. The thermal protection system, main engine, and avionics system were "outside the existing state of the art." Moreover, "a 200,000-pound glider with a very low lift-to-drag ratio at the end of a landing strip many miles from its normal operating base presents some interesting logistics problems." Finally, because it was designed for piloted landings, there could be no unmanned test flights as there had been for all previous space vehicles.[21]

TABLE 10. Requested and Authorized Apollo and Shuttle Budgets for First 10 Program Years (millions $)

Apollo Requested/Programmed[1]			Shuttle Requested/Programmed		
1962	160	75.6	1970	9	12.5
1963	617.2	1,184.0	1971	110	78.5
1964	1,207.4	1,147.4	1972	100	100.0
1965	2,677.5	2,614.6	1973	200	198.6
1966	2,997.4	2,941.0	1974	475	475.0
1967	2,974.2	2,922.6	1975	800	797.5
1968	2,606.5	2,556.3	1976	1,527	1,206.0
1969	2,038.8	2,025.0	1977	1,288.1	1,413.1
1970	1,691.1	1,684.4	1978	1,349.2	1,349.2
1971	956.5	913.7	1979		

[1]Programmed funds are the amounts actually allocated for expenditure after congressional appropriations.
Note: Apollo program funding ceased after a 1973 program allocation of $56.7 million.
Source: Linda Neuman Ezell, NASA Historical Data Book, II, p. 128; III, pp. 69, 71.

Previous spaceflight experiences, however, provided technological precedents that helped solve Shuttle problems. There was now a much higher experience base in the government/industry complex. Research airplanes (X series), Dyna-Soar and Gemini Earth landing studies contributed to Shuttle design. Tremendous advances in electronics and computer design facilitated the development of guidance and control systems and reliable pilot-controlled avionics systems. Both solid and liquid rocket engine history contributed to the design of the sophisticated engines. Silica-fiber based tiles protected with a boro-silicate material were chosen to cover the orbiter's bottom and sides and thus help resolve the thermal protection problem which existing metals could not solve at acceptable weight levels. The leading edges of the Shuttle wings and body were covered with heat-resistant "carbon-carbon" developed from a rayon material. There were historical precedents (with flying model and wind tunnel verification) for piggyback transport of Shuttles on the Boeing 747, but whether the method of transporting would work remained unproven until it actually happened.[22] While the Shuttle system was something new and different, development could now draw upon a reservoir of aerospace technology which was virtually nonexistent even a decade earlier.

A considerable part of that technology was located at JSC in the form of the expertise of its personnel and its laboratory and testing facilities. Building a space vehicle, or perhaps any engineered machine, involved conceptualizing the final product and each component of that machine, designing the parts, testing the parts (sometimes redesigning them), then assembling the parts and testing the whole (and sometimes redesigning the parts or the whole). Often the design or even conceptualization of a device could not proceed until sufficient testing had occurred to indicate how one might proceed. For example, a Langley Research Center-designed Manned, Upper stage, Reusable Payload (MURP) craft described as a "lifting body fuselage with variable geometry wings in a horizontal recovery, land landing vehicle" influenced the design of the Space Shuttle. MURP experimental vehicles of the early 1960's provided the first real experiences with hypersonic and supersonic flight and ablative heat shields. Wind tunnel tests, simulated reentry trajectories and, finally in 1966 and 1967, the firing of model reentry vehicles called the ASSET and PRIME (34 and 55 inches in length, respectively) from high suborbital trajectories into the atmosphere, produced real reentry test results for bodies and wings that would have the configuration of the Space Shuttle.[23]

A part of the built-in system of redundancy in spacecraft had to do with the thorough testing of all materials, parts, and systems. Much of what JSC was and is has to do with its laboratory and testing facilities. Engineers could neither manage nor be engineers without such facilities. Those unique attributes which JSC contributed to Shuttle development included its past experiences, but most significantly its unique Mission Control Center and operations expertise, and its special laboratories that were not replicated anywhere else.

Those laboratories included the vibration and acoustics facilities, the Shuttle Avionics Integration Laboratory, the atmospheric reentry materials and structures laboratory, the thermal-vacuum laboratories, the electronic system test laboratory, the space environment simulation laboratory, the life sciences laboratory, and the Shuttle mission simulator. Less unique, but still reflecting the state of the art in those fields, were the radiation instruments laboratory, geology and geochemistry laboratory, photographic technology laboratory, geophysics and applied physics laboratories, a meteoroid simulation laboratory, thermochemical test areas, optical labs, and antenna and anechoic chambers. Instruments included magnetometers, spectroscopes, a Van de Graaff facility to calibrate radiation detectors, radiometric counting equipment, and vacuum chambers—among other devices. Universities, laboratories, or other public and private agencies might have had some of the specialized laboratory equipment, but its assemblage and concentration at JSC was in itself unique. The center possessed the most advanced electronic, radio, and radar equipment available, and created some that was previously nonexistent.[24] The center provided the engineer a veritable cornucopia of testing and laboratory equipment, all with very serious and ineluctable purposes.

Development of spacecraft such as Apollo and the Shuttle required unusual and "unworldly" laboratories such as this anechoic chamber at JSC. The chamber replicates space by absorbing radio and radiation emissions.

Engineering Facilities at JSC
Comment by Henry O. Pohl, Director, Engineering

To understand JSC and its facilities, it is necessary to understand the background of the men and women who moved from Langley Field to Houston. They came from a research background and were accustomed to operating with very small budgets. They did only what they themselves could accomplish to expand the envelope of flight. It is interesting to note that they never designed an airplane, yet every airplane of that era flew with Langley wings. They wrote the criteria and developed the formulas used by the various aircraft companies to design every wing. They often worked in crummy offices but had very fine laboratories.

They understood the value of good, accurate data. They understood the necessity to test their theories. They understood the absolute requirement for each engineer to understand his or her discipline and to understand firsthand the limitations and accuracy of the data. As flight speeds increased and missiles entered the scene, wind tunnels were no longer capable of providing the velocities, temperatures and pressures required to validate their theories. As a result, they turned their attention to placing models on top of rockets designed to place the model at the desired velocity and altitude. These models were precision devices and fairly expensive. Unlike models used in wind tunnels, these could be used only once. Therefore, it was essential that good, accurate data be obtained from each test. As a result, reliability also became important.

It was this experience base that the MSC personnel, charged with landing a man on the Moon within 8 years, brought with them from Langley Research Center. They realized they would have to train several thousand new employees in the exacting science of spaceflight. They realized that in a very few years they would have to depend upon these engineers and scientists to provide good, accurate data, and that many new theories would have to be validated. They realized that if they were to accurately supervise the multitude of contractors and systems, these new employees would have to develop confidence in themselves. The only way they knew to do this was to provide the facilities necessary to give employees the opportunity to verify their theories. This attitude was reflected by Hugh Dryden in an article published in *NASA Activities* (August 1976): ". . . each employee at NASA should be a doer of things, not just a watcher."

Therefore, when they laid out MSC facilities in a cow pasture south of Houston, they took into account all the disciplines required to build and operate manned spaceships. These were the finest laboratories money could buy. The facilities were built, not because they could afford it, but because they knew they could not afford to be without. It was no accident that MSC contained facilities that specialized in the health and well-being of humans as well as operations in life support, materials, metallurgy, structural dynamics, acoustics, guidance, control, communications, tracking, propulsion, power, explosives, data acquisition, data reduction, fabrication shops, machine shops, and large computers. One should remember that at the time MSC was designed, computers were mostly thought of as humans with Gerber Scales, slide rules, and mechanical calculators converting squiggly lines from oscillograph traces and strip charts to engineering units. Yet they knew that much more computing power would be an absolute necessity.

Moreover, the Mission Control Center was no accident. Langley/MSC engineers knew from their experience in flying models that it would be essential that they keep track of the status of the spacecraft and be able to reprogram the course of the missions if problems developed. Apollo 13 is a good example of where this forethought paid off. Similar examples can be cited in every discipline. Through the use of these facilities many of our current employees became true international experts in their disciplines.

This expertise, developed during the Apollo era, saw the Shuttle program through its developmental phase. Over the years, some of these facilities have changed very little. Much of the original equipment still functions and is in use. Some facilities are unique. Perhaps the greatest change in the facilities has been in the use of computers. We spend vast sums on new computing equipment and software each year—yet we no longer have state-of-the-art computers for spaceflight.

Another area in which NASA/JSC pioneered was in materials and metallurgy. Through better materials processes, stronger, lighter, and more consistent materials have been obtained. By thoroughly understanding the behavior of materials, it became possible to tailor materials for a specific application, resulting in lighter, more reliable systems. These processes have largely been adopted by the aircraft and auto industries.

There are over 70 laboratories and facilities listed in the Engineering Directorate Technical Facilities Catalog. These laboratories are located in 18 buildings, not including other laboratories used in Life Sciences and Operations. Sixty-four percent of the facilities were built in the 1960's, 10 percent during the 1970's, and 26 percent in the 1980's. During the 1980's facilities were added to accommodate robotics, microwaves, communications, data management, avionics integrations, and additional computer facilities. The replacement cost of test and evaluation facilities put in service during the 1960's would exceed $390 million. Replacement costs for those added during the 1970's is estimated at about $15 million, if the Shuttle Avionics Integration Laboratory (SAIL) were excluded. The $630 million SAIL facility was put in place to design and verify the hardware and software used in the electronic flight control system on the Space Shuttle. Construction costs for additional facilities built in the 1980's was about $16 million. In total, JSC technical and engineering facilities are among the most advanced and distinctive anywhere in the world and are central to the completion of the center's mission.

Suddenly, Tomorrow Came...

Johnson Space Center, 1977

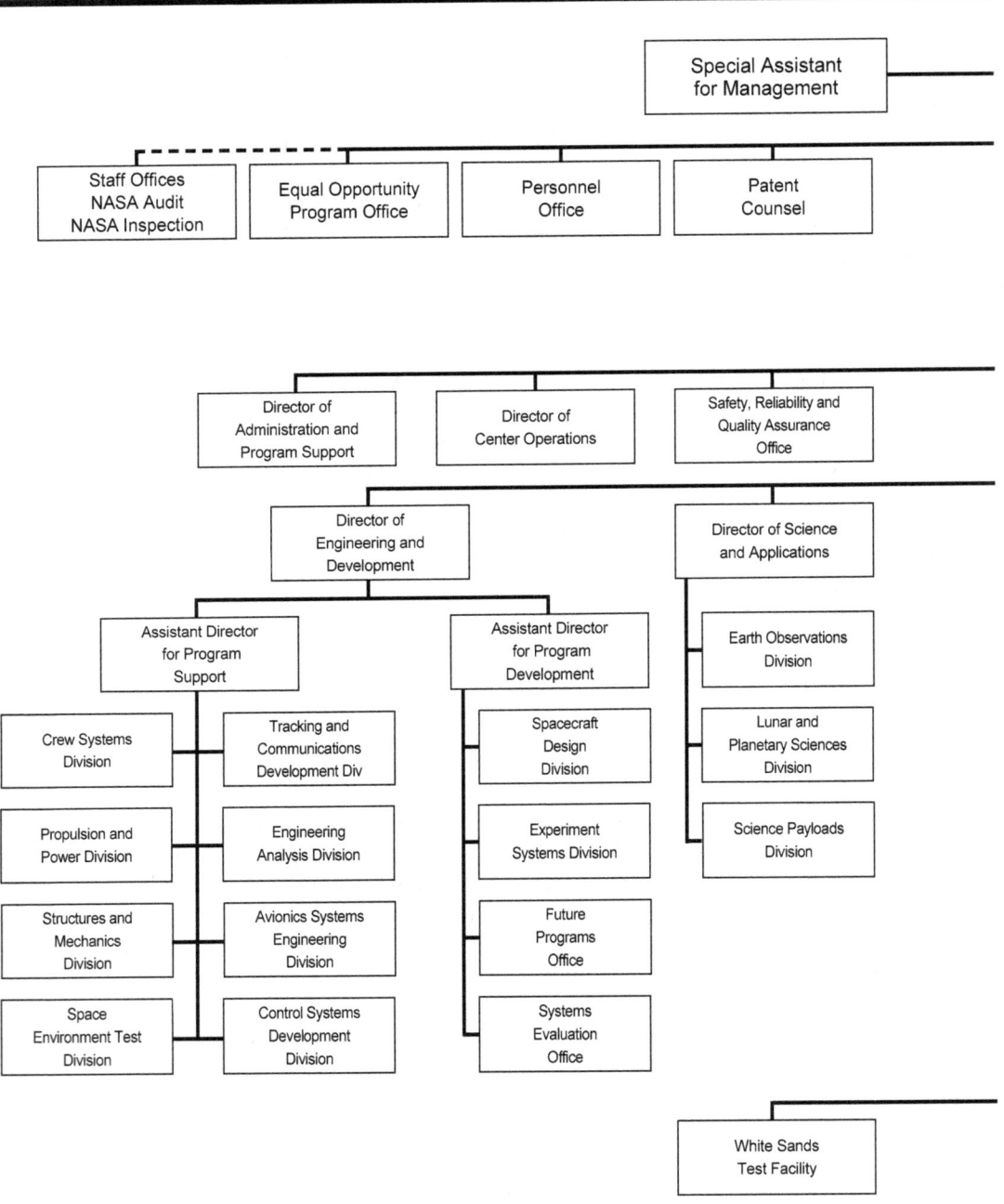

FIGURE 17. *Organization as of January 1977*

Lead Center

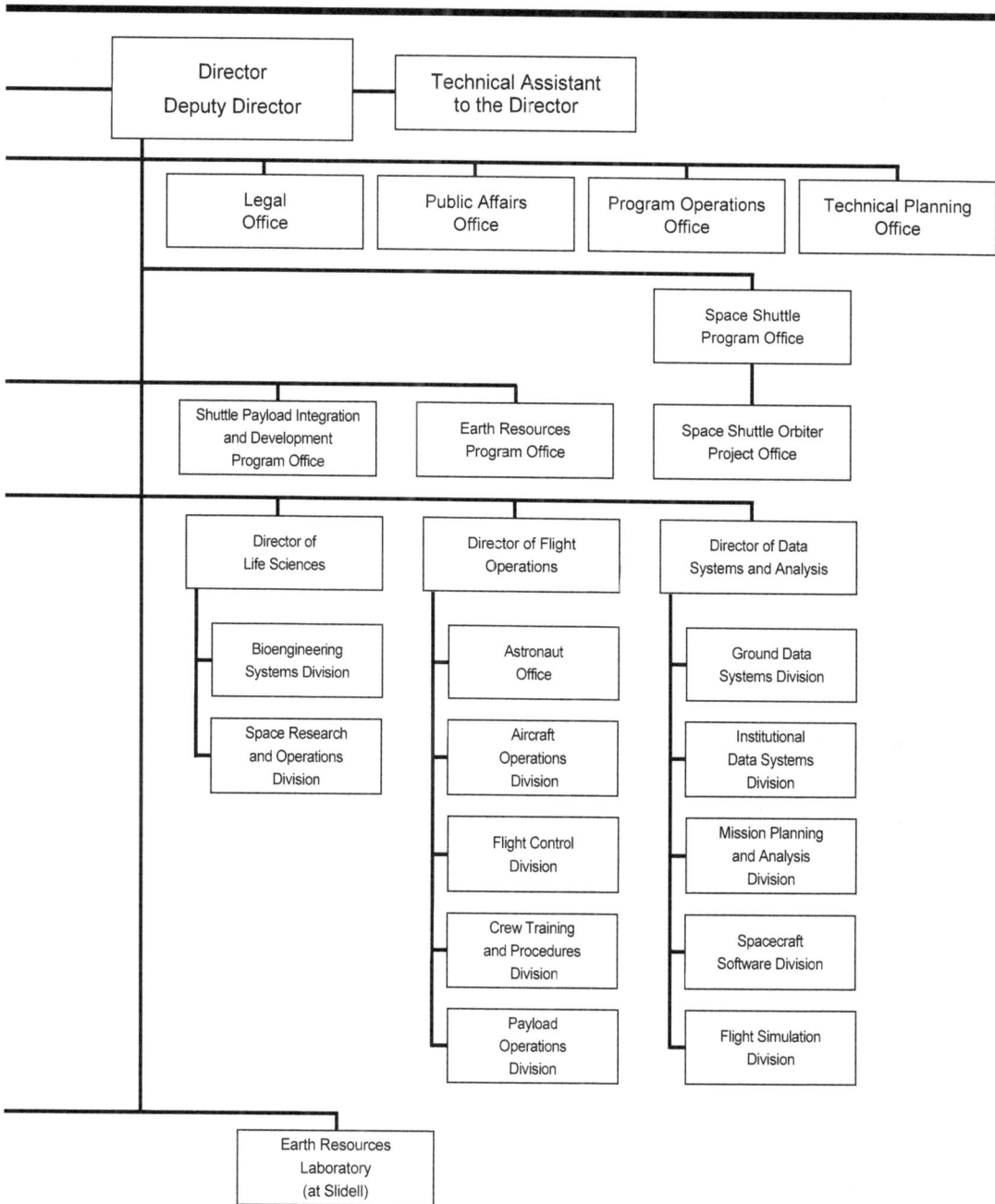

The history and culture of JSC is tied inextricably to its laboratories and tradition of hands-on engineering, but because much of its work could not be verified in Earth laboratories, JSC engineers relied heavily on analytical and computational methods of engineering. The culture of JSC engineers is described in the comment on engineering facilities at JSC by Henry Pohl who succeeded Aaron Cohen as head of Engineering and Development. Pohl began his career as a "rocket engineer" with the Army Ballistic Missile Agency in Huntsville, Alabama, in 1957. He transferred to MSC in 1962 as the senior propulsion engineer and became Chief of the Propulsion and Power Division in 1980 before being named Director of Engineering and Development in 1986.[25]

Within the testing and laboratory facilities at JSC, unseen and much of it literally underground, resided the heartbeat of the center. The laboratories operated within the Engineering and Development Directorate under a division chief and branch or section heads (figure 17). The Advanced Spacecraft Technology Division which became the Space Environment Test Division in the Shuttle era, managed the radiation and fields laboratories and test facilities. The Lunar Surface Technology Branch became part of the Science and Applications Directorate and managed the geology, geochemistry, cartographic, and geophysics laboratories. The old Meteoroid Technology and Optics Branch of Engineering, which directed the meteoroid simulation, and the applied physics (conducting optical experiments to determine methods of making space measurements which cannot be made from the Earth's surface) and planetary atmosphere laboratories became elements of Science and Applications Directorate, Earth Observations Division and Science Payloads Division.[26] Although the division (and laboratory designations) changed over time, figure 17 indicates the divisions within the JSC directorates, most of which operated laboratories and test facilities.

The Crew Systems Division, responsible for "establishing and validating the physiological design parameters for manned spaceflight," designed and tested everything having to do with life support systems. Food, spacesuits, water, waste, and health care fell under the division's supervision. A space environment could complicate the most innocuous human function. Pre-Shuttle toilet facilities, for example, comprised essentially strapped-on tubes, bags, and diapers. Perfecting a workable toilet in a zero-gravity environment proved difficult but "do-able."

A 20-foot and 8-foot diameter altitude chamber could simulate altitudes to 150,000 feet. A liquid nitrogen cold-trap tests heat exchangers and thermal characteristics. An envirotron chamber replicates a full range of vacuums, temperatures, and pressures. There are materials labs, chemistry and instrumentation laboratories, a waste management and microbiology laboratory, and a crew performance laboratory associated with the altitude chambers. The latter crew performance laboratory examines crew behavior under varying simulated extraterrestrial and flight conditions. An impact test facility determines what would happen to pieces of equipment such as gauges, lights, or cameras when subjected to shock. The flight acceleration facility, or centrifuge, was used to train crews and test their equipment through the Apollo era, but then was closed in the mid-1970's, and the building became a Weightless Environment Test Facility.[27]

The world of computers changed markedly and rapidly during the first three decades of NASA's existence. Generally, computer technology evolved from the massive corporate-owned IBM-style 7090 computers using vacuum tubes into the new-generation microchip mainframe computers such as the IBM 7044/7094. Subsequently, these mainframe units became more sophisticated, but also less costly, as microchip and printed circuitry evolved. The IBM 360s and 370s characterized this second phase of the computer industry. At this point, new corporations and new products began to challenge IBM dominance. In the 1980's, Amdahl, Packard, and Texas Instruments, among others, became formidable contestants in the rapidly expanding computer market.

Personal computers became popular in the late 1970's and became indispensable in the 1980's. Initially, engineers regarded personal computers as toys or for typing. Not only were the PCs unsuitable for running the engineer's FORTRAN programs, but they tended to be beneath the dignity of the professional engineer. In time this changed, but some tension continued to exist between the advocates of mainframe computer systems and personal computers, and often between administrators who opted for less costly and often less current computer equipment. In this, JSC compared to most large organizations caught in the throes of rapidly changing computer technology.

Digital Electronics Corporation introduced its powerful and affordable VAX computers that brought forth a new generation of division and department-owned computers that could be directed to specific uses. Meanwhile, personal computers became more powerful. Networking and individual workstations began to blur the old lines between mainframe and personal computing. Once essentially a "corporate" operation, computing for NASA engineers (and engineers everywhere) increasingly became a personal operation despite the later introduction of yet another generation of "super" mainframe computers in the 1980's. JSC's "computer" culture tends to reflect this overall pattern of development.

The branches of the Engineering (previously Computation) and Analysis Division at JSC specialized in work on engineering analysis, flight mechanics and applications, data processing, programming, and data systems development. Major computers in use during the 1960's and 1970's included an IBM 7044/7094, two IBM 7094s, A Univac 1108, two Control Data Corporation (CDC) 3600s, several hybrid systems, and a number of lesser units. An idea of what this means is suggested by the fact that an IBM 7094 could calculate in 5 seconds what would otherwise involve 86 person-years of labor and the UNIVAC 1108 was three times faster than the IBM 7094.[28] Even these systems were outmoded almost as soon as they were in place. Both computer hardware (the equipment) and software (the program) are constantly upgraded.

VAX 11/785 computers began to handle such engineering design graphics packages (not available in the 1960's and 1970's) as PLAID and TEMPUS, which can give multidimensional views of design models. The pencil became almost obsolete in engineering design. Even the Apple Macintosh and other desktop computers offered state-of-the-art graphics packages in the 1980's that greatly advanced the frontiers of engineering design. NASA's first new-generation supercomputer, a Cray 2 located at Ames Research Center, can do a quarter of a billion computations per second. At the heart of JSC Mission Control Center in the 1980's were five IBM 308X class computers. JSC completed installation of its own supercomputer, an "Engineering Computation Facility (ECF) Class VI," using Cray and Amdahl data processors in 1990. The Space

Shuttle itself housed five interconnected (modified IBM AP101) computers, each about the size of a breadbox, which surpassed NASA's total computer capability of the Apollo era (but which by today's standards are slow). A special testing device, built specifically for the Shuttle, used 32 computers (primarily Perkin-Elmer 8/32 computers) which are tied together through two "host" UNIVAC 1100/44 computers.[29] These are the brains of the Shuttle mission simulator (SMS).

The SMS, which became operational in 1978 and has been updated periodically, marked a significant enhancement of the Apollo simulation laboratory complex. Managed by the Flight Simulation Division, SMS became the primary training device for Shuttle crews and for the Mission Control Center with which the systems are integrated. Built by Singer Company's Link Division, the major components of the simulator are a fixed base crew station, a network simulation system, and a motion base crew station. The device provides real-time mission simulation: prelaunch, ascent, orbit operations, deorbit, entry, approach and landing. The crew's orbital operations, visual scenes, and aural cues are "rigorously" simulated. In the motion base crew station, displays, control responses, and inputs are indistinguishable from those aboard the actual Shuttle. An instructor at a remote station can initiate over 3800 malfunction situations requiring crew responses. A network simulation system simulates the ground spaceflight tracking and data network which provides telemetry, tracking and communications with the actual Shuttle that is tied to the Mission Control Center through the Goddard Space Flight Center.[30] By the time astronauts complete their mission training on the SMS, they think they have already flown the mission—many times over.

Testing applies not only to machines and spaceflight equipment, but to the men and women who would fly those machines and use that equipment. A new-generation space vehicle, the Shuttle would be manned by new generation astronauts. NASA had last recruited astronauts in 1966 and 1967. The class of 1966 comprised 19 pilot astronauts, and in 1967, 11 scientist astronauts were selected for the lunar science missions. In 1978, both pilot astronauts and astronaut mission specialists were recruited. From over 8000 qualified applicants, NASA chose 15 pilot astronauts and 20 mission specialists. Pilot astronauts needed a minimum B.S. or B.A. degree in engineering, biological, or physical sciences with an advanced degree preferred and a minimum of 1000 hours of high performance jet aircraft experience. Mission specialists (and payload specialists) did not need flight time or pilot experience and were subjected to a less rigorous physical examination than the Class I flight physical, but academic credentials and psychological testing weighed heavily in the selections.[31] Reflecting the growing national concerns about opportunities for minorities, affirmative action, and civil rights, NASA's first Shuttle class included six women and four minority candidates.

The Shuttle astronauts, once admitted to candidacy, completed 12 months of rigorous training and testing before a final review and full acceptance into the program. Studies included guidance and navigation, astronomy, meteorology, math, physics, and computer programming. They spent a lot of time in the SMS. Astronaut Robert Crippen told Henry S.F. Cooper, author of *Before Lift-Off*, that "if it weren't for the mission simulators, flying in space at all would probably be impossible." A spacecraft crew flies only once, and when it does, it must be a crew in all that the term implies. The crew is created by the simulator.[32]

Astronauts Class of 1978[33]

Bluford, Guion S.	Brandenstein, Daniel C.*	Buchli, James F.
Coats, Michael L.*	Covey, Richard O.*	Creighton, John O.
Fabian, John M.	Fisher, Anna L.	Gardner, Dale A.
Gibson, Robert L.*	Gregory, Frederick D.*	Griggs, David S.*
Hart, Terry J.	Hauck, Frederick H.*	Hawley, Steven A.
Hoffman, Jeffrey A.	Lucid, Shannon W.	McBride, Jon A.*
McNair, Ronald E.	Mullane, Richard M.	Nagel, Steven R.*
Nelson, George D.	Onizuka, Ellison S.	Resnik, Judith A.
Ride, Sally K.	Scobee, Francis R.*	Seddon, M. Rhea
Shaw, Brewster H., Jr.*	Shriver, Loren J.*	Stewart, Robert L.
Sullivan, Kathryn D.	Thagard, Norman E.	Van Hoften, James D. A.
Walker, David M.*	Williams, Donald E.*	

*Pilot astronauts

After their first flight, astronauts reported that the Shuttle simulations were so accurate that they felt they had indeed flown the mission many times. Mission specialists and payload specialists also trained with the remote manipulator system (RMS), which loaded or off-loaded cargo from the Shuttle bay during orbital missions. Built at the SPAR Aerospace Plant in Toronto, Canada, the RMS recalled perhaps a much earlier Anglo-Canadian contribution to the American space program—the 25 or so AVRO aerospace engineers who joined NASA and the original Space Task Group shortly after they were first formed. Crew specialization "sharply reduced training resources required (equipment, personnel, money) per mission."[34] Specialization eased the training problem, but not the overall task or the essential development of teamwork and trained responses.

By the time the Shuttle astronauts began training, the old centrifuge at JSC, which had whirled astronauts around at often unbearable speeds to simulate gravitational forces equivalent to launch pressures, was no longer needed. Whereas Apollo flights subjected astronauts to forces 15 times the pull of gravity, the Shuttle only subjected them to forces 3 times the pull of gravity. The centrifuge was replaced with a device designed to simulate zero gravity. A gravity-free environment cannot be achieved on Earth, but the conditions can be closely approximated by immersing the suited astronaut into as much as 25 feet of water. A Weightless Environment Training Facility (WETF), actually a large swimming pool 76-foot long, 35-foot wide, and 25-foot deep, approximated some aspects of a weightless environment. Astronauts trained intensively to do those tasks which required EVAs while submerged in the WETF. They also, as did the Apollo astronauts before them, got a momentary sense of weightlessness as passengers aboard a KC-135 (comparable to a large commercial jet) during flight over a parabolic curve.[35]

Astronauts began their Shuttle pilot training in a substantially modified Gulfstream II aircraft adapted to mimic the flight characteristics and instrumentation of the Shuttle. When Al Pacsynski, who was stationed at JSC's White Sands Test Facility for radar and propulsion

Suddenly, Tomorrow Came . . .

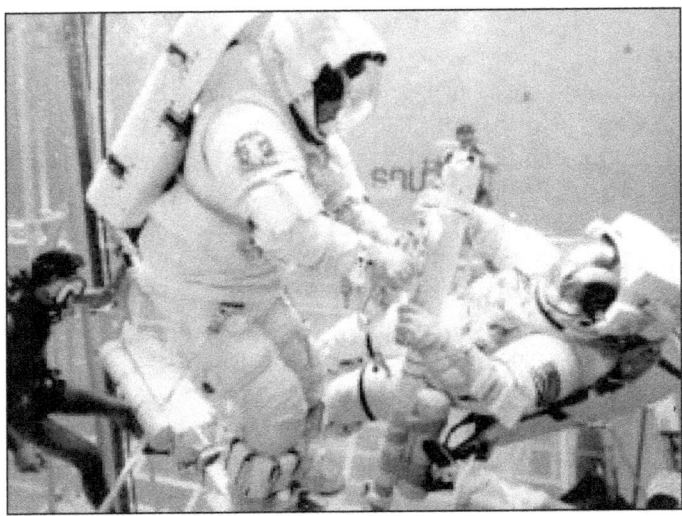

Astronauts trained to work in the weightless environment of space in JSC's WETF. Here, astronauts George D. Nelson and James D. van Hoften, STS 41-C mission specialists, go through a drill preparing them to repair the damaged Solar Maximum Satellite in orbit about Earth.

tests, heard that the astronauts needed a practice landing strip, he suggested the relatively unused 10,000-foot Northrup Strip located on the test facility grounds. Originally a part of the Army's White Sands Missile Range before being transferred to NASA in 1962 to test the Apollo command module and lunar descent engines, the Northrup Strip was used by Northrop Aviation to land target drones. (A typographical error in an early press release transformed Northrop to Northrup—and the latter stuck.) JSC reactivated the field in the summer of 1976, built a second runway in 1978, and expanded both runways to 35,000 feet in 1979, establishing the "White Sands Space Harbor" as both a training field and an alternative landing site for future Shuttle flights.[36]

The 60,000-acre White Sands Test Facility continued to operate propulsion and materials testing laboratories in the Shuttle era as it had for Apollo. Engine test stands at the laboratories include altitude chambers that simulate the vacuum of space during engine firings. Materials test laboratories are constantly adapted to simulate unique space conditions and to examine such things as space debris impact, the performance of high pressure pumps and valves, and metals flammability. Originally staffed largely by Grumman and North American contractor personnel, the White Sands Test Facility has been staffed in the Shuttle era by approximately 60 JSC (civil service) personnel and 500 or more Lockheed Engineering and Sciences Company personnel and staff. Although used only on one occasion, in March 1982, as an alternate shuttle landing field (for STS-3), the Space Harbor and laboratories of the White Sands Test Facility are an important adjunct of JSC's training, testing, and recovery capabilities.[37] Although they never flew, the SMS and its complementary SAIL located at JSC in Houston provided an almost real-time flight simulation experience.

The SAIL developed by the Avionics Systems Engineering Division (formerly the Instrumentation and Electronic Systems Division) after 1974, examined physical and electronic components and their integration into the Shuttle systems.[38] The whole concept of applying electronics to aviation, and the word avionics, evolved in the post-Sputnik aerospace industry. The SMS, and more especially the Shuttle orbiter itself, represented as of the 1970's and 1980's the leading edge in aerospace and computer electronics. NASA space programs, through the contractor affiliates, facilitated the diffusion of an advanced state of electronic, engineering, medical and geophysical technology throughout the socioeconomic system.

Changes within what might be considered the more traditional areas of engineering were no less startling than in the areas considered new and innovative. Thus the Propulsion and Power Division of the Engineering and Development Directorate was primarily concerned with thermochemical and pyrotechnic tests. The Structures and Mechanics (or Engineering Analysis Division) laboratories examined spacecraft materials, gaseous helium and liquid nitrogen systems. The Space Environment Simulation Laboratory (SESL), completed in 1965 and 1966, included two vacuum chambers. Chambers A (65-foot diameter and 120-foot height) and B (35-foot diameter and 43-foot height) were built to provide simulated space and lunar surface environments for Apollo training. Thermal vacuums are to space systems what aerodynamics is to aircraft; that is, the vacuum pressures, internal pressures, and heat create the stress on the structure. Temperature control in the chambers ranged from 80 to 400 degrees K. The Apollo lunar science experiments, Skylab, and Shuttle missions all relied heavily on data from the SESL tests. Built at a cost of $35 million, SESL was one of the most unique of all JSC laboratories, a machine tool designed to do things that otherwise could not be done on Earth.[39]

Although manned lunar flights ended with Apollo 17, new applications of space technology were already enriching human knowledge. Here a test is conducted on an Applications Technology Satellite's 30-foot umbrella-shaped antenna in Chamber A of the SESL at JSC.

Although the technology of structures and mechanics, thermochemical reactions, electronics and computers often skirted the mainstream of American interests, Moon rocks, and environmental issues did attract considerable public interest as the lessons learned from Apollo began to unfold. Those lessons evolved from the work completed in the Lunar and Planetary Sciences and Earth Observations Division laboratories of the Science and Applications Directorate. The division began with the establishment of the Lunar Receiving Laboratory in 1968. The laboratory is equipped with two-way biological safeguards to provide a sterile environment and protect the lunar samples and Earth from mutual contamination. It contains instruments for physical, chemical, petrographic, and mineral analysis—and radiation detection. The laboratory not only conducted its own investigations, but also distributed samples and information throughout the United States and to foreign countries.[40]

JSC made significant contributions to scientific knowledge about the heavens and the Earth. "Along with technology, national confidence, and human spirit," Jack Schmitt, the scientist astronaut aboard Apollo 17 recalled at a public review of Apollo science programs during the 20th Anniversary celebrations of the Apollo 11 lunar landing, "science benefited permanently from our exploits. Remembering just how little we knew about the Moon before Apollo serves to emphasize both how far we have come and how far we now may go."[41]

John Wood, with the Smithsonian Institute's Astrophysical Laboratory put it more succinctly: "What we thought we knew about the Moon before Apollo was wrong." People thought of the Moon as a "strange, weird, scary place." The Lunar Receiving Laboratory, Wood commented, was built to quarantine astronauts coming from the Moon as much as it was to analyze Moon rocks or data. What we found out was that there were no unknown pathogenic organisms on the Moon, and it was not, in fact, a strange, weird, scary place. As a result, people began to think of other planets as not so scary after all.[42] For a time, during the Apollo era, the Moon became one of those very special NASA laboratories.

William David Compton's study of the Apollo lunar exploration missions examines "how scientists interested in the Moon and engineers interested in landing people on the Moon worked out their differences and conducted a program that was a major contribution to science as well as a stunning engineering accomplishment." Although the analysis of lunar and planetary data is still in progress and conclusions are still tentative, one of the most important results of Apollo science and explorations was to change our understanding and knowledge of the Earth.[43]

Dr. Wendell Mendell, with the Solar System Exploration Division of JSC, commented that Apollo changed the American mind-set from the idea that space was difficult and expensive to the understanding that it is real and possible. We need to think of space as an evolving sector of the Earth's and the United States' economy and society. And the Moon, Mendell thought, had some real opportunities in terms of scientific research, resource utilization, and colonization.[44]

For Americans and the world, one of the indelible memories of Apollo included the photographs and videotapes of the lunar surface. Accepting the rubric that a picture is worth a thousand words, the photographs brought back from the Moon by the astronauts dispelled the old mysteries of the Moon. But the Apollo lunar voyages actually taught humans more about themselves than about the Earth, the Moon or the planets. There was a new awareness of Earth, of Earth resources, and of the total human environment. It was this new awareness of Earth and Earth people that in time became the central thrust of the Space Shuttle program as scientific experimental planning for Shuttle missions began to focus on Earth resources and the environment. Greater public support for the Shuttle emerged from the decision to begin to apply Shuttle capabilities to the more immediate and direct benefit of humankind on planet Earth.

Nevertheless, although the Shuttle staggered through a prolonged and agonizing developmental phase, it remained in the minds of NASA planners as one component of a broader space program, a facilitator for the construction of a permanent space station circling Earth, and subsequently a Space Shuttle to ferry passengers and cargo to the station and to waiting interplanetary vehicles. The development of the Shuttle helped make that rather ancient and distant dream become very real and, in the construct of human history, imminent.

By 1977 the complex elements of the Space Shuttle began to mesh. The four basic elements of the Shuttle system include two massive solid rocket boosters (SRBs) (each generating 2.65 million pounds of thrust at lift-off), an external fuel tank, and the manned Shuttle or orbiter. The three main engines fire in rapid sequence. Then the twin SRBs ignite and the Shuttle lifts off. When the SRBs burn out, they separate and parachute into the Atlantic for retrieval and reuse. The main engines continue to burn, using fuel from the external tank, until just before orbital insertion. After main engine cutoff, the external tank detaches and the orbiter moves away through short burns of its reaction control system thrusters. The tank is the only expendable element of the Shuttle stack—it tumbles and disintegrates during reentry. The Shuttle orbiter completes orbital insertion by using its two orbital maneuvering system engines. Usually one burn puts it into orbit; a second burn puts it into a stable circular orbit. Reentry into the atmosphere begins with a deorbit engine burn after which the Shuttle engines are shut off and the space vehicle glides to a landing at Edwards Air Force Base, Kennedy Space Center, or one of the alternative landing areas such as White Sands, New Mexico. The Shuttle, if it landed elsewhere than Kennedy, was then returned to its launch site at Cape Canaveral piggyback aboard a Boeing 747.[45]

The first test of the external fuel tank was completed by Marshall Space Flight Center in January, and a solid rocket booster engine was first fired in July 1977. Rockwell International Corporation, the orbiter prime contractor, rolled the first orbiter *Enterprise* (sans engines) from its plant in Downey, California, on January 30, 1977, for a year of testing. An unmanned *Enterprise* flew piggyback aboard a 747 in February and March, while control and guidance systems were being installed. The first manned unpowered flight of the Shuttle was made on June 18. NASA scheduled its first free flight with a separation from the 747 for August. Five free flights were made during the year. Shuttle main engines, rocket booster engines, and Shuttle orbital engines were being tested. The shuttle main engines misfired on occasion, causing NASA reviews and a National Academy of Science inquiry in 1978. The 31,000 shuttle tiles, each individually crafted and glued to the orbiter's bottom and sides, required an estimated 335 person-years of labor to install on the orbiter *Columbia,* which would be the first of the orbiters to fly.[46] Development was slow but sure—and costly.

In 1977, the *Enterprise* neared completion at a cost of $500 million; the *Columbia* was expected to cost somewhat more. Three additional Shuttles scheduled for production were now being estimated at $550 to $600 million each. NASA estimated that a 2-year funding delay for the construction of additional Shuttles might elevate costs as high as $1 billion each. Before the decade was out, Congress had authorized construction of five completed orbiters or shuttles: *Enterprise, Columbia, Challenger, Discovery,* and *Atlantis.* By 1981, the first shuttle was ready to fly—2 years behind schedule and $1 billion over costs anticipated in 1975.[47]

JSC worked on the Shuttle, but also developed a greater social consciousness in these years as it began to apply and to teach others the lessons learned from Apollo and from work on the Shuttle. Astronaut Charles M. Duke, Jr., left NASA in 1978 and later became a lay minister, while James B. Irwin (the eighth man to walk on the Moon) founded a religious organization after his retirement in 1972. Many engineers founded consulting companies. Others joined space-related corporations. Eugene Horton left JSC and became involved in industry and environmental programs. Horton, formerly the center's education officer, organized the Earth Awareness Foundation dedicated to uniting the industrialist and the environmentalist in their common cause. Each year the center hosted a week-long lunar and planetary science conference in cooperation with the neighboring Lunar and Planetary Institute of Houston. The Lunar Receiving Laboratory developed national and international associations. In 1977 the center hosted a junior and senior high school symposium for 2000 Texas students—including tours, lectures and visits with astronauts, scientists and engineers.[48]

Its outreach program extended to far distant west Texas. The center established an Emergency Medical Communications Console at Odessa, Texas, designed to provide emergency medical care for the sparsely populated but vast distances of the Texas Permian Basin area. The facility represented a direct application of Apollo and Shuttle technology to a present, on-Earth problem. Through the network, physicians and nurses could consult paramedics in the field, doctors could receive electrocardiograms on the telephone lines (now common), telephones and radios could be interconnected, and physicians and hospital staff members within a 17-county area could be paged directly through the central console.[49] Texans began to think of Texas not just as a traditional cotton, oil, and cattle kingdom, but as a center for new opportunities derived from the application of space technology and research.

The decade of the seventies began with the triumphs of the successful Apollo lunar landings, the construction and launch of Skylab with three long-duration manned flights, and the Apollo-Soyuz docking between a Soviet and an American spacecraft. During these years, the Shuttle program, officially designated the Space Transportation System, struggled for survival. Funding was short. NASA and contractor personnel declined precipitously. There were unanticipated technical difficulties. By 1975, the time of glory, what Gerald Griffin had earlier referred to as the "scarf in the wind" era of spaceflight, had ended. The last 5 years of the decade, during which time JSC and its personnel devoted most effort to the Shuttle program and the development and testing of the orbiter, were difficult years requiring hard work and perseverance. NASA learned how to do more with less, and discovered sometimes hidden costs. JSC people learned a lot during these years about aerospace engineering and lunar science, about people in general, and about themselves in particular.

When the Boeing 747 carrying the Shuttle *Enterprise* landed in March 1978 at Ellington Field near Houston and JSC, the center released its usual press release announcing the event. What happened was an overwhelming show of curiosity and support of the center and the space program as 240,000 people came out to view the Shuttle. A tour guide perhaps explained the situation better than the most enlightened engineer or scientist: "What I think is that this is different than Apollo and going to the Moon. I think the Shuttle is coming closer to the people. It is something they can relate to. They wanted to know when they can go on it."[50]

Although its primary role as a lead center for the Shuttle would continue through the 1970's and 1980's, JSC had developed a subordinate role as a research and development center leading Texas and the Southwest into a new era of economic and industrial development.

CHAPTER 13: Space Business and JSC

Houston and the Texas of the 1980's became something different than they had been in the 1950's, in part because NASA and JSC gave them the opportunity to be different. Admittedly, many factors were involved, including unplanned events. Oil, cattle, cotton, real estate and the Port of Houston were never displaced in the public mind or in commercial realities by aerospace or related industries. At first from necessity but then with real enthusiasm, particularly when oil prices plummeted and land prices collapsed in the 1980's, Texas and Houston turned to space industries as a new opportunity. However, when oil prices quickened and the economy improved at the close of that decade, Owen Morris (a former JSC aerospace engineer and a cofounder of Eagle Engineering) recalled that Houston interest in space declined correspondingly.[1] Nevertheless, space and related industries established under the NASA/JSC stimulus created a more diversified technology and industrial base in Texas than previously existed.

During President Richard M. Nixon's administration (1969-1974), the Republican Party's support of space programs, and specifically the Shuttle, was based on the premise that the Shuttle represented an economy measure and that, by providing a routine and economical access to near-Earth space, the Shuttle would support the commercialization of space. During President Jimmy Carter's administration, Democratic Party leaders, including former President John F. Kennedy's brother Senator Edward Kennedy and Vice President Walter Mondale, advised more funding for domestic programs and less for space. Upon assuming the presidency in 1981, Ronald Reagan promised to continue the space program, but reduced the NASA budget set by President Carter by $600 million.[2] Problems in the oil patch, privatization sentiments held by Republican administrations, affirmative action programs, and even the Shuttle's tightening budgets between 1970 and 1980 all contributed to the expansion of the private sector of space business, particularly in Texas.

Under the impetus of the marketplace and congressional programs seeking to stimulate minority and small businesses, many new small firms began to join the ranks of the giant aerospace corporations, such as Grumman, McDonnell Douglas, North American Rockwell, Ford Aerospace, RCA, Lockheed, Singer-Link, and IBM, as contractors and businesses which provided goods and services to NASA and to domestic and international aerospace consumers. Houston began to develop a space business complex and a new mind-set about modern technology.

From 1962 through September 1989, JSC disbursed $37.3 billion and most of that ($33 billion) to its contractors located throughout the United States for Research and Development work on the various space programs. Another $2.6 billion was spent in civil service salaries, $3.9 billion for research and program management, and $351 million for the construction of facilities. Most of these latter expenditures were made in the local economy.[3]

Through civil service and contractor-related employment, JSC was solely responsible for an average of 10,000 jobs in the Houston economic community after 1963. The average employee income exceeded that of the traditional petrochemical or agribusiness areas. A history of civil service and contractor employment levels at JSC from 1963 through 1990 (figure 18) illustrates both fluctuation levels and peak employment levels associated with the Apollo lunar landing in 1969, and the increases 20 years later associated with Shuttle and Space Station development. Although not always obvious to the casual observer, by any measure JSC impacted significantly on the Houston and Texas economy.

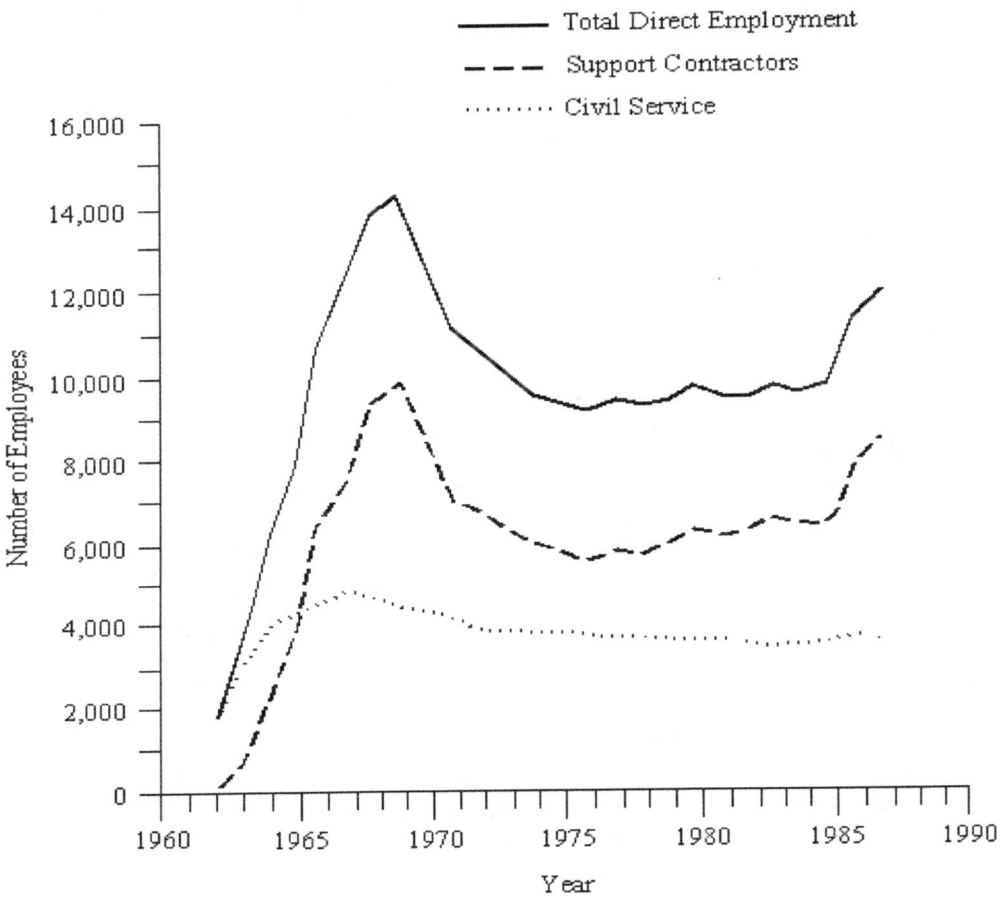

1987-1988 increase in support contractor MYEs primarily due to increases occurring in Space Station and Shuttle operations.

1988-1989 increase in support contractor MYEs primarily due to increases occurring in Space Station and Shuttle operations.

1989-1990 increase in support contractor MYEs primarily due to increases occurring in Space Station.

MYE: Man year equivalent

Source: Administration Directorate, JSC, Houston, Texas [1990].

FIGURE 18. *Civil Service and Support Contractor Employment History*

Moreover, as time passed, more and more of JSC's expenditures, including those to contractors for research and development work, were made to Texas-based firms. JSC estimated that its impact on the local economy rose from slightly less than $400 million per year to approximately $1 billion between 1979 and 1989.[4] That increase in local spending reflected the growth of a supporting industrial and technological base in the Houston area and in Texas. Much of that growth occurred during the Shuttle era rather than during the Apollo program.

Most of the contract spending in the Houston area through the 1960's and 1970's had been for construction work and for services provided by major aerospace firms which, by the mid-1960's, had begun to locate corporate installations and branches in proximity to JSC. As mentioned in earlier chapters, Congressman Olin E. Teague in his role as Representative from the Sixth Congressional District (stretching to but not through the Houston and Dallas metropolitan areas) and as Chairman of the NASA Oversight Subcommittee of the House Committee on Science and Astronautics, kept careful surveillance of NASA spending in Texas. His records include the distribution of contract funding within Texas by city and by firm. The 11 Texas cities receiving the largest cumulative total of NASA contracts from 1958 to 1968 are indicated on table 11.[5]

Texas firms produced relatively little domestic industrial technology useful to the space program in the 1960's. Most of the firms that had particularly relevant expertise were located in the Dallas metropolitan area and included LTV (Ling-Temco-Vought) Aerospace

TABLE 11. Total NASA Awards to Business and Nonprofit Institutions in Texas by City to 1968

City	$/Thousands
Amarillo	$813
Austin	7,846
College Station	2,563
Dallas	151,721
Denton	545
El Paso	1,128
Fort Worth	7,496
Houston	643,545
Richardson	45,363
San Antonio	2,210
South Houston	2,039

Note: In Amarillo, Union Carbide received the $813,000 in NASA contracts. The University of Texas accounts for all but a few hundred thousand dollars of the work in Austin, and Texas A&M University was the sole recipient of the $2.5 million in College Station. In Dallas, LTV Aerospace received most of the $151.7 NASA contract dollars. Texas Instruments and the Southwest Center for Advanced Studies accounted for about $20 million. In El Paso, the University of Texas-El Paso and Globe Exploration held contracts totaling about $850,000. General Dynamics received the lion's share of the contracts in Fort Worth. Most of the contract dollars going to Houston before 1968 were awards to divisions of national aerospace and technology companies such as General Electric ($135.5 million), IBM ($112.4 million), Philco Ford ($118.5 million), TRW ($62.7 million), and Lockheed Aircraft ($51.3 million). Brown and Root, a local construction firm, in a joint venture with Northrop Aviation received a total of $25.7 million by 1968. Other major Texas recipients included Southwestern Bell Telephone ($6.3 million) and Rice University ($13.6 million-primarily land sales).
Source: Olin E. Teague Papers, Texas A&M University Archives, College Station, Texas.

Corporation, Collins Radio, and Texas Instruments. By 1963, *Houston* magazine estimated that 103 national space-related corporations had opened 137 offices (including some with multiple division offices) near Houston. Philco, Lockheed Aircraft, Goodyear Tire and Rubber, and Hayes International Corporation were mentioned as some of the early arrivals. Dave W. Lang, Chief of JSC's Procurement and Contracts Division, told the Houston Chamber of Commerce in 1964 that the center had become a catalyst for a "NASA boom." Friendswood Development Corporation had begun Clear Lake City, a 15,000-acre, $500 million, residential-research-industrial and commercial complex. Five new shopping centers, two motels, ten office buildings, three banks, a savings and loan association, and five service stations were already under construction. The older established villages, such as Webster and Seabrook, began to stir or be stirred from their somnolence.[6]

Houston Mayor Louie Welch was elected in 1963, assumed his office on January 2, 1964, and served as mayor through the NASA boom, until January 2, 1974. He was born in the small community of Lockney and attended public schools in Slaton, Texas, before completing studies in history (magna cum laude) at Abilene Christian University. He became a realtor in Houston and served several terms on the city council before making his successful bid for mayor. He attributed the decision to locate JSC near Houston largely to the work of Albert Thomas and George Brown and to Rice University.[7]

Houston had a very good relationship with Dr. Robert Gilruth and the center, Welch recalled, but during the 1960's there was considerable confusion and conflict involved in the rapid growth of the south Houston area. Houston was somewhat hesitant to annex the Clear Lake area because it was then some distance from the city, and because annexation involved an enormous expense for utilities, roads, and public services. Understandably, the city wanted the tax revenues and utility income from the growth areas. So did the neighboring incorporated communities of Pasadena and Webster. There was also some question in the minds of many as to whether JSC and the supporting space community were "here to stay." In short, the City of Houston's posture at first really was more of a wait-and-see attitude. Houston, Mayor Welch said, did not want to annex the area, but neither did it want anyone else to do so. The result was politically very confusing as city boundary disputes, suits, and legislative bills stymied any final boundary and utility decisions until about 1968-1969.[8] By this time, Houston and Texas were on the edge of the "oil boom" caused by the OPEC embargo, and Houston's fascination with space business, as was true with the rest of the United States, began to diminish somewhat.

Although space began to lose its importance as a centerpiece of business activity, NASA research and fellowship funding to Texas universities, as well as contract dollars going to large firms operating in Texas, particularly the aerospace firms located in Houston, represented an investment in future technology that began to bear dividends of a different kind. New domestic, space, and technology business enterprises appeared in increasing numbers in the 1970's. By the 1980's, Houston, Fort Worth, San Antonio, and Austin, with Dallas, offered a considerably larger commercial and industrial complex that could and did exploit space-age technology. One of the NASA/JSC spin-offs of the past few decades involved a change both in the state-of-mind and in the state of Texas business.

Paradoxically, federal budget constraints for space programs contributed to the surge of space-related business development in the Houston economic community. NASA's budget problems had to do with continuing fiscal pressures on the federal government caused (among other things) by the Vietnam War, the OPEC oil embargo, double-digit inflation, and rising federal deficits. Fresh reductions in personnel levels at JSC were implemented between 1978 and 1980. Again, as in the early seventies, young engineers at JSC tended to be the first dismissed. But, in addition, many older personnel nearing retirement age were offered incentives for early retirement in theory to help preserve places for the younger employees. As a result, NASA and JSC experienced a sudden loss of expertise at two critical levels. One cadre lost were those relatively young, but now experienced engineers who had been with NASA for 6 to 8 years and were now qualified to assume positions as subsystem managers and intermediate program directors. Because most NASA engineers came on board during the expansive sixties and few were hired during the doldrums of the seventies, NASA had few young or mid-career civil servants. The experienced older hands with 20 or more years of NASA longevity, who had matured through the entire NASA experience were beginning to retire. Thus, the mid-level technical managers left at JSC as the decade of the eighties began (the era of Shuttle operations) were left without the supporting technical infrastructure at the bottom of the system and at the top. [9]

This deficiency, particularly that of top-level engineering expertise, was ameliorated somewhat by the formation of private Houston-based consulting and contracting firms such as Eagle Engineering, Barrios, Inc., and Hernandez Engineering, Inc., among others. These firms often provided some of the basic engineering services enabling NASA engineers to concentrate on program and project management. Many former NASA engineers, in effect, continued to work for NASA, but as contractors. Finally, the threat of government RIFs and a cap on salaries stimulated a flow of employees from the government sector into the private sector under early retirement incentives. In addition, as major aerospace contractors such as Lockheed, Grumman, McDonnell Douglas, and North American Rockwell in the Houston/JSC economy reduced employee levels throughout the seventies, many of these people, rather than leave the Houston area where they had worked for the past 10 to 15 years, moved to smaller service-oriented firms that became independent NASA contractors or subcontractors to the major aerospace firms. [10]

Others moved into positions in the then burgeoning petroleum and petrochemical industries. These people began to apply the knowledge learned in the space business to the petrochemical, communications and electronics, and medical industries and thus aided in the transfer of space technology. The petrochemical industry particularly benefited from NASA experience with systems engineering and electronic and computer applications. Communications industries drew heavily on NASA expertise. Mayor Louie Welch was particularly struck, he recalled, when President Nixon dialed extension 713 (the Houston area code) from the White House to reach the astronauts on the Moon. And he credited space technology with many of the advances in medicine and pointed to the Texas Medical Center in Houston as a special beneficiary of that new technology. [11]

In the Shuttle era, the business of space tended to become more fractured and more broadly based both as a product of the marketplace and as a result of congressional policies providing small-business and minority-owned business incentives. This meant, in part, that while JSC lost many of its younger apprentice managers and some of its older tenured and tested engineers in the late seventies, it could still draw upon that expertise within the

Houston/JSC economic community (and elsewhere) through the consulting, contract, and subcontract systems. Thus, former NASA/JSC employees such as Owen Morris, Miguel A. Hernandez, Jr., Emyre Barrios Robinson, Max Faget, Jerry Hammack, Deke Slayton, and many others helped establish a commercial sector that not only began to provide technical support to NASA, but also created a broader Texas-based private commercial space and technology sector. For example, Eagle Engineering and Barrios Technology, Inc. both began in the Houston area in 1980 and became successful multimillion dollar space and engineering firms. Both drew heavily on the expertise of retired or separated JSC or JSC contractor employees. But each began for markedly different reasons and each offered a different, but complementary expertise.

Eagle Engineering began one Sunday afternoon in 1978 in Owen Morris' home in Clear Lake City, Texas, when he and his wife with Hubert P. and Mary Davis, Carl Petersen, John Hanaway, R.E. Johnson, and William A. Bland met to discuss a plan for a private consulting business initially developed by Davis and an attorney friend, Art Dula. Davis recalled that when he added up the numbers on his Hewlett Packard HP57 pocket calculator, the early-out retirement opportunity that had to be elected prior to December 31, 1979, simply meant that a large number of civil service employees such as himself had been given a very strong incentive to leave government employment. He submitted his papers in September for retirement to be effective on December 7, 1979. His options, he said, were to take a lower level position with a local contractor, seek an equivalent position with a major aerospace firm out of state, go to the "oil patch" as many elected to do, or become an independent consultant. Being a native of San Antonio and a graduate of Texas A&M University, he wanted to remain in Texas and elected consulting work. He was able to obtain two personal consulting contracts, but as he became involved in setting up his office and putting the numbers back into his calculator for overhead expenses, he began to realize that this would not be a highly profitable venture. But there were also many, he knew, who shared the same predicament. The problem was to band these people together to achieve economies of scale—i.e. shared office spaces, secretaries, telephones, attorney fees and such. A corporate consulting entity, he thought, needed someone with a "higher profile" than himself to head it, and this led him to Owen Morris.[12]

Morris, a native of Shawnee, Oklahoma, who graduated from the University of Oklahoma in 1948 with an undergraduate and master's degree in aeronautical engineering, took his first job with NACA at Langley, Virginia, in 1948. His first assignment was to help design a large supersonic wind tunnel at Langley, and when the tunnel was built he had charge of the calibration of the tunnel as it began operations. Most of his work after that focused on stability and control problems of the new jet fighter series, such as the North American F-100, McDonnell 101, or Republic F-103 (which never became operational). He was working on the Atlas and Redstone missile development and doing research on hypersonic heating when he asked to transfer to MSC closer to home.[13]

Upon arrival in Houston in January 1962, Morris joined the Apollo Project Office under Bob Piland. Over the next 10 years he directed the Reliability and Assurance Division, became Chief Engineer on the lunar module, Project Manager of the lunar module after Apollo 11, and Apollo Program Manager for Apollo 17. He worked briefly for Aaron Cohen as deputy manager in the Shuttle orbiter office, and then moved to the Shuttle

Program Office to manage the Systems Engineering and Integration Division. Morris made important contributions to the Apollo program and to the design and inception of the Shuttle.[14]

Morris' NASA association with the Shuttle ended upon his retirement, but continued at least tangentially through his work as a NASA contractor. By the time Morris retired, Hugh Davis and Carl Petersen had opened an office for Eagle Engineering (the name was derived from the Apollo LM) on January 1, 1980, in a 600-square foot office on El Camino Real near JSC. Within a short time, Eagle became an association of approximately 30 of the 275 NASA engineers who had elected early retirement. The consultants comprised a pool of technical expertise from which individuals or a special mix of individuals could be recruited to accomplish a particular task. The firm offered special competency in systems engineering, production management, contract proposal review and preparation, and computer programming and "debugging." Most of the associates had 30 or more years of technical aerospace experience. They wanted to do something different, at their own choice and time, and to remain current and active in their fields. In addition, Eagle wanted to create a comfortable environment ("as comfortable as an old shoe") so that retirement would be easy. For example, the company paid cash advances to people traveling at government rates, as had been done by NASA. And the social aspects of their previous government work were also duplicated.[15]

A break came when Sun Oil Company recruited Eagle Engineering to work on its computer software systems. Soon other petrochemical companies, such as Exxon and Champlin Oil, began drawing upon Eagle expertise. Martin Marietta used Eagle engineers as subcontractors on a NASA project, but almost 5 years passed before Eagle embarked upon a direct contract for a NASA project. The reason for this delay was in part (JSC Assistant Director Joseph P. Loftus commented) that the center and NASA sensed an image problem (accusations of a brother-in-law deal) if NASA awarded contracts to Eagle Engineering and similar start-up firms that were developing in Houston and in proximity to other NASA centers. And in fact, Owen Morris and Hugh Davis, who were the primary shareholders in Eagle Engineering, went out of their way to pursue nonspace contracts.[16]

Within a relatively few years, Eagle began to experience growing pains as its work expanded to include systems engineering and consulting projects throughout the United States and in England. Within 5 years, its clients included Marathon Le Tourneau, Plessy Radar Ltd., RCA, Superior Oil, Tenneco, General Electric, and the U.S. Food and Drug Administration, among others. Within a few years of its organization, Max Faget retired as head of JSC's Engineering and Development Directorate and joined Eagle as a technical vice president. Other technical vice presidents included Robert E. Johnson (formerly the chief of JSC's Structures and Mechanics Division), Thomas Chambers (formerly chief of the Guidance and Navigation Division), and Burnell Bennett (former head of the Facilities Development Department of the Exxon Baytown [Texas] Refinery).[17]

Eagle accommodated its first decade of rapid growth by hiring a large staff of full-time young engineers and office personnel, expanding its list of associates to include several hundred people, and reorganizing under a parent umbrella company called Eagle Aerospace and a number of subsidiaries that provided more sharply defined services. Eagle Engineering focused on concepts, management, and marketing. Eagle Technical Services provided mainstream but cost-effective engineering services to a broad clientele and to Eagle Engineering. After a period of tension between Morris and Davis over administrative control of the firm, Davis withdrew from active participation and organized his own independent consulting firm, Davis Aerospace, which he runs from his Texas hill country home.[18] Eagle Engineering, also in the process of growth, helped create and spin-off yet other independent technical enterprises.

One of these was Space Industries, Inc. which Max Faget organized in 1982 in cooperation with Eagle, and with the support of Westinghouse, Boeing, and Lockheed, for the purpose of developing an industrial space facility—a permanent workplace in space for private enterprise and profit. Space Industries completed the design of the facility in 1988, and in 1989 won NASA payload contracts providing the space agency a "bridge" facility for its work and experiments during the construction and deployment of Space Station *Freedom*. The 35-foot-long industrial space facility was designed to maintain a permanent orbit, operate in conjunction with the Shuttle, and accommodate removable auxiliary modules to give it greater flexibility. The $30 million investment in the venture by 1990 represented a very serious step in the "privatization" of space. But by then, Space Industries was only one of many private American and foreign firms which, as the firm's slogan professed, sought to bring "the promise of space down to Earth."[19]

Barrios Technology, Inc., founded by Emyre Barrios Robinson and conceived in the same month of 1979 as Eagle Engineering, began providing basic engineering services to NASA, its contractors and other firms requiring cost-efficient technical services. Emyre Barrios was born in El Paso, Texas. Her father practiced medicine. She and her two brothers were reared in a traditional Mexican home and because she did not speak English, she "flunked kindergarten." She completed elementary and high school in El Paso, attended the University of California-Los Angeles for 2 years, married, had three children, divorced, and married Donald M. (Mack) Robinson (a Rockwell International engineer). They had one child, and after her family began to leave home, Emyre went back to school and completed a degree at the University of Houston in 1971 with a major in Spanish and a minor in business. She entered the business world for the first time in 1973 as an Associate Editor for Kentron International, a firm providing technical support services to JSC under contract. She became Manager of Data Services and then Business Manager for Kentron in 1978.[20]

She and her supervisor, Ray Perkins, became aware of an opportunity to bid on a NASA support contract designated as a "small business set aside" which precluded Kentron from bidding. The contract involved training nonengineering personnel to do the repetitive engineering tasks required to generate flight design data. Perkins decided to stay with Kentron, but encouraged Emyre to pursue the opportunity. She recruited Gary Zoerner to assemble a technical team that could meet the contract specifications. The team wrote the proposal between October and December of 1979, received notification of the $1.8 million 2-year contract award in May 1980, incorporated Barrios Technology, Inc. that month, and began work on July 1 with 15 employees.[21]

Within 10 years Barrios grew from its JSC dependency to become a regional and national aerospace services firm with 1989-90 business receipts of $22 million and 525 employees. It became one of the Nation's largest Hispanic-owned businesses, and with 51 percent of the stock, Emyre was one of the few female chief executive officers in an aerospace firm. But by the end of the decade, Barrios had reached that awkward stage: "neither small, and really not big." To help make that leap to the midsize arena and enter the international scene and also to better diversify into commercial services, in 1990 Emyre Robinson sold a controlling interest in the company to H. Ray Barrett and investors Charles Whynot and Lyle Anderson. Barrett, previously President of Hi-Port Industries, became the chief executive officer of Barrios Technology, Inc. Emyre continued as president of Barrios and a "space activist." She accepted an appointment to the Governor's Texas Space Commission which was charged to develop a strategy to establish the State as an industrial, academic and scientific leader in space. The commission elected her its chairman. It is imperative, she believes, that Texas and the Nation become firmly committed to space. "If we let this go we will become a Third World nation."[22]

Miguel (Mike) A. Hernandez, Jr., a refugee from one of those Third World nations, founded one of Houston's NASA/JSC spin-off space firms in 1983. Hernandez fled Fidel Castro's regime in Cuba in 1960 and settled in Florida. He graduated from the University of Florida in 1966 and accepted a job with JSC, in part because he was promised the opportunity to locate with the Apollo program flight crew training group stationed at Kennedy Space Center. "We were something of a JSC island in the midst of Kennedy," he recalled, "but we worked very closely with our Link contractors and with Kennedy personnel." When Apollo ended, that segment of astronaut training was transferred to Houston and the training branch was relocated there. Hernandez recalls some difficulty in selling his home near Cape Canaveral. Contractor and NASA personnel layoffs, beginning in 1969, severely depressed the housing market in and around the Kennedy Space Center. He finally sold his house in 1972 when people from northern states migrated into Florida to Cocoa Beach because of the bargains in the housing market. His Cuban origins and the United States embargo did create some personal problems for Hernandez. For one thing, he was grounded from commercial flying by NASA because of the threat of being hijacked to Cuba.[23]

One of the last Apollo training episodes, he recalls, was during a simulated training mission for Apollo 16. Astronaut Charles M. Duke, who was the LM pilot, failed to make the session. John Young, the commander for Apollo 16, asked Mike Hernandez to handle the LM controls for the simulation exercise. Edward Mitchell, the CAPCOM in charge of the exercise, was unaware of the substitution of Hernandez for Duke. When the communication checks began, Young responded, and then Mitchell called for a response by the LM pilot. What he got was a distinctly Latin accent on the communications network, rather than the Carolina drawl of Charles Duke. There was silence on the entire communications system. Finally, John Young broke in and reported to Mitchell: "Houston, this is Apollo 16. We've been hijacked to Cuba."[24]

Hernandez decided to leave NASA in 1980, just before Shuttle flights began, when he was recruited by Scott Science and Technology, Inc. to work on Air Force astronaut training programs. That company was itself a NASA/JSC spin-off. David R. Scott was among the third group of astronauts recruited by NASA (1963) and piloted Gemini 8 and Apollo 9. He commanded the Apollo 15 lunar mission and was the seventh man to walk on the Moon. He became special assistant for Mission Operations during the Apollo-Soyuz flight, and then in 1975 was appointed Director, Dryden Flight Research Center in Edwards, California. In 1977 he resigned to establish the company that became Scott Science and Technology, Inc., based in Los Angeles.[25] After only 2 years with Scott, Mike Hernandez decided to establish his own company and pursue opportunities elsewhere.

He and Scott Millican organized Hernandez Engineering in 1983, and opened their office in Houston with "two and one-half" employees—including himself and Millican. He began with a consulting job overseas. The European Space Agency (comprising members of the European Economic Community) had been organized and Hernandez saw an opportunity to provide training, payload integration expertise, and control center operation experience. The company won a small consulting contract with a German space company located in Bonn. Hernandez then won contracts with General Electric providing them support on Shuttle payload integration. At the same time they added another engineer. Next an engineer was hired to remain permanently stationed in Germany to work on the expanding European business.[26]

After 1985 the company's business grew rapidly. Jerome B. Hammack retired from JSC as Safety and Reliability officer on August 1, 1987, and began similar duties with Hernandez on August 3. The firm won contracts at Goddard and NASA Headquarters. William R. Holmberg, an engineer from the University of Texas with a graduate degree in fine arts, had taught for a few years before joining McDonnell Douglas and becoming one of the first contract (as opposed to civil service) mission control front room leaders at JSC. He joined Hernandez in September 1987 as manager of Operations and Logistics. Millican left the firm and Hernandez spun off his overseas operation as an independent subsidiary incorporated in Germany. Hernandez then doubled the size of the company adding over 200 new employees, with the assumption of a technical information and public affairs support services contract with JSC in 1989—his first with the center.[27]

Unlike most space firms, Hernandez Engineering began its business as an international operation and then later developed activities within the United States. Mike Hernandez believes that space exploration and space business for the foreseeable future are inextricably tied, wherever it is, to government programs and government funding. Unlike the aircraft industry, it does not yet have an independent private sector. But the aircraft industry did not truly become an industry until would-be airplane manufacturers began to receive orders for military aircraft and would-be airlines began to receive mail delivery contracts.[28] Space, international business, and doing business with governments are individually difficult spheres in which to operate. To do all three requires a very special business expertise.

Space Services Inc. of America, founded by a group of Texas investors headed by Houston real estate developer David Hannah, Jr. and directed by Donald K. (Deke) Slayton, Director of Flight Crew Operations who retired from JSC in 1982, made the first commercially licensed rocket launch by an American firm on March 29, 1989. The Starfire rocket was lofted from the White Sands Missile Range in New Mexico on a contract with the University of Alabama for a suborbital flight to test medicines and materials in a weightless environment.[29] Despite congressional incentives, the fledgling space company encountered frequent difficulties with technical malfunctions, financing problems in the face of Texas banking woes, and competition for elusive federal space contracts.

Congress did, however, attempt to facilitate and encourage private ventures in space and other high-tech industries in a variety of stratagems throughout the 1980's. The Small Business Innovation Development Act of 1982 established contracting guidelines for all federal agencies that spent $100 million or more annually on research. The program encouraged scientific and technical feasibility research projects by small businesses (having fewer than 500 employees) with $50,000 (Phase I) investigative grants. Phase II awards to successful investigators could qualify for funding of $100,000 to $500,000 per year for the development of prototypes of the project. In addition, Congress directed NASA to provide more specific support for space-related projects by small businesses in a 1984 amendment to the National Aeronautics and Space Act. The amendment directed NASA to "Seek and encourage to the maximum extent the commercial use of space."[30] The measures reinforced the public-oriented attitude and a procurement process that encouraged smaller subcontractors to provide goods and services to larger, primary contractors.

TABLE 12. Distribution of JSC Procurements

| Fiscal Year | Business | Net Value of Obligations (Millions of Dollars) | | | Total |
		Educational/ Non-Profit	Government Agencies	Outside U.S.	
1988	1708.2 (94.5)*	34.2 (1.9)	41.7 (2.3)	23.7 (1.3)	1807.8
1987	1540.3 (94.6)	34.2 (2.1)	42.4 (2.6)	11.8 (0.7)	1628.7
1986	1296.4 (95.0)	30.4 (2.2)	33.6 (2.5)	4.2 (0.3)	1364.6
1985	1651.1 (96.0)	30.8 (1.8)	28.9 (1.7)	9.2 (0.5)	1720.0
1984	1561.1 (96.0)	23.1 (1.4)	28.1 (1.7)	13.7 (0.9)	1626.0
1983	1664.0 (95.7)	23.9 (1.4)	32.8 (1.9)	17.7 (1.0)	1738.4
1982	1627.6 (96.3)	18.7 (1.1)	26.7 (1.6)	17.6 (1.0)	1690.6
1981	1567.7 (96.4)	20.0 (1.3)	21.8 (1.3)	16.7 (1.0)	1626.2
1980	1380.4 (96.6)	19.0 (1.3)	20.8 (1.4)	9.5 (0.7)	1429.7
1979	1154.9 (96.4)	16.2 (1.3)	22.4 (1.9)	4.6 (0.4)	1198.1
1978	980.9 (96.6)	17.6 (1.7)	17.4 (1.7)	.2 (<.1)	1016.1
1977	1083.4 (97.1)	16.8 (1.5)	15.4 (1.4)	<.1 (<.1)	1115.6
1976	991.7 (96.7)	16.9 (1.7)	16.3 (1.6)	<.1 (<.1)	1024.9
1975	801.1 (96.3)	17.5 (2.1)	13.3 (1.6)		831.9
1974	633.3 (93.6)	24.1 (3.5)	19.4 (2.9)		676.8
1973	449.9 (91.4)	24.9 (5.0)	17.6 (3.6)		492.4
1972	391.6 (87.1)	35.4 (7.9)	22.4 (5.0)		449.4
1971	542.0 (89.0)	45.1 (7.4)	21.9 (3.6)		609.0
1970	995.5 (94.0)	36.6 (3.5)	26.8 (2.5)		1058.9
1969	1101.4 (95.3)	28.9 (2.5)	25.8 (2.2)		1156.1
1968	1174.1 (95.2)	31.6 (2.6)	26.8 (2.2)		1232.5
1967	1408.4 (94.7)	26.6 (1.8)	51.9 (3.5)		1486.9
1966	1396.9 (90.3)	22.0 (1.4)	127.8 (8.3)		1546.7
1965	1280.5 (86.1)	21.5 (1.4)	185.4 (12.5)		1487.4
1964	1234.6 (85.2)	18.9 (1.3)	195.1 (13.5)		1448.6
1963	560.8 (76.1)	15.4 (2.1)	161.0 (21.8)		737.2
1962	169.2 (83.0)	3.1 (1.0)	32.2 (16.0)		204.5

*Numbers in parentheses are percentages.
Source: JSC Annual Procurement Reports (FY 1963-1988).

The Technology and Commercial Projects Office at JSC under the New Initiatives Office, provided special assistance to small businesses. In the Administration Directorate, the Procurement Operations Branch of the Procurement Support Division actively assisted in the identification of small business subcontracting opportunities. A Small and Disadvantaged Business Office and an Industry Assistance Office offered special services to firms needing assistance in developing proposals or tailoring their firms to best fit NASA procurement requirements. In 1987, small and small disadvantaged businesses received $729.2 million in total direct NASA prime contract awards and approximately $815 million in subcontract awards. Of the total $4,807 million in contracts awarded by JSC in 1988, $92.1 million (or 5.4 percent) went to small businesses.[31] These efforts helped broaden both the space services and the space technology base within the Houston economic community and in Texas.

That impact is documented in part by analyses prepared by JSC and by the University of Houston-Clear Lake, Bureau of Research. The following tables denote the distribution of JSC procurement dollars since it began operations among business, educational, and government agencies and the percentage of distributions to small business and minority firms. Table 12 illustrates the severe contraction of space expenditures in fiscal years 1971 through 1974 and the expansion concurrent with Shuttle operations and space station program development in the 1980's. As indicated in table 15, NASA/JSC budgets for 1980 through 1987 generated an estimated 21,000 to 28,000 jobs in the Houston area and pumped some $560 to $859 million into the local economy. Although much more difficult to measure than dollars, the greater impact of JSC on Houston and Texas had to do with conversion from a dominantly agricultural and petroleum economy to a more diversified economy with a strong base in aerospace industries, communications, electronic and computer technology, and sophisticated medical technology. This had as much to do with a changing world-view by Texans and with new educational and employment opportunities and incentives, as it had to do with specific business developments or technology. Although JSC was by no means uniquely or singly responsible for the substantive transformation in the Texas economy, it was a catalyst in that change.

Hard times brought about by reductions in personnel and contract spending by JSC between 1971 and 1974 resulted in a considerable absorption of personnel and talent by the domestic petroleum industry, which was then enjoying boom times, and by academia. Subsequently, civil service retirements, start-up ventures by former NASA engineers and scientists, federal incentives to small businesses and minority business, and an overriding initiative to transfer much of the government's business to the private sector affected the transfer of space technology to the Texas economic order. Finally, during much of the 1980's, Texas capital and Texas business turned increasingly (and sometimes desperately) to space ventures and new technology, as petroleum and real estate entered into a severe recession. Thus, change in the Texas economy became much more pervasive and significant in the dawning Shuttle era than it had been during the previous several decades. The development, production, and experimental flights of the Shuttle were coterminous with the changing economic order.

During its first two decades, JSC became the home of human spaceflight and a significant contributor to the development of a new regional business culture that took space and high technology in its stride, and that learned to do business not only with the federal government, but also with governments and businesses throughout the world. Particularly during 1972 through 1982, which witnessed the development and orbital test flights of the Space Shuttle, JSC became a conduit for the transfusion of people and their experiences and know-how into a changing Texas and national economy.

TABLE 13. Small Business and Minority Business Participation in JSC Procurement Activity

Fiscal Year	Small Business Value of Obligations (millions of dollars)	Minority Business Value of Obligations (millions of dollars)	Value of Obligations (millions of dollars)
1988	1708.2	92.1 (5.4)*	24.1 (1.41)
1987	1540.3	89.3 (5.8)	27.5 (1.79)
1986	1296.4	79.3 (6.1)	23.7 (1.83)
1985	1651.1	74.4 (4.5)	20.4 (1.24)
1984	1561.1	58.9 (3.8)	15.3 (0.98)
1983	1664.0	57.1 (3.43)	13.5 (0.81)
1982	1627.6	49.5 (3.04)	16.4 (1.01)
1981	1567.7	40.7 (2.6)	10.0 (0.64)
1980	1380.4	33.4 (2.4)	2.1 (0.15)
1979	1154.9	28.2 (2.4)	2.4 (0.21)
1978	980.9	25.3 (2.6)	2.4 (0.25
1977	1083.4	25.1 (2.3)	2.7 (0.25)
1976	991.7	19.6 (2.0)	2.3 (0.23)
1975	801.1	22.1 (2.8)	1.78 (0.22)
1974	633.3	20.8 (3.3)	1.4 (0.22)
1973	449.9	19.8 (4.4)	0.431 (0.09)
1972	391.6	21.3 (5.4)	0.611 (0.16)
1971	542.0	29.3 (5.4)	
1970	995.5	27.8 (2.8)	
1969	1101.4	22.8 (2.1)	
1968	1174.1	27.3 (2.3)	
1967	1408.4	28.9 (2.1)	
1966	1396.9	17.2 (1.2)	
1965	1280.5	23.3 (2.0)	
1964	1234.6	21.5 (2.0)	
1963	560.8	11.6 (2.1)	

*Numbers in parentheses are percentages.
Source: JSC Annual Procurement Reports (FY 1963-1983).

TABLE 14. *Geographical Distribution of JSC Procurement Excluding Intragovernmental Actions*
(thousands of dollars)

Fiscal Year	Far West	Mountain States	Mid-West	Texas	Great Lakes States	Southeast	Mid-Atlantic	New England	Alaska	Hawaii	Outside U.S.
1988	817,100	22,586	21,453	746,552	5726	55,154	35,305	38,479	3	—	23,705
1987	758,582	21,941	5410	608,252	8313	82,698	50,730	38,561	1	—	11,767
1986	1,572,304	18,396	3585	608,252	7759	72,151	38,002	51,627	30	39	4179
1985	928,363	15,671	4148	513,797	5330	68,433	72,647	73,444	3	—	9199
1984	922,942	23,681	2074	453,315	3508	54,418	65,799	58,952	—	1	13,289
1983	1,060,422	30,553	2785	407,972	7861	58,131	52,704	67,446	1	—	17,715
1982	1,146,372	12,390	1955	344,822	5603	46,892	43,722	44,572	3	—	17,624
1981	1,144,176	23,471	2220	266,274	7726	35,159	68,204	40,443	—	10	16,663
1980	978,514	33,594	2536	246,268	4634	30,166	65,573	37,973	75	9	9543
1979	810,162	4341	1568	218,901	5002	23,631	56,719	50,852	<1	9	4568
1978	679,861	21,226	7948	197,161	4959	8932	51,602	26,807	10	36	211
1977	827,093	14,857	4350	182,124	4944	10,978	41,866	13,971	-(23)	7	47
1976	770,773	8766	2267	141,394	5500	10,958	61,527	7451	-(3)	<1	-(18)
1975	607,405	7252	7849	128,873	6001	11,661	42,328	7068	108	2	80
1974	433,411	15,945	13,041	109,159	5450	9101	58,218	12,971	30	1	132
1973	257,618	22,657	9682	103,945	10,370	5189	44,578	20,617	45	2	119
1972	140,366	28,971	17,495	116,605	26,357	3697	57,394	35,991	145	<1	12
1971	153,588	20,717	14,233	118,974	60,967	3717	154,992	59,420	192	33	287
1970	403,248	55,258	18,919	139,283	42,688	13,723	304,118	54,529	21	—	275
1969	494,057	14,411	17,801	83,403	49,790	53,330	375,966	40,613	422	—	443
1968	504,871	8143	27,395	84,240	71,204	40,309	433,867	35,265	103	—	278
1967	591,218	7666	16,714	74,681	98,342	27,832	578,907	39,367	—	—	348
1966	694,966	7368	58,795	50,566	134,649	24,862	405,612	41,785	89	—	—
1965	654,503	7261	177,478	30,360	76,624	23,998	295,982	38,375	—	—	—
1964	991,226	3046.6	654,659.3	19,158.5	55,810	209,133.3	79,535.9	—	—	—	—

Source: JSC Annual Procurement Reports (FY 1964 - 1988)

TABLE 15. *Impact of the NASA Budget on JSC and Houston Area Economy*

	Fiscal Year (dollars in millions)										
	1980	1981	1982	1983	1984	1985	1986	1987*	1988	1989	1990
Total NASA Budget	5,243	5,522	6,020	6,836	7,248	7,552	7,764	10,774	7,800	10,897	12,296
Total JSC Budget	1,557	1,702	1,789	1,745	1,662	1,617	1,445	2,909	1,445	1,935	2,505
JSC's Share of NASA Budget	29.7%	30.8%	29.7%	25.5%	22.9%	21.4%	18.6%	27.0%	18.5%	17.8%	20.4%
Dollars Spent in Houston	404	475	502	559	564	645	706	788	797	973	1,182
Houston's Share of JSC Budget	25.9%	27.9%	28.1%	32.0%	33.9%	39.9%	48.9%	27.1%	55.2%	50.3%	47.2%
Houston's Share of NASA Budget	7.7%	8.6%	8.3%	8.2%	7.8%	8.5%	9.1%	7.3%	10.2%	8.9%	9.6%

*In 1987, the NASA/JSC budget reflects a one-time appropriation for a replacement orbiter. Production of the new orbiter was primarily done in California.

Source: JSC Almanac, "Economic Impact," 1991.

CHAPTER 14: *Aspects of Shuttle Development*

Although the Shuttle flights, beginning with the four orbital test flights in 1981 and 1982, took Americans back into space after an absence of 6 years, the ground rules or "Earth rules" for spaceflight had changed. The Shuttle had a different commitment and different purposes than previous programs. The Shuttle was exclusively an Earth-to-orbit transportation system. Defense and earth sciences loomed proportionately larger in its development and operation. Costs remained critical. Benefits were of the essence. "Payloads" became a Shuttle euphemism for payoff. Popular enthusiasm for space waned. National prestige was no longer so threatened as it had been before Apollo. Americans in the Shuttle era no longer mobilized for space as though preparing for a hopefully short and determinate war. They began to learn to accept space, with its technology, its benefits and its costs, as a part of everyday life.

The enthusiasm, the commitment, and the funding for space ventures declined perceptibly after Apollo. Apollo had the national spotlight. It was a prestigious program, was popular, and seemed to have unlimited backing. Money was always available for necessary work. The Shuttle was conceived under that same aura, but developed and flown under different circumstances. When the Shuttle began, it was to be one element of a grand design which included a space station, unmanned planetary missions, and a manned flight to Mars. The Johnson Space Center was to become a multiprogram center.[1] But the Shuttle ended up being the only program.

The designation of JSC as lead center, effectively transferred Level II or technical control of Shuttle development from Headquarters to JSC. During Apollo, although Headquarters nominally exercised technical control, technical management was actually dispersed among the spaceflight centers which operated under very strong leadership. Thus, the designation of one center as lead center put technical control where NASA had in-depth technical support. Headquarters exercised less technical management on the Shuttle than it had on Apollo, in part because of its relatively smaller technical staff.[2] The lead center management style made most efficient use of NASA's personnel and resources.

Owen Morris, previously identified as a cofounder of Eagle Engineering after he retired from NASA as head of the Systems Engineering and Integration Division in the Shuttle Program Office, had primary responsibility for integrating the orbiter in the overall Shuttle system. He believed that under lead center management the work and coordination among the centers went quite well. The Shuttle represented a challenge to systems engineering. It was a much more complex machine. There had been a progressive increase in the complexity in the interfaces involved in each program from Mercury through the Shuttle. The Apollo manned capsule interface with the propulsion system was accomplished with 96 bolts and 93 wires. The Shuttle was an integrated vehicle and much more complex. There were 3200 separate wires leading from the propulsion system of the Shuttle. The forthcoming space station has yet a "much, much more complex interface."[3]

Despite budget and personnel cuts, the Shuttle was a relatively well designed and managed program. The budget cuts were most damaging in that NASA could never plan for lean years and good years. Cuts were always followed with promises of better funding ahead; thus NASA always tried to rebuild and gear up for another productive surge, only to have the funds cut at the last minute—often by OMB rather than by Congress. OMB, which functioned under the authority of the Executive Office, tended to be less supportive of NASA programs than did Congress. But the real problem with lean years was that invariably research funding suffered first, and as development and construction began, budget constraints often translated into reducing spare parts or redundant systems. Rodney G. Rose, who headed a special flight operations planning group for Apollo, believed that budget constraints meant that the Shuttle became operational with far fewer spares than the Air Force, for example, considered adequate.[4]

Moreover, inasmuch as the program office made research allocations, research funding tended to be directed to specific purposes and lacked the broad base and diversity needed. In addition, when basic research and developmental work was delayed or slipped to meet a launch deadline, it meant that down the line some of the essential "dirty-handed" engineering would not be available when needed.[5] The costs of Shuttle development were exacerbated by the delays. The technical losses (largely in the area of basic research and development) were long-term rather than immediate.

One unique element in Shuttle development had to do with mission operations planning, which had evolved to a considerably higher level of sophistication compared to that of the Apollo and earlier programs. Mercury had begun with a fairly simplistic aircraft flight operations approach. The process matured during Gemini operations when a systems handbook and direct interface between flight control teams and the crew provided real-time ground-to-space interaction. Gemini EVA heightened the relationship between the astronaut, the task, and the working environment. During Apollo, the operations team "worked in an integrated fashion on all issues involving flight systems, flight design, science, and manned operations."[6]

Shuttle flights, however, had greater and more diversified capabilities and more participants in terms of federal agencies, institutions, and even foreign nations. Skylab flight operations were much more of a learning experience for Shuttle flight operations than had been Apollo. During Skylab, systems engineering and integration processes began to be applied to flight operations in a formal context. Skylab, as the Shuttle would have, had a complex flight program involving the designer and builder of the craft, the science experiment user, the crew, and mission control. Critical engineering support for Skylab flights was derived through the creation of a joint JSC/Marshall Space Flight Center review team which screened the systems engineering and integration processes as they related to the flight plan. In other words, it tested prior to flight the compatibility of the men and the machine and anticipated the ability of both to accomplish the mission. This was done through formal systems operations compatibility and assessment reviews.[7]

Whereas one might remember that the Mercury capsule was built almost oblivious to the fact that it would carry a person (almost by accident did an astronaut discover that the original design had no visor plate), Shuttle design and construction involved close support from the mission operations team. The Shuttle was built with the understanding that good flight operations required something of a symbiotic relationship between the human occupants and the machine—and that this relationship must extend to its ground support systems. For flight operations, systems engineering and integration is a process by which "the technical, operational, economic and political aspects of programs are integrated to support the program objectives and requirements consistent with sound engineering, design and operations management principles."[8]

Shuttle flight, STS-5 *(Columbia)* crewmember Joseph P. Allen observed, is technologically complex, cooperatively challenging, and personally exhilarating. Each launch is unique and one of the "richest events" of a lifetime. It was a sentiment generally shared by NASA personnel at each of the centers, and especially by the mission control personnel linked to the Shuttle through the invisible threads of radio, electronics, and human spirit.[9] It was at the Mission Control Center during Shuttle flight that the rich mixture of crew, machinery, engineering, scientific and support structures melded.

Fewer operators worked the Mission Control Center at JSC than in the days of the Apollo lunar missions, but Shuttle flight operations required a networking of the support team composed of the flight control room, the multipurpose support rooms with the payload operations control centers located at JSC or elsewhere. A payload operations control center at Goddard Space Flight Center, for example, monitored all free-flying (satellite) systems delivered, retrieved or serviced by the Shuttle, including the two communications satellites delivered into orbit by *Columbia* on the STS-5 mission. Both satellites were built by Hughes Aircraft Company under contract—one for a private company, Satellite Business Systems, and the other for Telesat of Canada. Hughes engineers, as well as technicians representing the contractors, monitored the satellite launches from remote payload operating control centers. In the event the Shuttle was delivering satellites for interplanetary exploration, as would be true in later flights, a payload operating control center at the Jet Propulsion Laboratory in Pasadena, California, managed the payload.[10]

Beneath the primary flight control rooms at JSC (on the second floor of the Mission Control Center or for DoD missions on the third floor), the network interface processor provided an intermittent flow of real-time information coming from the Shuttle and other operating centers and fed it to flight control. Also on the first level, the data computation complex compared tracking and telemetry data with Shuttle flight progress.[11] Although the facade and apparatus of the Mission Control Center had changed little since the days of the Apollo lunar flights, flight control systems were enhanced substantially due to the advances in electronics and computer technology—advances which were in part derived from previous NASA spaceflight experiences.

Shuttle flight control became much more streamlined than during Apollo flights, and depended on advanced information systems and computer programs (although the external hardware in the Mission Control Center was much the same). The Shuttle required all new computer software—adjusted and reconfigured for each Shuttle mission. Development and ownership of software was a big challenge in the design of Shuttle operations. Improved information systems, derived from more sophisticated computer hardware and sophisticated programming, at least in part, facilitated the reduction in the numbers of flight control personnel. Better systems engineering and integration also helped. Mission Control teams for

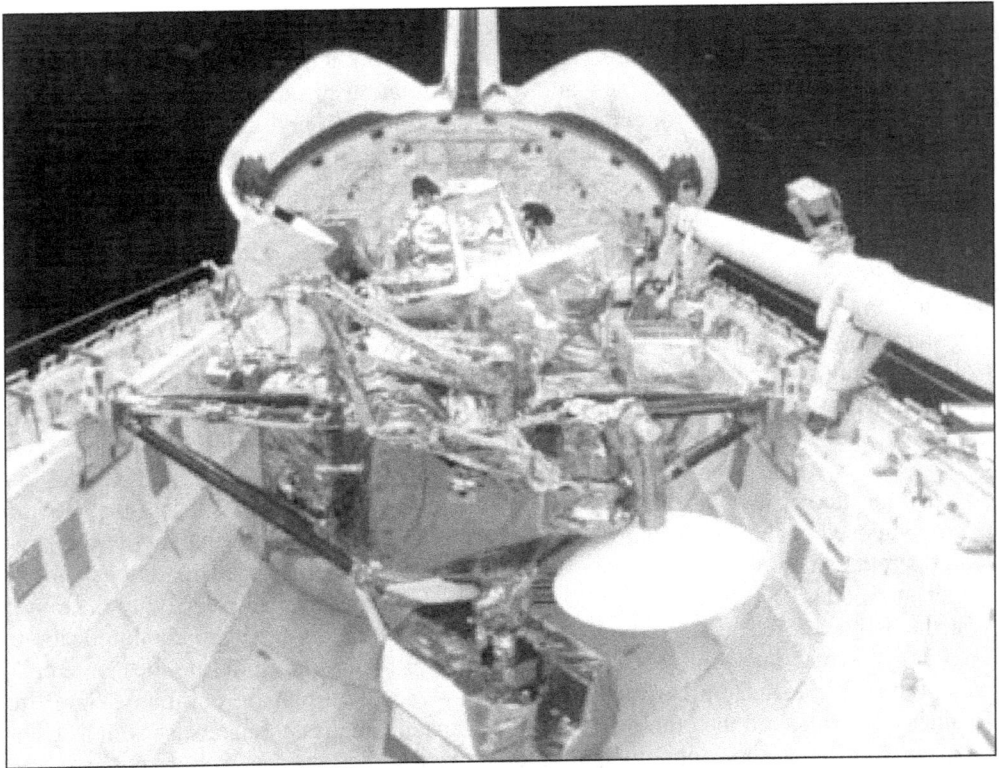

This photograph of the Upper Atmosphere Research Satellite in the payload bay of the Earth-orbiting Shuttle Discovery *shows the large size and versatility of the new space vehicle designed for near-Earth operations.*

the Shuttle were pared to one-half the size of Apollo teams with 80 people on each flight control team and 3 teams for each mission. The 22 controllers stationed in the flight control room of the Mission Control Center managed a host of technical advisors in multipurpose support rooms in the Control Center and had access to support groups stationed throughout the United States—and indeed in other countries. As mentioned earlier, Wayne Hale, a Shuttle flight director, likened the operation of the Shuttle to the operation of a battleship, except that instead of thousands of crewpersons aboard the ship, there were only six or seven on the orbiter and the other thousands were stationed on Earth.[12]

Mission planning for early Shuttle missions began 3 or 4 years before each mission. Approximately one-third of the Shuttle flights developed problems during flight which required adjustments in the mission flight plan. Shuttle flying time usually consumed only about 10 percent of the total hours that went into each mission with most of the hours and work related to flight planning, simulation, training, and preflight preparation. Payload planning and payload mission planning often consumed as much time and energy as did Shuttle flight preparations. Whereas during Mercury, Gemini and Apollo years the flight was itself the essence of the mission, for the Shuttle (as had been true of the Skylab missions), the payload was the most important element of the mission.

Payload decisions had to do with commercial competition, foreign governments, competition for payloads between NASA centers, as well as broader scientific and earth-resource interests. NASA began to tackle the problem very early in the Shuttle's development by creating one of those interim, ad hoc, shadow organizations that appeared, made a critical decision or contribution, and then disappeared. Such was the Ad Hoc Shuttle Payload Activities Team organized in January 1974 under the tutelage of Charles Donlan, who returned to NASA for one of the frequent periods of activity following his retirement some years earlier. The Payload Team met at each of the NASA centers during January 1974 to discuss the establishment of management policies affecting Shuttle payloads.[13]

The issues raised at those meetings, more than the resolution of the problems, denoted the complexity and sensitive aspects of payload decisions. A JSC contingent attending Payload Team meetings held in Houston on January 10 and 17, for example, believed that because Shuttle involved both manned and unmanned operations (the unmanned being the payloads), the traditional manned versus unmanned program definitions within the NASA management system should be abandoned. Lewis Research Center personnel called for a radical change in thinking because payload operation had to be separated from operation of the transportation system. Kennedy Space Center observers stressed that a single center needed to be responsible for the sustained engineering of the Space Shuttle to accommodate each payload. Vehicle preparation for certain payloads could take years. Some individuals thought that there might be greater efficiency and payoff in missions if planning were subordinated to letting the crew and specialists go up and "klunk around" in space.[14]

JSC managers noted that an Agency guideline specifying that mission decisions should consider "minimum practical total cost to attain mission and program objectives" was a "state of mind" rather than real criteria. The Agency needed a double standard for the Shuttle: transportation systems would have to meet high standards and rigid criteria, while payloads would have to meet varying but generally lower standards. And JSC people wanted to squelch a recurring suggestion that Mission Control for the Shuttle be placed somewhere other than at JSC. "There is no viable alternative to doing this job at JSC," they said, "and entertaining alternatives is divisive and inhibits developing harmony within NASA.[15]

The issues discussed ranged from the very broad to the very finite. Mission planning would have to be separated from flight planning. Low-cost payloads could not come in the form of de-emphasizing sophistication in scientific instruments. Langley proposed to assign the Marshall Space Flight Center responsibility for maintenance and operation of the experiment modules and pallets, and it wanted to manage its own payload operations, but do so through the Mission Control Center at JSC. Ames Research Center experiences argued against giving the Shuttle or transportation operator any responsibility for payload decisions. Jet Propulsion Laboratory personnel advised that mission specialists (as part of the flight crew) should operate the payloads and come from the payload organization. They also thought that Shuttle crew operations and payload maintenance could properly be combined under the authority of the Kennedy Space Center. And Ames advised Headquarters to stick to policy decisions regarding payloads and avoid operational decisions.[16] It thus became clear long before the Shuttle began to fly that it was mechanically much more complex than previous spacecraft, and socially (or procedurally and politically) much more complicated and integrated.

The following year, 1975, JSC established a Shuttle Payload Integration and Development Program Office under Glynn S. Lunney. Lunney, it might be recalled, left Lewis Research Center in 1958 to join the Space Task Group and served as Chief of the Flight Director's Office during the Apollo flights. The Payload Integration Office (redesignated the STS Operations Program Office in 1980) had responsibility for planning and integrating JSC-sponsored payloads to include engineering and operations interface and integration responsibilities.[17]

Lunney believed that the Shuttle was a dramatically different program than Apollo, and that it particularly drew upon one of the greater strengths of JSC—the interaction between the design engineers and the flight crews. It involved a complete change of "mind-set" by JSC, from the tradition of pouring every energy into every single flight to the idea that the flight itself was peripheral to the payload. NASA and JSC ceased being inventors and became producers. Production, for example, required the development of standard connectors in the cargo bay for payloads, and assimilating the customer's emphasis on containing costs—that is doing that which was necessary, but no more. Shuttle missions related to payloads and to external relations with people around the Nation and in other countries (which is one reason, Lunney added, that the Apollo-Soyuz planning group which he headed became the payload integration team for the Shuttle).[18]

Shuttle development thus involved social as well as technical engineering. As the machinery was perfected, techniques of using the machine were honed by constant exercises involving the crew and ground control teams on the shuttle mission simulator (SMS). The SMS provided integrated training for flight crews, the Mission Control Center and mission operators, payload support groups, and tracking and telemetry systems. Built under a contract awarded in 1976 to Singer Company's Link Division, the SMS provided real-time simulation capability for all phases of Shuttle orbital flight. Its three basic stations, including a fixed-base crew station, a network simulation system, and a motion-base crew station, became operational in 1978. The machinery (and the station operating crews) could simulate every phase of the flight including motion and directional simulation, vehicle dynamics, orbital environment, visual scenes and aural cues. The SMS neatly blended the complex human and mechanical elements of Shuttle spaceflight.[19]

Owen Morris' conviction that the Shuttle was a well conceived, well engineered, and well built vehicle was certainly supported by the 24 Shuttle flights made between April 1981 and January 1986. During the first 5 years of Shuttle operations, NASA almost doubled the cumulative hours compiled by astronauts during the 10 years of Mercury, Gemini and Apollo flights. The critical moment in Shuttle development came, according to Morris, when for the first time NASA put the engine, the fuel tanks and the orbiter together on the test stand at the Mississippi Test Facility near Biloxi about March 1977. The Space Shuttle main engines experienced serious problems during tests until late 1980. Those engines (with three mounted on each Shuttle) were the most powerful hydrogen-oxygen engines yet built, and the technical problems proved to be considerably greater than anticipated in the design stage. Components, such as valves, seals, and pumps, had to be redesigned to withstand the pressures. One high-pressure fuel pump on an engine was so powerful it could empty an "Olympic-sized swimming pool in just 25 seconds."[20] Thus the engine itself represented a major technological advance, rather than a simple adaptation of existing knowledge.

One of the most frustrating Shuttle problems involved not just the development, but the adhesion of silicate fiber tiles to heat-bearing surfaces on the Shuttle. NASA engineers, working with Lockheed, McDonnell Douglas, Battelle/Columbus and university scientists and engineers, developed and tested a tile that could withstand the 2700 degree F (1428 degrees C) reentry. The featherweight tile "could throw off heat so quickly that a white-hot tile directly out of an oven, with a temperature of 2300 degrees F (1260 degrees C) could be held in a bare hand without burning or causing other injury." Although JSC engineers protest that the "bare hand" rhetoric overstated the case, the tiles could handle temperatures no previous man-made substance could withstand. But initially the tiles were very susceptible to meteorite and impact damage. Most tiles failed stress tests until they were thickened and redesigned with a ludox (silicon-boron) base. Next, each of the 31,000 tiles, with no two alike, had to be glued to the Shuttle surfaces—a job requiring an estimated 670,000 hours or 335 person-years of labor![21]

Despite the intensive tests of tiles and engines, the Shuttle's space-worthiness would be proved only by piloted flights. Unlike the earlier Gemini or Apollo spacecraft, there would be no automated tests of the Shuttle. The Shuttle was designed for manned flight. Thus, when John W. Young and Robert L. Crippen made the first orbital test flight of the new Space Shuttle (STS-1) *Columbia* on April 12, 1981, they were truly "man-rating" America's first aerospace vehicle. Crippen, a native of Beaumont, Texas, and a graduate of the University of Texas, came to the astronaut corps by way of the Air Force Manned Orbiting Laboratory Program. Among other things, he had immersed himself in Shuttle computer software problems. It was his first flight. Young, a Georgia Tech aeronautical engineer who was born in California, was a veteran astronaut who made his first flight as pilot of Gemini 3 and was command module pilot for Apollo 10 and commander of the Apollo 16 flight.[22] They were the real guinea pigs.

The flight was, by traditional aeronautical standards, an unusual one. About one million people were on hand to watch the launch from the Kennedy Space Center—millions more watched on worldwide real-time television. Nine minutes after launch the astronauts were in orbit. Once in orbit about the Earth they flew their craft tail-forward and upside down (to get a better view of Earth and its horizon). During their 2-day, 6-hour and 21-minute flight, they changed their orbit apogee (high point) by some 172 statute miles and checked out the computers, jet thrusters, cargo bay doors, and control systems. The flight marked, according to the official NASA Mission Report, "a new era in space promising countless benefits for people everywhere."[23] It was "top billing" in theatrical or PR (public relations) terms, but *Columbia*'s almost flawless voyage held enormous promise and infused great optimism in a NASA and JSC cadre that had lived on the edge during the past decade.

The Shuttle *Columbia* returned "hardly worse for wear" after its searing atmospheric reentry through temperatures that reached 3000 degrees F (1650 C). The two solid rocket

Suddenly, Tomorrow Came . . .

boosters were recovered in the Atlantic off Daytona Beach and could be refurbished at a fraction of the cost of building new expendable rocket engines. And the Shuttle, but for the damage to 12 tiles on the tail section, returned intact and fully reusable. For a time the old surge of Apollo lunar mission pride and national resolve returned after a long American absence from space. President Ronald Reagan told Congress that the return to space "did more than prove our technological abilities. It raised our expectations once more. It started us dreaming again."[24]

A few months later *Columbia* was ready to fly again. Joe H. Engle and Richard H. Truly, who had flown the *Enterprise* in landings after drops from a Boeing 747 in 1977, manned *Columbia's* second flight into space on November 12. Engle was unique. He entered the astronaut program already an astronaut, having completed 16 flights in the X-15 experimental aircraft during one of which he reached an altitude of 280,600 feet. Neither he nor Truly previously had flown an orbital mission. The STS-2 mission not only further flight-tested the readiness of the Shuttle, but also carried a test package of scientific experiments prepared by the Office of Space and Terrestrial Applications and a robotic arm for managing the Shuttle cargo built by the Canadian Government as its participation in Shuttle flights and technology. The pallet or container housing the five experiments carried by STS-2 was designed and developed by the Spacelab Program Office at Marshall Space Flight Center

The new Shuttle, designed as a reusable spacecraft, made its maiden voyage on April 12, 1981. After an absence of 6 years, America had returned to space. Eugene Kranz, Chris Kraft, and Max Faget monitor the return to space.

and built by the British Aerospace Corporation for Zentral Gesellschaft VFW-Fokker mbH on behalf of the European Space Agency.[25] The second flight helped identify the scientific and Earth resources orientation of Shuttle missions, and denoted the broadening of American space programs to include European and Canadian participation.

The experimental package included a Shuttle Imaging Radar-A designed to send and receive microwave radiation to create maplike images of the Earth. Satellites carrying such equipment previously recorded ancient Mayan canals, ice flow and ocean wave patterns, and, it was surmised, could aid in the study of Earth's geological formations leading to the discovery of oil and mineral deposits. A Shuttle multispectral infrared radiometer supplemented the radar surveillance by detecting the best spectral bands to be used in remote sensing. A feature identification and location experiment activated the infrared and radar experiments when atmospheric conditions were best suited for observation. Another package, called the Measurement of Air Pollution from Satellites, checked particularly for the distribution of carbon monoxide in the middle and upper troposphere (7.5 to 11 miles above Earth's surface). An ocean color experiment identified chlorophyll and other pigments in the oceans, thus providing locations for schools of fish and pollution.[26]

During the interim between the flight of STS-1 and STS-2, the House Subcommittee on Space Science and Applications held public hearings on "Future Space Programs" based on what the members considered "a look at the space program from the point of view of society." The intimation was that in previous hearings the focus had been on what society could do for the space program. The subcommittee hearings considered how space might relate to the "Nation's wealth, broadly defined," and what programs might be considered "fruitful investments." No NASA personnel were interviewed, although several of the guests had previous NASA associations. The comments from business and academia (e.g., Dr. Melvin Kranzberg, Professor of History of Technology at Georgia Institute of Technology; David Hannah, President of Space Services, Inc.; Dr. Donlin M. Long, Johns Hopkins Hospital; General Thomas Stafford (retired), former Head of USAF Space Programs) ranged widely but focused on the general theme of costs versus possible benefits. The committee finally called for broad debate on the civil space program and the definition of national goals.[27]

The cost-benefit theme had indeed preoccupied NASA through much of the Shuttle development phase. A critical element in those deliberations had to do with the establishment by NASA of a Shuttle cargo or payload policy. The kind of cargo or payload the Shuttle was to carry would be very critical in the cost-benefit evaluation. Payload decisions also had to do with commercial competition, foreign governments, and competition for payloads between NASA centers, as well as broader scientific and Earth-resource interests. NASA, as previously mentioned, began to tackle the problem very early in the Shuttle's development by creating one of those interim, ad hoc, shadow organizations that appeared, made a critical decision or contribution, and then disappeared.[28]

Headquarters ultimately approved Shuttle payloads. Each center, or other government agency such as DoD, and sometimes foreign governments and institutions might sponsor payload proposals. NASA solicited proposals from universities and the general public. The final payload decision ultimately rested on the technical feasibility and outfitting costs as judged by JSC's Payload Integration Office. Outfitting and redesign of the Shuttle to accommodate the varying payloads affected payload scheduling.[29]

The Payload Integration Office (redesignated the STS Operations Program Office in 1980) had responsibility for planning and integrating JSC-sponsored payloads to include engineering and operations interface and integration responsibilities. In January 1982, following the successful flights of STS-1 and STS-2, JSC Center Director Chris Kraft announced the merger of the STS Operations Office with the Space Shuttle Program Office. It meant that with payload management and integration problems resolved, after two more orbital test flights the Shuttle was ready to become fully operational.[30]

On schedule, with its turnaround time cut from 102 to 69 days, the Shuttle *Columbia* blasted off for its third successive flight on March 22, 1982. Commander Jack R. Lousma, a veteran of 59 days aboard Skylab, was a colonel in the Marine Corps who joined NASA in 1966 with the fifth group of astronauts. His pilot, C. Gordon Fullerton, had not flown in space but had manned three of the *Enterprise* glide flights during its initial tests in 1977. They now pushed the orbiter a bit closer to its flight limits. The scientific experiments approved by NASA's Office of Space Science investigated space plasma physics, solar physics, astronomy, life sciences, and space technology. The orbiter also carried Todd Nelson's "Insects in Flight Motion Study" experiment. Nelson, an 18-year-old high school student won a NASA and National Science Teachers Association competition for the opportunity. It denoted yet another broadening of Shuttle applications into the everyday world. Engineers performed additional tests on the Shuttle's control systems and its aerodynamic performance during launch and reentry, and accomplished a cold start of the orbital maneuvering engines. As mentioned previously, STS-3 was the first and only shuttle to land at the Northrup Strip at White Sands, New Mexico.[31]

Finally, the fourth flight of *Columbia* (STS-4), the first DoD secret mission, ended auspiciously on July 4, 1982, when flight commander Thomas K. (Ken) Mattingly and pilot Henry W. Hartsfield, Jr., returned from a 7-day orbital flight to the concrete runway (another first) at Edwards Air Force Base, California. It was the end of the beginning for the Shuttle program. "Its on-time launch, near-flawless completion of all assigned tasks, and perfect landing ushered in a new era in the Nation's exploration of space—a fully operational, reusable spacecraft now set to begin its job in earnest."[32]

The payload carried the Shuttle's first commercial package, an electrophoresis experiment (involving the separation of biological materials in fluids) developed by McDonnell Douglas Astronautics Company with the Ortho Pharmaceutical Division of Johnson and Johnson. Two high school students, Amy Kusske of California and Karla Hauersperger of North Carolina, submitted separate medical experiments testing chromium levels and cholesterol levels in the astronauts from urine and blood samples. Nine experiments by Utah State University students were funded under NASA's low-cost Getaway Special, a small, self-contained payload experimental program. The crew also tested the Shuttle skin and tiles for prolonged exposure to extreme heat and cold by changing the position of the spacecraft relative to the sun.[33] Spaceflight, many thought, had now become routine.

The business of space had changed markedly in that brief decade between the inception of the Shuttle program and the completion of four Shuttle orbital test flights.

The Shuttle was a much more complex mechanism. Its management systems had become much more integrated and involved political and economic decisions as well as technical decisions. Space appeared to be an evolving sector of the national economy and an increasingly significant element in the local and regional economy of Texas and the Southwest. Now that the engineering and developmental phase had ended, it was time to put the Shuttle to work.

CHAPTER 15: *The Shuttle at Work*

The launch of STS-5 *(Columbia)* on November 11, 1982, brought to an end the politically and fiscally sensitive interlude between the Apollo-Soyuz mission in 1975 and the first Shuttle test flight in 1981. The focus of manned spaceflight shifted from earlier emphases on development and exploration to operations. With the return to flight, the old verve and vitality returned to JSC and throughout NASA. At JSC the collective pulse quickened as personnel turned to mission planning, flight software development, payload integration and crew training. It was time to put the Shuttle, and space, to work. Named for America's first Navy ship to circumnavigate the globe (1836) and for the command module that carried Neil Armstrong, Michael Collins and Edwin (Buzz) Aldrin to the Moon in 1969, *Columbia* began in earnest the pursuit of a new space goal—not simply getting there, but using the resources one found in space.

A reusable aerospace vehicle, the Shuttle is launched like a rocket, orbits like a spacecraft, and lands like a glider (albeit a very heavy 150,000-pound glider). Designed to reduce the cost of spaceflight and to accommodate civilian passengers, the Shuttle promised economies because of its reusable orbiter, its recoverable solid rocket boosters, its lower launch costs as compared to the Atlas-Centaur and Delta rocket alternatives, and most especially because of its cargo delivery and retrieval capacities. Congress defined the Shuttle as the key element in making space an *extension of life on Earth*. The Shuttle was to be part of a "total transportation system linking Earth with space" including "vehicles, ground facilities, a communications net, trained crews, established freight rates and flight schedules—and the prospect of numerous important and exciting tasks to be done." In 1980 NASA projected that within 5 years the Shuttle might fly weekly round-trip missions between the United States and Earth orbit.[1]

Instead, upon completion of the four orbital test flights, the Shuttle averaged closer to one flight every 2 months for the next 3 years. Between November 1982 and mid-January 1986, the Shuttle flew 20 operating missions during which 117 crewmembers accumulated a total of 17,576 hours or 2 years of flight in space. Payloads launched included 29 satellites (most of them communications and navigational satellites such as Telstar, Palapa, WESTAR), 5 of which were retrieved and repaired after operational failures. There were four spacelab missions, a number of science laboratory packages deployed or operated during flight, and two classified DoD missions. During the same interval, another 29 satellites were launched aboard unmanned and expendable Delta, Centaur, and occasionally Scout rockets. The European Space Agency had also begun to launch satellites on the French-built Ariane rocket. Independent consultants estimated in 1982 comparative satellite launch costs for the Delta at $37 million, the Ariane at $31 million, and the Shuttle at $17.5 million, but noted that the Ariane launches often offered price concessions and that Shuttle manifests were already full through 1985.[2]

Estimates of comparative launch costs vary widely, depending on the elements included in the base. Costs are, however, considerably higher than those estimated in 1982, and the cost advantage of the Shuttle considerably less than that estimated in 1982. A 1986 NASA study, for example, estimated Shuttle launch costs at $31 million and Delta launch costs at $38 million (1986 dollars). Costing has been a very problematical thing, but the following data (table 16) provided by JSC for 1990 represents a recent analysis. The analysis indicates an $8.9 billion launch cost for Apollo and a $1.8 billion cost for the Shuttle.

Although the public perceived Shuttle flights to have become largely routine by 1985, the Shuttle never became and never could be fully operational in the traditional sense of aircraft. Each flight was unique. Each mission required unique and comprehensive training of the astronauts, operations crew, launch crew, and payload operations team. Each payload required modifications to the flight operations. Each flight required a reconstitution of the computer software and information systems. Although reusable, after each flight the orbiter had to be carefully examined, repaired and reconstructed. The Shuttle required "an incredible amount of tending."[3]

TABLE 16. Comparative Shuttle and Apollo Launch Vehicle Launch Costs

	Apollo					Shuttle		
	RY $	Infl. Factor	1991 $			RY $	Infl. Factor	1991 $
1962	75.7	6.063	459.0		1970	12.5	4.094	51.2
1963	1,184.0	5.857	6,926.4		1971	78.5	3.851	302.3
1964	2,273.0	5.605	12,740.2		1972	100.0	3.643	364.3
1965	2,614.6	5.421	14,173.8		1973	198.6	3.447	684.6
1966	2,992.2	5.114	15,302.1	Shuttle Prog. R&D	1974	475.0	3.215	1,527.1
1967	3,002.6	4.875	14,637.7		1975	797.5	2.902	2,314.4
1968	2,556.0	4.625	11,821.5		1976	1,206.0	2.662	3,210.4
1969	2,025.0	4.376	8,861.4		1977	1,291.1	2.403	3,102.5
1970	1,684.4	4.094	6,895.9		1978	1,401.0	2.229	3,122.8
1971	913.7	3.851	3,518.6		1979	1,707.8	2.036	3,477.1
1972	601.2	3.643	2,190.2		1980	2,054.9	1.839	3,779.0
1973	56.7	3.447	195.1		1981	2,301.8	1.657	3,814.1
					1982	2,689.6	1.526	4,104.3
Total	19,979.1	—	97,721.9		1983	3,357.5	1.430	4,801.2
*NASA Pocket Facts shows 20,444.0 total program costs.					1984	3,068.9	1.352	4,149.2
					1985	2,786.7	1.301	3,625.5
Apollo				SFCDC (Prod/Ops only)	1986	2,987.9	1.259	3,761.8
97,721.9 ÷ 11 = 8.9 billion/launch (8.884)					1987	5,138.3	1.213	6,232.8
					1988	2,917.9	1.153	3,364.3
Shuttle					1989	3,500.3	1.097	3,839.8
63,599.6 ÷ 36 (through Sep. 30, 1990) = 1.8 billion/flight (1.766)					1990	3,818.2	1.040	3,970.9
Ratio: Apollo/Shuttle per flight = 5:1					Total	41,890.0	—	63,599.6

Source: Papers of Joseph P. Loftus, Assistant Director (Plans), JSC.

Columbia returned from the first working Shuttle flight after 5 days in orbit, having delivered its payload of the two commercial communications satellites. But the flight, like its successors, was not routine. A planned EVA was scrapped due to malfunctions of a ventilation motor in one space suit and a pressure regulator in the other.[4] A characteristic of spaceflight from the earliest days of Mercury seemed to be that no flight was uneventful or routine. Launch and flight control usually involved troubleshooting, either of problems that had developed or that might be anticipated.

Similarly, the maiden flight of *Challenger* (STS-6), April 4, 1983, became a continuing test in problem-solving. North American Rockwell delivered the *Challenger* (named for a Navy vessel which conducted extensive exploration in the Atlantic and Pacific Oceans between 1872 and 1876) to the Kennedy Space Center on July 5, 1982. Originally scheduled for its first flight on January 20, hydrogen gas leaks required the removal, repair and reinstallation of two main engines and the replacement of the third. During repair work, a severe storm broke seals on the payload changeout room and particles contaminated the payload, which had to be removed and cleaned. The realities of flight preparation and training, payload planning, and launch costs gradually changed the idea that there would be frequent or even weekly Shuttle flights with returns for repairs and reoutfitting of payloads. NASA began considering 12 Shuttle flights per year, and as the problems and realities of flight continued to unfold, administrators decided to strive for 5 to 8 launches per year and stress flight duration and mission success.[5]

At last, in April, *Challenger* successfully placed NASA's first Tracking and Data Relay Satellite (TDRS, or "Teadras") in orbit. The second stage booster of the 5000 pound satellite ceased its burn 33 seconds earlier than scheduled. Over a period of several months, ground controllers succeeded in nursing the satellite into a satisfactory orbit. The first of three planned tracking satellites, the TDRS-1, virtually made the old ground control network used for Mercury, Gemini and Apollo obsolete and considerably improved the control, communications, and response between Shuttle flights and ground control. Each TDRS satellite could maintain communications with the orbiter for nearly one-half of the globe.[6]

During the flight of STS-6, mission specialists F. Storey Musgrave, an M.D. and Ph.D. (Physiology), and Donald H. Peterson, an Air Force transfer to NASA, tested the new space suits especially designed for Shuttle EVA use.[7] The 5-day mission provided more experience for crew and flight controllers and helped build new confidence in the Space Transportation System.

That confidence seemed to be wholly warranted by the almost flawless launch and performance of *Challenger* (STS-7) on its 6-day mission beginning June 18, 1983. *Challenger* carried a shuttle pallet satellite (SPAS) built in West Germany and placed it in orbit from the cargo bay by the remote manipulator (Canadarm). Mission specialist Dr. Sally K. Ride, the first American woman in space, managed the remote manipulator arm that placed the SPAS in orbit, and then, after the Shuttle was maneuvered around the satellite, plucked it from its orbit and returned it to the orbiter cargo bay.[8]

Ride's flight finally quelled a festering public relations problem that had plagued NASA since the early 1970's when women activists began to perceive NASA and the astronaut corps as a macho, male only, antifeminist organization. When astronaut James

A. Lovell responded to a reporter's inquiry in 1972 about why there were no women being sent into space by saying that there had thus far been no reason to, but that "in the near future we will fly women into space and use them the same way we use them on Earth—for the same purpose," a storm of protest understandably swept the press, Congress, NASA administrators, and James Lovell. Whether it can be attributed to Lovell's gaffe, social consciousness, public relations consciousness, political pressures, or all of the above, once astronauts began to be picked from the science community and were no longer defined by the pool of pilots with test pilot experience, women were included in the candidate pool.[9]

In preparation for Skylab and post-Apollo operations, NASA, as early as 1973, began to encourage applications from women and minority candidates who met the established criteria for pilot astronauts and mission specialists. None were admitted, however, since there were no astronauts chosen between 1970 and 1978. Earlier classes drew from candidates with test pilot experience (but for the scientist-astronauts selected in 1967 who were required to attend jet pilot school for one year). There were few if any women with hypersonic test pilot experience before 1973. In 1978, from 659 astronaut pilot applicants (including 8 women and 10 minority applicants) NASA selected 15 finalists only one of whom came from a civilian background and none of whom were women or minority candidates. But of the 5680 who applied as mission specialists (including 1251 women and 338 minority candidates), NASA selected 20 astronaut candidates including 6 women and a number of candidates of African, Asian, and Hispanic heritage. NASA selected two additional women mission specialists in the 1980 astronaut class. Certainly by the time Sally Ride made her historic flight aboard *Challenger,* it had become well established that the "right stuff" for spaceflight included men and women of many professional, cultural and racial backgrounds. "In short, the Shuttle opened the door for a vast broadening of the human experience in space."[10]

Sally Ride obtained advanced degrees in physics (M.A. and Ph.D.) from Stanford. She received her doctorate in 1978, and that year was selected for the astronaut corps. She met her future husband, Steven A. Hawley, in the 1978 astronaut class. Another 1978 astronaut, Robert Lee (Hoot) Gibson, met and married a classmate, Margaret Rhea Seddon, an M.D. from the University of Tennessee College of Medicine. Interestingly, Anna Fisher, an astronaut classmate of Steven Hawley, met her husband, William F. Fisher in the 1980 astronaut candidate pool and became with the Hawleys and Gibsons, America's first "partners in space." Ride, after serving as CAPCOM in Mission Control for the STS-2 and STS-3 missions, joined the crew of STS-7 commanded by Robert L. Crippen, who had flown the first orbital shuttle flight (STS-1) aboard *Columbia.*[11]

The STS-7 flight crew, which included Crippen and Ride, pilot Frederick C. Hauck, and mission specialists John M. Fabian and Dr. Norman Thagard, also placed in orbit a Palapa B satellite for the Indonesian Government and a Telesat satellite. The satellites made possible the first modern communications system for the 3000 islands comprising that country. Getaway Special experiment packages included studies of an ant colony in zero gravity and the germination of radish seeds in space. Dr. Thagard conducted studies on the effects of Space Adaptation Syndrome which causes nausea and sickness during the initial hours of spaceflight.[12]

Getaway Special

Officially titled "Small Self-Contained Payloads," the Getaway Special program is offered by NASA to provide to anyone the opportunity to fly a small experiment aboard the Space Shuttle. The experiment must be of a scientific research and development nature. The Getaway Special experiments are flown on Shuttle missions on a space-available basis. A Getaway Special Flight Verification Payload was first flown aboard the STS-3 mission. The test payload, a cylindrical canister 61 cm (24 in.) in diameter and 91 cm (36 in.) deep, measured the environment in the canister during the flight. The first private sector payload was flown on STS-4. The Getaway Specials are available to industry, educational organizations, and domestic and foreign governments for legitimate scientific purposes.

Challenger (STS-8) rose in a fiery arc from Kennedy Space Center in the first night launch of the Shuttle on August 30, 1983. The mission, commanded by Richard H. Truly, carried America's first black astronaut, Guion Bluford (Ph.D. in aerospace engineering), an INSAT IB satellite for India, and 12 Getaway Special canisters—4 contained scientific experiments and 8 held U.S. first-day postal stamp covers.[13] Perhaps the lessons learned from the controversy that erupted when Apollo 15 astronauts carried unauthorized first-day covers to the Moon had not been lost on U.S. postal authorities.

More significantly, the communication satellite payloads being carried by Shuttle flights were quietly revolutionizing communications and the quality of life around the world. Twenty years earlier, on July 10, 1962, NASA had launched the Telstar 1 communications satellite built by American Telephone and Telegraph. Telstar carried the first transatlantic television broadcast between the United States and Europe. In 1964 NASA's successful placing of Syncom 3 in a geosynchronous orbit marked the beginning of a satellite communications network that would provide the capability of real-time voice and television communications between most points on Earth.[14]

Underscoring the international aspects of Shuttle missions, the next Shuttle flight would be a world-class flight with a European Space Agency-sponsored Spacelab payload. Contractors delivered the orbiter *Discovery*, named both for Henry Hudson's ship which sought the Northwest Passage in 1610 to 1611 and for Captain Cook's vessel which discovered Hawaii, to the Kennedy Space Center on November 9, 1983, while technicians readied STS-9 *(Columbia)* for a November 28 lift-off.[15]

Spacelab 1, bolted into the cargo bay of *Columbia,* carried experiments relating to atmospheric and plasma physics, astronomy, solar physics, material sciences, technology, life sciences, and Earth observations. The European Space Agency selected Ulf Merbold from West Germany as its payload specialist for the mission; and MIT (Massachusetts Institute of Technology), which managed the American experiments aboard Spacelab, selected Bryon Lichtenberg. Lichtenberg was the first non-NASA astronaut to fly in space, and Merbold the

TABLE 17. Space Shuttle Missions in Brief, 1985 to 1986

Mission Name	Astronauts	Launch Date	Orbiter	Primary Payload	Launch Pad	Result
STS 51-C	Mattingly, Shriver, Buchli, Onizuka, Payton	1-24-85	*Discovery*	DoD	39A	S
STS 51-D	Bobko, Williams, Seddon, Griggs, Hoffman, Garn, Walker	4-12-85	*Discovery*	Anik C 1/ SYNCOM IV-3	39A	S
STS 51-B	Overmyer, Gregory, Lind, Thagard, Thornton, van den Berg, Wang	4-29-85	*Challenger*	Spacelab 3	39A	S
STS 51-G	Brandenstein, Creighton, Lucid, Nagel, Fabian, Baudry, Sultan Al-Saud	6-17-85	*Discovery*	Arabsat-1B/ Telstar 3-D/ Morelos 1	39A	S
STS 51-F	Fullerton, Bridges, Musgrave, England, Henize, Acton, Bartoe	7-29-85	*Challenger*	Spacelab 2	39A	S
STS 51-I	Engle, Covey, van Hoften, Lounge, Fisher	8-27-85	*Discovery*	AUSSAT 1/ ASC 1/ SYNCOM IV-4	39A	S
STS 51-J	Bobko, Grabe, Stewart, Hilmers, Pailes	10-03-85	*Atlantis*	DoD	39A	S
STS 61-A	Hartsfield, Nagel, Buchli, Bluford, Dunbar, Furrer, Ockeis, Messerschmid	10-30-85	*Challenger*	Spacelab D-1	39A	S
STS 61-B	Shaw, O'Connor, Cleave, Spring, Ross, Vela, Walker	11-26-85	*Atlantis*	Morelos-2/ AUSSAT-2/ RCA Satcom Ku-2	39A	S
STS 61-C	Gibson, Bolden, Chang-Diaz, Hawley, Nelson, Cenker, Nelson	1-12-86	*Columbia*	RCA Satcom Ku-1	39A	S
STS 51-L	Scobee, Smith, McNair, Resnik, Onizuka, Jarvis, McAuliffe	1-28-86	*Challenger*	TDRS-B	39A	U

first non-American to fly aboard an American spacecraft. A special payload operations control center at JSC became operational and tied science managers at the center to the Shuttle crew and to remote stations at MIT, to the European Space Agency in Bonn, Germany, and to the Goddard Space Flight Center in Maryland. For the first time, the TDRS satellite became fully operational and relayed an enormous volume of data from the Shuttle and its Spacelab payload.[16]

Hans Mark, NASA Deputy Administrator, visiting the payload operations control center at JSC, said one could actually watch the crew members performing their experiments on television monitors while scientists on the ground discussed results and suggested changes in the procedures in real time. Watching those people at work, he said, removed any doubts one might have had about the necessity of having human intelligence and judgment in space. During the flight, President Ronald Reagan and Chancellor Helmut Kohl of Germany talked to the crew, and Reagan and Kohl talked to each other through the *Columbia's* communications loop. *Columbia* landed at Edwards Air Force Base on December 8, 1983, after more than 10 days and 166 orbits about the Earth.[17] In those few days the world had somehow grown smaller and more interdependent.

NASA confused the historical record thereafter by changing the designation of Shuttle flights from the simple numerical sequence (STS-1, 2, etc.) to a formula by which flights were numbered first by the year of launch, second by the launch site (1 for KSC, 2 [in theory] for Vandenberg Air Force Base), and the final alphabetical designation representing the original order in which the flights were assigned to fly. By this time NASA knew that Shuttles would not fly in the order assigned. Shuttles, as it turned out, were temperamental machines and required "enormous tending." So did their payloads.

Thus STS 41-B (*Challenger*, [4] from 1984, [1] from KSC, [B] the second flight of the year) lifted off as planned on February 3, 1984. That is about as far as actual events conformed to the planning. A commercial Westar VI communications satellite carried for Western Union was deployed from the cargo bay, but when its booster engine malfunctioned after firing for only a few seconds, the satellite coasted into a useless 600-mile-high orbit instead of the intended 22,300-mile geosynchronous orbit, where it would have maintained a fixed position above the Earth's surface. A similar Palapa satellite, also built by Hughes Aircraft Company, was scheduled for deployment for the Indonesian Government. Hughes engineers decided that the misfiring of the Westar booster was an anomaly and the Indonesian Government agreed to release the Palapa from the cargo bay the next day. But the Palapa duplicated the Westar. "Almost impossibly," the Palapa rocket "sputtered and died just as Westar's had done." Palapa too was lost. Tens of millions of dollars of communications hardware orbited uselessly about the Earth.[18]

But the *Challenger's* work was not yet done. Bruce McCandless, who joined the astronaut corps in 1966 but had never previously flown in space, tested the new manned maneuvering unit (MMU)—a device which he helped design—in the world's first untethered flight that took him some 300 feet from the Shuttle. Bob Stewart took the "Buck Rogers" device out the next day for another flawless flight. The crew also tested techniques designed to rescue and refurbish the Solar Max satellite during a subsequent Shuttle mission. Solar Max, a $235 million scientific satellite launched in February 1980 from a Delta Rocket and designed for the study of solar flares in an effort to better understand "the violent nature of the Sun and its effects on Earth," now orbited uselessly in space with blown fuses in its attitude control box. Success with the MMU convinced the crew and mission planners that not only Solar Max but the errant Westar and Palapa satellites might be successfully rescued, repaired and returned to orbit using the MMU.[19]

Subsequently, STS 41-C lifted off from Cape Kennedy on April 6, 1984. At its first orbital stage, *Challenger's* Canadarm lifted the long duration exposure facility (LDEF) into orbit some 288 miles above the Earth. LDEF weighed 21,300 pounds and carried 57 different experiments developed by 200 researchers from 8 different countries. One of those

Astronaut George Nelson practices an EVA with a mockup of the MMU.

experiments proposed to test the fertility of 12 million tomato seeds after exposure to space for one year, when the LDEF facility was to be retrieved. Another tested the ability of honey bees to build a honeycomb in a space environment. After a few unsuccessful misshapen tries, the honeybees corrected for the microgravity environment and built a very comfortable Earth-like honeycomb.[20] As it turned out, LDEF had a much longer stay in space than planned.

At a second orbital plane 300 miles high, *Challenger* "parked" some 200 feet from the Solar Max satellite while mission specialists James D. van Hoften and George D. Nelson fitted into their space suits. Nelson then piloted the MMU out to Solar Max, but after three tries was unable to secure a specially designed locking device to the satellite. Scientists and engineers at an enhanced payload operations control center (interfaced with the Mission Control Center at JSC) at Goddard Space Flight Center directed the work. Nelson's attempts to grapple the satellite caused it to begin more erratic tumbling motions. During the night, Goddard payload control crews were able to stabilize Solar Max, and the next morning *Challenger* moved in closer for a try at grappling the satellite with the Canadarm. They succeeded on the first effort and brought the satellite into the cargo bay where Van Hoften and Nelson, working in their spacesuits, replaced the attitude control module and the main electronics box for one of the instruments. Solar Max was put back into orbit and to work the next day. The mission demonstrated both the ability to retrieve and repair a satellite in orbit, and the importance of having people in space who could use their intelligence, judgment and imagination in problem solving.[21] Rather than routine, Shuttle flights had become a continuing exercise in problem solving and improvisation.

In June, after three previous launch delays, onboard computers aborted lift-off of *Discovery* 4 seconds before launch due to a fuel valve problem. An August 29, another launch attempt failed due to problems with computer software for the main engine controllers. Meanwhile, NASA canceled a previously scheduled Shuttle mission (41-E) because delays and cost overruns required pruning the schedule to conserve funds. Engineers then refitted *Discovery* to carry some of the planned cargo for mission 41-F, and canceled that mission as well.[22] A Shuttle cargo manifest, as Henry S.F. Cooper explained in his story of flight crew training, *Before Lift-Off*, "was almost a living thing." Launch problems caused "periodic massive overhauls of the manifest" through many subsequent scheduled flights.

Even after a mission had been decided on and a launch date, a cargo, and a crew assigned to it, a dozen things could change, including its cargo, its crew, its launch date, the landing site, and its duration; indeed, entire missions were canceled just before launch and their cargoes and crews combined with future missions. The reshuffling could upset a year's planning and send ripples far down the manifest.[23]

Finally, all systems were go. On August 30, *Discovery* blasted off, and orbiting high above the Earth successfully deployed three large commercial communications satellites. Mission specialists conducted a variety of tests and experiments while a payload specialist, McDonnell Douglas employee Charles D. Walker, manufactured a pharmaceutical product for his company in the installed contractor-furnished laboratory.[24]

As additional evidence of NASA's confidence in the reliability and flight-worthiness of the Shuttle, President Ronald Reagan announced a "Teacher in Space Project" on August 27, 1984. The project, intended to generate a sense of wide public participation and to expand Shuttle flight opportunities to a wider range of private citizens, generated applications from some 11,000 public school teachers. From these a special panel selected 104 individuals for a nomination list from which 10 were chosen for final interviews and testing. Two candidates, Christa McAuliffe and runner-up Barbara Morgan, were chosen to begin training for flight at JSC in September 1985. The teacher in space would telecast live classroom lessons to school children throughout the United States. In preparation for the flight, school children participated in "space" lessons, and selected teachers helped prepare the final lesson plans.[25] The program evoked widespread public interest and helped rekindle a somewhat flagging public interest in Shuttle flights for the remainder of 1984 and 1985.

In part because of the wide publicity given the Teacher in Space Project, the October 1984 *Challenger* (STS 41-G) flight seems to have attracted a bit more of the public's interest, enthusiasm, and imagination than had been true in the flights made earlier in the year. *Challenger* carried a crew of seven on a heavily science-oriented mission that included the deployment of an Earth Radiation Budget Satellite (ERBS) designed to measure the amount of energy received from the Sun and reradiated into space and to study seasonal changes in the Earth's energy levels. During the mission, astronaut Kathryn D. Sullivan made the first spacewalk by an American woman. Payload specialists Paul Scully-Power, with the Naval Research Laboratory, conducted oceanographic experiments and Marc Garneau, from Canada, managed a package of Canadian-sponsored experiments having to do with medicine, climate, materials and robotics. An experimental package prepared by NASA's Office of Space Science and Applications (OSTA 3—an acronym for the original office title which was the Office of Space and Terrestrial Applications) included radar imaging experiments and Earth air pollution measuring devices. High-resolution cameras and experiments with refueling orbiting satellites comprised part of the workload.[26] There were, as usual, problems, but they were resolved in flight.

NASA provided yet another window to the public during the STS 41-G flight by permitting author Henry S.F. Cooper to live and work with the *Challenger* crew as they trained for their mission. Cooper's book, *Before Lift-Off: The Making of a Space Shuttle Crew*, published several years after the flight, described the "human dimensions" of training a Shuttle crew for a space mission. The flight itself, Cooper pointed out, was only the tip of the iceberg of spaceflight. For every hour in space the crew spent thousands of hours in every imaginable and many unimaginable training regimes.[27] And he might have added that, for every hour of Shuttle flight, thousands of other individuals at JSC and throughout the NASA system as well as countless others in the contractor's offices, laboratories, and factories labored uncounted hours.

Within weeks of *Challenger*'s return, *Discovery* took off for a rescue mission to salvage the wayward Westar and Palapa communications satellites stranded since February. Insurers, faced with a $180 million loss, decided to fund the rescue with an additional $10.5 million payload investment. Mission specialists deployed two communications satellites (an Anik D2 and SYNCOM IV-1) before maneuvering into position alongside the Palapa B-2 satellite. Joe Allen approached the satellite in his MMU. While Hughes payload controllers on the ground slowed the satellite's rotation, Allen inserted a "stinger," tying the Palapa to his MMU.[28] As he recalled:

> The tip of the stinger was just beginning to penetrate the nozzle. The lights on my helmet flooded the empty volume inside the nozzle; I let the stinger drift further in, then pulled the lever that opened the toggles. I could see the fingers pop out, called "soft dock," as they expanded, and then shortened the stinger with the crank until the ring had pressed tight against the satellite: hard dock. "Stop the clock! I've got it tied!" That was all I could think to say, but the capture had been far easier than rodeo calf-roping would be.[29]

He and Gardner then attempted to secure the satellite to the Canadarm, being operated from within the Shuttle by Anna Fisher. But when a special A-frame built to hold the satellite failed to fit, Allen and Gardner manually fixed an adapter and hauled the satellite into its berthing in the cargo bay during a physical struggle of more than 5-1/2 hours. The exhausted crew and Mission Control personnel and payload controllers on the ground spent the next day planning how to best handle the Westar salvage operation.[30] It was decided to improvise once again.

Discovery rendezvoused with Westar. Dale Gardner flew out in the MMU, captured the satellite and brought it alongside. Joe Allen, riding on the end of the Canadarm, held the satellite while Gardner fitted the adapter to it. Then Anna Fisher lowered Allen, still holding Westar, into the cargo bay and the adapter with the Westar satellite slid smoothly into the guides. "We learned later," Allen recalled, "that when the news reached officials at Lloyd's of London, one of the principal underwriters of the satellites, they ordered the ringing of the Lutine bell, the insurers traditional signal of a successful recovery . . ."[31] Space was no longer just adventure, innovative engineering, or exciting science, but in part good business—at least for insurers when there were no losses.

Space Shuttles also involved government business—that is, secret Air Force government business. STS 51-C *(Discovery)* carried a classified DoD cargo. It lifted off on January 24, 1985, and returned 3 days later. It was the first NASA mission dedicated wholly to defense. It was a milestone in the course of a long and very difficult debate and tenuous relationship between two very different government agencies and two somewhat incompatible directives. The 1958 Space Act created two separately managed space programs, one civilian (NASA) and one military. The Air Force, which had the "responsibility for conducting the national security related space program" held strong reservations about the 1972 decision that the Shuttle would eventually replace all other launch vehicles. The

Air Force regarded the Shuttle as a "truck" with the cargo a separate entity from the vehicle and flight crew. The Air Force also persisted in retaining a capability to "launch national security related payloads on conventional expendable launch vehicles . . . until such time that the Shuttle proved to be completely reliable."[32]

In 1977 and 1978, when the Shuttle program faced new cutbacks from initiatives in the OMB and Congress, Air Force concerns about the impact of Shuttle budget reductions on national security led to a study headed by Hans Mark, then Undersecretary of the Air Force, and Thomas P. Stafford, who resigned from the astronaut corps in 1975 to become commander of the Air Force Flight Test Center at Edwards Air Force Base. Stafford became Deputy Chief of Staff of the Air Force for Research and Development in 1978. The classified study entitled "The Utility of Military Man in Space" attempted to offer Air Force options should the Shuttle program be truncated or canceled.[33]

Meanwhile, Hans Mark (Director of Ames Research Center, 1969 to 1977), acting in the interests of the Air Force, helped throw Pentagon support to NASA and the Shuttle program. Gerald Griffin, who in the late seventies and early eighties left JSC for administrative positions at Dryden Flight Research Center, Kennedy Space Center, and NASA Headquarters before replacing Chris Kraft as Director of JSC in 1982, said that the Shuttle program was in deep jeopardy. That the Air Force and Hans Mark stood up and said, "we have got to have the Shuttle" had a big impact on the future of the program. Subsequently, Griffin said, the Reagan administration was 100 percent sold on NASA and the Shuttle. And Hans Mark, in his fascinating personal account of *The Space Station*—which actually tells much more about NASA, space personalities, and the intricacies of space, government agencies, administrations and Congress than it does about the Space Station—explains how that relationship with the Reagan administration came to be.[34]

Nevertheless, the alliance between the DoD and NASA behind the Shuttle was an imperfect union, if not an unholy alliance, in part because of the Air Force's necessity for security which contradicted NASA's "open door" policy, and in part because the physical packages sometimes delivered by the Air Force did not fit the dimensions of the Shuttle. Thus, the Shuttle *Discovery* completed in 1983 and *Atlantis* delivered on April 3, 1985, had been redesigned during construction to meet DoD requirements. The dependency of national defense and intelligence operations on the Shuttle became even more critical when Shuttle flights were grounded for 2 years in consequence of the loss of the *Challenger* and its crew in January 1986.[35]

Yet, in the ensuing 12 months between the flight of STS 51-C and STS 51-L, nine Shuttle missions made successful flights into space, placed satellites in orbit and completed Spacelab and Getaway Special experiments. *Discovery*, in April 1985, carried a Congressman into space (and returned him to Earth)—Senator E.J. "Jake" Garn (R-Utah), chairman of the Senate committee with oversight responsibilities for NASA's budget. Although there were astronauts who went to Congress following a career with NASA (Senator John Glenn of Ohio, Senator Harrison Schmitt of New Mexico, and Representative Jack Swigert of Colorado), Garn was the first Congressman to reverse the procedure. At the age of 52, Garn trained about 200 hours for the flight and maintained his own rigorous physical fitness program. His project as a payload specialist was to be a medical specimen for a variety of tests.[36]

Once in orbit the *Discovery* crew worked unsuccessfully on a Leasat-3 satellite whose booster stage failed to fire. Upon landing, the Shuttle blew a tire. Before the end of the month, *Challenger* (STS 51-B) carried 2 monkeys, 24 rodents, Spacelab 3, 2 Getaway Special experiments, and placed a NUSAT (Northern Utah Satellite) into orbit. A Global Low Orbiting Message Relay Satellite (GLOMR) failed to deploy and was retrieved and returned to Earth for repairs and a later try.[37]

In June, *Discovery* (STS 51-G) successfully launched three communications satellites—one for Saudi Arabia, one for Mexico, and one for AT&T. Prince Sultan Salman Al-Saud flew as payload specialist for the Arabsat 1B satellite built by Aerospatiale of France. Patrick Baudry, from France, managed a package of French biomedical experiments. Mexico's Morelos 1 satellite provided "educational and commercial television programs, telephone and facsimile services, and data and business transmission services to even the most remote parts of Mexico." AT&T's Telstar 3-D could handle 86,400 two-way telephone calls at one time. Mission specialists also deployed and later retrieved a 2223 pound Spartan carrier with astronomy-related experiments.[38] There were problems, but this time they were precious few.

Space Shuttle *Atlantis blasts toward orbit on two powerful solid rocket boosters. After they are dropped, the three main engines continue to fire, fueled by the large external fuel tank.*

Challenger experienced problems with a main engine coolant valve that delayed its scheduled lift-off for several weeks. There were then minor problems with the orbiter that delayed lift-off for 1 hour and 37 minutes on July 29, 1985, and before the planned orbit was achieved, the Shuttle's No. 1 engine shut down. But the crew nursed the craft from a 124- to a 196-mile orbit using OMS burns. Once in a satisfactory orbit, the Spacelab 2 experiments conducted cooperatively by the crew in space and scientists on Earth exceeded all expectations. *Challenger* (STS 51-F) landed at Edwards Air Force Base on August 6, after 7 days and 22 hours in space.[39]

Approximately 2 months after its return from its previous mission, *Discovery* first encountered a launch delay caused by a local thunderstorm at Cape Canaveral and then another when a computer had to be replaced, but the STS 51-I mission got underway on August 27. The crew deployed with great difficulty an AUSSAT satellite. The satellite sunshield got tangled in the antenna, but Canadarm and the crew came to the rescue. The AUSSAT provided communications services including television, radio, data transmission, and air traffic control to Australia and its offshore islands. Another satellite deployed, the ASC-1 built by RCA, provided voice, data, facsimile and videoconferencing services to U.S. businesses and government agencies. The crew deployed a SYNCOM IV-4 satellite serving DoD, and rescued and repaired SYNCOM IV-3 which had been inoperative since it was put in orbit by *Discovery* in April 1985. When Canadarm failed, mission specialists James van Hoften and William F. Fisher accomplished the retrieval and repair in some unplanned EVAs.[40]

STS 51-J, the first flight of the Shuttle *Atlantis*, lifted off on October 3, on a classified DoD mission. *Challenger* (STS 61-A) left at the end of the month with a crew of eight and a payload largely financed and operated by West Germany and the European Space Agency. German and Dutch mission specialists, working with science controllers in the German Space Operations Center at Oberpfaffenhofen, near Munich, managed the payload experiments (Spacelab Deutsch 1) while mission controllers at JSC worked with the flight crew on Shuttle operations. The crew also deployed the GLOMR satellite that had previously been deployed but retrieved after an operating failure in April.[41] The mission demonstrated the radical changes that had occurred in worldwide communications and NASA's developing ability to service space communications systems. It generated an enormous store of data relating to human physiology, biology, chemistry and physics.

Three satellites, an AUSSAT-2 (for Australia), a Morelos-2 (Mexico), and a SATCOM Ku-2 (RCA American Communications) were placed in orbit by *Atlantis* (STS 61-B) which blasted off from Kennedy Space Center on November 26.[42] Astronauts also experimented with assembling large structures in space—a permanent space station having been given the President's blessings in the State of the Union Speech delivered by Ronald Reagan on January 25, 1984:

> We can follow our dreams to distant stars, living and working in space for peaceful, economic and scientific gain. Tonight, I am directing NASA to develop a permanently manned space station and to do it within a decade.[43]

The EASE/ACCESS construction experiments in space made a Space Station seem eminently feasible—and within the decade.

Columbia (STS 61-C), however, seemed to defy the logic of it all by being exceedingly uncooperative during repeated launch attempts. Launches were scheduled variously for December 18 and 19, 1985, January 6, 7, 9, and 10, 1986, but valves, hydraulic systems, and weather, among other things, prevented a launch until January 12, 1986. Mission and payload specialists (the latter including Congressman Bill Nelson, Chairman of the Subcommittee on Science and Technology) deployed an RCA SATCOM Ku-1 satellite and conducted 13 Getaway Special and other experiments, but failed to get photographs of Halley's Comet as planned. The mission returned on January 18.[44]

Suddenly, Tomorrow Came . . .

January 28, 1986, Challenger *lifted off.*

Ten days later "*Challenger* (STS 51-L) lifted off from Pad B, Launch Complex 39, Kennedy Space Center, at 11:37 a.m. on January 28, 1986. At just 73 seconds into the flight an explosion occurred, which caused the loss of the vehicle and its crew."[45] It was a terrible end to life and to a time of optimism and innocence.

Deep grief, personal trauma, and a kind of paralysis swept through the NASA community. Americans everywhere felt a sense of loss and confusion. Astronauts made space, science, and the incredible machines used in spaceflight more human, understandable and comfortable. The *Challenger* accident not only touched the heart but somehow signaled a loss of control of man over the machine. Condolences arrived from all

This photograph of 51-L shows the flame developing near the O-Ring on the solid rocket booster. Moments later the entire craft exploded in a fiery ball.

over the world. Japan, Germany, Indonesia, Russia, Australia, Arabia—all had become through the Shuttle not just observers but participants in American spaceflight. And the children in American schools who waited eagerly for Christa McAuliffe's lessons on "The Ultimate Field Trip," and "Where We've Been, Where We're Going, Why?" would, with the rest of the American people, ponder those untaught lessons deeply.[46]

The President appointed an independent investigative commission headed by William P. Rogers, former Attorney General (1957-1961) and Secretary of State (1969-1971) to "establish the probable cause or causes of the accident" and to recommend corrective actions. Members included Neil A. Armstrong (vice-chairman), the first astronaut to walk on the Moon (who left NASA in 1971 for a career in academia) and Brigadier General

In Memoriam

The future is not free: The story of all human progress is one of a struggle against all odds. We learned again that this America, which Abraham Lincoln called the last, best hope of man on Earth, was built on heroism and noble sacrifice. It was built by men and women like our seven star voyagers, who answered a call beyond duty, who gave more than was expected or required and who gave it with little thought of worldly reward."

—President Ronald Reagan *January 31, 1986*

Francis R. (Dick) Scobee   Michael John Smith
Commander   *Pilot*

Ellison S. Onizuka   Judith Arlene Resnik   Ronald Erwin McNair
Mission Specialist One   *Mission Specialist Two*   *Mission Specialist Three*

S. Christa McAuliffe   Gregory Bruce Jarvis
Payload Specialist One   *Payload Specialist Two*

Suddenly, Tomorrow Came...

The Nation deeply mourned the loss of the Challenger *astronauts. Thousands gathered at JSC for memorial services on January 31, 1986.*

Charles (Chuck) Edward Yeager, who as an Air Force test pilot helped pioneer hypersonic flight. Sally K. Ride, physicist and the first American woman in space; Dr. Albert D. Wheelon, then senior vice president of the Space and Communications Group of Hughes Aircraft Company and a former Deputy Director of Science and Technology for the Central Intelligence Agency; Robert W. Rummel an aerospace engineer, aerospace consultant, and former vice president of Trans World Airlines for 35 years; and Arthur B. C. Walker, Jr., a professor of applied physics at Stanford University served. Dr. Richard P. Feynman, professor of theoretical physics and a Nobel prize winner, and Dr. Eugene E. Covert, professor of aeronautics at Massachusetts Institute of Technology and a consultant to NASA on rocket engines, provided their special expertise. Robert B. Hotz, an author, journalist and editorial consultant for McGraw-Hill offered a broad perspective. David C. Acheson, an attorney, author, and former senior vice president for Communications Satellite Corporation, with Major General Donald J. Kutyna, Director of Space Systems for the Air Force and a much-decorated pilot, completed the panel.[47] NASA and JSC initiated their own internal investigations. Individuals officially and unofficially reached their own conclusions.

John Young, Chief of the Astronaut Office at JSC, argued that the Shuttle is "not airline machinery," but is "an inherently risky machine to operate." Young, rather bitterly, argued that there were conditions and situations existing with the Shuttle programs that were as potentially catastrophic to the program as the 51-L accident. Flight safety, he believed, was not being given preeminence, and he faulted management for inadequate testing and poor safety standards and priorities. "If the management system is not big enough to STOP the Space Shuttle Program *whenever* necessary to make flight safety corrections, it will NOT survive and neither will our three Space Shuttles or their flight crews."[48]

Richard L. (Larry) Griffin, Commander of the Space Command at Falcon Air Force Station in Colorado felt impelled to respond to Young's memorandum, which had been disseminated throughout the NASA and Air Force space community, with the observation that Young was part of management and that spaceflight and aviation were inherently risky businesses. He enjoined a witch hunt and advised fair and responsible investigations.[49]

By the end of March 1986, NASA began, at the direction of Headquarters, an intensive and exhaustive examination of virtually every element of spaceflight associated with the Shuttle. During a 2-year study, NASA and each spaceflight center reassessed its program management structure. A special panel focused on the solid rocket motor joint design. Study groups reviewed all testing, checkout, and assembly processes involving flight hardware. Launch and abort rules and systems were completely reconsidered. Kennedy Space Center and JSC cooperated on a study of Shuttle flight manifest procedures and the impact of manifest changes on launch and flight operations. A "First Stage Abort Options Group," comprising largely JSC and Kennedy Space Center personnel, examined the entire history of Shuttle first-stage failures, launch aborts, and crew safety systems and concepts.[50]

Tommy W. Holloway, Chief of the Flight Director Office, headed a Mission Planning and Operations Team which reviewed all of the mission planning processes, flight design, scheduled crew activities, training, manifests, and safety procedures relating to the 51-L mission. The JSC team met intermittently with its counterpart, the Planning and Operations Panel of the Presidential Commission led by astronaut Dr. Sally Ride, during March and April. Presentations by many and diverse JSC personnel provided information and insight into basic Shuttle operations, payload integration processes, flight rules, safety procedures and training. Panelists engaged in intense discussions regarding flight schedule pressures, the rationale for landing at Kennedy Space Center, the history of aborts during solid rocket booster thrust experiences, and crew escape systems. There were sessions focusing on program development, and evaluations for "Failure Mode and Effects Analysis" and criticality ratings. The O-Ring seals on the Shuttle rocket booster engines were given careful attention, as were the procedures for monitoring and inspecting Shuttle and engine system components. Leonard Nicholson, in JSC's Space Shuttle Integration and Operations Office, explained payload manifest processes and relationships between the payload and the launch window. Panelists discussed payload safety, manifest changes, and the integration of changes, payloads, and crew safety procedures.[51] Congress, science panels, industry groups, and every NASA center participated in the exhaustive review and study process that emerged in the June 1986 *Report of the Presidential Commission on the Space Shuttle* Challenger *Accident*.

The Report concluded that the destruction of STS 51-L was "an Accident Rooted in History": "The Space Shuttle's Solid Rocket Booster problem began with the faulty design of its joint and increased as both NASA and contractor management first failed to recognize it as a problem, then failed to fix it, and finally treated it as an acceptable flight risk."[52]

The Commission believed that "cost" considerations had been preeminent in the selection of Morton Thiokol, Inc. as the contractor for the development of the solid rocket boosters, and that NASA managers explicitly considered the dual O-Ring seals designed by Thiokol to have increased reliability and decreased operational procedures at the launch site, "indicating good attention to low cost . . . and production."[53]

TABLE 18. The Shuttle in Flight, 1981 to 1989

Mission	Crew	Date	Mission elapsed time, hr:min:sec	Cumulative U.S. manhours in space, hr:min:sec
Space Transportation System				
STS-1 (OFT)	Young, Crippen	Apr. 12 to 14, 1981	54:20:53	22612:30:02
STS-2 (OFT)	Engle, Truly	Nov. 12 to 14, 1981	54:13:13	22720:56:28
STS-3 (OFT)	Lousma, Fullerton	Mar. 22 to 30, 1982	192:04:45	23105:05:58
STS-4 (OFT)	Mattingly, Hartsfield	Jun 27 to Jul 4, 1982	169:09:40	23443:25:18
STS-5	Brand, Overmyer, J. Allen, Lenoir	Nov. 11 to 16, 1982	122:14:26	23932:23:02
STS-6	Weitz, Bobko, Peterson, Musgrave	Apr. 4 to 9, 1983	120:23:42	24413:57:50
STS-7	Crippen, Hauck, Ride, Fabian, Thagard	Jun. 18 to 24, 1983	146:23:59	25145:57:45
STS-8	Truly, Brandenstein, D. Gardner, Bluford, W. Thornton	Aug. 30 to Sep. 5, 1983	145:08:43	25871:41:20
STS-9	Young, Shaw, Garriott, Parker, Lichtenberg, Merbold	Nov. 28 to Dec. 8, 1983	247:47:24	27358:25:44
41-B	Brand, Gibson, McCandless, McNair, Stewart	Feb. 3 to 11, 1984	191:15:55	28314:45:19
41-C	Crippen, Scobee, van Hoften, Nelson, Hart	Apr. 6 to 13, 1984	167:40:07	29153:05:54
41-D	Hartsfield, Coats, Resnik, Hawley, Mullane, Walker	Aug. 30 to Sep. 5, 1984	144:56:04	30022:42:18
41-G	Crippen, McBride, Ride, Sullivan, Leestma, Garneau, Scully-Power	Oct. 5 to 13, 1984	197:23:33	31404:27:09
51-A	Hauck, Walker, D. Gardner, A. Fisher, J. Allen	Nov. 8 to 16, 1984	191:44:56	32363:11:49
51-C	Mattingly, Shriver, Onizuka, Buchli, Payton	Jan. 24 to 27, 1985	73:33:23	32730:58:44
51-D	Bobko, Williams, Seddon, Hoffman, Griggs, Walker, Garn	Apr. 12 to 19, 1985	167:55:23	33906:26:25
51-B	Overmyer, F. Gregory, Lind, Thagard, W. Thornton, van den Berg, Wang	Apr. 29 to May 6, 1985	168:08:46	35083:27:47
51-G	Brandenstein, Creighton, Lucid, Fabian, Nagel, Baudry, Al-Saud	Jun. 17 to 24, 1985	169:38:52	36270:59:51
51-F	Fullerton, Bridges, Musgrave, England, Henize, Acton, Bartoe	Jul. 29 to Aug. 6, 1985	190:45:26	37606:17:53
51-I	Engle, Covey, van Hoften, Lounge, W. Fisher	Aug. 27 to Sep. 3, 1985	170:17:42	38457:46:23
51-J	Bobko, Grabe, Hilmers, Stewart, Pailes	Oct. 3 to 7, 1985	97:44:38	38946:29:33
61-A	Hartsfield, Nagel, Buchli, Bluford, Dunbar, Furrer, Messerschmid, Ockels	Oct. 30 to Nov. 6, 1985	168:44:51	40296:28:21
61-B	Shaw, O'Connor, Cleave, Spring, Ross, Neri-Vela, C. Walker	Nov. 26 to Dec. 3, 1985	165:04:49	41452:02:04

TABLE 18. The Shuttle in Flight, 1981 to 1989 (concluded)

Mission	Crew	Date	Mission elapsed time, hr:min:sec	Cumulative U.S. manhours in space, hr:min:sec
Space Transportation System				
61-C	Gibson, Bolden, Chang-Díaz, Hawley, G. Nelson, Cenker, B. Nelson	Jan. 12 to 18, 1986	146:03:51	42474:29:01
51-L	Scobee, Smith, Resnik, Onizuka, McNair, Jarvis, McAuliffe	Jan. 28, 1986	00:01:13	42474:37:32
STS-26	Hauck, Covey, Lounge, G. Nelson, Hilmers	Sep. 29 to Oct. 3, 1988	97:00:11	42959:38:27
STS-27	Gibson, G. Gardner, Mullane, Ross, Shepard	Dec. 2 to 6, 1988	105:05:35	43485:06:22
STS-29	Coats, Blaha, Bagian, Buchli, Springer	Mar. 13 to 18, 1989	119:38:52	44083:20:42
STS-30	D. Walker, Grabe, Thagard, Cleave, Lee	May 4 to 8, 1989	96:57:31	44568:08:17
STS-28	Shaw, Richards, Adamson, Leestma, Brown	Aug. 8 to 13, 1989	121:00:09	45173:09:02
STS-34	Williams, McCulley, Chang-Díaz, Lucid, E. Baker	Oct. 18 to 23, 1989	120:39:24	45776:26:02
STS-33	F. Gregory, Blaha, Musgrave, Carter, K. Thornton	Nov. 22 to 27, 1989	120:06:46	46371:54:46
STS-32	Brandenstein, Wetherbee, Dunbar, Low, Ivins	Jan. 9 to 20, 1990	261:00:37	47676:57:51
STS-36	Creighton, Casper, Mullane, Hilmers, Thuot	Feb. 28 to Mar. 4, 1990	106:18:22	48208:29:41
STS-31	Shriver, Bolden, Hawley, McCandless, Sullivan	Apr. 24 to 29, 1990	121:16:06	48814:50:11
STS-41	Richards, Cabana, Shepherd, Melnick, Akers	Oct. 6 to 10, 1990	98:10:03	49305:40:26
STS-38	Covey, Culbertson, Springer, Meade, Gemar	Nov. 15 to 20, 1990	117:54:27	49895:12:41
STS-35	Brand, G. Gardner, Hoffman, Lounge, Parker, Durrance, Parise	Dec. 2 to 10, 1990	215:05:07	51400:48:30

U.S. Manhours in Space

Program	Mercury	Gemini	Apollo	Skylab	ASTP	STS
Manhours in space	54	1940	7506	12,352	652	20,124
Number of manned flights	6	10	11	3	1	28
Crew members	1	2	3	3	3	2 to 8

Cumulative U.S. manhours in space: 51,400 hours 48 minutes 30 seconds

Source: Shuttle Flight Data and In-Flight Anomaly List, Revision Q, July 1989.

Despite evidence provided initially by tests and later indicated by actual flight, Thiokol and NASA engineers regarded leakages or faults in the O-Ring seals on the Shuttle rocket booster engines as "not desirable" but "acceptable." Engine tests run in 1977, showed that the seal housing the O-Ring opened rather than closed (as expected in the design) under extreme pressures, thus increasing the pressure on the actual O-Ring (Viton rubber seals). Thiokol engineers called this "joint rotation," and in their reports to NASA managers at Marshall Space Flight Center noted that they did not anticipate significant problems resulting from it. Marshall engineer Leon Ray, however, through his chief, John Q. Miller, advised a redesign of the joints "to prevent hot gas leaks and resulting catastrophic failure." Although Marshall engineers pursued the problem through 1978, and Leon Ray and Glenn Eudy personally visited Precision Rubber Products and Parker Seal Company (which manufactured the O-Rings) and informed them of the test results, the O-Ring design was not changed.[54]

Inspections following the flight of STS-2 in 1981, demonstrated serious O-Ring erosion. Marshall decided that the secondary O-Rings used as a backup, or redundant system for the first set of O-Rings, actually ceased to function under a certain set of pressure conditions—but they agreed with Thiokol engineers that those conditions would be "exceptional" and that the primary O-Ring seal was reliable. Subsequent inspections and tests turned up no serious O-Ring problem until STS 41-B returned in February 1984 with O-Ring erosion damage. But laboratory tests indicated that such erosion should not constrain future launches. In January 1985, following a launch of STS 51-C under unusually cold conditions at Cape Canaveral (51 degrees F), O-Ring erosion showed on both boosters. Although some thought that cold temperatures at launch may have exacerbated the O-Ring problem, O-Ring blowby had occurred on six flights on which temperatures ranged from 51 degrees F to 80 degrees F. Each of the next four flights during 1985 showed evidence of joint seal leakage. In July, Roger Boisjoly, a Thiokol engineer, advised that unless the seal problem were resolved a flight failure might be expected—"a catastrophe of the highest order—loss of human life."[55] But continuing successful flights, perceived pressures by NASA to perform, and the unspoken elements of costs seemed to mitigate the sense of danger or the urgency to redesign and remanufacture. Thus, on a very cold day in January 1986, *Challenger* left the Earth and met disaster.

The accident traumatized NASA, left a legacy of personal sorrow, and instigated an almost unremitting public and media investigation and reinvestigation of the accident. Was the Shuttle program a policy failure? Was it a "free fall" to disaster? Was there a cover-up by NASA? Why did the Shuttle blow up? Why were they dead? And then the nagging question: "Whither America in Space?" There was a NASA internal investigation, an official Presidential Commission investigation, a Congressional Science and Technology Committee study, an official NASA implementation or response to Commission recommendations, and an "After Challenger, What Next?" study headed by Dr. Sally K. Ride that focused on goals and future programs.[56]

JSC, as the Shuttle lead center, established status review teams to implement and monitor recommendations by the Presidential Commission. They included a Design Team headed by Aaron Cohen, a Shuttle Management Structure Team under Richard (Dick) Kohrs, a Criticality Review and Hazard Analysis group led by Bill J. McCarty, a Safety Organization under Martin L. Raines, an Improved Communications Team also led by

Kohrs, and a Landing Safety Team under Bryan O'Connor. Cliff Charlesworth directed the Launch Abort and Crew Escape Team, Leonard Nicholson's Flight Rate Group helped establish new criteria for establishing the parameters of flight frequency, and Gary Coultas's team monitored new maintenance safeguard procedures. These teams reported directly to Headquarters on the progress of implementing Commission recommendations. NASA submitted a preliminary report to the President in June 1986, and a follow-up status report the following June 1987.[57]

Paradoxically, in the short term, the *Challenger* disaster appeared to strengthen support for the Shuttle program. Discussion, concentrated media attention, indications of haste and carelessness by NASA and its contractors, and some very outspoken public criticism and hostility seemed to sharpen the public's generally positive view and confidence in NASA. By January 1987, the public was expecting an early return to flight. In that, they were disappointed. Funding attitudes, however, changed markedly. Perhaps in an effort to assuage the *Challenger* trauma, Congress seemed to become more, rather than less, willing to fund new programs.[58] Although always qualified, always conditioned by costs, time, expediency, and alternative views, NASA and the Shuttle survived the *Challenger* crisis. The Agency and JSC began to mend the wounds. NASA revised its Shuttle organization. Managers and contractors worked to correct and improve the spacecraft. Ultimately, the *Challenger* disaster forced a reappraisal of national objectives in space and helped build a stronger foundation for new initiatives in space.

CHAPTER 16: *New Initiatives*

The 32 months between the fiery destruction of *Challenger* and its crew and the return to flight marked an intense "downtime" for NASA and especially for the Johnson Space Center. Individuals coped as best they could with their own grief and sense of responsibility. The organization was traumatized. Despite an initial outburst of public support and confidence in NASA, media coverage became more critical and hostile. NASA's old self-confidence and "can-do" attitude seemed to erode. Administrative changes at Headquarters and within the centers, made to compensate for perceived weaknesses, contributed to a sense of insecurity. Relations between the centers, and particularly between JSC and Marshall Space Flight Center, deteriorated as each became more territorial. Relations between JSC and Headquarters became increasingly strained.[1]

The downtime involved several phases. In the aftermath of *Challenger* came trauma, characterized by personal grief and indecision within NASA. Almost every person at JSC had a personal or professional relationship with one or more of the astronauts. Public opinion samples by private pollsters indicated an initial show of public support for the agency and spaceflight. Administrative reactions, already described, included the internal NASA investigation of the causes for the disaster and the Presidential Rogers Commission inquiry. But events continued to unfold. A National Commission on Space headed by Thomas Paine had begun its work prior to the *Challenger* accident. Its report encompassed *Challenger* but focused on the next 50 years of *Pioneering the Space Frontier*. The report, published in May 1986 and dedicated to the crew of the Space Shuttle *Challenger*, now somewhat incongruously offered "an exciting vision of our next 50 years in space." NASA launched a special leadership and management review—a response both to *Challenger* and to the report of the National Commission on Space. Newspapers, magazines, and television conducted an almost uninterrupted debate of the causes of the disaster and the reasons for spaceflight. "The task of understanding the accident," commented one NASA historian, "became confused with the excitement of assigning blame . . ."[2] NASA survived as an agency, but with credibility tarnished, confidence shaken, morale low and leadership completely displaced.

The loss or disruption of leadership at almost every level of the NASA organization affected the style of management in place at the time of the *Challenger* disaster, and almost certainly impeded recovery. In a 12-month period surrounding the *Challenger* accident, NASA lost or replaced almost every top manager. Many retired, some went to business, others simply changed from one program to another, as, for example, from shuttle work to space station work. A sizable and relatively successful program in remote sensing at JSC was phased out between 1984 and 1985 because of costs. Glynn S. Lunney, manager of the National Space Transportation System Program Office and an experienced flight director, announced his resignation from NASA/JSC in April 1985. Gerald Griffin, who headed the center since 1982 and had spent most of his professional career with JSC, retired from NASA to become director of the Houston Chamber of Commerce on January 14, only weeks before the *Challenger* launch. It was, Griffin thought, an opportune time for a career change. The Shuttle was going great! The space station program had been approved. JSC had been designated the lead center for the space station. It was a good time for a change!" Griffin wanted to do something different, "something I didn't know a damn thing about."[3] Griffin left JSC when morale and optimism were at their highest. There had, in fact, been a considerable turnover of top NASA administrators in 1985 before the *Challenger* launch. Even more resignations and administrative changes followed the *Challenger* accident.

Hans Mark left his post as NASA Deputy Administrator in September 1985 to become Chancellor of the University of Texas. William Graham came to NASA from private industry as the new deputy administrator in November 1985. On December 2, a federal grand jury indicted General Dynamics Corporation and three of its former executives—one of whom happened to be NASA Administrator James Beggs—for fraud relating to a Defense Department contract on an antiaircraft gun. On December 4, Administrator Beggs took an indefinite leave of absence. President Reagan appointed Graham, who now had 8 days of NASA work experience, as acting administrator. Graham then appointed NASA's experienced associate administrator Phil Culbertson to serve as "general manager." Culbertson was a 20-year NASA veteran and headed the Office of Space Station at Headquarters. Beggs was later fully exonerated and received a letter of apology from the Attorney General. The impact of his untimely indictment and departure from NASA will ever be debated.[4]

When Gerald Griffin left JSC, Robert C. Goetz, formerly with Langley Research Center who came to JSC as deputy director in 1983, succeeded Griffin as acting director. On January 23, just days before the scheduled launch of *Challenger* on January 28, Jesse W. Moore, Associate Administrator for Space Flight, was named Director of JSC. He joined the Headquarters staff in 1978 as Deputy Director of the Solar Terrestrial Division in the Office of Space Science and became Director of the Space Flight and Earth and Planetary Exploration Division, before moving in 1983 to the Office of Space Flight were he served variously as deputy associate administrator, acting associate administrator and then associate administrator, from August 1984 until January 1985. Goetz, who had, on his own initiative and in the context of JSC's role as lead center for the Shuttle, implemented NASA's accident contingency plan when the new acting administrator declined to do so without higher authorization, concentrated on internal reviews and investigations. Concurrently, Jesse Moore made what came to be a rather transient move to JSC. By October of that year Moore was back in Washington D.C. and JSC had yet another director.[5]

Shuttle management became entangled not only in the abnormal rate of personnel changes but also in the spill-out from the *Challenger* accident. It became increasingly clear outside of NASA that the Shuttle was one of the most complex machines ever built. It required "the world's largest pit crew"—tens of thousands of highly skilled workers from a dozen NASA facilities, thousands more from large companies such as Lockheed and Rockwell, and yet more from small subcontractors scattered throughout the country. Its payloads involved the work of still more scientists and technicians, including often the work of those in other nations.[6] Neither fish nor fowl, the Shuttle was not truly an experimental device, nor was it—contrary to its early portrayal by NASA—a wholly operational spacecraft or "shuttle" that could routinely fly people and cargo into space and return to Earth.

Its prime purpose was not national defense, nor was it science. Other than for the placement of communications satellites into orbit, opportunities for profitable business ventures seemed esoteric and distant. Neither was the Shuttle essentially a research and development tool. The Shuttle could carry the American flag and American astronauts into space—but not to the Moon—and the latter had already been done. Mistakenly compared to expendable launch vehicles, the Shuttle was unique in its ability to return from space and in being reusable. The American public, however, was not sure just what the Shuttle was, or should do. Inasmuch as JSC was most closely associated in the public mind with Shuttle flights, the center lost stature because of the *Challenger* accident and the declining reputation of the Shuttle. Moreover, the Shuttle's various constituents, such as space-related businesses and contractors, scientists, engineers, national defense interests, and diverse governmental agencies held often conflicting purposes.

The Shuttle seemed somehow emasculated from a broadly conceived space program or policy. It seemed to be an interim device filling the void between the program that had been (Apollo) and that which was to be (the space station). But what was to be was not yet fully known. The purposefulness and clarity of vision for the space program suffered, consensus waned, and confusion persisted.[7]

Indecision, change, and confusion extended to JSC's role in the space station program. At NASA Headquarters, John Hodge, regarded by some as a "controversial figure with respect to his relationship with the Johnson Space Center" became acting associate administrator for the space station in December 1985. In 1969, while Manager of the Manned Spacecraft Center's Advanced Missions Program, Hodge presented an overview and justification for the development of a large space station in Earth orbit coupled with a low-cost Earth-to-orbit transportation system to the House (Committee on Science and Astronautics) Subcommittee on Manned Space Flight. That presentation reflected then current NASA and MSC thought about post-Apollo programs. Later, Hodge directed the initial space station task force planning group and favored centralizing space station (Level B) management at the Headquarters level. Following the *Challenger* accident, Hodge (with the concurrence of Acting Administrator Graham) asked General Samuel Phillips to head a special committee to study the space station program management system.[8]

The existing three-tiered space station management plan provided that Level A Headquarters would establish policy and guidelines and protect Level B JSC Program Office from "the political environment and allow them to do the technical job," while Level C would comprise the space station project offices located at the various centers which managed the specific contracts related to the space station (figure 19). As the space station management study began in Washington, D.C., Neil Hutchinson, a former JSC flight director who headed the program office of the space station at JSC and managed the developmental (Level B) phase of the space station, stepped down as program manager in February 1986. John Aaron, formerly special assistant to the Director of JSC, replaced Neil Hutchinson as Acting Manager of the Space Station Program Office at JSC for the remaining months of JSC's Level B lead center role. Finally, in March of that year, James C. Fletcher, who held the position between 1971 and 1977, reassumed the post of NASA Administrator, replacing William R. Graham.[9]

Suddenly, Tomorrow Came...

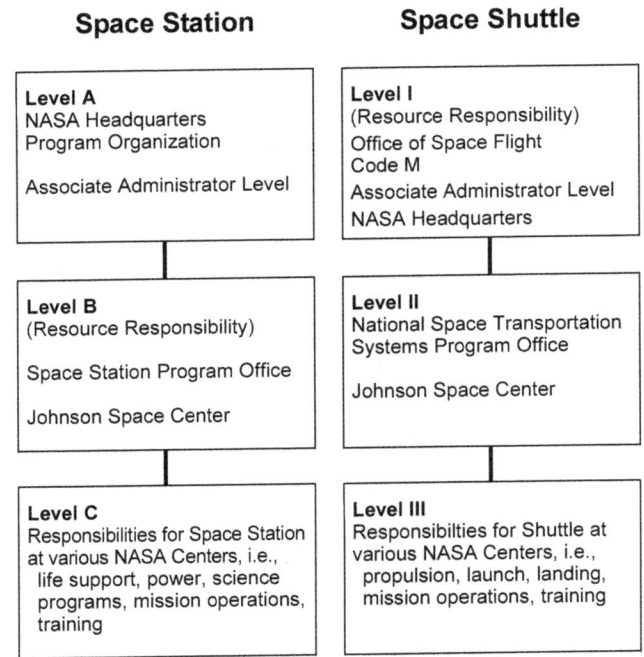

**Space Station Phased Development
as of 1986**

Block I:
    Phase A   Concept definition
    Phase B   Definition (Specifications for hardware, etc.)
    Phases C & D   Construction and launches—for initial capacity

Block II:
Advanced capacity and technology development

FIGURE 19. *Shuttle and Space Station Compared*

Within the 12-months surrounding *Challenger*, NASA had three administrators (Beggs, Graham [acting], and Fletcher). During calendar year 1986, JSC had four directors (Gerald Griffin, Robert Goetz [acting], Jesse W. Moore, and Aaron Cohen). The space station had a new associate administrator at Headquarters and two different program managers at JSC. Aaron Cohen described the post-*Challenger* era as a time "worse than chaos." JSC had no spokesman and NASA no leadership. Morale was low, it was imperative that there be a return to flight, and the space station was "upon us."[10] The leadership changes created uncertainty and confusion, delayed an effective response to the *Challenger* accident, and contributed to a realignment of the space station program.

In 1986 the space station received a new directive which removed the Level B program office from JSC and placed it at Headquarters. As Hodge anticipated, Phillips' group recommended that the lead center management system be abandoned for the space station and that Level B management be concentrated at the Headquarters level. In addition, the committee recommended that Marshall Space Flight Center assume project management for the life habitat module systems, JSC manage the truss or support structures, Goddard Space Flight Center manage the platforms and servicing facility, and Lewis Research Center supervise the power systems.[11]

As the incoming NASA Administrator, Fletcher accepted the Phillips committee recommendations and began reorganizing along the lines suggested. Concurrently, Jesse Moore left JSC to accept the position of Special Assistant to the General Manager at Headquarters without having communicated to JSC personnel the full content of the Phillips report or the organizational changes being made.[12]

As the space station reorganization developed, personnel at JSC, now somewhat adrift in the throes of management changes, viewed the elimination of the center's "lead" position both as punishment for the *Challenger* accident and the product of Machiavellian machinations at Headquarters. In addition, JSC engineers considered crew systems and flight capsules to be under their domain and wholly foreign to the expertise at Marshall Space Flight Center. Moreover, many believed that Headquarters lacked the technical depth, expertise, and experience to manage the space station Level B program. JSC engineers associated effective program and project management with hands-on engineering and in-house laboratory support systems. They believed that successes in the earlier Mercury, Gemini and Apollo programs derived in part from the fact that center engineers duplicated much of the developmental engineering work in their own shops and laboratories, and thus were better and more experienced contract and project managers.[13]

Leadership changes, the *Challenger* trauma, and reorganization combined to drop morale at JSC to a new low. With the space station reorganization an apparent *fait accompli*, Administrator Fletcher set about to restore order and direction to NASA and within JSC. He addressed the issue of order by creating a special task group headed by astronaut Dr. Sally K. Ride to report on NASA's status, review the long-term goals of the civilian space program, and consider the Agency's ability to meet those goals. The confusion and turbulence at JSC were met in part, and perhaps somewhat belatedly, by naming a new director to replace Jesse Moore. On October 12, 1986, Aaron Cohen became JSC's director.[14]

A Texan by birth, Cohen completed engineering studies at Texas A&M University and an advanced degree in applied mathematics at Stevens Institute of Technology. He served in the Army and had a brief stint as a microwave tube design engineer with RCA where he developed patents on a microwave tube and a color TV tube before joining General Dynamics as a senior research engineer. He joined the Apollo Program Office at MSC in Houston in 1962, managed the Command and Service Modules Offices of the Apollo spacecraft program from 1969 to 1972, and headed the Space Shuttle Orbiter Project Office from 1972 to 1982. Prior to his appointment as Director of JSC, Cohen managed the Engineering and Development Directorate. Cohen felt a deep sense of personal frustration at the seeming inability of NASA to resolve the Shuttle problems and wanted to get things moving.[15]

Cohen and others sensed that Headquarters and perhaps history were nudging JSC into the role of being exclusively an operations center. The Chris Kraft-Gerald Griffin/

Apollo-Shuttle era had imprinted a JSC operations mode into the collective mind of the general public and administrators at Headquarters. Other centers, such as Marshall Space Flight Center, might well have encouraged that trend as a way of allocating a greater share of research and development (R&D) or project management to those respective centers. From this point of view—that is, that JSC should be preeminently concerned with operations—it made perfectly good sense at Headquarters and at Marshall Space Flight Center that the major technical management contracts related to the space station go elsewhere than to JSC. Moreover, Marshall's experience as a "lead center" for Skylab seemed to point to its expertise in heading the space station's habitable modules. Cohen believed that the unique strength of JSC was in fact its technical project management expertise relating to the design and development of manned spacecraft and the integration of that expertise with flight operations.[16]

Since the days of Mercury, JSC's unique capability had in fact been the design, integration and technical management of manned space vehicles, astronaut training, and flight operations. Historically, among the manned spaceflight centers, JSC managed the spacecraft design, development, and use (including crew training and operations). Marshall's expertise was in propulsion and launch vehicles, Kennedy's in launch operations, and Goddard's in communications and data acquisition. Or so it seemed to those at JSC.

The traditional center roles were changing in response to space station planning. The "lead center" management style, always under some duress, became suspect after the *Challenger* accident. In addition, whereas the Mercury, Gemini, and Apollo programs each had a finite life, the long-term programs envisioned in the Shuttle, Space Station, and Space Telescope required that each center have more diverse capabilities but also that each interface more closely with the others in program development.[17] Thus Headquarters became more involved in program implementation as well as oversight, and centers increasingly began to overlap and compete with each other for the same projects. Although JSC actually concentrated most of its energies and personnel on the design and development of spacecraft, its public reputation derived from flight operations.

Cohen sought to reestablish engineering and technical management as the center's fulcrum. It was important, he thought, that JSC avoid becoming wholly an "operations" center, and that it preserve its unique "design, fly, and use" capability. JSC managers and planners began looking not only at the problem of returning the Shuttle to flight, but also at JSC's role under the new space station management plan. William Huffstetler, who began his space career with the Army Ballistic Missile Agency before joining NASA's Marshall Space Flight Center (from which he transferred to MSC in 1963), recalled "serious conversations" with Aaron Cohen beginning in 1986 that eventually led to the establishment of the center's New Initiatives Office. That office sought to maintain the center's technical diversity in research, development and operations, to institutionalize long-range planning, and specifically to respond to perceived challenges and competition from the Marshall Space Flight Center.[18]

Indeed, the New Initiatives Office, which Cohen appointed Huffstetler to head, unabashedly imitated the Program Development Directorate established years earlier at the Marshall Space Flight Center which was originally headed by Eberhard Rees. James Murphy currently headed the office which gave Marshall an edge in program planning and in organizing its resources for new projects—such as the space station. Huffstetler's New Initiatives Office assumed the functions of the Chief of Technical Planning (Joseph P.

Loftus) in the JSC director's office, and broadened that capability by providing a full project-level office and support system. As a project-level office (rather than a directorate) the New Initiatives Office avoided direct competition with the line offices or directorates at JSC. It could draw upon the resources of the directorates and provide a mechanism for coordinating planning activities and future programs.[19]

The New Initiatives Office sought to maintain JSC's leadership in spacecraft design and development, and to learn to package and market new ideas to potential space customers in accord with the Agency's growing emphasis on technology transfer and efforts to realize practical benefits for Earth from space exploration. By 1990 the office had become the incubus for six major "future" projects. One related to Shuttle middeck experiments in the life sciences. Personnel also managed the design of an assured crew return vehicle (ACRV) for the space station. Others worked on the development of LifeSat, an unmanned satellite for low-level biological studies. One division focused on the development of a commercial Shuttle (or space station) module designed to carry up to 75 experiments for paying customers. Another group worked on studies and projects related to space servicing for the maintenance and repair of vehicles in space. Another division concentrated on the design of human and robotic spacecraft for future missions—such as a Mars mission. The work focused on "touch and feel" engineering, that is, in studies of hardware and project management capabilities.[20]

The office also was charged with helping JSC establish ties to the regional academic community. Major universities had become a neglected resource for new ideas and new

Through their conceptions, artists have enabled the New Initiatives Office to learn to package and market new ideas such as the ACRV (left) and space station robotics (right).

initiatives. After Apollo, JSC concentrated its energies and its diminishing funding on building and flying the Shuttle. By 1986, the Shuttle absorbed 80 to 85 percent of the center's resources and dollars.[21] One fallout of the *Challenger* was that NASA and JSC were forced to look beyond the Shuttle.

Although JSC had a long informal association with Texas universities, programs involving those universities later tended to be underfunded, neglected, and sometimes simply cosmetic. Many JSC employees came from Texas universities, and in that sense the center was very dependent upon them. Because of this and the fact that space was a powerful stimulus, JSC had a profound impact upon higher education in Texas and upon the public's perception of education and technology. Space helped jar Texas and the Nation from its yesterdays into a new tomorrow.

One of the factors leading to the location of the MSC in Houston, Texas, had been the presence of strong educational institutions—prominently Rice University and the University of Houston within the city, and the University of Texas and Texas A&M University in Austin and College Station, respectively. In 1963, Center Director Robert R. Gilruth announced an Aerospace Summer Intern Program that would give 30 junior/senior or graduate students drawn from throughout the United States (20 from science and engineering and 10 from public and business administration) the opportunity to work as salaried interns at the center.[22] The internship program was an effective device both for tapping the intellectual resources of the university, and also for assisting in the transfer of technology from NASA to the private sector. The program generally flourished through the 1960's, declined due to budget problems in the 1970's, and was reinvigorated as a cooperative educational program in the 1980's.

The location of the space center in Houston particularly provided a stimulant to the University of Houston to accelerate its development as a major university. Under the leadership of President Philip G. Hoffman, the University of Houston upgraded its academic offerings, faculty, and facilities. The University of Houston received direct fellowship funding from MSC totaling $365,000 by 1965 which provided 20 graduate fellowships in business, and space-related science and technology fields. In 1969, the university established a graduate campus adjacent to MSC, which became the University of Houston-Clear Lake campus, offering upper-division undergraduate and graduate courses. Paul Purser was assigned by JSC's director to assist in the creation of a graduate center at the University of Houston-Clear Lake. In the 1970's, both the Clear Lake and main University of Houston campuses offered specially devised management programs to help NASA administrators cope with shrinking personnel and budgets.[23]

Texas A&M University, with relatively strong programs in engineering and sciences, proposed to establish an Activation Analysis Laboratory and Space Research Center in 1961, and did implement several NASA-related predoctoral traineeships as early as 1962. Although the Texas Engineering Experiment Station (TEES), administered under the authority of Texas A&M University, did establish a Space Technology Division in 1963, the proposed Space Research Center failed to materialize for several decades. By 1965 Texas A&M University offered 32 scholarships funded by $614,000 in NASA grants. The Dean of Engineering, Fred J. Benson, informed Congressman Olin E. Teague that by 1966, the university had received $2.9 million in NASA-funded grants, including almost $1 million for traineeship grants for graduate students, over $300,000 for independent interdisciplinary research designed to "support faculty ideas which are in the area of the space effort," and $1 million for a Space Research Center building grant. Despite progress, Benson thought that NASA had funded non-Texas schools more generously, and that NASA had handled a number of Texas A&M research proposals "badly," resulting in interminable delays in decisions, funding and real progress, all the while giving verbal encouragement that the university should continue its space-related research programs.[24]

NASA's "Sustaining University Program" included the predoctoral training grants, internships, research grants, and facility grants designed to help rapidly enlarge the production of scientists and engineers in space-related fields. Physical and life sciences, mathematics, physics, biology and developing fields of bioengineering were greatly stimulated by America's space programs.[25] Texas A&M University and the University of Texas, for example, having nationally recognized programs in mechanical and petroleum engineering, began a buildup of faculty and facilities in chemistry, mathematics, physics, and electrical and aerospace engineering under the "space-race" impetus. Generally, however, Texas universities at the beginning of the space era were teaching oriented rather than research oriented, and their expertise lay in practical applications of engineering, agriculture, business, and the sciences.

By the 1980's, those same universities had developed a much broader research base, a more diversified academic program, and in many cases, space-oriented research capabilities. The latter is exemplified by the belated establishment of a Space Research Center at Texas A&M University in 1985, and the Texas Space Grant Consortium organized in 1989, which included 21 state universities, 18 Texas companies, and 2 state agencies. Progress in the development of space programs at Texas universities had, however, been spasmodic and in part deflected by the strong resurgence of traditional emphasis on petroleum production and petrochemical development. The oil boom, in a sense, slowed Texas conversion to diversified industries and to space, but once that boom ended, universities, like businesses, sought new opportunities in space, medical, and computer technologies. There was, in the latter part of the decade of the eighties, a conscious attempt by universities and by JSC to "reach out" to each other, thus ending a long period when relations between the two had been rather quiescent.

The TEES, located at Texas A&M University, initiated an effort to develop more direct research contacts with JSC. David J. Norton, a professor of aerospace engineering and then Assistant Director for Research at TEES, and Oran Nicks, who had joined TEES as a research engineer in 1980 after 20 years of NASA experience ranging from Director of Lunar and Planetary programs to Associate Administrator of Advanced Research and Technology (1960-1970) and Deputy Director of the Langley Research Center (1970-1980), took the initiative in attempting to develop a more permanent relationship between the university and JSC. The two actively began to explore the possibilities with JSC managers.[26]

Conversations with Director Gerald Griffin in the fall of 1984 led to more intense discussions with Michael B. Duke, Chief of the Solar System Exploration Division in the Space and Life Sciences Directorate of the center. Meanwhile, there was within JSC a growing realization that "things were changing," and that the center should become more actively involved in Texas business and university interests and look beyond the Shuttle.

Subsequently, Space and Life Sciences Director Carolyn L. Huntoon approved a response to Texas A&M inquiries by which Duke offered to fund (for $25,000) a proposal for a Space Research Center which would both "define" such a center and identify at least 10 specific tasks related to space and the work of JSC that TEES and Texas A&M could accomplish. Moreover, Duke charged Norton and Nicks to support their proposals by finding specific advocates among relevant JSC engineers and scientists.[27]

While this work progressed, Norton, with the direction and assistance of W. Arthur Porter, Director of TEES, developed a more broadly constructed proposal that would develop within TEES a statewide structure for coordinating cooperative space-related research efforts by Texas universities and businesses. In November 1984, Norton authored "Space Technology Consortium: A Concept Paper," which began circulating within NASA and through Texas universities. In this document, TEES proposed to "provide for an industry-university partnership which would enhance the mission of each through the development of new technology."[28] Both the Texas A&M University-based Space Research Center idea and the statewide Space Technology Consortium developed proponents and critics.

Some faculty and administrators at Texas A&M particularly believed that a consortium of universities under the auspices of TEES might divert research funding from their university to other universities. Although it was in fact a state agency, Texas A&M held a proprietary interest in TEES. Other universities regarded TEES as a competitor rather than an independent state agency. There was also concern that the Space Research Center might deflect research funding from academic departments already involved in space-related research to other departments, duplicate work already being done by the Texas A&M University Research Foundation, or simply impose another intermediary agency between the faculty members and their research. In addition, there was some concern that contract research and an academic-business alignment might jeopardize the integrity of academic research.

Nevertheless, in January 1985, the Board of Regents of Texas A&M University approved the creation of the Space Research Center under the jurisdiction of TEES, and appointed Oran Nicks its director. Nicks discussed the Space Research Center concept with a number of department heads, particularly those in aerospace, electrical, nuclear, and chemical engineering, and held a general meeting of interested faculty and administrators to discuss programs and possibilities. Nicks stressed that the center would serve as a broker between the university and NASA and that it would focus on stimulating interdisciplinary research. Research was to be funded by grants rather than through contracts. There was an important distinction in that proposals were to emanate from the researcher and were not necessarily driven by formal requests from NASA. Thus, research would tend to be generic and preserve academic integrity. Nicks explained that the center sought to promote space-related research, encourage interdisciplinary projects within the university, help attract accomplished scholars and researchers to the campus, and train future scientists and engineers for space activities. Subsequently, the faculty developed 31 rather than the 10 specific research proposals requested by Mike Duke, and the Space Research Center forwarded all of those to NASA through its designated contact, Aaron Cohen, then Director of Research and Engineering at JSC. Nicks commented that the large list of proposals was intended to demonstrate to JSC Texas A&M's capability and relevance in space research.[29]

Gerald Griffin, representing JSC, and the new Dean of Engineering and Director of TEES at Texas A&M, Herbert H. Richardson, approved a Memorandum of Understanding in February 1985, for the announced purpose of "permitting beneficial contact between JSC and TEES, providing opportunities for dissemination of information concerning the activities of the National Aeronautics and Space Administration, and providing opportunities for the scientific community to participate in problem-solving endeavors concerning the space program." Next, JSC funded from the Director's "Discretionary Account" six of the research proposals submitted by the Space Research Center.[30]

Concurrently, a special Space Advisory Board established by the Texas A&M University System, whose members included Gerald Griffin and Aaron Cohen from JSC (both former students of Texas A&M), actively considered establishing a Space Technology Consortium and six NASA Commercial Centers of Excellence to handle proprietary industrial and public domain government research, including prospective Texas projects related to President Ronald Reagan's "Star Wars" or Strategic Defense Initiative, and other independent research initiatives. Norman R. Augustine, Vice President of Martin Marietta Aerospace and President of Denver Aerospace, was also included on this consultative body. (In July 1990, Vice President Dan Quayle appointed Augustine to chair an Advisory Committee on the Future of the U.S. Space Program.) W. Arthur Porter, past Director of TEES, participated as the new director of what became the Houston Advanced Research Center (HARC—a privately funded institution located in The Woodlands, Texas, adjacent to north Houston). George Jeffs, a Rockwell International vice president; Dr. W.H. Pickering, former Director of the Jet Propulsion Laboratory and president of a Pasadena-based research company; Edward "Pete" Aldrich, an Undersecretary and later Secretary of the Air Force; Dr. Lloyd Lauderdale with E-Systems in Dallas, Texas; C.H. McKinley, an LTV Aerospace vice president; and John Yardley, president of McDonnell Douglas Astronautics participated in discussions regarding new initiatives in university-business space technology. Texas A&M University Chancellor Arthur G. Hansen believed that in the future, government/industry/university collaboration would be vital to the state's economic health and well-being—particularly in light of the state's sharply declining oil revenues.[31] Economic misfortunes, coupled with NASA's own *Challenger* misfortune, compelled Texas businesses and universities to seek new research and funding opportunities on the one hand, and JSC to more actively reach out to these constituents on the other hand.

By early 1986, Houston had come to know the depths of recession resulting from the collapse of petroleum prices. Industry began looking for strategies to diversify. When Gerald Griffin assumed the position of President of the Houston Chamber of Commerce, he immediately established an Aerospace Task Force that might seek ways to better integrate JSC's "space economy" into the economy of the city. The task force, led by Walter Cunningham (a former astronaut), included Aaron Cohen, Richard Van Horn, President of the University of Houston, Richard Wainerdi, head of the Texas Medical Center, David Norton, Director of the Space Technology and Research Center of HARC, and Eleanor Aldridge, coordinator for the study. In addition, local business leaders founded a Space Foundation to promote research and private space ventures. It in turn stimulated the organization of a Space Business Roundtable which facilitated monthly discussions by bankers, entrepreneurs, engineers, accountants and business representatives of space business and opportunities. The Space Roundtable idea soon spread from Houston into seven major cities, including Washington, D.C.[32]

When he became Director of JSC, Aaron Cohen established a special $2 million research package to be allocated on the basis of grant applications to the major research institutions in Texas, including Rice University, Texas A&M University, the University of Texas, and the University of Houston. Meanwhile, Frank E. Vandiver, then President of Texas A&M University, proposed to the Space Advisory Board at its first meeting in 1985, that it support an effort to have Congress create national Space Grant universities, as it had already established and funded Land Grant and Sea Grant institutions. Subsequently, Senator Lloyd Bentsen of Texas sponsored legislation authorizing a National Space Grant College and Fellowship program to be administered by NASA. Congress approved the measure in 1987. Texas A&M University and the University of Texas received "Space Grant" status in 1989, bringing together a "Texas Space Grant Sonsortium" involving 21 Texas universities. The two universities, which historically worked in competition rather than in consort, joined in the formation of a broader Texas Space Grant Consortium operating under a Space Grant Program Office located on the University of Texas campus. Dr. Byron Tapley became program director, and Dr. Sallie Sheppard of Texas A&M University and Dr. Steven Nichols of the University of Texas became associate directors. Oran Nicks, with the Texas A&M Space Research Center, chaired the Space Grant Board of Directors, composed of university, industry, and state agency representatives.[33] The space consortium marked the development of a new relationship between Texas universities, and between those universities, JSC, and space industries.

There were other symptoms of those changing relationships. W. Arthur (Skip) Porter, who had participated in the inception of the Space Consortium and Space Research Center, left Texas A&M University to head what became HARC. Created by Houston oilman George Mitchell (President of Mitchell Energy & Development Corporation) in 1984, HARC was a rather unique public-private endeavor (richly funded by George Mitchell) and with a yet undefined mission to stimulate research and technological development through university-business cooperative projects. HARC was yet another umbrella organization within which Texas universities—including Rice University, Texas A&M University, the University of Texas, and the University of Houston—could cooperate without forsaking their familiar turf. It provided for Porter an opportunity to develop the kind of cooperative enterprises he had sought to build through TEES, but in what he thought would be a more conducive and independent environment.[34] Although much of HARC's initial work related to locating a supercollider and supercomputer in Texas, one of HARC's first initiatives was to develop a space research component within HARC.

To do this, Porter recruited his former assistant director at TEES, Dr. David J. Norton, to head the Space Technology and Research Center at HARC. Norton now continued to build from the private sector that which he had worked to develop within the public sector—a liaison between industry and education for research on space-related projects—specifically directed to the needs of NASA and developing private space enterprises. One of the mechanisms for promoting this new liaison was Partners in Space, a nonprofit organization which emerged from the work of the Houston Aerospace Task Force and HARC. Partners in Space helped integrate NASA/JSC into the intellectual and economic fabric of Texas through education, conferences and other forums.[35]

Complementary developments occurred within individual universities and in collaborative efforts by business and civic organizations. NASA's Headquarters Office of Commercial Programs which had launched a program for the establishment of "Centers for the Commercial Development of Space," approved and provided funding, in cooperation with industry, for a Space Vacuum Epitaxy Center at the University of Houston in 1986 and a Center for Space Power at Texas A&M University in 1987. The Epitaxy Center began under the direction of Dr. Paul C.W. Chu, who had become famous for his work on superconductors. Epitaxy, "the atomically ordered growth of a thin film on a substrate in an atom-by-atom, layer-by-layer manner," if conducted in a vacuum such as space, provided the means for producing electronic, magnetic, and superconducting thin film materials of extremely high quality.[36] Like HARC in The Woodlands and the Space Research Center at Texas A&M University, the Epitaxy and Space Power Centers helped build a bridge between private business and NASA.

More directly to the point of the business-space-NASA connection, the University of Houston-Clear Lake, in cooperation with JSC, established a Space Business Research Center in 1986. The Clear Lake center offered library and data base information services relating to NASA and space industries, conducted educational seminars, and provided economic analyses and market forecasts.[37] In addition, beginning in 1987, The Space Foundation (a nonprofit Washington, D.C. organization), *Space Magazine*, and *Washington Technology* (a biweekly business newspaper) in cooperation with JSC, and area chambers of commerce and businesses, sponsored an annual Space Technology Commerce and Communications Conference and Exhibition at the George R. Brown Convention Center in Houston. People in the community and at JSC began serious discussions about building a large visitor center that might better explain what had been done in the past and could be done in the future.[38]

In this general context of public enthusiasm, concern, or even desperation, Texas business and political interests took a very dim view indeed of NASA Headquarters' plan to reduce JSC's participation in the space station program. As information regarding the Phillips report began to filter through JSC and its external support groups, a sense of alarm began to build. Congressman Jack Brooks and Gerald Griffin met in Washington with members of the "Texas delegation" and business leaders to discuss an appropriate response to the threatened program transfers. One immediate result was a telephone call from House Speaker Jim Wright (D-Texas) to Administrator James Fletcher suggesting that if NASA persisted in the planned reallocation of space station spending away from JSC "no future NASA funding legislation would ever be enacted."[39] NASA did begin to reevaluate the planned transfers, and Congressman Jack Brooks subsequently became a self-appointed watchdog to safeguard Texas (and JSC) interests. Vice Admiral Richard H. Truly, who succeeded James Fletcher in the Administrator's post in 1989, commented that Brooks was "one tough customer."[40] As a result, space station management and JSC participation in the program remained relatively unsettled through 1989, when NASA awarded its major space station contracts.

All of this meant, in part, that NASA and JSC had begun to take a different look at the world, and that conversely, the world outside, especially that in Texas, was beginning to change its role from being largely an observer to becoming an active participant in space and new technology. JSC had begun to move from its more narrowly defined role as a project manager, to incorporate a broader perspective of its role as an arbiter of changing technology in society. Nevertheless, the center was aware, as executive assistant Daniel A. Nebrig explained, that there are "some things we cannot do and should not do." Aaron Cohen added that JSC's leadership is best demonstrated by doing those things it can do well.[41] Whatever its intentions, over three decades, Texas society had clearly changed under the influence of space in general and JSC in particular. There had been not only a technology transfer, but also the development of a new attitude and perspective by people in business, education and government. Texans had taken one small step beyond the old frontiers of cattle, cotton, and oil, into the new frontier of space.

NASA too was adapting to change. In May 1986, the President's National Commission on Space charged to help formulate an "aggressive civilian space agenda to carry America into the 21st Century," submitted its report entitled *Pioneering the Space Frontier*. The report offered an "exciting vision of our next 50 years in space." Although predicting the future can be hazardous, the authors noted that Wernher von Braun and Chesley Bonestell predicted a reusable launch vehicle, a space telescope, and a rotating space station as early as 1951. Space technology, the commissioners concluded, "has freed humankind . . . from Earth . . . to expand to other worlds." They advised stimulating individual initiative and free enterprise in space, harnessing solar energy, creating a sustained space program using "a small but steady fraction of the national budget," and fostering international cooperation and American leadership in space (with a critical lead role by the U.S. Government).[42]

The commission recommended an aggressive space science program that first, stressed an understanding of the structure and evolution of the universe, the galaxy, the solar system and planet Earth; second, applied this understanding to forecast future phenomena "of critical significance to humanity"; and third, used space to study basic properties of matter and life. NASA should work on those projects related to sustaining human life in space, including robotic prospector missions, robotic and human exploration and surveying, and the establishment of human outposts in the inner solar system.[43]

The report recommended that NASA should encourage new space enterprises. The commissioners recommended that 6 percent, rather than 2 percent, of NASA's total budget be allocated for research and development in basic technologies, specifically those relating to aerospace plane propulsion and aerodynamics, advanced rocket vehicles, aerobraking for orbital transfer, long-duration closed ecosystems, electric launch and propulsion systems, nuclear-electric space power, and space tethers and artificial gravity.[44]

The commission advised that work on post-Shuttle transportation systems should include an economical cargo transport for low Earth orbit, a passenger transport to and from low Earth orbit, and a reusable round-trip multipurpose vehicle for destinations beyond Earth orbit. Technological milestones envisioned included the initial operation of a space station (projected for 1994), the initial operation of "dramatically lower cost transport vehicles to and from low Earth orbit," the establishment of spaceports in low Earth and lunar orbit, Mars orbital flights first by robotics and then by humans, and finally human exploration and prospecting on Mars. Because "the American public has become uncertain as to our national space objectives . . . We must strengthen and deepen public understanding of the challenges and significance the space frontier holds for 21st-century America . . ."[45] Although the report of the National Commission on Space did offer an "exciting vision," it tended to be ignored by the media and the public who were concentrating on the *Challenger* accident, economic and federal budgeting problems, and other domestic concerns. The report did, perhaps, help restore a sense of clarity and purpose to NASA's own vision for its future.

Civilian space goals needed to be carefully chosen and consistent with NASA's capabilities, that is, to be technically and financially feasible. The National Commission on Space offered a vision of new frontiers, while the report of the Rogers' Commission offered a very sobering view of that which was possible or desirable:

> . . . in the aftermath of the *Challenger* accident, reviews of our space program made its shortcomings starkly apparent. The United States' role as the leader of spacefaring nations came into serious question. The capabilities, the direction, and the future of the space program became subjects of public discussion and professional debate.[46]

Could NASA adopt a major, visionary goal or did the Shuttle and the space station already represent an overcommitment?

NASA attempted to confront the issues presented in resolving the futuristic goals of the National Commission on Space with the hard realities proffered by the Rogers Commission. A *Leadership* study, headed by astronaut Dr. Sally K. Ride and presented to Administrator Fletcher in August 1987, if it did nothing else, highlighted the complexity of civilian space program policy decisions—and it urged the institution of better long-range planning mechanisms. Somewhat reminiscent of the Sputnik scare and a reminder that the cold war had not ended, the study noted that unmanned Mariner and Viking missions visited Mars in the 1960's and 1970's, but none had done so since 1976, and the Soviets were beginning an extensive robotic exploration of the Martian surface in 1988. The American space station Skylab was last visited in 1974, whereas the Soviets' six space stations had been visited in orbit since the mid-1970's, and the Mir, launched in 1986, would give the Soviets a very long presence in space. The United States "has clearly lost leadership in these two areas and is in danger of being surpassed in many others during the next several years."[47]

The National Space Policy Act of 1982 obligated the United States to maintain "space leadership." And what constitutes leadership? Leadership is something that cannot be announced but must be earned. It involves the development of capabilities and the demonstration of those capabilities. It requires the ability to set and meet goals and achieve objectives. Space leadership requires the development and sustenance of strong programs in scientific research and technology development, and the demonstration of tangible accomplishments. Once achieved, leadership generates national pride and international prestige, and enhances "the human spirit's desire to discover, to explore and to understand."[48] How, then did one extrapolate leadership to the mundane world of NASA research, engineering, and contract development?

Very briefly, the *Leadership* report viewed NASA as the agency designated to implement national civilian policy in aeronautics and space. Those goals are to advance scientific knowledge of the Earth, the solar system, and the universe beyond; expand the human

Suddenly, Tomorrow Came . . .

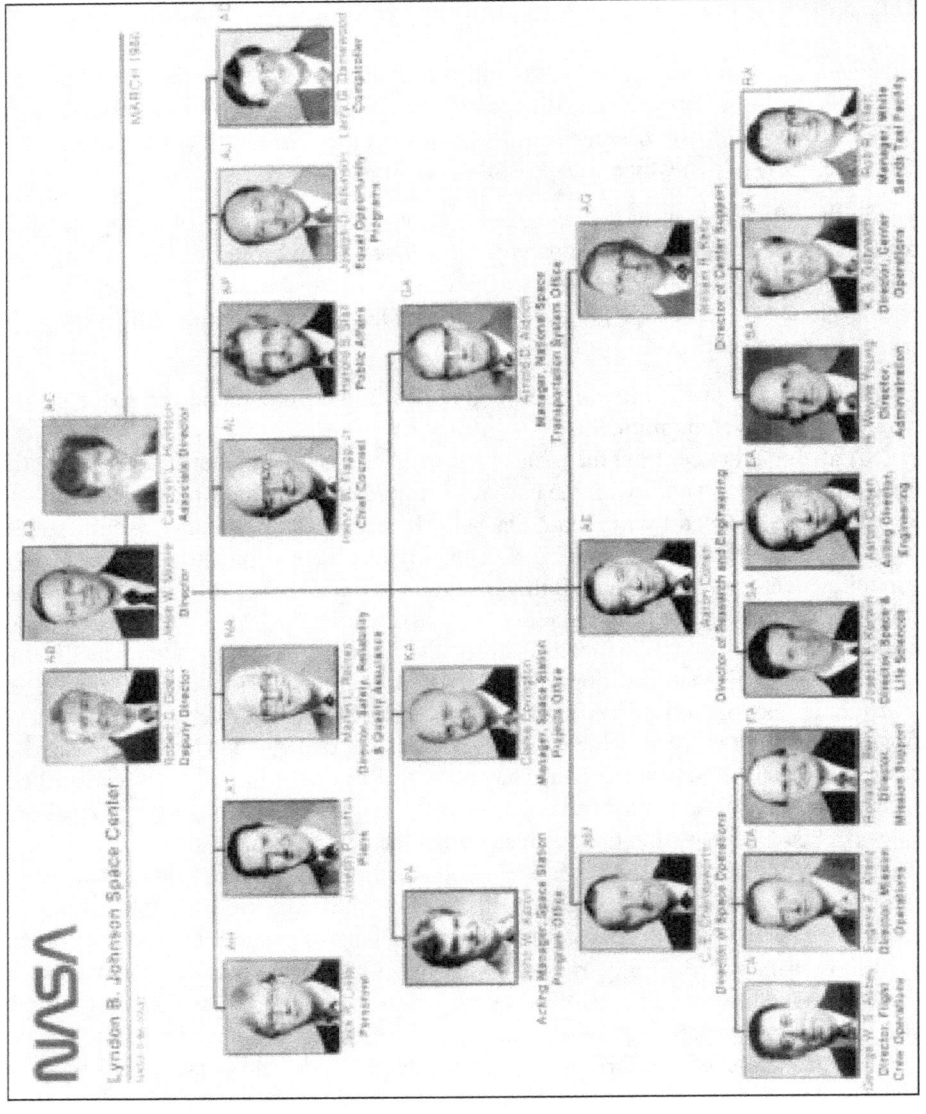

FIGURE 20. *Organization as of March 1986.*

New Initiatives

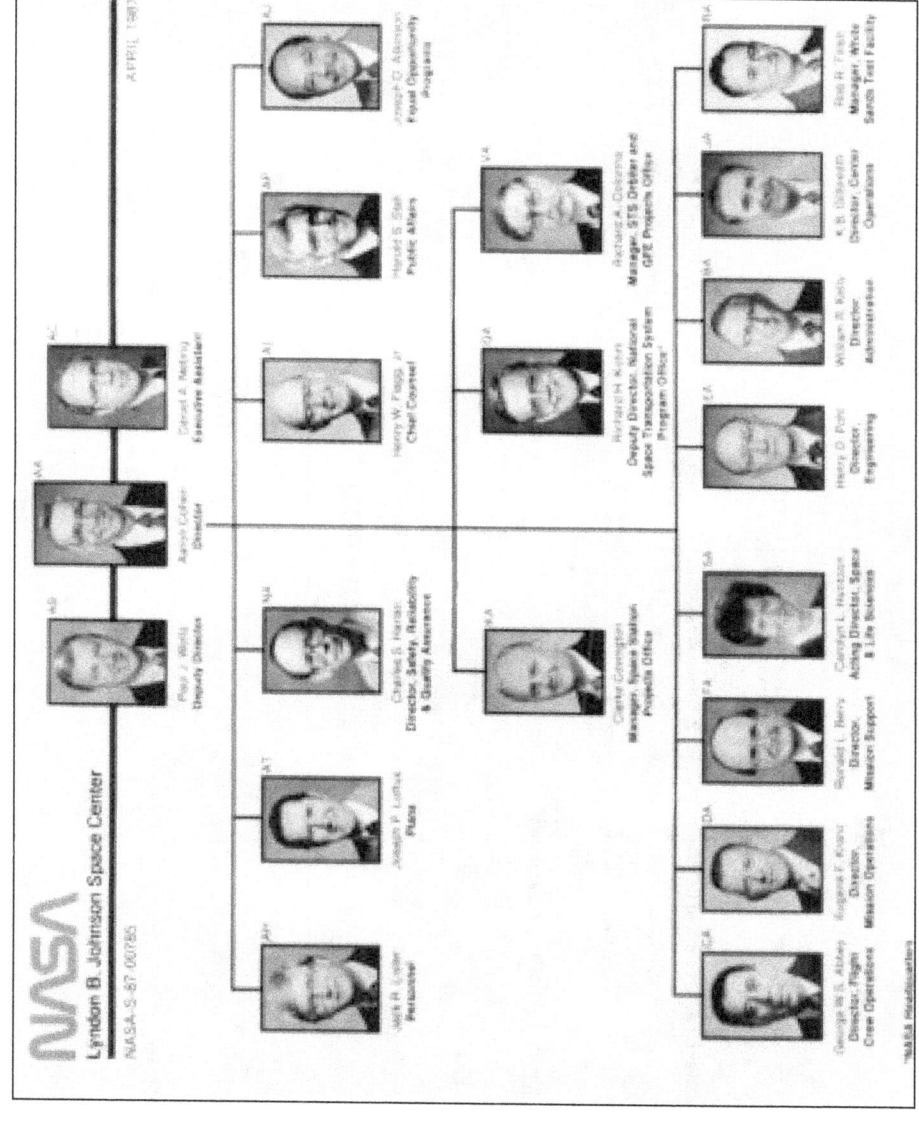

FIGURE 21. *Organization as of April 1987*

Johnson Space Center, 1988

Office of the Director (AA)
Aaron Cohen — Director
Paul J. Weitz — Deputy Director
Daniel A. Nebrig — Executive Assistant
Clifford E. Charlesworth — Special Assistant
John W. Young — Spec Asst for EO & S
Joseph P. Loftus — Assistant Dir (Plans)
John W. Aaron — Technical Assistant
George W. S. Abbey — Spec Asst (Programs)

NASA Inspector General Field Office
W. Preston Smith — Chief
Ned Eckerd — Deputy

Legal Office (AL)
Henry W. Flagg — Chief Counsel

Equal Opportunity Program Office (AL)
Joseph D. Atkinson — Chief

Human Resources (AH)
Jack R. Lister — Director
Harvey L. Hartman — Deputy

Public Affairs Office (AP)
Harold S. Stall — Chief
Douglas K. Ward — Deputy

Space & Life Sciences Directorate (KA)
Carolyn L. Huntoon — Director
Donald E. Robbins — Deputy
Lawrence F. Dietlein — Asst Dir for Life Sci

- Medical Science Division
- Life Sciences Project Division
- Solar System Exploration Div
- Man-Systems Division

Orbiter & GFE Projects Office (VA)
Richard Colonna — Manager
Daniel M. Germany — Deputy
Gary A. Coultas — Asst Mgr
Robert V. Battey — Asst Mgr for Plans & Sch

- Orbiter Project Resident Office – Downey
- JSC Resident Office – KSC
- Flight Data and Evaluation Office
- Orbiter Avionics Systems Office
- Flight Support Equipment Office
- Orbiter Engineering Office

Space Station Projects Office (KA)
Clarke Covington — Manager
Carl B. Shelley — Deputy
Robert E. Bobola — Mgr for Dev
Jesse F. Goree — Mgr for Integ
W. E. Rice — Asst Mgr for Policy

- Information Systems Office
- Flight Elements Office
- Management Integration Office
- External Integration Office
- Manufacturing and Test Office
- Systems Engineering and Integration Office

Mission Support Directorate (EA)
Ronald L. Berry — Director
Elric N. McHenry — Deputy
John R. German — Asst to Dir
Samuel M. Keathley — Asst to Dir

- Data Processing Systems Div
- Mission Planning and Analysis Div
- Spacecraft Software Division
- Systems Development Division

Flight Crew Operations Directorate (KA)
Donald R. Puddy — Director
Henry W. Hartsfield — Deputy
Richard W. Nygren — Asst to Dir
Karol J. Bobko — Asst to Dir

- Vehicle Integration Test Office
- Aviation Safety, Chief
- Crew Integration Office
- Astronaut Office
- Aircraft Operations Division

National Space Transportation Systems (GA)
Richard H. Kohrs — Mgr/Dep Dir
Jay F. Honeycutt — Deputy
Bryan D. O'Connor — Asst Mgr for Ops

- Management Integration Office
- NSTS Integration and Operations Office
- Space Station Integration Office
- Flight Production Management Ofc
- Mission Integration Office
- Cargo Engineering Integration Ofc
- Customer Integration Office
- NSTS Engineering Integration Ofc
- Project Integration Office
- Avionics Systems Office
- Systems Engineering Office

Administration Directorate (BA)
William R. Kelly — Director
R. Wayne Young — Deputy

- Office of Director of Procurement
- Office of the Comptroller
- NSTS Program Control Office
- Orbiter and GFE Projects Control Ofc
- Management Analysis Office
- Space Station Projects Control Ofc

Mission Operations Directorate (DA)
Eugene F. Kranz — Director
John W. O'Neill — Deputy
James D. Shannon — Asst to Dir

- Reconfiguration Management Ofc
- Administrative Office
- Flight Director Office
- Management Integration Office
- Flight Design and Dynamics Div
- Operations Division
- Systems Division
- Schedules and Flow Office
- Training Division
- Facility and Support Systems Div
- Space Station Mission Ops Ofc

Center Operations Directorate (JA)
Kenneth B. Gilbreath — Director
Grady E. McCright — Deputy

- Logistics Division
- Security Division
- Management Services Division
- Photography and Television Technology Division
- Technical Services Division
- Plant Engineering Division
- Facilities Design Division

Safety, Reliability & Quality Assurance Office (NA)
Charles S. Harlan — Director
Gary W. Johnson — Deputy

- NSTS SR&QA Office
- Quality Assurance, Reliability, and Safety Office – Downey
- Advanced Projects Definition Ofc
- Technology and Commercial Projects Office
- Flight Projects Office
- Lunar and Mars Exploration Ofc

New Initiatives Office (IA)
William J. Huffstetler — Manager

- Crew Emergency Return Vehicle (CERV) Office
- Advanced Projects Definition Ofc
- Technology and Commercial Projects Office
- Flight Projects Office
- Lunar and Mars Exploration Ofc

Engineering Directorate (EA)
Henry O. Pohl — Director
Max Engert — Deputy

- Engineering Management Office
- Advanced Programs Office
- Program Engineering Office
- Crew and Thermal Systems Div
- Systems Development and Simulation Division
- Avionics Systems Division
- Tracking and Communications Div
- Propulsion and Power Division

NSTS Operation Integartion Office
Richard A. Thorson — Manager

White Sands Test Facility (RA)
Rob R. Tillett — Manager

- Administration Office
- Propulsion Office
- Laboratories Office

FIGURE 22. Organization as of May 1988.

presence beyond the Earth; and strengthen aeronautics research. Its immediate commitments were to return the Space Shuttle to flight status, develop advanced space transportation capabilities, and build the facilities and pursue the science and technology needed for the Nation's space program. In the process, NASA would promote the application of aerospace technologies to improve the quality of life on Earth, and conduct cooperative activities with other nations.[49] It seemed to be a reasonable and clearminded charge, attainable, and more practical than visionary. It was something that NASA and JSC could attack and Congress and the public could understand. The *Leadership* report, delivered to James Fletcher in August 1987, helped resolve doubts and uncertainties.

By the close of 1987, most of the reports, studies, and organizational changes relating to the space station and the Shuttle had been completed. Program management, both for the Shuttle and the space station, shifted to NASA Headquarters. Aaron Cohen revised JSC's administration by having each directorate report directly to the center director, rather than through a division head. Thus, whereas Flight Crew Operations, Mission Operations, and Mission Support had reported through C.E. Charlesworth to the director, now each reported directly (figures 20 through 22). JSC relinquished the National Space Transportation System Program Office to Headquarters. Arnold D. Aldrich became director and Richard H. Kohrs deputy director. Robert L. Crippen, one of the Air Force manned orbiting laboratory trainees who transferred to the astronaut corps in 1969, became deputy director of the Space Transportation System Operations Office stationed at Kennedy Space Center in Florida.[50] Administrative arrangements for the space station and the Shuttle, however, continued to change.

In addition to the regular line directorates, the three project offices at JSC included the New Initiatives Office, the Space Station Projects Office managed by Clarke Covington, and an STS Orbiter and GFE Projects Office under Richard A. Colonna.[51] The Space Station Projects Office concentrated on planning and designing the proposed station until 1989 when major development contracts were let. The Orbiter Office readied the three remaining Shuttles for flight, and, in July 1987, began to award contracts for a replacement vehicle for the *Challenger*.

The replacement vehicle (Orbiter 105) was to be markedly improved over the three existing Shuttles. Orbiter 105, Colonna said, "would have stronger

On August 2, 1991, STS-43 soared toward space to begin a 9-day mission to deploy a Tracking and Data Relay Satellite (TDRS-E) similar to that lost aboard Challenger. *The TDRS-E had waited more than 5 years while NASA investigated the* Challenger *accident and the Nation reassessed its space programs.*

wings, midbody, and tail, better brakes and tires on stronger axles, improved electronics, bigger and better computers, improved auxiliary power units, a safer heat exchanger, a parachute brake, fewer tiles, and new fuel disconnect valves." Rockwell, the Shuttle prime contractor, had major structural parts for a new Shuttle on hand which speeded construction by an estimated 2 years. Subcontractors such as Grumman (wings), Fairchild Republic (tail), and IBM (computers) manufactured their components in their own plants and delivered them for assimilation at the Rockwell plant in Downey, California, or for final assembly to a Rockwell hangar in Palmdale. The new Shuttle was scheduled for delivery in April 1991. In addition, the solid rocket booster joints had been completely redesigned under the supervision of Marshall Space Flight Center and Morton Thiokol engineers.[52]

The Orbiter Project Office also managed an extensive refurbishment of the existing Shuttle fleet. There were 210 modifications to the orbiters, including a rudimentary escape system, new pressure suits to be worn during launch (versus the old shirtsleeve uniform), inflatable life rafts and emergency signals, structural reinforcements in the wings and fuselage, better brakes, and enhanced computers. In January 1987, NASA selected a crew (Frederick Hauck, Richard Covey, George Nelson, Mike Lounge, and David Hilmers), a ship (*Discovery*), and a launch date (February 18, 1988) for the return to flight. That launch date slipped to September 29, when fuel tank tests, and then solid-fuel rocket booster test failures, leaky seals, and valve malfunctions caused delays.[53] NASA's response to *Challenger* and the Rogers Report was better engineering and enhanced safety and quality control checks.

But NASA now seemed to be plagued with problems that lay beyond engineering expertise and quality control. Moreover, problems of any sort in the post-*Challenger* era had become matters of public concern and alarm. Although it had long labored under the public eye, NASA's performance was now being viewed through a magnifying lens. In March 1987, a NASA Atlas-Centaur carrying a Navy communications satellite broke up after being struck by lightning only moments after launch. In June, three rockets being readied on the Wallops Island launch pad were struck by lightning and launched in premature and "hopeless" flight. In July, the upper stage of an Atlas-Centaur rocket on the launch pad at Kennedy Space Center was destroyed in an accident. In August, valve malfunctions on the orbiter and hydrogen leaks in the ground support equipment forced postponement of a *Discovery* launch. And in September a hurricane off the Texas coast threatened to disrupt operations in Mission Control and forced a delay in setting a firm launch date. Finally, the countdown began on September 25. An estimated one-half million people gathered at the Cape to view the launch. "After 2-1/2 years of anguish, resolve, and accomplishment," on September 29, 1988, *Discovery*—and the United States returned to space.[54] The downtime ended.

Out of this time of chaos came a reassessment of America's role in space. The results of these deliberations were remarkably consistent with visions and programs delineated in the 1950's and 1960's. A space shuttle, unmanned missions to the planets and beyond, a space station, and habitation by humans of other solar bodies still comprised the general plan for an American presence in space. The dream was still alive. There was now a new space station initiative, a better and safer Shuttle, and a "New Initiatives" management group at JSC. The structures for administering the Shuttle and the space station had changed and would continue to do so as the programs themselves changed and as the new styles of management were field-tested. There were new initiatives by NASA, JSC, universities, and businesses that would fabricate a new extension and a new permanence of space in the affairs of people on Earth, and particularly among those in that community associated with JSC.

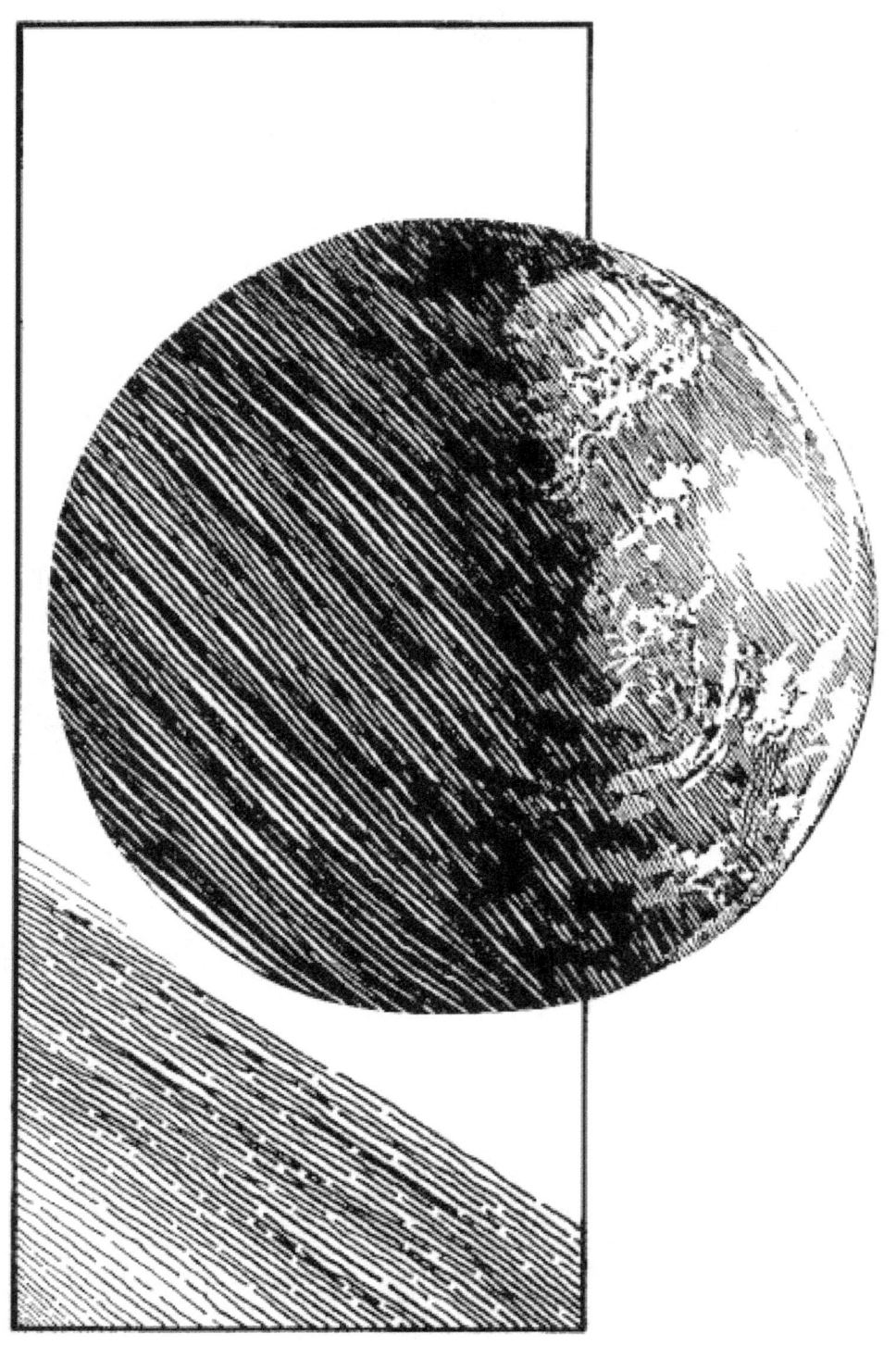

CHAPTER 17: Space Station Earth

Children in kindergartens in Texas in the 1980's began to sing new lyrics to the old song "Farmer in the Dell." They went like this: "We're blasting off to space, We're blasting off to space, Fly high away from Earth, We're blasting off to space." And the round continued—"We're going to the Moon . . . We're going to the planets . . . We're going to the stars . . . Fly high away from Earth, we're going to the stars." It meant that human spaceflight had become a part of the folklore. Whatever the pros and cons of putting humankind in space might have been over the past three decades, or the costs, or the real or imagined benefits for present and future generations, spaceflight had become as ingrained in the American mystique as "Farmer in the Dell."[1]

People could leave the planet Earth. They could fly in space. They had flown in space. They could walk on the Moon. They had walked on the Moon. They could fly a vehicle into an Earth orbit and return safely to Earth. They had done so many times. And they were now building a permanent space station for working and living in space. These things had been conceived, designed, constructed, and flown in part through the work of the National Aeronautics and Space Administration and Lyndon B. Johnson Space Center and its contractors. Because of these things Earth and its peoples would never be the same again.

With the return to flight, activity and energy-levels at NASA and JSC quickened. Congress began firming up a NASA budget for 1989 that contained a 30-percent increase over the 1988 budget, including a $6 billion, 3-year commitment to space station funding and a 27-percent increase in Shuttle program funding. The improved budget rejuvenated work on the space station, which had experienced budget cuts late in 1988 that resulted in delaying plans to have a fully operational station in Earth orbit from 1995 to 1997. Budget and design problems would continue to cause "slippages" in the space station program, but it had for a time developed a new life and a new name—Space Station *Freedom*.[2]

While *Discovery* orbited the Earth in September 1988, NASA began negotiating development contracts for the four, 10-year space station work packages to be handled respectively by Boeing Aerospace, McDonnell Douglas, General Electric's Astro Space Division, and Rockwell International's Rocketdyne Division. JSC, as manager of the McDonnell Douglas $2.6 billion package, had responsibility for the space station's integrated truss structure, mobile servicing system transporter, airlocks, and hardware and software data management (relating to guidance, navigation and control, and communications and tracking). Boeing, under Marshall Space Flight Center direction, would develop the laboratory and habitation module, while Goddard Space Flight Center managed contracts for an unmanned polar-orbiting platform and a flight telerobotic system, and Lewis Research Center directed work on an electrical power and distribution system. JSC anticipated increasing its civil service employees from a near all-time low of about 3340 to about 3460, with support contractor employment remaining stable at approximately 9000. Human Resources Director Jack Lister noted that since 1970, JSC's civil service force had been in a precipitous decline which strained the center's "in-house, civil service technological strength." Things were now looking up.³ Employee and contractor morale began to rise more than proportionately.

But long-term budget and space policy awaited decisions by a higher authority. Governor Michael Dukakis of Massachusetts, the Democratic candidate for President, ran on a ticket with Texas Senator Lloyd Bentsen. The latter, if not professedly pro-space, had a proud history of being supportive of government programs that helped Houston and Texas. George Bush, the Republican candidate for President, who claimed Houston as his hometown, supported an invigorated space program, as did incumbent Republican President Ronald Reagan. Reagan visited the crew of *Discovery* and personnel at JSC just days before the countdown began for the return to space. "You don't launch rockets," he told the crowd of government employees and contractors, "you launch dreams. . . . America's going to space again, and we are going there to stay." STS-26 commander Rick Hauck told the President they would like to take him with them on the flight, but being unable to do that, would carry his personal name tag and a flight patch for the jacket he then presented the President.⁴

Certainly one spin-off of the *Discovery* flight was to help focus the voters' attention on NASA and space. It was, in fact, an unusually well-informed electorate who over the past 2-1/2 years had been exposed to numerous blue-ribbon committee studies and reports, and to a relentless media scrutiny that, in the words of Administrator James Fletcher, "sought to question every action and to uncover its every perceived blemish and wart."⁵ For example, *Apollo: The Race to the Moon* by Charles Murray and Catherine Bly Cox suggested that the Apollo successes derived from a brilliant, "old-boy" network that produced an organization based on reliability and trust, but which in time became an "incestuous buddy-buddy bureaucracy" that had experienced "hardening of the arteries." Political scientist Howard E. McCurdy observed that NASA "lost touch with many Apollo precepts, including the importance of testing, the need for hands-on activity, and a commitment to recruitment of exceptional people." Historian Alex Roland thought that the Agency had fallen into a vicious cycle in repeatedly trying to outdo Apollo. "NASA simply wouldn't face the evidence that their plans were too expensive."⁶

A *New York Times* article, reprinted in a NASA newsletter, suggested that while the dream might still be alive, it was an impoverished dream. "Like the Shuttle, the space station is not an end but a means, infrastructure, built for when a President someday decides what to do with it. No wonder the space program has become a yawn." The writer thought that the space station should be scrapped and that the United States should seek a joint mission with Russia to Mars.⁷

And while *Discovery* orbited the Earth, James Van Allen, the professor of physics who in the Explorer experiments discovered the "Van Allen" radiation belt, raised the old cry that NASA's emphasis on manned spaceflight prevented spending on constructive scientific work that might be done with unmanned vehicles. An unidentified NASA "top administrator" is supposed to have said that "we made Van Allen famous, and he's been kicking us in the butt ever since."⁸

Although a poll by the Associated Press in August 1988, just before the return to flight, indicated that a large portion of the population had lost confidence in NASA over the past few years, a remarkable 58 percent said that the Agency had done a good or excellent job, with most of those polled favoring a budget at least equal to or greater than that for the past years.

But a similar Gallup Poll a year later suggested that public commitment to the space program was lukewarm. Roughly one-fourth of those polled favored raising the NASA budget, while an equal number would cut the budget. There was little sense of urgency about being the first to land on Mars, and no strong opinions about whether space exploration should concentrate on manned or unmanned missions.[9]

In October 1988, as NASA celebrated its 30th birthday and the return to flight, most Americans could not recall a time when there had been no NASA. For them, the Apollo program and lunar landing were history, nothing more. And the cold war which had triggered the turn to space was dissipating rapidly and would virtually evaporate. The polarized cold war world helped keep the space race alive. Could peace and international cooperation do the same? Did space ventures, in fact, offer an institutional and intellectual framework for international cooperation that might in and of itself justify a continued commitment to space? Whereas the cold war and international competition brought Soviets and Americans into space, and in the process threatened to destroy them all either through military conflict or economic collapse, could peace and international cooperation sustain an effort in space, or was technological advancement somehow dependent upon international rivalry and war or the threat of war? American goals in space, whatever the motivations of the past decades, entered a new conceptual framework in the last decade of the 20th century because of the changing world order reflected by the demise of the cold war. For those under the age of 30, the rationale for an American presence in space would not be quite the same as for those of the cold war generation.

In the realization that NASA had not fully communicated its accomplishments and its purposes to the general public, JSC and individuals in the private sector for some years had been interested in improving the center's visitor complex and educational information facilities. In October 1988, Aaron Cohen signed a Memorandum of Understanding with the Manned Space Flight Educational Foundation for a privately endowed visitor center on site. Approximately one million visitors came through the gates of JSC each year; and although there were Mercury and Apollo rockets, lunar modules and astronaut flight suits on display, the center was really not equipped, nor did it have the personnel or time, to commit the kind of energies to public education and information that did seem to be justified. In cooperation with community leaders and retirees, JSC officers—prominently William R. (Bill) Kelly (Director of Administration) and Harold S. Stall (Director of Public Affairs)—organized a nonprofit corporation as a vehicle for raising funds to operate a planned $40 million (or more) visitor center.[10]

Kelly, who signed the Memorandum of Understanding with Cohen, was Chairman of the Foundation. Stall was President and Chief Operating Officer. Six more JSC officials including Dr. Carolyn Huntoon (Director of Space and Life Sciences), Harvey Hartman (Deputy Director of Human Resources), Paul J. Weitz (JSC Deputy Director), John W. O'Neill (Assistant Director of Mission Operations), and Grady McCright (Deputy Director of Center Operations) joined five board members from the private sector. In November, the Foundation kicked off an $8 million fund-raising campaign for "Space Center Houston," seeking to procure one-half of the total from aerospace contractors and the remainder from Houston-area civil and philanthropic organizations. An additional $42 million was to be raised from revenue bonds funded by admission charges. The Foundation also entered into a preliminary

Space Center Houston: Built to be an "adventure of the mind," the visitor center is expected to relieve JSC of the difficult obligation of hosting the general public and at the same time produce a better informed, more aware, and more supportive public.

agreement with Walt Disney Imagineering for the design and development of the visitor complex.[11]

The economic down-turn, oil price collapse, and banking and savings and loan crises resulted in long delays and failed commitments to the Foundation and difficulty in obtaining funding for the proposed revenue bonds; but by early 1991, "Space Center Houston" was back on track. The complex had been enlarged to a 180,000 square foot facility that would cost about $70 million and host an anticipated 2.3 million visitors per year. Hal Stall said that the new center sought to dispel the myth of space and explain its realities. "It is not to be," he said, "a theme park or a museum, but an experience center where visitors can see, touch and feel. It is to be an adventure of the mind."[12] The visitor center is expected to have a great impact on the local economy, relieve JSC of the difficult obligation of hosting the general public, and produce a much better informed, more aware, and presumably more supportive public. Thus, September, October, and now November 1988, marked a turning point in the affairs of NASA and JSC in yet another way.

The newly elected President, George Bush, interpreted his victory at the polls as at least in part a vote for a sustained and somewhat enhanced space program. A transition team began the search for a new NASA Administrator. Among those being considered were Gerald Griffin and Chris Kraft, former JSC directors; H. Ross Perot, a businessman from Dallas; Hans Mark, Chancellor of the University of Texas; Richard H. Truly, NASA's Associate Administrator and a former astronaut; Frank Borman, a former astronaut and President of Eastern Airlines; and former astronauts Bill Anders, Tom Stafford, and Frederick H. (Rick) Hauck, who commanded the *Discovery* in its September flight.[13] Although the list of candidates suggested a strong JSC representation and certainly the President's personal interest in space (and in Houston) boded well for JSC, center personnel were preoccupied with more pressing things than the politics of succession.

Technicians and flight crews began final preparations for the launch of *Atlantis* (STS-27) scheduled for late November or early December. In November, JSC unveiled its new $4.8 million space station mockup and trainer building and facilities, and soon after, the *Atlantis* crew flew to Kennedy Space Center for a "dry" test launch. On December 2, after a delay for bad weather, the Shuttle lifted off on a classified DoD mission. There were now two successful returns to flight. Following *Atlantis*' return, JSC employees and contractors received a total of 158 individual and 43 group return-to-flight NASA awards. JSC remembered that 20 years earlier Apollo 8 made the historic translunar orbital flight. Relatively few of the personnel on board then were still at the center. It seemed appropriate to begin planning a reunion—a 20th anniversary celebration of the Apollo 11 lunar landing to be held at JSC in July, 1989.[14]

A mockup of the new Space Station Freedom *is housed in Building 9 at JSC to provide training and design experience for the space station scheduled to be placed in permanent orbit about Earth near the close of the decade of the 1990's.*

Discovery entered the Vehicle Assembly Building at Kennedy in late January 1989, for mating with the STS-29 Shuttle components in preparation for a March 1989 lift-off. On March 13, pretty much on schedule, *Discovery* rode into orbit. In a telephone call to the orbiting vehicle, President Bush congratulated the crew and NASA: ". . . you have our strong support," he said, "We're living in tough budgetary times, but I am determined to go forward with a strong, active space program." And on the ground at JSC, in his farewell tour of NASA facilities, Administrator Fletcher congratulated JSC employees and predicted revisits to the Moon, lunar bases, and manned missions to Mars.[15]

As a reflection of his commitment to space, President Bush resurrected the National Space Council which had fallen into disuse during previous administrations. Soon after the return of *Discovery*, its head, Vice President Dan Quayle, visited JSC in company with Rear Admiral Richard H. Truly, NASA Associate Administrator for Space Flight. The Vice President told a lunch crowd in the Building 11 cafeteria that President Bush told him there were three things he would love about Houston, "the weather, the barbecue, and the Johnson Space Center." The National Space Council, he said, would look beyond the traditional divisions of space interests (the civil, commercial and national security interests), and would seek to formulate policies regarding the privatization of space and the promotion of educational opportunities "to ensure an abundant supply of qualified scientists and aerospace engineers."[16]

Suddenly, Tomorrow Came . . .

Within days, President Bush nominated Admiral Truly to the post of NASA Administrator. The Senate approved the nomination in late June, and in July Richard H. Truly became the first administrator with astronaut experience. A Navy pilot, Truly joined the Air Force Manned Orbiting Laboratory Program in 1965 and transferred to NASA's astronaut corps in 1969 when the Air Force program was canceled. He piloted the STS-2 flight and commanded the STS-8 flight. He then served as commander of the Naval Space Command before going to NASA Headquarters as Associate Administrator for Space Flight. Henry Hartsfield, Deputy Director of Flight Crew Operations at JSC, said that because Truly had flown in space and "managed the return to flight (of the Shuttle), he understands how to sell the budgets and how to develop sound ideas." But not all were so supportive. George Henry Elias, author of *Breakout Into Space: Mission for a Generation*, thought that Truly was too much the specialist (an astronaut, military officer and technician) when what NASA needed was a "generalist with broad vision and deep experience."[17] Generally, however, JSC felt good about having one of its own at the NASA helm.

It felt good too to get *Atlantis* back into orbit on May 4, carrying the unmanned Magellan spacecraft to be launched from the Shuttle for an orbital exploring mission to the planet Venus. A month later the Magellan probe had traveled 3.735 million miles from Earth and was moving at a velocity of 5500 miles per hour.[18] The space program seemed to be back on track.

NASA scheduled 4 more Shuttle flights for 1989, 9 for 1990, 8 in 1991, and 12 in 1992 (including the introduction to the Shuttle fleet of the new *Endeavour*—replacing the *Challenger*). The flight manifest planned 14 Shuttle flights in 1993, 13 in 1994, and 10 through September of 1995. In addition, NASA began scheduling launches using Titan IV expendable rockets. Like Magellan, two more unmanned planetary probes, Galileo to Jupiter and Ulysses to the Sun, were scheduled for 1989 and 1990, respectively. The program office slipped the scheduled launch of the Hubble Space Telescope from December 1989 to March 1990 in order to retrieve a Long Duration Exposure Facility (LDEF) deployed in 1984 and originally scheduled for retrieval in 1985. That retrieval time was now long past due and the LDEF satellite was in danger of plunging back to Earth. There were to be many Earth science missions and experiments, a number of cooperative missions and experiments with foreign nations, and in 1995, the first assembly missions for components of Space Station *Freedom*.[19] There was much to do—much to look forward to.

Aaron Cohen remarked: ". . . Our number one job (at JSC) is still to fly the Shuttle and fly it safely." In the Technical Services Division, machining, sheetmetal and welding fabrication, sculpturing, electronics and computer devices were produced on order for the Shuttle and space station. Here JSC engineers fabricated the mockups, models, and government-supplied equipment where ideas and designs became tangible artifacts, and technicians manufactured everything from "soup to nuts." Here hands-on management began. Many of the specialty fabrications for the new ship *Endeavour,* scheduled for a 1992 maiden flight, came out of the JSC shops. Much of the center's energy and talent continued to be directed to maintaining and flying the Shuttle fleet.[20] JSC also made a number of administrative changes affecting the management of both the Shuttle and the space station.

Daniel M. Germany assumed direction of the Orbiter and GFE (Government-Furnished Equipment) Projects Office at JSC, replacing Richard A. Colonna who headed for the outback to become NASA's representative to Australia. Work on *Endeavour* now consumed more of the orbiter office's attention. Rockwell International's Space Construction Division increased its workforce on the *Endeavour* construction project from 600 to 850. Roger Hicks, JSC's orbiter project operations officer stationed at the construction site in Palmdale, reported that everything was on schedule, if not a bit ahead, but that the crucial work would come in 1990 when the various components and fuselage, wings, tail, and crew modules were assembled and electronic systems were integrated and tests began.[21]

Leonard Nicholson replaced Richard Kohrs as Deputy Director of the Space Shuttle Program Office, and that office was moved from Headquarters to JSC to better mesh the technical work with management—thus ameliorating (in the minds of JSC engineers) the separation of Level II management from the center technical expertise that had occurred with the scrapping of the lead center system and the reorganizations following the *Challenger* disaster. Nicholson, who joined the Spacecraft Integration Branch in the Engineering and Development Directorate at JSC in 1963, rose through the ranks to become technical assistant to the Manager of the Apollo Program Office, then manager of the Space Transportation System (STS) Operations Office, and manager of STS Integration and Operations. Jay Greene, who had a diverse background at JSC in flight dynamics, as chief of the Mission Operations Branch, flight director, and chief of the Safety Division, became deputy manager of the National Space Transportation System Program Office (which would soon be renamed the Space Shuttle Program Office). Larry Williams, who joined NASA in 1962, became manager of the Engineering Integration Office. C. Harold (Hal) Lambert, a 1957 Langley Research Center veteran who went to JSC's Propulsion and Power Division in 1962, became manager of the Shuttle Integration and Operations Office.[22]

There were also management changes and program changes for the space station. Richard Kohrs left JSC for Headquarters where he would direct the Space Station Freedom Program Office. Kohrs, like Nicholson, began his NASA career at MSC in 1963 and was STS systems integration manager and deputy manager of the STS Program Office before moving to space station work. Kohrs assigned Richard A. Thorson to JSC as deputy manager for Space Station Freedom Program Integration. This assignment helped reestablish the essential association between the Level B (Level II after 1987) program management and technical expertise.[23] Thus, the new space station Level B and Shuttle Level II management structures attempted to establish a bridge between the old lead center system and the post-*Challenger* organization that had effectually isolated the technical integration management from its technical resources.

Another bridge between project and program management involved simply the transfer of personnel between Headquarters and JSC. Thus, Arnold D. Aldrich, who joined the Langley Space Task Group soon after its formation and became a member of the staff of MSC, moved to Headquarters in 1987 as Director of the National STS Office. Aldrich managed the STS through his deputy director and colleague, Leonard Nicholson at JSC. Nicholson, of course, had replaced Richard Kohrs as deputy director. About the time that Kohrs went to Headquarters, John W. Aaron, who was managing the Lunar and Mars Exploration Activity Office in Washington, D.C., transferred to Houston to head the Space Station Projects Office. Aaron, who joined the JSC task force in 1964 as a flight controller, had been assistant and then chief of the Spacecraft Software Division (1979-1984) and a special assistant to Aaron Cohen. Clarke Covington, who formerly headed the Space Station Projects Office, now became a technical assistant to the JSC Director. But because the project office reported directly to the Center Director, rather than through the program office as had occurred under the lead center style of management, the interface between the program office and the project office was not as close as it had been under the lead center system.[24] However, no management system had been perfect and the vital ingredient in effective project/program management involved the proper "people" and experience mix. It did appear that insofar as JSC and the Shuttle were concerned, the old "collegial" management mix that had provided an interface between Headquarters and JSC and had served NASA so well during the Mercury, Gemini, and Apollo/George Low era had been reinstated.

Aaron Cohen defined project management as "the business of creating—through a sensible sequence of efforts that utilize to best advantage the resources available—a product that achieves the objective." JSC's product:

> ... is putting men and women into space, keeping them alive and productive while they're there and returning them safely to Earth. We design, develop and operate manned spacecraft and train the crews that use them. We conduct scientific and medical experiments that help us understand how space affects our astronauts and spacecraft . . .

Cohen thought that the key to effective project management was to nurture the environment and culture that motivated people to strive for technical excellence above all else. After intensive in-house studies, JSC initiated a Total Quality Management program that sought to continually enhance performance at all levels through cooperative contractor-manager team planning and collaboration. Regarded as a strategic approach to change, the new processes sought to produce real savings and better performance (earning JSC a Quality Improvement Prototype award from OMB in 1990).[25]

Cohen had learned that hands-on experience was essential to controlling the three classical elements of project management—performance, cost, and schedule. Schedules drive costs, and costs determine what can be produced. Performance is a product of costs and schedules. Contract management and project control are as important to management as technical expertise. Decisions must not only be made, they must be timely. Compromise is both acceptable and necessary. Not all problems can be solved. Product development involves selecting that which is best or better, not that which is perfect. And finally, project management is a people-oriented business. Patience, communication, honesty and fair treatment are necessary elements of effective management.[26] Thus, space projects were people projects, and the culture and environment of space project management, to be sure, extended far beyond the confines of JSC. The President, Congress, Headquarters, all of the NASA centers, the contractors and their employees, and even the media and electorate contributed to the culture that had formed about space and its technology.

Space was a complex business that required a sustained level of activity, careful scheduling, continual testing and development, cost controls, a relentless attention to detail and quality in product and performance. An event such as the launch of *Columbia* (STS-28) in August 1989 on a DoD mission reflected not only a triumph in technology, but a significant accomplishment in very large-scale project and systems management, as well as a real achievement of the human spirit. Glenn Lunney sensed that the 25 years he spent at JSC had been a time memorable for extraordinary events, a "Camelot, a magic time, . . . when what we did was more than the sum of all of us."[27]

Although none said so in so many words, Lunney's sense of things seemed to reflect the sentiments of most of those gathered at JSC to celebrate the 20th anniversary of the lunar landing. "Were you there? . . . Yes, I was there." They meant, of course, not that they were on the Moon, but that they shared that time as one of the several thousand scientists, technicians, engineers, flight controllers, and staff of JSC. Comments from speakers included: "Think about what humans have done the past 100 years, when you think about the possibilities for the future." "There was a sense of trying to accomplish something that had not been done before." "What we did was nothing short of fantastic," another commented. "We had more responsibility at age 30 than most people have in a lifetime." "This was an enormously successful and dedicated organization." "The door's been opened . . ." "The things we thought were not important . . . really are." "This was a pause as one climbed the mountain . . ." "This was done by ordinary people!"[28] And with each Shuttle launch, each placement of a satellite, or each design, development and testing of a space station component, that magic continued.

While JSC and NASA celebrated the Apollo Lunar Landing 20th Anniversary, President George Bush announced a new "Space Exploration Initiative."

> We must commit ourselves anew to a sustained program of manned exploration of the solar system and, yes, the permanent settlement of space . . .
>
> First, for the coming decade—for the 1990's—Space Station *Freedom*, our critical step in all our space endeavors. Next, for the new century, back to the Moon, back to the future, and this time, back to stay. And then a journey into tomorrow, a journey to another planet, a manned mission to Mars.[29]

As Congress and the American people began to digest this proposed long-range continuing commitment in space, NASA selected a special study group, headed by JSC's Aaron Cohen, to frame the essential elements and guidelines affecting decisions about a lunar-Mars initiative. And the pace and excitement within NASA seemed to quicken.

Atlantis (STS-34) moved into the Vehicle Assembly Building at Kennedy Space Center in August, within days of the return of STS-28. Voyager 2, an unmanned planetary probe, began sending images of the planet Neptune 2.8 billion miles through space to Earth. *Atlantis* lifted off from Kennedy Space Center on October 18 "after being threatened by a court challenge, delayed five days by a suspect main engine controller, and one day by unfavorable weather." The crew, including commander Don Williams, pilot Mike McCulley, and mission specialists Ellen Baker, Franklin Chang-Díaz and Shannon Lucid, deployed the Galileo spacecraft for a 5-year journey to the planet Jupiter and a 1995-1997 orbital tour of the great planet.[30]

That week in the Gilruth Center at JSC, Dr. Robert L. Forward, a physicist, science consultant and author, speculated with JSC personnel about the feasibility of interstellar travel. In November, Congress approved a $12.4 billion NASA budget, providing an 11.9-percent increase over the previous year. JSC's space station work would increase markedly, and a construction program was scheduled that would add a new central computing facility, an auxiliary chiller for air-conditioning, additions to the atmospheric reentry materials and structures evaluation facility, a space station "high-bay" assembly building, a space station control center, and an improved simulator/training facility. The Hubble Space Telescope, scheduled for flight in March 1990, began instrumentation tests at Kennedy Space Center.[31]

Atlantis (STS-32), however, seemed poised interminably at the launch pad for favorable weather and a good launch, as the decade of the eighties drew to a close. The launch of *Atlantis*, said Aaron Cohen, "if we do our jobs well . . . will be the first successful Space Shuttle mission of a busy, challenging year. We also find ourselves working toward the well-defined, long-range goals of establishing a permanent base on the Moon, and then sending humans on to Mars and beyond. Separate, these efforts are extremely important. Together, they are the realization of dreams."[32]

The launch went well. Another decade in space began. *Atlantis* sped into space on January 8, 1990, and returned with a prize, the LDEF, a bus-sized satellite stranded in space for almost 6 years, which carried rich documentation for long-duration spaceflight and habitation. In the returning cargo were thousands of tomato seeds sent as part of the Space Exposed Experiment Developed for Students (SEEDS). After the seeds were returned and preliminary tests were completed, NASA sent seeds to schools and individuals throughout the United States and the world in response to 130,000 requests. Would space-exposed seeds germinate, grow, and bear fruit? Technical foreman Dan Alexander planted seeds outside building 326 at JSC. There was an 85-percent germination rate for the space seeds, and a 62-percent germination for a test batch of earthbound seeds![33]

The new year also began with the promulgation of the study of the President's Space Exploration Initiative. The study, a product of a comprehensive NASA effort and directed by Aaron Cohen, involved program associate administrators at Headquarters, center directors, technical study groups, and a report assembly team. Directed to Administrator Truly for the National Space Council, the report sought to provide criteria and framework for a determination of the necessary money, personnel, and materials that might be required for a "new and continuing course to the Moon and Mars and beyond." The Space Exploration Initiative Cohen defined as encompassing both robotic and human missions—but overall a "distinctly human adventure" in the broadest sense, in that human and robotic missions into space would extend into the solar system the "skills, imagination, and support of many thousands of people who will never leave Earth."[34]

The study addressed again the question raised when the United States began its space program. "Why fly into space?" And it addressed a more contemporary question, but one which echoed from NASA's own past—"after the Shuttle, what next?" It was not unlike the question posed as the decade of the 1960's ended, and America had indeed met the challenge of putting a man on the Moon "within this decade." Then it had been—"after Apollo, what next?" Like the answers given earlier, the answers given in 1989 and 1990 to those questions would never be wholly satisfactory.

> The imperative to explore is embedded in our history, our traditions, and our national character . . . Now, in the late 20th Century and the early 21st, men and women are setting their sights on the Moon and Mars, as the exploration imperative propels us toward new discoveries.
>
> To enrich the human spirit, to contribute to national pride and international prestige, to inspire America's youth, to unlock the secrets of the universe, and to strengthen our Nation's technological foundation: human exploration of the Moon and Mars will fulfill all these aspirations and more.[35]

Almost concurrently with the preparation and release of the report on the new American space initiatives, the world as it had been known began to unravel and change markedly. The demise of the cold war, the withdrawal of Russian troops from East Germany and Soviet satellite countries, and the seemingly incredible reunification of Germany and breakup of the Soviet Union cast the old questions about space, and their answers, in a totally new context. Could the American space program survive peace? Could international cooperation replace international competition? Would Congress and the American people commit their resources to new space initiatives now that the justification for space and even high technology had seemingly changed?

Although the existing Shuttle, space station initiatives, telecommunications, navigation and information management systems provided the basic infrastructure for a lunar-Mars initiative, the Shuttle and expendable launch vehicles would need to be enhanced. Cargo flights for extraterrestrial human exploration required a lift capacity of 60 metric tons for the Moon and 140 metric tons for Mars, compared to the 17.3 metric ton capability of the Shuttle. More work was needed in the life sciences, including medical care, life support systems, and studies of human behavior in an extraterrestrial environment. Space Station *Freedom* would be intrinsic to the development of extraterrestrial capabilities, and the Shuttle, in turn, vital to the construction of the space station.[36] But much more would be required in basic research and development.

NACA/NASA, with its 75 years of research, development and operational experience, provided the core capability for the new space initiative. The new programs would, however, require a "significant augmentation of civil service positions," and a "solid balance between in-house and contracted works." New exploration initiatives offered the potential and opportunity for international cooperation. They also created a favorable environment for scientific and technological research and development, and necessitated on the part of NASA a further nurturing of science and engineering in American educational institutions.[37]

Cohen's study group examined technical variables and scheduling, or the evolutionary processes of a Moon-Mars exploration program. The committee conducted a technology assessment that linked existing and projected capabilities with costs and schedules. Although existing technology could take people back to the Moon, long-term and regenerative life support systems were yet to be developed. For example, an Earth-Mars return flight was estimated at 14 months; and with surface operations, the trip would require self-sustained flight of at least 600 days. By comparison, the Shuttle was built for a nominal 7-day mission. New propulsion systems were advised, possibly expanding on nuclear thermal rocket technology developed in the NERVA rocket program between 1955 and

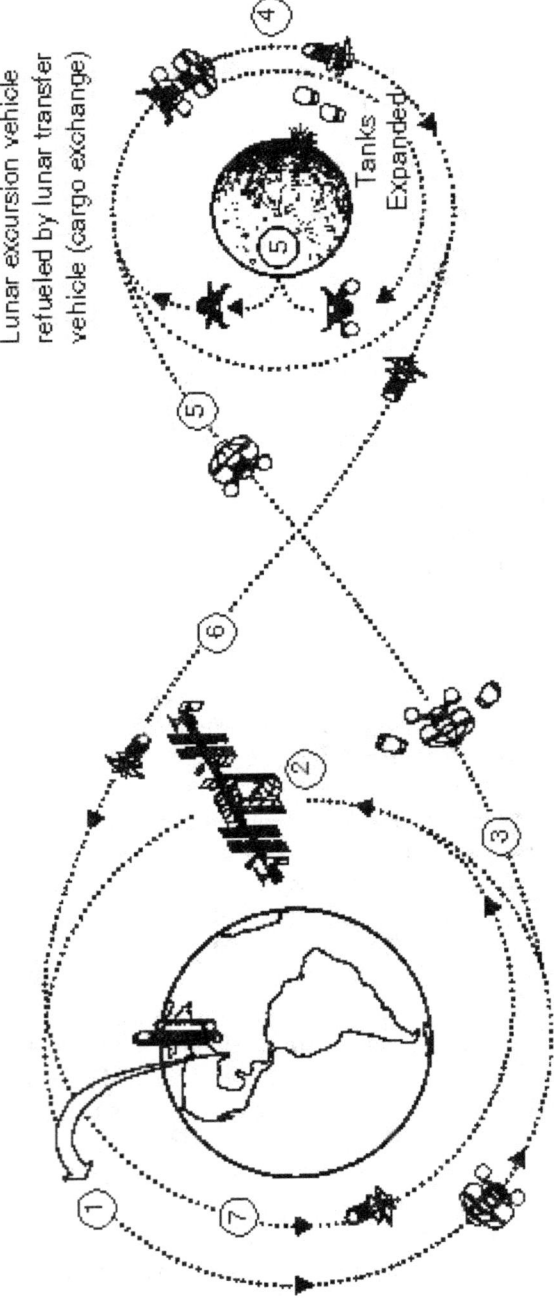

FIGURE 23. *Lunar Mission Profile*

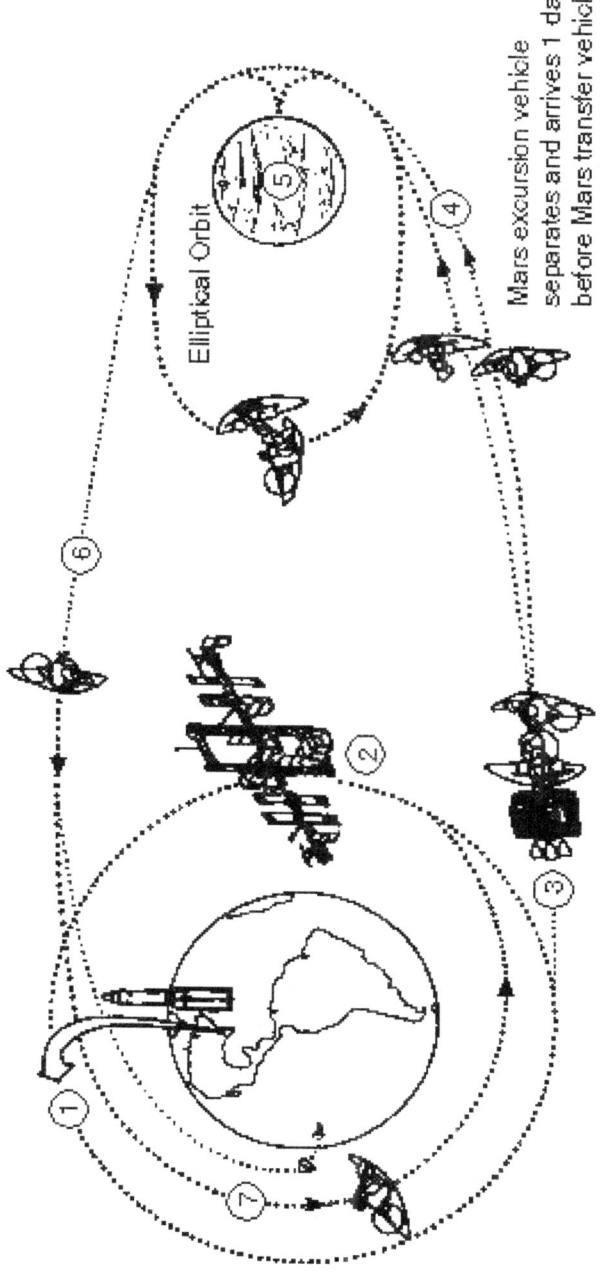

FIGURE 24. Mars Mission Profile

1973, or derived from ongoing work on electric propulsion thruster systems. Basic research and new technologies would be required to develop some of the essential service and maintenance systems.[38]

Thus, extended habitation on the Moon or a trip to and from Mars involved substantially more than launching another Apollo-type space vehicle. Apollo and the Shuttle represented relatively primitive machines and technology compared to the requirements for the Space Exploration Initiative. As was true with the decision to put a man on the Moon in the decade of the sixties, the engineering and the scientific community generally regarded the impediments, the difficulties, and the unknown as a challenge rather than a deterrent. No wonder, as Aaron Cohen noted, the 90-day study itself had generated a new enthusiasm, dedication, and excitement within the Agency.[39] But the proposed programs required a sustained, long-range, and continuing commitment. Whereas the Apollo, Shuttle, and even the space station were programs that could nominally be attained within a decade, a lunar base and a Mars expedition required many decades and more total resources than space programs had yet absorbed.

Despite what seemed to be real progress, uncertainties and doubts abounded. In the spring of 1990, NASA awarded the Operations Support Contract, including mission operations support, facility operations, and flight crew training for the space station and other programs to Rockwell Space Operations Company of Houston. Most of the contract management tasks for the 10-year $814 million contract were assigned to JSC Mission Operations Directorate. Rockwell subcontractors included Barrios Technology, Inc., Bendix Field Engineering Corporation, Omniplan Corporation, Science Applications International, Systems Management American Corporation, and UniSys-Air Defense and Space Systems Division. New jobs generated under the contract in the Houston-Clear Lake area were expected to rise from 200 in 1990 to 1450 by 1996. In anticipation that there would soon be other space vehicles added to the launch fleet, the Shuttle lost its old identity as the National Space Transportation System (NSTS) and became simply the Space Shuttle.[40]

In April, the Shuttle *Discovery* (STS-31) was mated to its tanks and rockets for a scheduled April launch. *Discovery* would carry the long awaited and much heralded Hubble Space Telescope into orbit. With the Hubble telescope, "a new era of astronomy and a new awareness of how humans fit in the cosmos will begin." After a number of "glitches," *Discovery* lifted from its pad on April 25 and placed the Hubble telescope in orbit. During the *Discovery* launch, *Columbia* (STS-35), carrying the ASTRO-1 ultraviolet astronomy telescope and a Broad Band X-ray Telescope, moved slowly aboard its crawler transporter to the adjoining launch pad for a May 16 launch. But problems with valves and freon coolant loops and hydrogen leaks forced repeated delays, until *Columbia* was rolled back to the Vehicle Assembly Building for more thorough checks. Meanwhile, the world waited expectantly for a new and brighter view of the cosmos while astronomers and technicians began targeting and focusing the Hubble telescope.[41]

By mid-summer the *Columbia* had not yet flown, and the source of its hydrogen leaks could not be located. Worse, the Hubble telescope simply could not focus the way it was supposed to focus—its primary mirror was flawed. Doubts and uncertainty grew greater. Congress stripped $300 million from the lunar-Mars initiative. But President Bush remained a supporter and asked Congress to raise NASA funding for 1991 to a record $15.2 billion, an increase of almost 25 percent over that of 1990. In an effort to help further resolve questions about America's future in space, in July, Vice President Dan Quayle, as head of the National Space Council, created a committee headed by Norman B. Augustine, Chief Executive Officer of Martin Marietta Corporation, to investigate and recommend to the Vice President, through the administrator, programs and approaches by which NASA might implement the U.S. space program in the years ahead.[42]

The Advisory Committee on the Future of the U.S. Space Program, as it came to be called, included scientists, engineers, former astronauts, business leaders and former Congressmen. Augustine's committee began its work in August, at a time when a number of external events began to intrude significantly on NASA's operations, and, indeed, on the world. In August, Iraqi armies directed by Saddam Hussein invaded and occupied neighboring Kuwait. President Bush and the United Nations responded by sending American and international forces to Saudi Arabia. The activation of reserve units immediately began to affect JSC employees. Of less traumatic but threatening proportions, Congress' failure to ratify a new fiscal year budget on time threatened to invoke the Gramm-Rudman-Hollings deficit reduction program which would require a 31.9 percent budget cut by NASA and nondefense government agencies, and more immediately result in the furlough of civil service employees until Congress did approve a budget bill.[43]

If this was not enough, NASA continued to be plagued with technical problems. Administrator Richard Truly tried to reassure NASA employees, who felt somewhat abused and confused by the problems, rising hostility in the press, the threat of foreign war, and budgetary and job uncertainties. It seemed to be something of an understatement when he explained to NASA employees in a radio broadcast from his office that "some things haven't gone right this summer." While engineers struggled to locate *Columbia's* hydrogen leak, *Discovery* (STS-41), with four previously scheduled launches already scrubbed because of such things as bent electrical connector pins and freon pressure losses, was readied for an October 5 lift-off. It made it—one day late—but on a near-perfect flight the flight crew launched the Ulysses probe bound for the planet Jupiter, conducted a variety of experiments, and returned to Earth.[44]

Crews now readied both *Atlantis* (STS-38) and *Columbia* (STS-35) on their launch pads at Kennedy Space Center for pre-Christmas launches. Payload and weather problems forced a week's delay of the *Atlantis* DoD mission, but the mid-November launch was routine. Finally, on December 2, after three previous failed launch attempts, *Columbia* carried her crew of seven into orbit and completed the long-delayed science missions. A special investigating team had discovered a crimped or damaged seal in two different engines, and tightened connections and checked all seals.[45] Since *Challenger,* the prelaunch checkout of each Shuttle, if it had not been so before, was meticulous, thorough, exhausting, time-consuming, and costly—but effective. There had been problems, but as Truly stated, those were problems uncovered by NASA.

Unfortunately, unlike the Shuttle's case, the problems with the Hubble telescope had not been uncovered prior to its launch. It turned out that the manufacturer of the Hubble mirror had tested the mirrors using an instrument which was itself defective; and NASA contract managers, who were concentrating on confining costs, forewent additional tests that could have revealed the flaws.[46] NASA began to actively consider a repair mission, perhaps as early as 1993. Meanwhile, the flawed Hubble did produce important new images and data of the universe. But the NASA image was blemished, as was the Hubble mirror.

On December 17, 1990, not long after *Columbia*'s return, Norman Augustine delivered the report of the Advisory Committee on the Future of the U.S. Space Program to Administrator Truly. There had obviously been many reports during NASA's 30 years of operation, such as *The Next Ten Years in Space, 1959-1969*, completed in 1959 by the staff of the Select Committee on Astronautics and Space Exploration, and more recently the Rogers report on the *Challenger* accident, the Ride report on *Leadership* in the post-*Challenger* era, and Cohen's report on a lunar-Mars initiative; but there was a growing perception within NASA and at JSC that the Augustine report might indeed be the charter for NASA's tomorrow. The report offered a very brief, candid, pragmatic, down-to-earth analysis of what the United States and NASA had done and might yet do in space.[47]

NASA's current problems needed to be set in the context of space history. The *Challenger* failure, hydrogen leaks aboard the *Columbia* and other Shuttles, cost overruns, and the Hubble aberration problem derived from errors or situations developing 5, 10, or more years ago. Spaceflight is not and has never been risk free. Of 37 satellite launches attempted before 1960, less than one-third were successful. Ten of the first eleven unmanned probes to the Moon failed. Three astronauts died in the AS-204 fire. A tank explosion on Apollo 13 damaged the spacecraft and jeopardized the mission. During the few months surrounding the *Challenger* accident, a Delta rocket, an Atlas-Centaur, two Titans, a French Ariane-2, and a Soviet Proton were lost.[48] Trouble-free, risk-free Apollo or Shuttle flights never existed.

There has been a distinct lack of consensus about what the goals of the American space program are and how they should be accomplished. Most people seemed to support a space program, but no two people agreed on what that program should be. Some urged robotic missions only as an efficient, low-cost approach; others argued that human involvement is the essence of exploration; still others advised commercialization of the space effort; and others stressed the pure scientific, research goals of spaceflight—"only to be challenged in turn to prove the tangible value of studies in astronomy." The committee agreed that NASA was trying to do too much—that it was overcommitted, perhaps in response to the very disparate pressures upon it. Changing project budgets demoralized both those doing the work and those paying the bills. Civil service personnel policies were incompatible with the need to maintain within NASA a "leading-edge, aggressive, confident, and able workforce of technical specialists and technically trained managers." The tendency for projects to grow in "scope, complexity, and cost," had to be countered. Space projects are very unforgiving of any neglect or human failures. Finally, the program was overly dependent on the Space Shuttle.[49] Even as that analysis was being drawn, conditions within the space industry and NASA were changing rapidly as alternative space programs developed and diminishing emphasis on defense industries turned engineer talents increasingly to civilian space interests.

Given these parameters, the NASA Advisory Committee concluded that the Nation's space effort must continue to be directed by NASA, because it contained "by far the greatest body of space expertise in any single organization in the world." And what should be the U.S. space program? "What *should* we afford?" During the Apollo program, NASA spending accounted for 0.8 percent of gross national product (GNP), 4.5 percent of total federal spending and 6 percent of discretionary spending. From 1975 through 1990, NASA spending, equaled about .25 percent of GNP, 1 percent of federal expenditures, and 2.5 percent of discretionary spending.[50]

With the caveat that NASA cannot do everything, the report stated that the Agency should give funding priority to the space science program as the "fulcrum of the entire civil space effort." Its mission-oriented programs should support two major undertakings—a Mission to Planet Earth and a Mission from Planet Earth. The former would focus on climate and environmental issues that affect the quality of life on Earth, the latter would be focused on the exploration of space. It was the latter that had represented the most costly part of the civil space program.[51] As previously mentioned, between 1975 and 1990, approximately 85 percent of JSC resources were allocated to the Shuttle.

During the past decade, the Shuttle and manned spaceflight had been central to the controversy surrounding a space program. The advisory committee rejected unanimously the option that a space program should dispense with human flight. But what should be the objectives or projects of a manned space program? Not the cargo flights of a Shuttle to and from near-Earth space to deliver satellites or cargoes that might better be carried by unmanned vehicles! The advisory committee concurred with President Bush's lunar-Mars initiative, but counseled a "significant new approach in the planning of human space exploration." Rather than schedules, such as a landing on the Moon in this decade, a program with the long-term objective of the human exploration of Mars should be tailored to the availability of funding. Moreover, such an initiative should be a shared program of a consortium of nations. Space Station *Freedom* cannot be justified as a (nonbiological) science laboratory or as an essential transportation mode. Rather its validity derives from the contributions it can make as a life sciences laboratory and as a microgravity experiment station. This being the case, the space station can be "simplified, reduced in cost, and constructed on a more evolutionary modular basis." (Subsequently, in fiscal year 1991 Congress cut the Space Station *Freedom* budget by $500 million, directed that the planned facility be scaled back in scope, and advised NASA to cut approximately $6 billion from proposed station spending through 1997.)[52]

The Augustine Report recognized the technology base and the existing space transportation system as the two fundamental building blocks for extending the human presence in space. "NASA simply must take those steps needed to enhance the Shuttle's reliability, minimize wear and tear, and enhance launch schedule predictability." Cost reductions are "desirable," but "secondary to the preceding objectives." Costs could also be contained and efficiencies gained by providing a predictable and stable funding—which is wholly outside the province of NASA and dependent upon the support of the administration and Congress. The committee suggested diverting the funds proposed for an additional Shuttle orbiter to the construction of a new unmanned heavy lift launch vehicle, support by NASA to help nurture a commercial space industry, and a continuing national effort to enhance the Nation's mathematics and science programs.[53]

But space activity is inherently difficult and complex, the committee admonished: "As we labor under such challenges, we should insist upon excellence" and "strive for perfection." But we should be prepared for the occasional failure. The Nation "has no

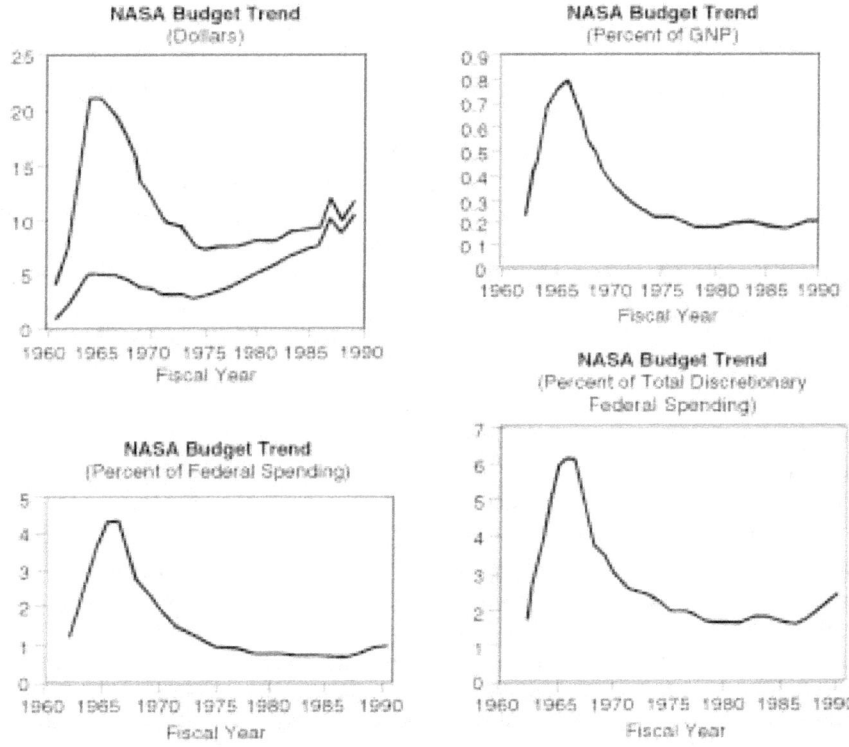

FIGURE 25. *NASA Budget Trends, 1960 to 1990.*

business in space if it places too great a premium on not making errors, and on "ridiculing those who strive but occasionally fail." The Augustine Committee made 15 fairly specific recommendations regarding goals, programs, costs, and management.[54]

It urged the establishment of an Executive Committee to the National Space Council including the Administrator of NASA, major civil service reforms (or specific exemptions affecting NASA employees), a review of the mission of each of the NASA centers so as to consolidate and refocus their efforts with a minimum of overlap, a reorganization of the Headquarters administrative structure, and the retention of an independent cost analysis group. The report recommended against multicenter projects, but recommended that when they could not be avoided, an "independent project office reporting to Headquarters be established near the center having the principal share of the work for that project"—and that this office have a systems engineering staff and full budget authority. This seemed to be an endorsement of the lead center management system which had been virtually abandoned for the Space Station *Freedom* program.[55]

Moreover, the report said "NASA should concentrate its hands-on expertise in those areas unique to its mission." Contract monitoring is best accomplished by systems managers with pertinent experience. NASA program offices, in effect, take the place of traditional prime contractors in defense and other government contracts, and contract work stresses performance rather than specifications—and can involve considerably more government people.[56]

Many JSC personnel believed that effective program management had more to do with attitude than organization. It involved primarily the tradition of advisory/participatory management inherited through the old NACA style and its intellectual predecessor, the National Academy of Sciences, which like NACA and then NASA had been conceived as a response to a real or imagined technological and economic threat. In the case of the National Academy of Sciences, founded in 1863, the threat was the Industrial Revolution in Europe. In NACA's case, it was competition in aviation technology; and in NASA's, a perceived national crisis engendered by Sputnik. NACA/NASA programs required the cooperation and participation of industry, academia and government, and because they also involved research and development rather than the fabrication of known structures, they further necessitated a cooperative, participatory management style.[57]

But there were other ramifications to space ventures that went beyond the role of a government agency producing a product—be it a shuttle or a space station. Exploration involves more than a product or development, but rather relates to the discovery of new frontiers and the use by society of the resources derived from that frontier. In this sense, the

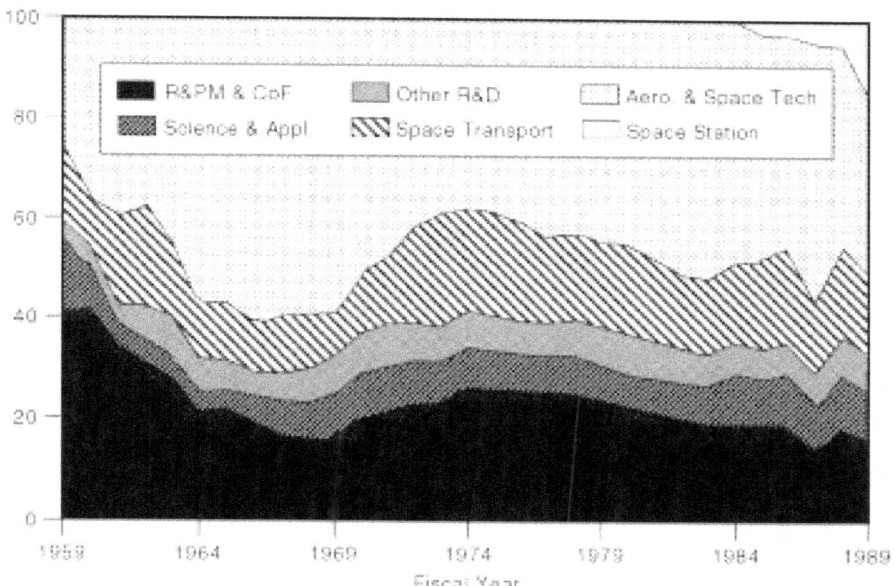

Source: Congressional Budget Office.

FIGURE 26. NASA Budget, 1959 to 1989

American frontier in space *did* reflect those earlier experiences by which the United States Government actively supported expansion into new frontiers. The Lewis and Clark Expedition (1803-1806), U.S. Coastal and Geodetic Surveys (1807, 1808, 1832), the U.S. Exploring Expedition (1838-1842), transcontinental railroad construction (c. 1863-1890), the Panama Canal and even the Federal Highways Act of 1915, the Interstate Highways Act of 1952, and the establishment of the Atomic Energy Commission fit the model of federal-private-scientific cooperation, and required the commitment of a considerable proportion of the federal budget.[58] Unlike the government's commitment to space, however, these earlier projects were considerably more finite in scope, purpose, resources used, and product derived. Space seemed to have no boundaries.

Epilogue

Earthrise! A phenomenon never before observed by humankind suggests the profound technical and philosophical impact of the U.S. manned spaceflight program upon people of the Earth.

Although the book ends, the story continues. The repercussions of America's ventures in space will ripple through time and space to affect life on Earth, perhaps for all time. The ripples will be somewhat greater or lesser depending upon future funding, continuing programs and achievements, but should NASA or JSC come to the end of their time, what has been done in the past will have a continuing effect on humankind. During the past three decades, spaceflight, thought about space under the leadership of JSC, and new technology engendered by space have contributed to epochal changes in human history.

Although there were preliminaries—and the luminaries such as Robert H. Goddard, Konstantin E. Tsiolkovsky, and Hermann J. Oberth—America's ventures in space really began as a response to the challenge or threat (real or imagined) caused by the Soviet's successful launch of Sputnik I. The Space Task Group headed by Robert R. Gilruth was formed soon afterward from Langley Aeronautical Laboratory's Pilotless Aircraft Research Division.

Upon the approval of the National Aeronautics and Space Act by Congress in 1958, the older National Advisory Committee for Aeronautics (NACA) and its research centers were redirected as the research arm for NASA, while the STG became the nucleus of a new multicenter manned spaceflight program. The Space Task Group was enriched with the addition of Canadian and British engineers from Canada's recently closed AVRO Aircraft, Ltd., a subsidiary of Britain's A.V. Roe Company. Subsequently, a large contingent of German rocket scientists, headed by Wernher von Braun and working with the Army Ballistic Missile Agency, joined the NASA organization. Air Force, Navy and civilian engineers soon swelled the ranks of NASA's civil service personnel.

STG personnel with the AVRO engineers became the core of the Manned Spacecraft Center in Houston, Texas. Von Braun's group comprised the essential ingredient of the Marshall Space Flight Center organized in Huntsville, Alabama, while elements from both Marshall and MSC, with Air Force and other military personnel, were the basic ingredients of what became the Kennedy Space Center. Goddard Space Flight Center, Stennis Space Center, MSC, Marshall Space Flight Center and Kennedy Space Center, with other laboratories and test operations, constituted the manned space and operations arm of NASA.

MSC, renamed the Lyndon B. Johnson Space Center in 1973 for one of Texas' leading statesmen and an architect of the national space program, became the lead center in developing the design, flight systems, and crew training for the Mercury, Gemini, Apollo and Shuttle programs. Apollo-Soyuz missions were distinctive, both in the use of diverse space hardware and in facilitating cooperation between two nations enmeshed in cold war. Johnson Space Center provided astronauts, life support systems, and flight operations for Skylab missions. In the 1990's, JSC supports Space Shuttle flight missions, Space Station *Freedom*, and the development of a lunar base that is integral to a manned mission to Mars.

During its three decades of existence, JSC developed a distinct culture and management style within the national aerospace community. That style heavily reflects the imprint of Dr. Robert R. Gilruth, the head of the original STG and first Director of MSC. It involved primarily the tradition of advisory/participatory management inherited from NACA. It was sharpened by a strong and continuing emphasis on hands-on engineering and technical accuracy, by the development of an effective and sophisticated structure of systems engineering, and by a deep sense of personal commitment to the program and to participants in the program.

NASA and space programs have directly affected the private sector of the national economy, while JSC has impacted heavily on the Houston and Texas economy. Space is largely a private business conducted by private contractors managed by NASA engineers. Pioneering the space frontier is inextricably tied to the expansion of those new, intensely earthbound frontiers in technology, organization, and management being pioneered by NASA and JSC in collaboration with American business and educational systems.

MSC/JSC Directors

Suddenly, Tomorrow Came...

MSC/JSC Directors

Dr. Robert R. Gilruth
1962 to 1972

Dr. Christopher C. Kraft, Jr.
1972 to 1982

Gerald Griffin
1982 to 1986

Jesse W. Moore
1986

Dr. Aaron Cohen
1986 to Present

Reference Notes

Chapter 1

1. Lyndon Baines Johnson, *Vantage Point: Perspectives of the Presidency, 1963-1969* (New York: Holt, Rinehart and Winston, 1971), 272; Memo, Excerpts from Speeches, Special Committee on Space and Astronautics, Senate Papers, Box 357, Lyndon B. Johnson Library.
2. Preliminary History of the National Aeronautics and Space Administration During the Administration of President Lyndon B. Johnson, November 1963 to January 1969. Final Edition, NASA, January 15, 1969, I-6, I-7.
3. Documents in the History of NASA: An Anthology, Preliminary Edition, NASA History Office, NASA, Washington, D.C., June 1975, 41,45.
4. Interview, T.H. Baker with Edward C. Welsh, July 18, 1969, Oral History Collection, AC74-268, LBJ Library; and see Walter A. McDougall, . . . *the Heavens and the Earth: A Political History of the Space Age* (New York: Basic Books, Inc., 1985), 141-145.
5. Robert H. Ferrell, ed., *The Eisenhower Diaries* (New York: W.W. Norton, 1981), 217-359; R. Alton Lee, *Dwight D. Eisenhower: Soldier and Statesman* (Chicago: Nelson-Hall, 1981), 217-291.
6. Memo, Papers Relating to the Armed Services Preparedness Investigating Subcommittee, Other People's Statements on Satellites, Senate Papers, Box 356, LBJ Library (hereafter cited Other People's Statements on Satellites).
7. Other People's Statements on Satellites.
8. Other People's Statements on Satellites; Letter to "My Dear Mr. Secretary," October 11, 1957, Papers Relating to the Armed Services Preparedness Investigating Subcommittee, Senate Papers, Box 355, LBJ Library; Enid Curtis Bok, "The Establishment of NASA: The Political Role of Advisory Scientists," Organization Subseries, JSC History Office.
9. Other People's Statements on Satellites; Memo, Horowitz to Johnson, October 11, 1957, Senate Papers, Box 355, LBJ Library.
10. Other People's Statements on Satellites; Preliminary History of NASA, I-7.
11. Other People's Statements on Satellites.
12. Other People's Statements on Satellites.
13. Inquiry into Satellite and Missile Programs, Hearings before the Preparedness Investigating Subcommittee of the Committee on Armed Services, 85th Cong., 1st and 2nd Sess., Part I: 1-2 (hereafter cited Preparedness Subcommittee Hearings).
14. Preparedness Subcommittee Hearings, I:1-2.
15. Preparedness Subcommittee Hearings, I:1-2.
16. Preparedness Subcommittee Hearings, I:3.
17. Preparedness Subcommittee Hearings, I:1-2.
18. Johnson, 1971, 271.
19. Preparedness Subcommittee Hearings, I:4-20, 21-22, 23-57.
20. Preparedness Subcommittee Hearings, I:57.
21. Preparedness Subcommittee Hearings, I:57-89.
22. Preparedness Subcommittee Hearings, I:111-141.
23. Neil McAleer (with introduction by Chuck Yeager), *The Omni Space Almanac: A Complete Guide to the Space Age* (New York: World Almanac, 1987), 11.
24. McAleer, 1987, 10-12; McDougall, 1985, 76-77.
25. McAleer, 1987, 10; McDougall, 1985, 76-77.

26. David S. Akens, *Historical Origins of the George C. Marshall Space Flight Center*, MSFC Historical Monograph No. 1, MSFC, NASA, Huntsville, Alabama, December 1960, 23-24; McDougall, 1985, 77-78.

27. Akens, 1960, 24-26.

28. Akens, 1960, 26-28; Preparedness Subcommittee Hearings, I:578-579.

29. Akens, 1960, 26-28.

30. Documents in the History of NASA: An Anthology, 1975, 42.

31. Documents in the History of NASA: An Anthology, 1975, 27-29; Jane Van Nimmen and Leonard C. Bruno with Robert L. Rosholt, *NASA Historical Data Book*, NASA SP-4012 (Washington, D.C.: 1988), I:423-424.

32. Van Nimmen, 1988, I:423-424; Preparedness Subcommittee Hearings, I:539-577.

33. Documents in the History of NASA, An Anthology, 1975, 29-39.

34. James R. Hansen, *Engineer in Charge: A History of the Langley Aeronautical Laboratory, 1917-1958*, NASA SP-4305 (Washington, D.C.: NASA Scientific and Technical Office, 1987), 271-309.

35. Hansen, 1987, 1-22.

36. Hansen, 1987, 9.

37. Hansen, 1987, 16-18.

38. Hansen, 1987, 301-302, 385-389.

39. Hansen, 1987, 385-389; and see NASA/JSC Biographical Data Files.

40. "Robert Rowe Gilruth," Biographical Data Subseries, JSC History Office; Interview, Robert B. Merrifield with Paul Purser, May 17, 1967, MSC, Houston, Texas, in Oral History Series, JSC History Office; Hansen, 1987, 262-267, 276-278, 385-389.

41. "Statement to the Wire Services, December 16, 1958," and Memorandum, "George to Senator Johnson," in Papers Relating to the Armed Services Preparedness Investigating Subcommittee, Senate Papers, Box 355, LBJ Library.

42. McAleer, 1987, 13; Preliminary History of NASA, I-9,10, and see notes, I-71 to 74.

43. Memorandum, Clinton P. Anderson, Papers Relating to the Special Committee on Space and Astronautics, Senate Papers, Box 357, LBJ Library.

44. Staff Report of the Select Committee on Astronautics and Space Exploration, *The Next Ten Years in Space, 1959-1969*, 86th Cong., 1st Sess., House Doc. No. 115 (Washington, D.C.: Government Printing Office, 1959); Arnold S. Levine, *Managing NASA in the Apollo Era*, NASA History Series, NASA SP-4102 (Washington, D.C.: 1982), 12-13.

45. Johnson, 1972, 276-277.

46. "Introduction to Outer Space, The White House, March 26, 1958," in Documents in the History of NASA, An Anthology, 1975, 51.

47. Introduction to Outer Space, 50-59.

48. Memorandum for the Secretary of Defense and Chairman, The National Advisory Committee for Aeronautics, April 2, 1958, Special Committee on Space and Astronautics, Senate Papers, Box 357, LBJ Library. (Note: NASA Historical Data Book, I:4, incorrectly notes that the bill was submitted to Congress on April 14.)

49. Senate papers, Box 357, LBJ Library.

50. NASA Act 1958, 85th Congress, Second Session, Report 2166, July 15, 1958.

51. Astronautics and Space Exploration, Hearings before the Select Committee on Astronautics and Space Exploration, 85th Cong., 2nd Sess., 401-410.

Suddenly Tomorrow Came . . .

52. Hansen, 1987, 385-86; Letter, Bill Stoney to Dethloff, March 1, 1990.
53. Memorandum, "Statement by the President," February 19, 1973; Wright Patman, "The Lyndon Baines Johnson Space Center," *Congressional Record*, Extensions of Remarks, January 26, 1973, E-467, in Organization Subseries, JSC History Office.

Chapter 2

1. Dr. Robert R. Gilruth, "Memoir: From Wallops Island to Mercury, 1945-1958," presented at the Sixth International History of Astronautics Symposium, Vienna, Austria, October 13, 1972, in Vertical Files, JSC History Office, 29 (hereafter cited Gilruth Memoir, 1972.)
2. Interview, Robert B. Merrifield with Abe Silverstein, Washington D.C., January 30, 1969, in Oral History Series, JSC History Office.
3. Gilruth Memoir, 1972, 36-39.
4. Gilruth Memoir, 1972, 34-38; Interview, Merrifield with Silverstein, 1969, 1-3; *Report on the Activities of the Committee on Science and Astronautics*, 86th Cong., 1st Sess., House Doc. No. 32, i-iii, 1; "Authorizing Construction for the National Aeronautics and Space Administration," *Hearings before the Select Committee on Astronautics and Space Exploration*, 85th Cong., 2nd. Sess., August 1, 1958, 17-21.
5. Lyndon B. Johnson, "The Space Age and the Engineer," *American Engineer* (August 1958) V. 28, no. 8:11-14.
6. "Statement of Dr. T. Keith Glennan, Administrator, National Aeronautics and Space Administration before the House Committee on Science and Astronautics," February 2, 1959, and "T. Keith Glennan," Biographical Data, in Olin E. Teague Papers, "Early Organization," Texas A&M University Archives, Box 222. Preliminary History of the National Aeronautics and Space Administration During the Administration of President Lyndon B. Johnson, November 1963 - January 1969, Final Edition, NASA, January 15, 1969, I-15 to I-16; Documents in the History of NASA: An Anthology, Preliminary Edition, NASA History Office, Washington, D.C., June 1975, 104-106.
7. Gilruth Memoir, 1972, 39-40. Note: Members of the joint committee included Gilruth (Chairman), Dr. S.B. Batdorf (ARPA), Dr. A.J. Eggers (NASA), Maxime Faget (NASA), George Low (NASA), Warren North (NASA), Walter Williams (NASA), and Robertson Youngquist (ARPA); Paul E. Purser, "Log for Week of October 6, 1958," Purser Logs to Gilruth Subseries, JSC History Office.
8. Glennan Statement, 1959, 6; Interview, Robert B. Merrifield with Wesley L. Hjornevik, March 9, July 1, and September 9, 1967, Oral History Series, JSC History Office.
9. Gilruth Memoir, 1972, 39-41 and Appendix B.
10. Gilruth Memoir, 1972, Appendix B; Purser, "Log for Week of October 20, 1958," and "Log for Week of October 28, 1958," Purser Logs to Gilruth Subseries, JSC History Office.
11. Gilruth Memoir, 1972, Appendix B.
12. Gilruth Memoir, 1972, 35-45; Interview, Merrifield with Silverstein, 1969; Interview, Robert B. Merrifield with George M. Low, JSC, Houston, Texas, January 9, 14, 1968, February 4, 1969, in Oral History Series, JSC History Office.
13. Interview, Merrifield with Low, 1968.
14. Gilruth Memoir, 1972, 42.
15. Interview Merrifield with Silverstein, 1969; James R. Hansen, *Engineer in Charge: A History of the Langley Aeronautical Laboratory, 1917-1958*, NASA SP-4305 (Washington, D.C.: 1987), 385.
16. Interview, Merrifield with Silverstein, 1969; Interview, Merrifield with Low, 1968; Gilruth Memoir, 1972, 36-41.

Reference Notes

17. Gilruth Memoir, 1972, 43-44; Biographical, "Charles J. Donlan, Associate Director, Langley Research Center," Biographical Data Subseries, JSC History Office.
18. Gilruth Memoir, 1972, 43.
19. Gilruth Memoir, 1972, 44; Preliminary History of NASA, 1969, I-18.
20. Gilruth Memoir, 1972, 42-43.
21. Gilruth Memoir, 1972, 44-45.
22. Gilruth Memoir, 1972, 47.
23. Interview, Henry C. Dethloff with R. Bryan Erb, JSC, Houston, Texas, February 15, 1989, Oral History Series, JSC History Office.
24. Interview, Dethloff with Erb, 1989; Interview, Henry C. Dethloff with Rodney Rose, Wimberly, Texas, February 24, 1989, Oral History Series, JSC History Office.
25. Interview, Dethloff with Rose, 1989.
26. Interview, Dethloff with Rose, 1989.
27. Interview, Dethloff with Erb and with Rose, 1989.
28. Interview, Dethloff with Rose and with Erb, 1989.
29. Interview, Dethloff with Rose and with Erb, 1989; Memorandum, "English/Canadians from AVRO" in Organization Subseries, JSC History Office.
30. Interview, Merrifield with Silverstein, 1969; and with Low, 1968; Interview, Henry C. Dethloff with Oran Nicks, College Station, Texas, February 17, 1989.
31. Interview, Merrifield with Silverstein, 1969.
32. David S. Akens, *Historical Origins of the George C. Marshall Space Flight Center* (Washington, D.C.: NASA History Office, December 1963), 67-83.
33. Akens, 1963, 67-83.
34. Preliminary History of NASA, 1969, I-23 - I-24; Interview, Oran W. Nicks with T. Keith Glennan, May 11, 1982, Reston, Virginia, personal files of Oran Nicks, College Station, Texas; and see Ken Hecnler, *Toward the Endless Frontier: History of the Committee on Science and Technology, 1959-1979* (Washington, D.C., U.S. House of Representatives: Government Printing Office, 1980), 1-1073.
35. Preliminary History of NASA, 1969, I-23 - I-24.
36. Lyndon B. Johnson, *The Vantage Point: Perspectives of the Presidency, 1963-1969* (New York: Holt, Rinehart & Winston, 1971), 278-279; Ken Belieu, Memorandum for Senator Johnson, "Governmental Organization for Space Activities," December 17, 1960, and "Space Problems," December 22, 1960, in Vice Presidential Security File, Box 17, LBJ Library.
37. Johnson, 1971, 278-279.
38. Johnson, 1971, 278.
39. Preliminary History of NASA, 1969, I-26; Interview, Dethloff with Nicks, 1989.
40. Johnson, 1971, 279-280.
41. Johnson, 1971, 280.
42. Johnson, 1971, 280-281; Preliminary History of NASA, 1969, I-28 - I-32; Interview, T.H. Baker with Dr. Edward C. Welsh, July 18, 1969, Oral History Collection, LBJ Library, 12-16.
43. Johnson, 1971, 280-281.
44. Preliminary History of NASA, 1969, I-30 - I-32.
45. Preliminary History of NASA, 1969, I-33 - I-34.

Suddenly Tomorrow Came . . .

46. Johnson, 1971, 280-281.
47. *The Practical Values of Space Exploration: Staff Study of the Committee on Science and Astronautics*, House of Representatives, 87th Cong., lst Sess., August 17, 1961 (Washington, D.C.: Government Printing Office, 1961), 1- 51.
48. Interview, Merrifield with Silverstein, 1969.
49. Preliminary History of NASA, 1969, I-35; Interview, Dethloff with Nicks, 1989; Arnold S. Levine, *Managing NASA in the Apollo Era*, NASA SP-4102 (Washington, D.C.: 1982), 18-20, 307-312.
50. Howard Benedict, *NASA: A Quarter Century of Achievement* (The Woodlands, Texas: Pioneer Publications, Inc., 1984), 10-11.
51. Benedict, 1984, 10-11; Gilruth Memoir, 1972, 49-50.
52. NASA News Release 61-207, September 19, 1961.

Chapter 3

1. Paul E. Purser, "Log for Week of December 4, 1958," Purser Logs to Gilruth Subseries, JSC History Office.
2. Interview, Robert B. Merrifield with Abe Silverstein, Washington, D.C., January 30, 1969; Interview, Robert B. Merrifield with George M. Low, JSC, January 9, 14, 1968, February 4, 1969, Oral History Series, JSC History Office; Robert R. Gilruth, Memorandum for Staff, Organization Subseries, JSC History Office; Courtney G. Brooks, James M. Grimwood, and Loyd S. Swenson, Jr., *Chariots for Apollo: A History of Manned Lunar Spacecraft*, NASA SP-4205 (Washington, D.C.: 1979), 7-19. Note: George Low recalled that the decision to create a separate center for the STG was made by Abe Silverstein in Silverstein's office, with he and Wesley Hjornevik present. Hjornevik indicates that the decision was initiated by Glennan and Dryden.
3. Interview, Merrifield with Low, 1968 and 1969, 3-4; Interview, Merrifield with Wesley L. Hjornevik, March 9, July 1, and September 9, 1967, 6-10; Memorandum for Associate Administrator, November 18, 1960, Organization Subseries, JSC History Office.
4. (Note: Hjornevik identifies Langley as one of the locations precluded by Glennan, but this does not appear in *Chariots for Apollo*); see Brooks, Grimwood, and Swenson, Jr., 1979, 7-8, 19-20. (Note: Members of the "Manned Lunar Landing Committee" now included Eldon Hall, Oran W. Nicks, Alfred M. Mayo, Ernest O. Pearson, Max Faget, and Heinz H. Koelle).
5. Brooks, Grimwood, and Swenson, 1979, 19.
6. Brooks, Grimwood, and Swenson, 1979, 19; Memorandum, "Manned Spacecraft Development Center, Organizational Concepts and Staffing Requirements," Organization Subseries, JSC History Office.
7. Interview, Merrifield with Low, 1968 and 1969.
8. N. Philip Miller, "Memorandum for the Record," Organization Subseries, JSC History Office.
9. Interview, Robert B. Merrifield with Max Faget, October 16, 1967; Interview, Merrifield with Paul G. Dembling, September 25, 1969; Interview, Merrifield with Col. John A. Powers, November 9, 1968, Oral History Series, JSC History Office; Interview, Henry C. Dethloff with Paul Purser, David Lang and Bill Petynia, January 18, 1989; and see correspondence in Olin E. Teague Papers, Texas A&M University Archives, Box 222.
10. Memorandum for All Mercury Personnel, November 23, 1960; Memorandum for Staff, September 16, 1960; T. Keith Glennan to Marvin G. Miles, August 26, 1960; "Astronaut Plans Well Conceived," *Los Angeles Times* (August 21, 1960), Organization Subseries, JSC History Office.
11. "Space Task Group, Langley Field, Virginia, Organizational History," Organization Subseries, JSC History Office.

Reference Notes

12. Memorandum for Staff, January 23, 1961, "Progress Report by the House Committee on Science and Astronautics," Organization Subseries, JSC History Office.
13. Olin E. Teague Papers, Box 222; James E. Webb, Memorandum for the President, September 14, 1961, Organization Subseries, JSC History Office.
14. Webb Memorandum for the President, 1961.
15. Interview, Robert B. Merrifield with I. Edward Campagna, August 24, 1967, Oral History Series, JSC History Office.
16. Interview, Merrifield with Powers, 1968.
17. Webb Memorandum for the President, 1961.
18. NASA News Release 61-207, September 19, 1961; NASA News Release, 61-189, August 24, 1961; "Memorandum on Manned Space Flight Laboratory," Olin E. Teague Papers, Box 222.
19. Interview, Robert B. Merrifield with Robert R. Gilruth, April 17, 1969, Oral History Series, JSC History Office.
20. Interview, Robert B. Merrifield with Martin A. Byrnes, Jr., December 12, 1967, 18, Oral History Series, JSC History Office.
21. Interview, Merrifield with Byrnes, 1967, 18; Interview, Merrifield with Hjornevik, 1967, 20, 35.
22. *Houston Press* (September 20, 1961); *Houston Chronicle* (September 20, 24, 1961).
23. Interview, Robert B. Merrifield with W.A. Parker, May 16, 1967, Oral History Series, JSC History Office.
24. Interview, Merrifield with Low, 1969.
25. Interview, Merrifield with Powers, 1968.
26. Interview, Robert B. Merrifield with Grace Winn, Houston, Texas, June 3, 1968, Oral History Series, JSC History Office.
27. Interview, Merrifield with Winn, 1968.
28. Interview, Merrifield with Winn, 1968.
29. Interview, Merrifield with Winn, 1968; Interview, Henry C. Dethloff with Max Faget, Houston, Texas, March 15, 1989, Oral History Series, JSC History Office.
30. Interview, Merrifield with Winn, 1968; MSC Announcement 44, April 18, 1962, General Reference Series, JSC History Office; "Comments on the Preliminary draft of Chapter 2," by Lester Sullivan, undated, JSC History Office.
31. Interview, Merrifield with Winn, 1968; Interview, Merrifield with Hjornevik, 1967.
32. "STG Renamed: Will Move," *Space News Roundup* (hereafter cited *Roundup*) (November 1, 1961), JSC History Office 1, 3; MSC Announcement 2, November 1, 1961, General Reference Series, JSC History Office.
33. Interview, Merrifield with Parker, 1967.
34. Interview, Merrifield with Parker, 1967.
35. Interview, Marrifield with Parker, 1967.
36. Interview, Merrifield with Byrnes, 1967.
37. Interview, Merrifield with Byrnes, 1967.
38. Interview, Merrifield with Byrnes, 1967.
39. Interview, Merrifield with Byrnes, 1967.
40. Interview, Merrifield with Byrnes, 1967; Interview, Robert B. Merrifield with James M. Bayne, December 9, 1968.

41. Interview, Merrifield with Hjornevik, 1967; Interview, Dethloff with Faget, 1989; Interview, Merrifield with A. Malcom Lovett, January 21, 1971; MSC Announcement 21, February 26, 1962, General Reference Series, JSC History Office.
42. Interview, Merrifield with Campagna, 1967.
43. Interview, Merrifield with Campagna, 1967.
44. Interview, Merrifield with Campagna, 1967; Interview, Dethloff with Faget, 1989; Memorandum, "File on Gemini for Apollo, June 2, 1962, Organization Subseries, JSC History Office; Interview, Merrifield with Gilruth, 1969.
45. Memorandum, Manned Spacecraft Center for the National Aeronautics and Space Administration, Clear Lake, Harris County, Texas, "Background and History of Development of Master Plan and Architectural Concept for Construction of Facilities," Organization Subseries, JSC History Office.
46. Memorandum, MSC.
47. Memorandum, Paul Purser, "Institutional Planning for the Manned Spacecraft Center," July 22, 1968, Vertical Files, JSC History Office.
48. *Roundup* (June 27, 1962).
49. *Roundup* (September 19, 1962); MSC Announcement 6, September 11, 1962, General Reference Series, JSC History Office.

Chapter 4

1. "Gilruth Cites MSC Progress Despite Difficult Relocation," JSC, Houston, Texas, *Space News Roundup* (hereafter cited *Roundup*) (July 11, 1962).
2. *Roundup*, July 11, 1962.
3. *Roundup* (New York: World Almanac, 1987), 17-18.
4. *Roundup* (May 30, 1962).
5. *Roundup (*May 30, 1962).
6. Barton C. Hacker and James M. Grimwood, *On Shoulders of Titans: A History of Project Gemini*, NASA SP-4203 (Washington, D.C.:NASA, 1977), 1-9.
7. Hacker and Grimwood, 1977, 1-46; Courtney G. Brooks, James M. Grimwood, and Lloyd S. Swenson, Jr., *Chariots for Apollo: A History of Manned Lunar Spacecraft,* NASA SP-4205 (Washington, D.C.: NASA, 1979), 1-39.
8. Interview, Robert B. Merrifield with George M. Low, January 9, 14, and February 4, 1969, Oral History Series, JSC History Office.
9. Interview, Merrifield with Low, 1969.
10. Interview, Merrifield with Low, 1969.
11. Interview, Merrifield with Low, 1969.
12. Interview, Henry C. Dethloff and Oran Nicks with W. Hewitt Phillips and Robert G. Chilton, College Station, Texas, March 24, 1989, Oral History Series, JSC History Office.
13. Enid Curtis Bok, "The Establishment of NASA: The Political Role of Advisory Scientists," presented to the American Association for the Advancement of Science, Philadelphia, 1962, Organization Subseries, JSC History Office.
14. Interview, Henry C. Dethloff with Max Faget, NASA/JSC, March 15, 1989.
15. Interview, Henry C. Dethloff with Paul Purser, January 18, 1989, Oral History Series, JSC History Office.

16. R. Wayne Young, The Utilization of Project Management Concepts at the NASA Johnson Space Center, Unpublished doctoral dissertation, Graduate School of Public Affairs, University of Colorado, 1983, 209-211; Interview, Henry C. Dethloff with Robert Piland, The Woodlands, Texas, April 13, 1989.

17. "Charles J. Donlon," "Walter C. Williams," "Kenneth S. Kleinknecht," Biographical Data Subseries, JSC History Office; James R. Hansen, *Engineer in Charge: A History of the Langley Aeronautical Laboratory, 1917-1958,* NASA SP-4305 (Washington D.C.: 1987), 300-301; "Space Task Group, Langley Field, Virginia, Organizational History," September 1961, 1-5, Organization Subseries, JSC History Office.

18. Interview, Robert B. Merrifield with Wesley L. Hjornevik, March 9, July 1, and September 9, 1967, Oral History Series, JSC History Office; Interview, Henry C. Dethloff with Wesley L. Hjornevik, Wimberley, Texas, February 23, 1989.

19. Interview, Merrifield with Hjornevik, 1967; Interview, Dethloff with Hjornevik, 1989; Interview, Henry C. Dethloff with Paul Purser, David Lang and Bill Petynia, JSC, January 18, 1989, Oral History Series, JSC History Office.

20. Interview, Dethloff with Hjornevik, 1989; and see Wesley L. Hjornevik, "NASA Programs and Their Management," delivered to the Harvard Business School Club of Houston, Texas, January 28, 1964, Organization Subseries, JSC History Office.

21. Interview, Dethloff with Hjornevik, 1989; Interview, Merrifield with Hjornevik, March 9, 1967.

22. Interview, Merrifield with Hjornevik, 1967.

23. MSC Announcement 7, January 15, 1962, General Reference Series, JSC History Office.

24. MSC Announcement 8, January 15, 1962, General Reference Series, JSC History Office.

25. Interview, Robert B. Merrifield with Joseph P. Loftus, Jr., March 16, 1971, Oral History Series, JSC History Office; Interview, Henry C. Dethloff with Joseph P. Loftus, Jr., January 6, 1989; Interview, Dethloff with Piland, 1989.

26. Interview, Merrifield with Loftus, 1971; Interview, Dethloff with Loftus, 1989; Interview, Dethloff with Piland, 1989; and MSC Announcement 8, January 15, 1962, General Reference Series, JSC History Office.

27. Interview, Dethloff with Piland, 1989; Brooks, Grimwood and Swenson, 1979, 1-29; Organizational charts, September 26, 1960, May 23, 1961, Organization Subseries, JSC History Office.

28. Interview, Dethloff with Piland, 1989; Brooks, Grimwood and Swenson, 1979, 7-11; Guidelines for Advanced Manned Space Vehicles prepared by Flight Systems Division, March 1960, Apollo Series, JSC History Office.

29. Interview, Dethloff with Piland, 1989; Brooks, Grimwood and Swenson, 1979, 37-39.

30. MSC Announcement 21, February 26, 1962, General Reference Series, JSC History Office.

31. Memorandum, "MSC Senior Staff Meetings," March 30, 1962; Memorandum, "Meetings," April 9, 1962; Memorandum, "Summer Vacations for Staff," June 20, 1962, Organization Subseries, JSC History Office.

32. Interview, Henry C. Dethloff with Max Faget, July 12, 1989; Interview, Dethloff with Piland, 1989; MSC Announcement 8, January 15, 1962, General Reference Series, JSC History Office.

33. Interview, Dethloff with Piland, 1989; Brooks, Grimwood and Swenson, 1979, 1-29; Organization charts, September 26, 1969; May 23, 1961, Organization Subseries, JSC History Office.

34. Interview, Dethloff with Piland, 1989; Brooks, Grimwood and Swenson, 1979, 7-11; Guidelines for Advanced Manned Space Vehicles prepared by Flight Systems Division, March 1960, Apollo Series, JSC History Office.

35. Interview, Dethloff with Piland, 1989; Brooks, Grimwood and Swenson, 1979, 37-39.

36. MSC Announcement 21, February 26, 1962, General Reference Series, JSC History Office.

37. Interview, Henry C. Dethloff with Dennis Fielder, JSC, July 17, 1989.

Suddenly Tomorrow Came...

38. Memorandum, "MSC Senior Staff Meetings," March 30, 1962; Memorandum, "Meetings," April 9, 1962; Memorandum, "Summer Vacations for Staff," June 20, 1962, Organization Subseries, JSC History Office.
39. Virginia F. McKenzie, "Living in a Space Community," *Space News Roundup* (hereafter *Roundup*) (July 14, 1989); Interview, Robert B. Merrifield with Grace Winn, Houston, Texas, June 3, 1968.
40. *Roundup* (October 3, 1962); *Roundup* (October 17, 1962).
41. *Roundup* (October 17, 1962).
42. *Roundup* (June 26, 1963).
43. Interview, Merrifield with Low, 1969; Interview, Dethloff with Piland, 1989; *Roundup* (March 7, 1962).
44. Interview, Merrifield with Low, 1969; Interview, Dethloff with Piland, 1989; *Roundup* (March 7, 1962); Note: Piland sees Shea as a key combatant with the Golovin Committee which reviewed modes of lunar flight including direct ascent, earth-orbit rendezvous, lunar-orbit rendezvous or lunar-surface mission.
45. Brooks, Grimwood and Swenson, 1979, 56-59; NASA, Office of Management, Management Processes Branch, "The Evolution of the NASA Organization" (March 1985), Organization Subseries, JSC History Office, 21-22.
46. Interview, Merrifield with Low, 1969; Brooks, Grimwood and Swenson, 1979, 127-128; Linda Neuman Ezell, *NASA Historical Data Book*, II, *Programs and Projects 1958-1968,* NASA SP-4012 (Washington, D.C.: 1988), 101-111.
47. *Roundup* (April 18, 1962); *Roundup* (July 10, 1963); Grumman Aircraft Engineering Corporation, Project Apollo Feasibility Study Summary, May 15, 1961, v. IV, Apollo Series, JSC History Office.
48. *Roundup* (May 2, 1962); Ezell, 1988, 56-58; Brooks, Grimwood and Swenson, 1979, 87-166; and see Roger E. Bilstein, *Stages to Saturn: A Technological History of the Apollo/Saturn Launch Vehicles,* NASA SP-4206, (Washington, D.C.: Government Printing Office, 1980); and Charles D. Benson and William B. Faherty, *Moonport: A History of Apollo Launch Facilities and Operations,* NASA SP-4204, (Washington, D.C.: Government Printing Office, 1978).
49. Ezell, 1988, 56-61, 171-194; and see Brooks, Grimwood and Swenson, 1979, 87-188.
50. MSC Announcement 268, November 5, 1963, MSC Announcement 274, November 21, 1963, General Reference Series, JSC History Office; *Roundup* (November 27, 1963).
51. *Roundup* (January 6, 1965).
52. Letter, George E. Mueller to Robert R. Gilruth (October 4, 1964); Memorandum, Robert R. Gilruth to MSC Senior Staff, October 5, 1964, w/enclosure "Presentation by Dr. George E. Mueller before the Senior Staff, MSC, Houston, Texas, October 5, 1964, Organization Subseries, JSC History Office.
53. Letter, Mueller to Gilruth, 1964.
54. Letter, Mueller to Gilruth, 1964.

Chapter 5

1. *Space News Roundup* (hereafter *Roundup*) (April 15, 1964).
2. Barton C. Hacker and James M. Grimwood, *On Shoulders of Titans: A History of Project Gemini*, NASA SP-4203 (Washington, D.C.: 1977); *Welcome to MSC Roundup* (1964).
3. *Welcome to MSC Roundup (*1964).
4. Interview, Robert B. Merrifield with Paul Haney, Houston, Texas, April 8, 1968, Oral History Series, JSC History Office.
5. Interview, Merrifield with Haney, 1968.
6. Interview, Merrifield with Haney, 1968.

Reference Notes

7. Interview, Merrifield with Haney, 1968.
8. *Welcome to MSC Roundup* (1964).
9. Telephone interview, Henry C. Dethloff with Max Faget, December 5, 1989.
10. Linda Neuman Ezell, *NASA Historical Data Book*, II, *Programs and Projects, 1958-1968*, NASA SP-4012 (Washington, D.C.: 1988), 149-170; Hacker and Grimwood, 1977, 1-50
11. Hacker and Grimwood, 1977, 3-5.
12. Hacker and Grimwood, 1977, 1-50; Ezell, 1988, 155-156.
13. James A. Chamberlin, "Biographical Data," Biographical Data Subseries, JSC History Office.
14. Chamberlin, Biographical Data Subseries.
15. Gemini Spacecraft Contract, Financial Management and Cost Documents, JSC History Archive, Woodson Research Center, Rice University, Houston, Texas.
16. Gemini Spacecraft Contract.
17. Hacker and Grimwood, 1977, 57, 64, 76, 95-98; Ezell, 1988, 121, 123, 126-127.
18. Interview, Robert B. Merrifield with Paul Purser, May 17, 1967.
19. Interview, Robert B. Merrifield with Robert G. Chilton, March 30, 1970; Interview, Henry C. Dethloff with Max Faget, March 15, 1989.
20. Interview, Robert B. Merrifield with Joseph P. Loftus, March 16, 1971.
21. Interview, Henry C. Dethloff with Henry Pohl, May 16, 1989.
22. Interview, Merrifield with Purser, 1967; Interview with Merrifield with Loftus, 1971; Interview, Robert B. Merrifield with Christopher C. Kraft, Jr., October 5, 1967.
23. *Roundup* (January 8, 1964).
24. *Roundup*, 1964.
25. Interview, Henry C. Dethloff with Dennis Fielder, July 17, 1989.
26. The interviews available in the Oral History Series of the JSC History Office that particularly relate to the development of the operations and flight control aspects of the center and focus on Mercury and Gemini flights include interviews, Henry C. Dethloff with Dennis Fielder, July 17, 1989; Dethloff with Christopher C. Kraft, Jr., July 12, 1989; Robert B. Merrifield with Fielder, March 21, 1968; Merrifield with Kraft, October 5, 1967; Merrifield with John D. Hodge, March 15 and 18, 1968.
27. Interviews, Dethloff with Fielder and Kraft, 1989; Interviews, Merrifield with Fielder, Kraft, and Hodge, 1967 and 1968.
28. Interview, Merrifield with Hodge, 1968.
29. Interview, Merrifield with Kraft, 1967.
30. Interviews, Dethloff with Fielder and Kraft, 1989; Interviews, Merrifield with Fielder, Kraft, and Hodge, 1967 and 1968.
31. Interviews, Dethloff with Fielder and Kraft, 1989; Interviews, Merrifield with Fielder, Kraft, and Hodge, 1967 and 1968.
32. Interview, Merrifield with Kraft, 1967.
33. Interview, Merrifield with Kraft, 1967.
34. Interview, Merrifield with Kraft, 1967.
35. Interviews, Merrifield with Fielder, Kraft, and Hodge, 1967 and 1968; Interviews, Dethloff with Fielder and Kraft, 1989.

36. Interview, Merrifield with Kraft, 1967; Interview, Merrifield with Hodge, 1968.
37. Interview, Merrifield with Kraft, 1967; Interview, Dethloff with Kraft, 1989; Interview, Merrifield with Hodge, 1968.
38. Interview, Merrifield with Kraft, 1967; Interview, Dethloff with Kraft, 1989; Interview, Merrifield with Hodge, 1968.
39. Interview, Robert B. Merrifield with James C. Elms, July 30, 1968.
40. Interview, Merrifield with Elms, 1968.
41. Interview, Merrifield with Elms, 1968.
42. Interview, Merrifield with Elms, 1968.
43. Interview, Merrifield with Elms, 1968.
44. Interview, Merrifield with Elms, 1968.; *Roundup* (November 13, 1963).
45. *Roundup* (November 13, 1963); and see MSC Announcements, 287-293, January 3, 1964; 64-113, August 11, 1964; 64-148, October 21, 1964; 64-21, February 20, 1964; 64-34, March 12, 1964; 64-52, April 14, 1964; 64-64, April 30, 1964; 64-65, May 1, 1964; 64-66, May 1, 1964, and other related announcements, General Reference Series, JSC History Office.
46. *Roundup* (January 22, 1964); MSC Announcement 64-2, January 17, 1964; 64-5, January 21, 1964, General Reference Series, JSC History Office. After the reorganization, Graves and Frick left MSC. Walter Williams also departed to become Operations Director at NASA headquarters under Dr. George E. Mueller.
47. *Roundup* (December 9, 1964), (January 20, 1965), (February 3, 1965), (March 19, 1965), (April 2, 1965); Ezell, 1988, 149-170.
48. *Roundup* (March 4, 1964), (June 10, 1964), (February 3, 1965).
49. Wesley L. Hjornevik, Assistant Director for Administration, "NASA Programs and Their Management," January 28, 1964, Organization Subseries, JSC History Office.
50. Gemini Spacecraft Contract.
51. "Introduction to Configuration Management for Management Interns," n.d., JSC Management Subseries, JSC History Office; "Apollo Spacecraft Program Office, Configuration Management Plan," March 19, 1965 (Revision B, dated March 15, 1966), Apollo Series, JSC History Office.
52. "Introduction to Configuration Management."
53. "Introduction to Configuration Management."
54. *Roundup* (June 11, 1965).
55. Telephone interview, Joey Pellarin, JSC History Office, with Bishop William Sterling, Episcopal Bishop of Texas, April 10, 1991.
56. Ezell, 1988, 149-170; *Roundup* (August 6, 20, September 3, 17, October 1, 29, December 10, 23, 1965); Interview, Dethloff with Kraft, 1989.
57. Ezell, 1988, 149-170, 184; *Roundup* (February 4, 18, March 4, 18, April 1, 1966).
58. Ezell, 1988, 149-170.
59. Presentation by Dr. Charles Berry, Apollo 20th Anniversary Celebration Speaker's Series, "Planning the Apollo Missions" (not transcribed), July 18, 1989, JSC.
60. Presentations by former flight directors Glynn Lunney, Stephen Bales, Gerald Griffin, and Cliff Charlesworth, Apollo 20th Anniversary Celebration Speaker's Series, "Flying the Apollo Missions" (not transcribed), July 19, 1989, JSC.
61. Hacker and Grimwood, 1977 xv-xx, 383-389.

Reference Notes

Chapter 6

1. John Mecklin, "Jim Webb's Earthy Management of Space," *Fortune* (August 1967), 5-12.
2. For an overview of NASA organization and history see Roger E. Bilstein, *Orders of Magnitude: A History of the NACA and NASA, 1915-1990*, NASA SP-4406 (Washington, D.C.: 1989), and basic center histories including James R. Hansen, *Engineer in Charge: A History of the Langley Aeronautical Laboratory, 1917-1958*, NASA SP-4305 (Washington, D.C.: 1987); Elizabeth A. Muenger, *Searching the Horizon: A History of Ames Research Center, 1940-1976*, NASA SP-4304 (Washington, D.C.: 1985); Richard P. Hallion, *On the Frontier: Flight Research at Dryden, 1946-1981*, NASA SP-4303 (Washington, D.C.: 1984); Clayton R. Koppes, *JPL and the American Space Program* (New Haven: Yale University Press, 1982); Virginia Dawson, *Engines and Innovation, Lewis Laboratory and American Propulsion Technology*, NASA SP-4306 (Washington, D.C.: 1991); and Joseph Adams Shortal, *A New Dimension, Wallops Island Flight Test Range: The First Fifteen Years*, NASA RP-1028 (Washington, D.C.: 1978); Edwin P. Hartman, *Adventures in Research: A History of the Ames Research Center, 1940-1965*, NASA SP-4302 (Washington, D.C.: 1970); and Interview, Henry C. Dethloff with Donald L. Hess, JSC, September 20, 1989; and see *NASA Historical Data Book*, I, *NASA Resources, 1958-1968*, NASA SP-4012 (Washington, D.C.: 1988), 245.
3. *NASA Historical Data Book*, I, 269-539.
4. *NASA Historical Data Book*, I, 269-539.
5. Alfred Rosenthal, *Venture into Space: Early Years of Goddard Space Flight Center* NASA SP-4301 (Washington, D.C.: NASA, 1968); William B. Faherty, *Moonport: A History of Apollo Launch Facilities and Operations*, NASA SP-4204 (Washington, D.C.: Government Printing Office, 1978); Interview, Dethloff with Hess, 1989.
6. James R. Hansen, 1987, 392-393; John Logsdon interview with George Mueller and others, July 21, 1989, JSC; Interview, Henry C. Dethloff with George Mueller, July 21, 1989; Interview, Henry C. Dethloff with Aleck C. Bond and Jerome B. Hammack, JSC, September 14, 1989.
7. Interview, Henry C. Dethloff with Dr. Chris Kraft, Houston, Texas, April 2, 1991.
8. Interview, Dethloff with Kraft, 1991; Interview, Henry C. Dethloff with Oran Nicks, College Station, Texas, October 20, 1989; telephone interview, Dethloff with Joseph P. Loftus, October 20, 1989; James E. Webb, lectures presented in the McKinsey Foundation Lecture Series at the Graduate School of Business, Columbia University, New York, New York; I, Doctrine and Practice in Large Scale Endeavors, May 2, 1968, 4; and see also Goal Setting and Feedback in Large Scale Endeavors, May 9, 1968; and Executive Performance and Its Evaluation, May 16, 1968; Vertical Files, JSC History Office.
9. Arnold S. Levine, *Managing NASA in the Apollo Era*, NASA SP-4102 (Washington, D.C.: 1982), 61-64, 175.
10. Dr. George E. Mueller, "Biographical Data," Biographical Data Subseries, JSC History Office.
11. Interview, Logsdon with Mueller and others, 1989.
12. Interview, Dethloff with Bond and Hammack, 1989.
13. Interview, Logsdon with Mueller and others, 1989; Levine, 1982, 174-175.
14. Interview, Logsdon with Mueller and others, 1989; Levine, 1982, 174-175; Interview, Henry C. Dethloff with Oran Nicks, College Station, Texas, August 22, 1989.
15. Levine, 1982. 174-175; Interview, Dethloff with Bond and Hammack, 1989.
16. Interview, Loyd Swenson and James M. Grimwood with Kenneth S. Kleinknecht, JSC, March 6, 1970, Oral History Series, JSC History Office; critique of preliminary Chapter 6 draft by Rodney Rose, undated, Intaglio Files and Papers.
17. "Apollo Program Development Plan, 15 January 1965," Apollo Series, JSC History Office, xiii, xiv.
18. "Apollo Development Plan," 1965, 1-12, 1-13/1-14.

19. "Apollo Development Plan," 1965, 1-12, 1-13/1-14.; Note: the Panel Review Boards included Crew Safety, Flight Evaluation, Mechanical Design, Instrumentation and Communications, Flight Mechanics, Electrical Systems, Launch Operations, and Flight Operations.
20. Interview, Dethloff with Nicks, 1989.
21. Interview, Dethloff with Bond and Hammack, 1989.
22. Julian Scheer, Washington, D.C., to Paul Haney, MSC Houston, December 2, 1968; George S. Trimble, Deputy Director, MSC, to Julian Scheer, Washington, D.C., December 17, 1968, Apollo Series, JSC History Office.
23. George E. Mueller to Robert R. Gilruth, December 16, 1968, Apollo Series, JSC History Office.
24. Interview, Robert B. Merrifield with George Low, January 9, 14, February 4, 1969, Oral History Series, JSC History Office; Interview, Swenson and Grimwood with Kleinknecht, 1970.
25. Interview, Swenson and Grimwood with Kleinknecht, 1970.
26. Interview, Robert B. Merrifield with Paul Purser, May 17, 1967, Oral History Series, JSC History Office.
27. Alex Roland, ed., *A Spacefaring People: Perspectives on Early Spaceflight,* NASA SP-4405 (Washington, D.C.: 1985), 57-58.
28. Roland, 1985, 57-58.
29. Roland, 1985, 57-58.
30. Interview, Swenson and Grimwood with Kleinknecht, 1970; Interview, Robert B. Merrifield with Max Faget, October 16, 1967, Oral History Series, JSC History Office.
31. Interview, Robert B. Merrifield with Robert R. Gilruth, April 17, 1969, Oral History Series, JSC History Office.
32. Interview, Merrifield with Gilruth, 1969; Interview, Merrifield with Low, 1969; Barton C. Hacker and James M. Grimwood, *On Shoulders of Titans: A History of Project Gemini,* NASA SP-4203 (Washington, D.C.: 1977), 111-114, 144, 161, 200.
33. Interview, Merrifield with Faget, 1967; Interview, Merrifield with Gilruth, 1969; and see William David Compton, *Where No Man Has Gone Before,* NASA SP-4214 (Washington, D.C.: 1989).
34. Interview, Merrifield with Gilruth, 1969; Interview, Merrifield with Low, 1969.
35. Interview, Merrifield with Gilruth, 1969; Interview, Merrifield with Low, 1969; Interview, Merrifield with Faget, 1967.
36. "Marshall May Take 2nd Apollo Control," *Houston Post* (October 14, 1965), 22; *Houston Post* (October 10, 1966), 1-6.
37. *Houston Post,* 1965, 22.
38. Memorandum, "Telecon from NASA, Office of Manned Space Flight, 10/14/66," Olin E. Teague Papers, Texas A&M University Archives, Box 222.
39. George E. Mueller, Associate Administrator for Manned Space Flight, to Olin E. Teague, Chairman, Subcommittee on Manned Space Flight, October 19, 1965, Teague Papers, Box 222.
40. Olin E. Teague, Washington, D.C. to William P. Hobby, Jr., Houston, Texas, November 10, 1966, Teague Papers, Box 222.
41. Courtney G. Brooks, James M. Grimwood, and Loyd S. Swenson Jr., *Chariots for Apollo: A History of Manned Lunar Spacecraft,* NASA SP-4205 (Washington, D.C.: 1979), 190-193.
42. Interview, Merrifield with Gilruth, 1969.
43. *The Kennedy Space Center Story* (Kennedy Space Center, Florida: NASA, John F. Kennedy Space Center, 1974), iii, 47-48.

Reference Notes

44. *The KSC Story,* 1974, 49-52, 63; Brooks, Grimwood, Swenson, 1979, 50-51; Interview, Swenson and Grimwood with Kleinknecht, 1970.
45. Brooks, Grimwood, and Swenson, 1979, 192-195; *The KSC Story*, 1974, 51.
46. Brooks, Grimwood, and Swenson, 1979, 208-212.
47. Brooks, Grimwood, and Swenson, 1979, 213-218.
48. Brooks, Grimwood, and Swenson, 1979, 218-222.
49. Brooks, Grimwood, and Swenson, 1979, 222-233.
50. Memorandum, "Mr. Teague . . . ," undated, Olin E. Teague Papers, Committee on Science and Astronautics, Texas A&M University Archives, Box 320.
51. Interview, Merrifield with Low, 1969.
52. Brooks, Grimwood, and Swenson, 1979, 223-230.
53. Interview, Dethloff with Bond, 1989; Interview, Henry C. Dethloff with Aleck Bond, February 1, 1990.
54. Interview, Dethloff with Bond, 1989 and 1990.

Chapter 7

1. Michael Collins, *Carrying the Fire, An Astronaut's Journeys* (New York: Farrar, Straus & Giroux, 1974), Reprinted by New York: Bantam Books, 1983, 274.
2. Collins, 1974, 407.
3. Linda Neuman Ezell, *NASA Historical Data Book*, III, *Programs and Projects 1969-1978,* NASA SP-4012 (Washington: 1988), 9, 61; Jane Van Nimmen and Leonard C. Bruno, with Robert L. Rosholt, *NASA Historical Data Book*, I, *NASA Resources 1958-1968* (Washington: 1988), 118.
4. D.H. Beyer and S.B. Sells, "Selection and Training of Personnel for Space Flight" (February 1957), 1-6, Flight Crew Operations Subseries, JSC History Office.
5. Beyer and Sells, 1957, 1-6.
6. Beyer and Sells, 1957, 1-6.
7. Beyer and Sells, 1957, 1-6.
8. John A. Pitts, *The Human Factor: Biomedicine in the Manned Space Program to 1980,* NASA SP-4213 (Washington, D.C.: 1985), 15-18; Jerry Bledsoe, "Down from Glory," *Esquire* (January 1973:83-86).
9. Charles A. Berry, M.D. "Biographical Data," Biographical Data Subseries, JSC History Office.
10. Pitts, 1985, 2.
11. *Astronauts and Cosmonauts, Biographical and Statistical Data* (Revised June 28, 1985), Prepared by the Congressional Research Service, Library of Congress for the Committee on Science and Technology, U.S. House of Representatives, 99th Cong., 1st Sess. (Washington: Government Printing Office, December 1985), 9; *First Semiannual Report to the Congress of the National Aeronautics and Space Administration*, House Doc. No. 187, 86th Cong., 1st Sess. (Washington: Government Printing Office, 1959), 10-11, 68; Robert B. Merrifield, "Men and Spacecraft: A History of the Manned Spacecraft Center, 1958-1969," unpublished mss., 2-29 to 2-33, JSC History Office; Memorandum, W. Randolph Lovelace II, A.H. Schwichtenberg, and Ulrich C. Luft, Selection Program for Astronauts for the National Aeronautics and Space Administration (November 5, 1959), 1-9 and attachments, Flight Crew Operations Subseries, JSC History Office.
12. *First Semiannual Report to the Congress,* 1959, 11.
13. *First Semiannual Report to the Congress*, 1959, 11; Interview, Henry C. Dethloff with Charles J. Donlan, Alexandria, Virginia, January 17, 1990. Note: NASA Press Release No. 59-113 from Bob Voas

dated April 9, 1959, indicates that 69 persons were interviewed rather than 63; and of these, 32 made the selection to a semifinal list.

14. *Astronauts and Cosmonauts,* 1985, 20; Interview, Henry C. Dethloff with Donald K. Slayton, JSC, Houston, Texas, December 5, 1989.
15. *Astronauts and Cosmonauts*, 1985, 20; Interview, Dethloff with Slayton, 1989; Collins, 1974, 27-28.
16. Collins, 1974, 27-28; Interview, Dethloff with Slayton, 1989.
17. Interview, Dethloff with Donlan, 1990; Memorandum, Press Conference, Mercury Astronaut Team, 9 April 1959, Vertical Files, JSC History Office.
18. Interview, Dethloff with Slayton, 1989; Interview, Robert B. Merrifield with John H. Glenn, Jr., Houston, Texas, March 15, 1968, Oral History Series, JSC History Office.
19. Interview, Robert B. Merrifield with John A. Powers, November 11, 1968, Oral History Series, JSC History Office.
20. Interview, Merrifield with Glenn, 1968.
21. Interview, Merrifield with Glenn, 1968; Interview, Merrifield with Powers, 1968.
22. Interview, Merrifield with Glenn, 1968.
23. Interview, Merrifield with Glenn, 1968.
24. Robert R. Gilruth to George E. Meuller, November 18, 1963, Flight Crew Operations Subseries, JSC History Office.
25. Memorandum from the Assistant Administrator for Public Affairs to Public Affairs Officer, MSC [1961]; Memorandum for Dr. Bundy (from Jerome B. Wiesner), The White House, March 9, 1961, Flight Crew Operations Subseries, JSC History Office.
26. Interview, Robert B. Merrifield with Paul G. Dembling, September 25, 1969, Oral History Series, JSC History Office.
27. Interview, Merrifield with Powers, 1968.
28. Interview, Merrifield with Dembling, 1969.
29. Interview, Merrifield with Powers, 1968.
30. Interview, Dethloff with Slayton, 1989; Interview, Merrifield with Glenn, 1968.
31. Interview, Merrifield with Glenn, 1968
32. Memorandum, Robert B. Voas, Human Factors Branch, "Outline of the NASA Astronaut Training Program," with appendices, typed May 5, 1959, Flight Crew Operations Subseries, JSC History Office.
33. Memorandum, Voas, 1959.
34. Memorandum, Voas, 1959.
35. Memorandum, Voas, 1959; Robert B. Voas, Harold I. Johnson, and Raymond Zedeker, *Mercury Project Summary Including Results of the Fourth Manned Orbital Flight, May 15 and 16, 1963*, NASA SP-45 (Houston: 1963), 171-197.
36. Voas, Johnson, Zedeker, 1963, 171, 173.
37. Interview, Dethloff with Slayton, 1989.
38. Interview, Robert B. Merrifield with Edwin E. Aldrin, July 7, 1970, Oral History Series, JSC History Office.
39. Interview, Merrifield with Aldrin, 1970.
40. Interview, Merrifield with Aldrin, 1970; Interview, Robert B. Merrifield with Donald K. Slayton, October 17, 1967; Interview, Dethloff with Slayton, 1989; Interview, Robert B. Merrifield with Warren J. North, May 1, 1970, 9-10, Oral History Series, JSC History Office.

Reference Notes

41. *Astronauts and Cosmonauts,* 1985, 9.
42. *Astronauts and Cosmonauts*, 1985, 9-10.
43. Memorandum, Plan for Integrating Scientists Into Flight Crews of Apollo Lunar Landing Missions, March 13, 1964, Apollo Series, JSC History Office.
44. Memorandum, Plan for Integrating Scientists; *Astronauts and Cosmonauts,* 1985, 10.
45. *Astronauts and Cosmonauts,* 1985, 10-11.
46. *Astronauts and Cosmonauts,* 1985, 10-11.
47. Aleck C. Bond, "A Review of Man-Rating in Past and Current Manned Space Flight Programs," Eagle Engineering/LEMSCO Report No. 88-193, May 20, 1988; Vertical Files, JSC History Office.
48. Bond, 1988, 5.
49. Bond, 1988, 7.
50. Bond, 1988, 4-27.
51. *Major Test Facilities of the Engineering and Development Directorate* (Houston: April 1966), 2-1 to 2-26, Engineering Subseries, JSC History Office.
52. *Major Test Facilities,* 1966, 2-1 to 2-26.
53. *Major Test Facilities,* 1966, 2-1 to 2-26.
54. *Major Test Facilities,* 1966, 7-16 to 7-41; Memorandum, Aleck C. Bond to Henry C. Dethloff, April 24, 1991.
55. Memorandum, Bond, 1991, and see Henry S.F. Cooper, Jr., *Before Liftoff, The Making of a Space Shuttle Crew* (Baltimore: Johns Hopkins University Press, 1987), X, 16, 32-33.
56. Interview, Robert B. Merrifield with Julian Scheer, July 20, 1967.
57. Interview, Merrifield with Scheer, 1967.
58. Interview, Merrifield with Scheer, 1967; Interview, Merrifield with Aldrin, 1970.
59. Interview, Henry C. Dethloff with Eugene Horton, February 15, 1989; Eugene Horton, personal papers and scrapbooks.
60. Courtney G. Brooks, James M. Grimwood, Loyd S. Swenson, Jr., *Chariots for Apollo: A History of Manned Lunar Spacecraft,* NASA SP-4205 (Washington, D.C.: 1979), 232-234.
61. MSC Announcement 67-183, December 14, 1967, General Reference Series, JSC History Office.
62. Brooks, Grimwood, and Swenson, 1979, 241-244.
63. Brooks, Grimwood, and Swenson, 1979, 247-250.
64. Brooks, Grimwood, and Swenson, 1979, 250.

Chapter 8

1. Arnold S. Levine, *Managing NASA in the Apollo Era,* NASA SP-4205 (Washington, D.C.: 1982), 68-70.
2. Levine, 1982, 68-70.
3. Memorandum signed by W.F. Rockwell, Jr., accompanying distribution of Beirne Lay, Jr., *Earthbound Astronauts: The Builders of Apollo-Saturn* (Englewood Cliffs, N.J.: Prentice-Hall, 1971); U.S. Manned Space Flight, The Manned Flight Program, The Astronauts (NASA memorandum), n.p., n.d. [1971], 0-32, Vertical Files, JSC History Office.
4. Courtney G. Brooks, James M. Grimwood, and Loyd S. Swenson, Jr., *Chariots for Apollo: A History of Manned Lunar Spacecraft,* NASA SP-4205 (Washington, D.C.: 1979), 88-90; Lay, 1971, 33; Ralph B.

Oakley, "Historical Summary, S & ID Apollo Program," North American Aviation, Inc., January 20, 1966, 1-50, in "Apollo Execs, January 27-28, 1967," NASA History Office, Washington, D.C., Drawer 15-3; Interview, Ivan D. Ertel with John W. Paup, June 7, 1966, Oral History Series, JSC History Office.

5. Interview, Ertel with Paup, 1966.
6. Interview, Ertel with Paup, 1966.
7. John Yardley, "McDonnell/Spacecraft," in "Apollo Execs, January 27-28, 1967," NASA History Office, Drawer 15-3.
8. Lyndon Baines Johnson, *The Vantage Point: Perspectives of the Presidency, 1963-1969* (New York: Holt, Rinehart and Winston, 1971), 270-271; Courtney G. Brooks, James M. Grimwood, and Loyd S. Swenson, Jr., *Chariots for Apollo: A History of Manned Lunar Spacecraft*, NASA SP-4205 (Washington, D.C.: 1979), 213-218; Memorandum, Robert C. Seamans, Jr. to George Mueller, November 21, 1966, in "Apollo Execs, January 27-28, 1967," NASA History Office, Drawer 15-3.
9. Yardley, 1967, Drawer 15-3.
10. Yardley, 1967, Drawer 15-3.
11. Interview, Ertel with Paup, 1966.
12. Interview, Ertel with Paup, 1966.
13. Interview, Ertel with Paup, 1966.
14. Documents in the History of NASA, An Anthology (Preliminary Edition), NASA History Office, Washington, D.C., June 1975, 149-51.
15. Documents in the History of NASA, An Anthology, 1975, 152-163.
16. Documents in the History of NASA, An Anthology, 1975, 149-151.
17. Interview, Ertel with Paup, 1966.
18. Oakley, 1966, 1-50.
19. Oakley, 1966, 1-50.
20. NASA News Releases, 89-97, 89-98, July 7, 1989, Langley Research Center.
21. NASA News Releases, 89-97, 89-98; Chris Kraft to Henry Dethloff, Comments on draft of Chapter 8.
22. Kraft to Dethloff on Chapter 8; Interview, Ertel with Paup, 1966, Oral History Series, JSC History Office.
23. Oakley, 1966, 21-23; and see Paul Purser to Henry Dethloff, Comments on draft of Chapter 8, March 14, 1990.
24. Oakley, 1966, 1-50.
25. Oakley, 1966, 1-50.
26. Olin E. Teague Papers, Texas A&M University Archives, Boxes 222, 223, 224.
27. Memorandum, "Mr. Teague from Pete G.," undated, Olin E. Teague Papers, Texas A&M University Archives, Box 222.
28. Memorandum, Teague Papers, undated, Box 222.
29. "Space" Billions - Now a Boom? *U.S. News and World Report* (July 20, 1964):8-12; Sylvia Doughty Fries, Evolution of the NASA Institution, Langley Research Center, January 17, 1990, unpublished paper.
30. Oakley, 1966, 3-12, 35-38.
31. Oakley, 1966, 3-12, 35-38, 13; *Historical Statistics of the United States, Colonial Times to 1970*, Part 2 (Washington, D.C.: Bureau of the Census, 1975), Series Y 457-465, 472-487, and 1114-16.
32. Oakley, 1966, 12-13; and see Fries, Evolution of the NASA Institution.

Reference Notes

33. Linda Neuman Ezell, *NASA Historical Data Book*, II (Washington, D.C.: NASA, 1988), 192-194.
34. Ezell, 1988, 192-194.
35. Ezell, 1988, 192-194.
36. Oakley, 1966, 21-29.
37. William David Compton, *Where No Man Has Gone Before: A History of Apollo Lunar Exploration Missions,* NASA SP-4214 (Washington, D.C.: 1989), 41-54.
38. Compton, 1989, 41-54.
39. Compton, 1989, 41-54.
40. Compton, 1989, 41-54.
41. Compton, 1989, 387-388; U.S. Manned Space Flight memorandum, Program, 1971, 9-11.
42. U.S. Manned Space Flight memorandum, 1971, 9-11.
43. Committee on Science and Technology, U.S. House of Representatives, *Astronauts and Cosmonauts: Biographical and Statistical Data* [Revised - June 28, 1985], Ninety-Ninth Congress (Washington, D.C.: Government Printing Office, 1985), 58, 61, 131.
44. "Returns from the Space Dollar," address by Edward C. Welsh, October 11, 1966, in Olin E. Teague Papers, Texas A&M University Archives, Box 222.
45. Address by Welsh, 1966, Box 222.
46. Address by Welsh, 1966, Box 222.

Chapter 9

1. Interview, Henry C. Dethloff with N. Wayne Hale, Jr., October 19, 1989, Mission Control, JSC, Houston, Texas.
2. *Encyclopedia Brittannica*, Eleventh Edition (Cambridge, England: The University Press, 1910), 184-186; Michael Collins, *Liftoff: The Story of America's Adventure in Space* (New York: Grove Press, 1988), 8.
3. Linda Neuman Ezell, *NASA Historical Data Book*, III, *Programs and Projects 1969-1978*, NASA SP-4012 (Washington, D.C.: 1988), 73-92.
4. Maxime A. Faget, "An Overview of United States Manned Space Flight From Mercury to the Shuttle," Preprint for distribution to the XXXII Congress, International Astronautical Federation, Rome, Italy, September 6-12, 1981, in Faget Subseries, JSC History Office.
5. Faget, 1981.
6. Interview, Henry C. Dethloff with Eugene F. Kranz, March 6, 1990, Houston, Texas, Oral History Series, JSC History Office.
7. John Hodge, Comments on preliminary draft of Chapter 9, undated.
8. Interview, Dethloff with Kranz, 1990.
9. Interview, Dethloff with Kranz, 1990.
10. Interview, Dethloff with Kranz, 1990.
11. Interview, Dethloff with Kranz, 1990; Hodge comments on Chapter 9.
12. Hodge comments on Chapter 9; Faget, 1981, 7.
13. Courtney G. Brooks, James M. Grimwood, and Loyd S. Swenson, Jr., *Chariots for Apollo: A History of Manned Lunar Spacecraft*, NASA SP-4205 (Washington, D.C.: 1979), 387-393.
14. *Space News Roundup* (December 16, 1988).

15. Brooks, Grimwood, and Swenson, 1979, 256-260; Frank Borman, Apollo 20th Anniversary Celebration Speaker's Series, "The Moon as Seen by Apollo Astronauts," July 21, 1989, Vertical Files, JSC History Office.
16. Brooks, Grimwood and Swenson, 1979, 256-260.
17. Brooks, Grimwood and Swenson, 1979, 256-260.
18. Brooks, Grimwood and Swenson, 1979, 272-274.
19. Brooks, Grimwood and Swenson, 1979, 273; Clifford E. Charlesworth and Glynn Lunney, Apollo 20th Anniversary Celebration Speaker's Series, "Flying the Apollo Missions" (not transcribed), July 19, 1989, JSC; Ron Berry, Apollo 20th Anniversary Celebration Speaker's Series, "Planning the Apollo Missions" (not transcribed), July 18, 1989, JSC.
20. Berry, 1989; Oran W. Nicks, *Far Travelers: The Exploring Machines* (Washington, D.C.: NASA, 1985), 93-94; and Oran W. Nicks, Comments on preliminary draft of Chapter 9, April 7, 1990.
21. Interview, Dethloff with Kranz, 1990; Gerald D. Griffin, Apollo 20th Anniversary Celebration Speaker's Series, "Flying the Apollo Missions" (not transcribed), July 19, 1989.
22. Charlesworth, 1989.
23. Charlesworth, 1989; Chris Kraft, Comments on preliminary draft of Chapter 9, undated.
24. Owen Morris, Comments on preliminary draft of Chapter 9, undated.
25. Interview, Dethloff with Kranz, 1990.
26. Interview, Dethloff with Kranz, 1990.
27. Brooks, Grimwood and Swenson, 1979, 98-113, 175-178; Project Apollo, Lunar Excursion Module Development Statement of Work, June 18, 1962, Apollo Series, JSC History Office.
28. Thomas J. Kelly, "Apollo Lunar Module Mission and Development Status," AIAA 4th Annual Meeting and Technical Display, Anaheim, California, October 23-27, 1967, Vertical Files, JSC History Office, 1-11.
29. Kelly, 1967; Thomas J. Kelly, "Technical Development Status of the Project Apollo Lunar Excursion Module" [AAS 10th Annual Meeting, May 4-7, 1964], Vertical Files, JSC History Office, 1-63.
30. MSC Announcements: 67-33, February 20, 1967; 67-34, February 23, 1967; 67-63, April 27, 1967, General Reference Series, JSC History Office.
31. Faget, 1981, 9; Brooks, Grimwood and Swenson, 1979, 178-179.
32. Eugene A. Cernan, Apollo 20th Anniversary Celebration Speaker's Series, "The Moon as Seen by Apollo Astronauts," July 21, 1989, Vertical Files, JSC History Office.
33. Michael Collins, Apollo 20th Anniversary Celebration Speaker's Series, "The Moon as Seen by Apollo Astronauts," July 21, 1989, Vertical Files, JSC History Office.
34. Interview, Dethloff with Kranz, 1990.
35. Steve Bales, Apollo 20th Anniversary Celebration Speaker's Series, "Flying the Apollo Missions" (not transcribed), July 19, 1989, JSC History Office.
36. Bales, 1989; Interview, Dethloff with Kranz, 1990.
37. Interview, Dethloff with Kranz, 1990.
38. Charlesworth, 1989.
39. Charlesworth and Lunney, 1989.
40. Collins, 1989.
41. Address by William B. Bergen (undated) in Eugene E. Horton, personal files.

42. Gerald Griffin and Glynn Lunney, Apollo 20th Anniversary Celebration Speaker's Series, "Flying the Apollo Missions" (not transcribed), July 19, 1989, JSC.
43. Griffin and Lunney, 1989; Ezell, 1988, 86.
44. Joseph P. Allen, Apollo 20th Anniversary Celebration Speaker's Series, "Flying the Apollo Missions" (not transcribed), July 19, 1989, JSC History Office.
45. MSC Announcement 67-7, January 10, 1967, General Reference Series, JSC History Office.
46. MSC Announcement 67-7, 67-4, January 13, 1967, General Reference Series, JSC History Office.
47. MSC Announcement 67-27, February 17, 1967, General Reference Series, JSC History Office.
48. Griffin and Allen, Apollo 20th Anniversary Celebration Speaker's Series, "Flying the Apollo Missions" (not transcribed), July 19, 1989, JSC History Office.
49. Interview, Dethloff with Kranz, 1990.
50. Interview, Dethloff with Hale, 1989.
51. James A. Lovell, Jr., Apollo 20th Anniversary Celebration Speaker's Series, "The Moon as Seen by Apollo Astronauts," July 21, 1989, Vertical Files, JSC History Office.
52. Interview, Dethloff with Kranz, 1990.
53. Interview, Dethloff with Kranz, 1990.
54. William David Compton, *Where No Man Has Gone Before: A History of Apollo Lunar Exploration Missions*, NASA SP-4214 (Washington, D.C.: 1989), 388, and see 386-393.
55. Interview, Dethloff with Kranz, 1990.
56. Interview, Dethloff with Kranz, 1990.

Chapter 10

1. "The Case for Space," Reprint of an Interview with Olin E. Teague published in North American Rockwell's *Skyline Magazine*, in Olin E. Teague Papers, Texas A&M University Archives, Box 224.
2. Courtney G. Brooks, James M. Grimwood, Loyd S. Swenson, Jr., *Chariots for Apollo: A History of Manned Lunar Spacecraft*, NASA SP-4214 (Washington, D.C.: 1979), 7-9.
3. Roland W. Newkirk and Ivan D. Ertel, with Courtney G. Brooks, *Skylab: A Chronology*, NASA SP-4011 (Washington, D.C.: 1977), 8-22.
4. Newkirk, Ertel, and Brooks, 1977, 8-22; E.H.Olling, Memorandum for the Office of the Director, Attn: Mr. Paul Purser, July 24, 1962, Post-Apollo Planning Subseries, JSC History Archive, Woodson Research Center, Rice University, Houston, Texas.
5. Olling Memo, 1962; Michael Getler, "Goodyear Shows New Inflatable Station," *Missiles and Rockets* (August 13, 1962), 24; Charles W. Mathews, Memorandum for Walter C. Williams, Associate Director (MSC), Houston, Texas, October 29, 1962 and J. Mockovciak, Jr., Grumman Aircraft Engineering Corporation (PDM-330-6) to Distribution, November 14, 1962, Post-Apollo Planning Subseries, JSC History Archive, Woodson Research Center, Rice University, Houston, Texas; Sylvia D. Fries, Space Station: Evolution of A Concept, NASA Headquarters History Division, February 29, 1984, 1-13, appendices.
6. Newkirk, Ertel, and Brooks, 1977, 23-38.
7. Olin E. Teague Papers, Committee on Science and Astronautics, February 3, 1965-April 20, 1965, Texas A&M University Archives, Box 370; Post-Apollo Planning Subseries, JSC History Archive, Woodson Research Center, Rice University, Houston, Texas.
8. U.S. Congress, 89th Cong., 1st Sess., Hearings Before the Committee on Aeronautical and Space Sciences, United States Senate, August 23, 24, and 25, 1965, 47-104; W. David Compton and Charles D. Benson,

Living and Working in Space: A History of Skylab, NASA SP-4208 (Washington, D.C.: 1983), 26-27.

9. Compton and Benson, 1983, 26.
10. Olin E. Teague to George E. Mueller, May 24, 1966, Apollo Series, JSC History Office; Arnold S. Levine, *Managing NASA in the Apollo Era*, NASA SP-4102 (Washington, D.C.: 1982), p. 189.
11. "Bold Decisions on Space Asked," *New York Times* (August 10, 1966).
12. P.A. Gerardi to Mr. Teague, October 6, 1966, Olin E. Teague Papers, Science and Astronautics, Texas A&M University Archives, Box 370.
13. Thomas G. Miller, Jr., *The National Space Program: Problems and Promise* (New York]: Arthur D. Little, Inc., September 1966), 1-31, Copy in Olin E. Teague Papers, under cover of letter from Miller to Teague, November 11, 1966, Texas A&M University Archives, Box 223.
14. Robert R. Gilruth to George E. Mueller, "Apollo Applications Program Approach to Future Manned Spaceflight," March 25, 1966, Shuttle Series, JSC History Office.
15. Walter G. Hall, Dickinson, Texas, to Olin E. Teague, January 16, 1967; Teague to Hall, January 24, 1967, in Olin E. Teague Papers, Texas A&M University Archives, Box 369.
16. "After Apollo, What Next?" Remarks by J.P. Rogan before the Washington, D.C., chapters of the American Institute of Aeronautics and Astronautics and the American Astronautical Society, January 5, 1967.
17. A Report of the President's Science Advisory Committee, *The Space Program in the Post-Apollo Period* (Washington, D.C.: The White House, February 1967), 1-99.
18. Levine, 1982, 185-204.
19. Levine, 1982, 187-188, 207-209, 233-234.
20. Position Paper for FY 1971 Budget Hearings, AAP Schedule Slippages, February 6, 1970, and (Draft) Post-Apollo Advisory Group Summary of Proceedings, July 25, 1968, Post-Apollo Planning Subseries, JSC History Archive, Woodson Research Center, Rice University, Houston, Texas.
21. Memorandum, Proposed Center Intentions on Institutional Planning for the Manned Spacecraft Center, by Paul E. Purser, July 22, 1968, Vertical Files, JSC History Office.
22. Letter, George Mueller to Kurt Debus, "Proposed Goals in Manned Space Flight and NASA in the Next Decade," September 26, 1969, Apollo Series, JSC History Office.
23. Space Task Group Report to the President, *The Post-Apollo Space Program: Directions for the Future* (September 1969), 1-37; Memorandum, Activities of President's Space Task Group and NASA Internal Planning Groups, April 1, 1969, Post-Apollo Planning Subseries, JSC History Archive, Woodson Research Center, Rice University, Houston, Texas; *America's Next Decades in Space, A Report for the Space Task Group* (NASA, September 1969), 1-87.
24. Position Paper for FY 1971 Budget Hearings, JSC History Archive, Rice University.
25. Robert F. Freitag to Distribution, April 29, 1970, with enclosure letters from respondents dated April 10 through 22, 1970, Apollo Series, JSC History Office.
26. (Typescript) E.M. Emme, *Preliminary History of NASA*, II, Supplement, October 1968 - January 1969, Washington, D.C., NASA, January 15, 1969, I-1 to V-25.
27. Interview, Henry C. Dethloff with Gerald D. Griffin, Houston, Texas, April 16, 1990.
28. Interview, Henry C. Dethloff with Aleck C. Bond, September 21, 1989.
29. Interview, Dethloff with Bond, 1989.
30. Interview, Dethloff with Bond, 1989.
31. Interview, Dethloff with Bond, 1989.
32. Jerry Bledsoe, "Down From Glory," *Esquire* (January 1973), 83-86, 94, 176-190; *Astronauts and*

Reference Notes

Cosmonauts, Biographical and Statistical Data [Revised June 28, 1985], Congressional Research Services, Library of Congress (Washington, D.C.: Government Printing Office, 1985), 139.

33. *Astronauts and Cosmonauts*, 1985, 114, 129.
34. *Apollo Program Summary Report* (Houston: NASA, JSC, April 1975), 2-41 to 2-45.
35. *Apollo Program Summary Report*, 1975, 2-45 to 2-48.
36. Press Release, July 18, 1972, "Aspin Blasts Apollo 15 Astronaut's Scheme"; Memorandum, Apollo 15 Crew Visit to the Capitol, September 9, 1971, Olin E. Teague Papers, Miscellaneous, Texas A&M University Archives, Box 224.
37. (Typescript) CBS News with John Hart, October 6, 1972, Olin E. Teague Papers, Miscellaneous, Texas A&M University Archives, Box 224.
38. *Astronauts and Cosmonauts*, 1985, 94, 135, 163.
39. *Apollo Program Summary Report*, 1975, 2-48 to 2-51, 3-1 to 3-105.
40. *Apollo Program Summary Report*, 1975, 2-48 to 2-51, 3-1 to 3-105.
41. Memorandum, Chief, Public Appearances Branch to Assistant Administrator for Public Affairs, "Apollo 17 Postflight," April 6, 1973, Flight Crew Operations Subseries, JSC History Office.
42. Olin E. Teague, "What Apollo 16 Mission Means," *Congressional Record*, 91st Cong., May 16, 1972, E5269.
43. Letters to Frank A. Bogart, re: JSC Organization Review, from Kenneth S. Kleinknecht (April 25, 1972), John Eggleston (April 27, 1972), C.E. Charlesworth (April 25, 1972), Joseph V. Piland (n.d.), Philip H. Whitbeck (May 9, 1972), Donald K. Slayton (n.d.), Bill Tindall (n.d.), and memorandum from Frank A. Bogart and George Low, in Loftus Subseries, JSC History Office.
44. Letters to Frank A. Bogart, 1972.

Chapter 11

1. "A Message from MSC Director Dr. Robert R. Gilruth," *Space News Roundup* (hereafter cited *Roundup*) (December 18, 1970).
2. Organizational Study, MSC (n.d.), Loftus Subseries, JSC History Office; Linda Neuman Ezell, *NASA Historical Data Book*, III, *Programs and Projects, 1969-1978*, NASA SP-4012 (Washington, D.C.: 1988), 93-113.
3. JSC Organization Review, 1972, Loftus Subseries, JSC History Office; and see R. Wayne Young, The Utilization of Project Management Concepts at the NASA Johnson Space Center, unpublished dissertation, University of Colorado, 1983, 284-295.
4. *Roundup* (January 21, 1972).
5. W. David Compton and Charles D. Benson, *Living and Working in Space: A History of Skylab*, NASA SP-4208 (Washington, D.C.: 1983), 19-112.
6. Compton and Benson, 1983, 29, 110.
7. Compton and Benson, 1983, 133ff.; Ezell, 1988, 100-103.
8. MSC Announcement 71-77, May 18, 1971, General Reference Series, JSC History Office.
9. Interview, Henry C. Dethloff with William R. Kelly, JSC, May 9, 1990; MSC Announcement 72-11, January 21, 1972, General Reference Series, JSC History Office.
10. Interview, Dethloff with Kelly, 1990.
11. MSC Announcement 72-26, February 22, 1972.

12. *Roundup* (February 2, 1973) and (August 17, 1973).
13. Compton and Benson, 1983, 119ff.; Ezell, 1988, 101-108; *Roundup* (May 25, June 8, 22, July 20, 1973).
14. *Roundup* (July 6, 20, August 3, September 14, September 28, October 12, 1973).
15. *Roundup* (October 12, 1973, January 4, February 15, 1974).
16. JSC Organization Review, 1972, Loftus Subseries, JSC History Office.
17. JSC Organization Review, 1972, Loftus Subseries, JSC History Office.
18. MSC Announcement 71-21, February 16, 1971, General Reference Series, JSC History Office.
19. MSC Announcement 71-23, February 16, 1971, General Reference Series, JSC History Office.
20. MSC Announcement 72-25, February 18, 1972, General Reference Series, JSC History Office.
21. MSC Announcements: 67-27, February 17, 1967, 67-4, January 13, 1967, 67-7, January 10, 1967, General Reference Series, JSC History Office; and see William David Compton, *Where No Man Has Gone Before: A History of Apollo Lunar Exploration Missions*, NASA SP-4214 (Washington, D.C.: 1989), 54, 73-90.
22. MSC Announcement 71-54, April 15, 1971, General Reference Series, JSC History Office.
23. MSC Announcement 72-10, January 20, 1972, General Reference Series, JSC History Office.
24. MSC Announcement 71-65, May 5, 1971, General Reference Series, JSC History Office.
25. George Low, Acting Administrator, Memorandum, International Participation in the Post-Apollo program," to Distribution, November 2, 1970, Apollo Series, JSC History Office; Walter Froehlich, *Apollo Soyuz*, EP-109 (Washington, D.C.: 1976), 29-36.
26. Froelich, 1976, 29-36; *Roundup*, (December 3, 1971; January 7, May 26, June 9, 1972).
27. Maxime A. Faget, Background and Planning for the Apollo/Soyuz Mission, August 26-31, 1974, Max Faget Subseries, JSC History Office, 1-21.
28. *Roundup* (October 13, 1972); MSC personnel accompanying him included Hugh Scott, Clarke Covington, M.P. Frank, Thomas P. Stafford, Robert Ward, Edgar C. Lineberry, Donald C. Wade, Robert White, William Creasy, Lawrence Williams, C.C. Johnson, John Schliesing, R.E. Smylie, W.W. Guy, Dr. W.R. Hawkins, Raymond Zedeker and H.E. Smith; Ezell, *NASA Historical Data Book*, III, *Programs and Projects, 1969-1978*, 108-112.
29. Ezell, 1988, 108-112; Faget, 1974; Froelich, 1976, 7-14.
30. Ezell, 1988, 108-112.
31. *Roundup* (August 15, 1975).
32. Memorandum, Max Akridge (PD-RV), Space Shuttle History, January 8, 1970, MSFC Reports Subseries, JSC History Office.
33. Akridge, Space Shuttle History.
34. Akridge, Space Shuttle History; and see Eagle Engineering, Inc., Report No. 86-125C, Technology Influence on the Space Shuttle Development, June 8, 1986, MSC/JSC Subseries, JSC History Office.
35. Akridge, Space Shuttle History, 25.
36. Akridge, Space Shuttle History, 26-35.
37. Akridge, Space Shuttle History, 36.
38. Akridge, Space Shuttle History, 36-46.
39. Akridge, Space Shuttle History, 46-47.
40. Akridge, Space Shuttle History, 47-48.

Reference Notes

41. Akridge, Space Shuttle History, 49; Ezell, 1988, 113-118.
42. Akridge, Space Shuttle History, 58-60.
43. Akridge, Space Shuttle History, 62-65; Interview, Henry C. Dethloff with John Hodge, Falls Church, Virginia, January 15, 1990.
44. Akridge, Space Shuttle History, 62-65.
45. Akridge, Space Shuttle History, 62-65.
46. Akridge, Space Shuttle History, 68-98.
47. Ezell, 1988, 127-362.
48. NASA News Release 70-4, January 8, 1970.
49. NASA News Release 70-4, January 8, 1970; Interview, Ivan Ertel and Ralph Oakley with Dale D. Myers, Downey, California, May 12, 1969, Oral History Series, JSC History Office.
50. Interview, Henry C. Dethloff with Charles J. Donlan, Alexandria, Virginia, January 17, 1990, Oral History Series, JSC History Office.
51. Memorandum from Joseph Loftus in response to an inquiry from Aaron Cohen [1990], "How did we manage integration in the Apollo and Shuttle Programs?" Loftus personal files.
52. *Roundup* (June 18, 1971); Dale D. Myers, "Space Shuttle Program Management," March 14, 1972, NASA Management Instruction Subseries, JSC History Office.
53. Memorandum, Loftus to Cohen (1990).
54. *Roundup* (June 18, 1971) and (January 21, 1972).
55. Roger E. Bilstein, *Orders of Magnitude: A History of the NACA and NASA, 1915-1990*, NASA SP-4406 (Washington, D.C.: 1989).
56. George Low, "Meeting with the President on January 5, 1972," memo for the record, January 12, 1972, Shuttle Series, JSC History Office.
57. *Roundup* (January 21, 1972).
58. MSC Announcement 71-21, February 16, 1971, General Reference Series, JSC History Office.
59. Ezell, 1988, 113-124.
60. *Roundup* (January 4, 1974).

Chapter 12

1. John F. Guilmartin (Management Analysis Office Administration Directorate), *A Shuttle Chronology, 1964-1973: Abstract Concepts to Letter Contracts* (Houston: NASA, JSC, December 1988), VII-3, 16, 48, 52 (hereafter cited *A Shuttle Chronology*, 1988); Interview, Henry C. Dethloff with Henry O. Pohl, JSC, May 9, 1990.
2. Myers to Distribution, FY 1974 Budget Program Guidelines, August 17, 1972, Shuttle Series, JSC History Office; *A Shuttle Chronology*, 1988, VII-3, 16, 48, 52; Interview, Dethloff with Pohl, 1990.
3. *Catalog of Center Roles* (Washington, D.C.: NASA, December 1976), 1-30, Loftus Subseries, JSC History Office.
4. *Catalog of Center Roles*, 1976, 1-30; *A Shuttle Chronology*, 1988, VI-43.
5. Robert F. Thompson, Biographical Data Subseries, JSC History Office.
6. *Catalog of Center Roles*, 1976, 1-30.
7. *Catalog of Center Roles*, 1976, 1-30.

8. Aaron Cohen, Biographical Data Subseries, JSC History Office.
9. JSC Announcement 73-24, February 15, 1973, and 73-53 and 73-54, March 27, 1973, General Reference Series, JSC History Office.
10. *Space News Roundup* (hereafter cited *Roundup*) (April 19, 1991).
11. *A Shuttle Chronology*, 1988, VII-3.
12. Clippings and memoranda, January 6-9, 1972, Shuttle Series, JSC History Office.
13. *A Shuttle Chronology,* 1988, see particularly Chapters V, VI-10-11, and VI-40: The suborbital staging plan is attributed to Jack Funk, John T. McNeely, Burl G. Kirkland, Stewart F. McAdoo, Jr., and Victor R. Bond of the Advanced Mission Design Branch, Mission Planning and Analysis Division of JSC. See "Development of Suborbital Staging for the Shuttle External Propellant Tank," Loftus History Committee Subseries, JSC History Office.
14. *A Shuttle Chronology*, 1988, VII-3 to 4, VII-48.
15. *A Shuttle Chronology*, 1988, VII-98 to 99.
16. *Roundup* (March 1, 1974).
17. Eleanor H. Ritchie, *Astronautics and Aeronautics, 1977* (Washington, D.C.: NASA, 1986), 52-93.
18. Interview, Henry C. Dethloff with Christopher C. Kraft, Houston, Texas, March 6, 1990.
19. Interview, Dethloff with Kraft, 1990.
20. *A Shuttle Chronology,* 1988, VII:61-62.
21. Robert F. Thompson, The Space Shuttle - Some Key Program Decisions, 1984 Von Karman Lecture, Houston, Texas, Vertical Files, JSC History Office; Aaron Cohen, Progress of Manned Space Flight from Apollo to Space Shuttle, AIAA 22nd Aerospace Sciences Meeting, January 9-12, 1984, Reno, Nevada (NY: American Institute of Aeronautics and Astronautics).
22. Thompson, 1984; Cohen, 1984.
23. *A Shuttle Chronology,* 1988, I:6-9.
24. *Major Test Accomplishments of the Engineering and Development Directorate* (Houston: MSC, April 1966), sections 1-8, Engineering Subseries, JSC History Office.
25. Henry O. Pohl, Biographical Data Subseries, JSC History Office.
26. Henry O. Pohl, 1:1-23.
27. Henry O. Pohl, 2:1-26.
28. Henry O. Pohl, 3:3-10; NASA Information Summary, *Computers*, PMS-016 (JSC), May 1987, 1-8.
29. NASA Information Summary, 1-8.
30. Charles Olasky, Shuttle Mission Simulator, in Proceedings of the 11th Space Simulation Conference, JSC, September 23-25, 1980 (Washington, D.C.: NASA, 1980).
31. Kathy Colgan, "Astronaut Selection and Training," June 18, 1982, 1-10, Flight Crew Operations Subseries, JSC History Office.
32. Henry S.F. Cooper, *Before Lift-Off: The Making of a Space Shuttle Crew* (Baltimore: Johns Hopkins Press, 1987), 33.
33. *Astronauts and Cosmonauts, Biographical and Statistical Data* (Washington, D.C.: Government Printing Office, December 1985).
34. Colgan, 1982; C.H. Woodling to Distribution, Shuttle Crew Training, June 8, 1973, Flight Crew Operations Subseries, JSC History Office.
35. Colgan, 1982.

36. *Roundup* (December 12, 1962), (June 10, 1964), (October 14, 1964), (September 1, 1989), and (September 22, 1989).
37. *Roundup*, 1962, 1964, 1989.
38. Colgan, 1982.
39. *Roundup* (April 9, 1971).
40. William David Compton, "NASA and the Space Sciences," *Science and Technology in Texas*, 122-134.
41. Harrison Schmitt, "The Moon Before Apollo" (not transcribed), Apollo 20th Anniversary Celebration Speaker's Series, July 17, 1989, JSC History Office.
42. John Wood, "What Did We Learn and Where Did it Lead?", Apollo 20th Anniversary Celebration Speaker's Series, July 20, 1989, Vertical Files, JSC History Office.
43. See William David Compton, *Where No Man Has Gone Before: A History of Apollo Lunar Exploration Missions*, NASA SP-4214 (Washington, D.C.: 1989).
44. Wendell Mendell, "What Did We Learn and Where Did it Lead?", Apollo 20th Anniversary Celebration Speaker's Series, July 20, 1989, Vertical Files, JSC History Office.
45. For a good and simplified description of the operation and components of the Space Shuttle, see Michael Collins, *Liftoff, The Story of America's Adventure in Space* (New York: NASA, Grove Press, 1988), 201-222.
46. Collins, 1988, 214-215; Ritchie, 1986, 45, 108, 118, 121, 143, 196, 206, 234.
47. Ritchie, 1986, 108-109; Collins, 1988, 203, 215.
48. Ritchie, 1986, 41-42; Eugene Horton, Personal Scrapbooks.
49. Ritchie, 1986, 74.
50. The NASA History Office, *Astronautics and Aeronautics, 1978*, 49-50.

Chapter 13

1. Interview, Henry C. Dethloff with Owen G. Morris, JSC, August 8, 1990.
2. See Audrey L. Schwartz, "Selling Space: The Product Life Cycle of the Manned Space Program 1957-1981," University of Houston-Clear Lake, Space Business Research Center (August 1989), Space Flight Justification Subseries, JSC History Office, an unpublished manuscript which relates the space program to a product "life cycle" for which the consumer appetite waned precipitously in the 1970's.
3. *Space News Roundup* (hereafter cited *Roundup*) (February 16, 1990). The Center spent some $1.1 billion in fiscal year 1989 with Texas firms and out-of-state companies that pay salaries to employees in Texas. Texas followed California ($2.7 billion) and Florida ($1.2 billion) in states receiving NASA funds for contracts or grants.
4. *Roundup*, 1990.
5. Olin E. Teague Papers, Texas A&M University Archives, Boxes 222 and 224.
6. *Houston Magazine* (September, 1962) 26; (September 1963) 19, 20, 24; (March 1964) 26; (September 1964) 19-21.
7. Telephone interview, Henry C. Dethloff with Mayor Louie Welch, Houston, Texas, May 30, 1990.
8. Interview, Dethloff with Welch, 1990.
9. Interview, Dethloff with Morris, 1990.
10. Interview, Dethloff with Morris, 1990.
11. Interview, Dethloff with Morris, 1990; Interview, Dethloff with Welch, 1990.

12. Interview, Dethloff with Morris, 1990; Telephone interview, Dethloff with Hubert P. Davis, August 8, 1990.
13. Interview, Dethloff with Morris, 1990.
14. Interview, Dethloff with Morris, 1990
15. Interview, Dethloff with Morris, 1990; Telephone interview, Dethloff with Davis, 1990.
16. Telephone interview, Dethloff with Davis, 1990.
17. Telephone interview, Dethloff with Davis, 1990; Howard Benedict, *NASA: A Quarter Century of Space Achievement* (The Woodlands, Texas: Pioneer Publications, Inc., 1984), 209.
18. Eagle Aerospace, Informational Brochure (1990).
19. Space Industries, Inc., "Bringing the Promise of Space Down to Earth" (1990).
20. Interview, Henry C. Dethloff with Emyré Barrios Robinson, Clear Lake, Texas, July 17, 1990; Emyre Barrios Robinson (biographical).
21. Robinson (biographical); *Houston Chronicle* (August 21, 1988) and (July 20, 1990).
22. *Houston Chronicle*, 1988 and 1990.
23. Interview, Henry C. Dethloff with Miguel A. Hernandez, Jr., Jerome B. Hammack, and William R. Holmberg, Clear Lake City, Texas, March 6, 1990.
24. Interview, Dethloff with Hernandez, Hammack, and Holmberg. 1990.
25. Interview, Dethloff with Hernandez, Hammack, and Holmberg. 1990; Committee on Science and Technology, *Astronauts and Cosmonauts, Biographical and Statistical Data* (Washington, D.C.: Government Printing Office, December 1985), 135.
26. Interview, Dethloff with Hernandez, Hammack, and Holmberg, 1990.
27. Interview, Dethloff with Hernandez, Hammack, and Holmberg, 1990.
28. Interview, Dethloff with Hernandez, Hammack, and Holmberg, 1990.
29. *Houston Chronicle* (May 17, 1990).
30. Bryan-College Station *Eagle* (August 21, 1990); NASA *Spin-off* (August 1988), 4, 42-49.
31. NASA, JSC, Procurement Operations Branch, *Subcontracting Opportunities* (March 31, 1989), 1-19; NASA, JSC Almanac (May 1989), 109.

Chapter 14

1. Interview, Henry C. Dethloff with Owen Morris, Houston, Texas, August 8, 1990.
2. Interview, Dethloff with Morris, 1990.
3. Interview, Dethloff with Morris, 1990.
4. Interview, Dethloff with Morris, 1990; Rodney Rose, Comments on preliminary draft of Chapter 13.
5. Rose, Comments on Chapter 13.
6. Eugene F. Kranz and Christopher C. Kraft, Systems Engineering and Integration Process of NASA, JSC, Mission Operations Directorate, Final dated December 19, 1990 (scheduled for publication by the NASA Alumni League).
7. Kranz and Kraft, 1990.
8. Kranz and Kraft, 1990.
9. Interview, Henry C. Dethloff with N. Wayne Hale, Jr., JSC, October 19, 1989.
10. Interview, Henry C. Dethloff with Dennis Webb, JSC, July 19, 1990.

Reference Notes

11. Interview, Dethloff with Webb, 1990; Interview, Dethloff with Hale, 1990.
12. National Space Transportation System Program in New Employee Handbook, JSC Management Subseries, JSC History Office.
13. Memoranda, Ad Hoc Shuttle Payload Activities Team, January 4 - January 31, 1974, Loftus Subseries, JSC History Office.
14. Memoranda, Ad Hoc Shuttle Payload Activities Team, 1974.
15. Memoranda, Ad Hoc Shuttle Payload Activities Team, 1974.
16. Memoranda, Ad Hoc Shuttle Payload Activities Team, 1974.
17. JSC Announcement 81-43, May 7, 1981, General Reference Series, JSC History Office; Glynn S. Lunney, Biographical Data Subseries, JSC History Office.
18. Interview, Henry C. Dethloff with Glynn S. Lunney, Houston, Texas, October 25, 1990, Oral History Series, JSC History Office.
19. Charles Olasky, "Shuttle Mission Simulator," NASA Conference Publication 2150, *11th Space Simulation Conference*, Houston, Texas, September 23-25, 1980, 113-125.
20. Interview, Dethloff with Morris, 1990; Roger E. Bilstein, *Orders of Magnitude: A History of the NACA and NASA, 1915-1990* (Washington, D.C.: 1989), 110-111; Neil McAleer, *The Omni Space Almanac; A Complete Guide to the Space Age* (New York: World Almanac, 1987), 69-70.
21. Bilstein, 1989, 69-70.
22. U.S. House of Representatives, Committee on Science and Technology, *Astronauts and Cosmonauts: Biographical and Statistical Data*, 99th Cong., 1st Sess. (Washington, D.C.: Government Printing Office, December 1985), 56, 164.
23. NASA Mission Report, "A Free People Capable of Great Deeds,"- The Story of STS-1, The First Space Shuttle Mission (MR-001), Vertical Files, JSC History Office.
24. NASA Mission Report (MR-001); House Subcommittee on Space Science and Applications, *Future Space Programs Report 1981*, 97th Cong., 2nd Sess. (Washington, D.C.: Government Printing Office, 1981), 4.
25. *Astronauts and Cosmonauts: Biographical and Statistical Data*, 1985, 63, 154; NASA, JSC, STS-2, Second Space Shuttle Mission Press Kit, September 1981, 1-35. STS-2 Documents Subseries, JSC History Office.
26. Second Space Shuttle Mission Press Kit, September 1981, 35-43.
27. *Future Space Programs*, 1981, 1-45.
28. Memoranda, Ad Hoc Shuttle Payload Activities Team, 1974.
29. Memoranda, Ad Hoc Shuttle Payload Activities Team, 1974.
30. Memoranda, Ad Hoc Shuttle Payload Activities Team, 1974.
31. NASA, JSC, STS-3, Third Space Shuttle Mission Press Kit, March 1982, 1-91, STS-3 Documents Subseries, JSC History Office.
32. NASA, "STS-4 Test Mission Simulates Operational Flight—President Terms Success "Golden Spike" in Space," Mission Report (MR-004), Vertical Files, JSC History Office.
33. NASA Mission Report (MR-001).

Chapter 15

1. Howard Allaway, *The Space Shuttle at Work* (Washington, D.C.: NASA, 1979), Foreword, 21-27, 51-63.

2. NASA Information Summaries (PMS-020A/JSC) March 1989; National Space Transportation System Program in New Employee Handbook, JSC Management Subseries, JSC History Office; Comparison between STS, Delta and Ariane by Coopers & Lybrand, July 1982, STS Management Subseries, JSC History Office.
3. Interview, Henry C. Dethloff with Dennis Webb, JSC, July 19, 1990.
4. JSC, NASA Facts, "Mission Control Center" (n.d.); NASA Information Summaries, "Space Shuttle Mission Summary, 1981-1983." See Joseph P. Allen with Russell Martin, *Entering Space, An Astronaut's Odyssey* (New York: Stewart, Tabori & Chang, 1984, Revised 1986), 59.
5. "Shuttle Mission Summary, 1981-1983."
6. "Shuttle Mission Summary, 1981-83." Note: A TDRS can handle up to 300 million bits of information each second from a single user spacecraft. It operates on Ku-band and S-band frequencies.
7. "Shuttle Mission Summary, 1981-83."
8. "Shuttle Mission Summary, 1981-83"; JSC News Release 80-038, May 29, 1980.
9. *Astronauts and Cosmonauts, Biographical and Statistical Data*, 13-14; Correspondence and Memorandum, Flight Crew Operations Subseries, JSC History Office. Note: See Correspondence from 1973 to 1978. The six women astronaut candidates recruited in 1978 were Anna Fisher (M.D.), Shannon Lucid (Ph.D., biochemistry), Judith Resnik (Ph.D., electrical engineering), Sally Ride (Ph.D., physics), Margaret Seddon (M.D.), and Kathryn Sullivan (Ph.D., geology). And see Hans Mark, *The Space Station, A Personal Journey* (Durham, North Carolina: Duke University Press, 1987), 57-58.
10. Mark, 1987, 57-58.
11. "Shuttle Mission Summary, 1981-1983"; JSC News Release 80-038, May 29, 1980.
12. JSC News Release 80-038, 1980.
13. "Shuttle Mission Summary, 1981-1983."
14. Neil McAleer, *The Omni Space Almanac: A Complete Guide to the Space Age* (New York: World Almanac, 1987), 14,78.
15. McAleer, 1987, 14,78.
16. McAleer, 1987, 14,78.
17. Mark, 1987, 57-58, 188-90.
18. NASA Facts, "Space Shuttle Mission Summary 1984, STS Missions 41-B Thru 51-A," Kennedy Space Center; Allen with Martin, 1984, 223-224.
19. "Shuttle Mission Suumary, 1984"; McAleer, 1987, 84-86; NASA, Headquarters, Washington, D.C. and Goddard Space Flight Center, Greenbelt, Maryland, Press Release 80-16, February 6, 1980.
20. "Shuttle Mission Summary, 1984."
21. "Shuttle Mission Summary, 1984"; Mark, 1987, 205.
22. "Shuttle Mission Summary, 1984."
23. Henry S.F. Cooper, Jr., *Before Lift-Off, The Making of a Space Shuttle Crew* (Baltimore: Johns Hopkins University Press, 1987), 17.
24. Cooper, 1987, 17.
25. NASA, "Teacher in Space Project," 1985, STS-51L Documents Subseries, JSC History Office.
26. Cooper, 1987.
27. Cooper, 1987, 17.
28. Cooper, 1987, 17; Allen with Martin, 1984, 223-236.

29. Cooper, 1987, 17; Allen with Martin, 1984, 231.
30. Allen with Martin, 1984, 231-236; "Shuttle Mission Summary, 1984."
31. "Shuttle Mission Summary, 1984."
32. Mark, 1987, 67.
33. Mark, 1987, see especially 68-88.
34. Mark, 1987, see especially 68-88.; Interview, Henry C. Dethloff and David Norton with Gerald Griffin, Houston, Texas, April 16, 1990.
35. Mark, 1987.
36. "Shuttle Mission Summary 1984"; *Washington Post* (April 11, 12, 1985).
37. "Shuttle Mission Summary, 1984."
38. "Shuttle Mission Summary, 1984."
39. "Shuttle Mission Summary, 1984."
40. "Shuttle Mission Summary, 1984."
41. "Shuttle Mission Summary, 1984."
42. "Shuttle Mission Summary," 1984."
43. Mark, 1987, 195.
44. "Shuttle Mission Summary, 1984."
45. "Shuttle Mission Summary, 1984."
46. NASA Press Kit, "Space Shuttle Mission 51-L," (January 1986) STS 51-L Documents Subseries; and see, 51-L Investigation Subseries, JSC History Office.
47. Presidential Commission on the Space Shuttle Challenger Accident, Biographies, n.d., STS 51-L Investigation Subseries, JSC History Office.
48. Memorandum, John W. Young, Chief, Astronaut Office to Director, Flight Crew Operations (cc: R.H. Truly, Hqs, and others) March 4, 1986; Letter, Richard L. Griffin to John W. Young, March 12, 1986, STS 51-L Investigation Subseries, JSC History Office.
49. Memorandum, Young, 1986; Letter, Griffin, 1986.
50. STS 51-L Investigation Subseries, JSC History Office.
51. See Summaries of Meetings with the Mission Planning and Operations Panel of the Presidential Commission on the Space Shuttle Challenger Accident, under cover dated May 15, 1986, STS 51-L Investigation Subseries, JSC History Office.
52. Report of the Presidential Commission on the Space Shuttle Challenger Accident, Washington D.C.: June 6, 1986, 19.
53. Report on Challenger Accident, 1986, 120-48.
54. Report on Challenger Accident, 1986, 120-48.
55. Report on Challenger Accident, 1986, 120-48.
56. *Houston Chronicle* (September 25, 1988, October 8, 1988, November 13, 1988, January 22, 1989, February 23, 1990); Bruce Murray, "Whither America in Space?" *Issues in Science and Technology* (Spring 1986), 22-45; John M. Logsdon, "The Space Shuttle Program: A Policy Failure?" *Science* (30 May 1986), 1099-1105; Sally K. Ride, *Leadership and America's Future in Space, A Report to the Administrator* (Washington, D.C.: August 1987), 1-63.
57. Memorandum, Jesse W. Moore, JSC to NASA Headquarters, Response to Recommendations of the Presidential Commission on the Space Shuttle Challenger Accident, June 13, 1986, STS 51-L Mission

Subseries, JSC History Office; *Report to the President, Actions to Implement the Recommendations of the Presidential Commission on the Space Shuttle Challenger Accident: Executive Summary* (Washington, D.C.: NASA, July 14, 1986), 1-5.

58. Jon D. Miller, The Impact of the Challenger Accident on Public Attitudes Toward the Space Program, A Report to the National Science Foundation, Public Opinion Laboratory, Northern Illinois University, DeKalb, Illinois, January 25, 1987, STS 51-L Mission Subseries, JSC History Office.

Chapter 16

1. Interview, Henry C. Dethloff with William Huffstetler, JSC, November 8, 1990.
2. Sylvia Doughty Fries, "Recovering from Challenger," *Action* Volume 10, Number 10 November/December, 1990, 11-13.
3. Interview, Henry C. Dethloff with Gerald D. Griffin, Houston, Texas, April 16, 1990.
4. Interview, Dehtloff with Griffin, 1990; NASA, *Astronautics and Aeronautics, 1985: A Chronology*, NASA SP-4025 (Washington, D.C.: NASA History Series, 1988), 176-185.
5. *Astronautics and Aeronautics, 1985: A Chronology,* 1988, 176-185.
6. John M. Logsdon, "The Space Shuttle Program: A Policy Failure?" *Science* (30 May 1986), 1099-1105; Alcestis R. Oberg, "After the Parades," *Final Frontier* (September/October 1990), 43-59.
7. The lack of purpose and direction particularly affected development of the space station program.
8. See note 7; Thomas J. Lewin and V.K. Narayanan, *Keeping the Dream Alive: Managing the Space Station Program, 1982-1986* (Washington, D.C.: NASA, 1990), 111, 116-17.
9. Lewin and Narayanan, 1990, 107-23, 177, 179, 181.
10. Lewin and Narayanan, 1990, 107-23, 177, 179, 181, 1990; Interview, Henry C. Dethloff with Aaron Cohen, JSC, January 3, 1991.
11. Interview, Dethloff with Cohen, 1991.
12. Interview, Dethloff with Cohen, 1991.
13. These are conclusions reached by the author from comments both explicit and implicit in interviews with various JSC persons including prominently, Gerald Griffin, Aleck Bond, Christopher Kraft, Owen Morris, and Max Faget.
14. Sally K. Ride, *Leadership and America's Future in Space, A Report to the Administrator* (Washington, D.C.: NASA, August 1987), 1-63; Aaron Cohen, Biographical Data Subseries, JSC History Office.
15. Aaron Cohen, Biographical Data Subseries, JSC History Office.
16. Interview, Dethloff with Huffstetler, November 11, 1990.
17. See National Academy of Public Administration, The NASA Organization and Its Needs for the Future (June 1985), 1-12, NASA Management Subseries, JSC History Office.
18. Interview, Dethloff with Huffstetler, 1990.
19. Interview, Dethloff with Huffstetler, 1990.
20. Interview, Dethloff with Huffstetler, 1990.
21. Interview, Dethloff with Huffstetler, 1990.
22. *Space News Roundup* (February 20, 1963).
23. *Houston Magazine* 35(September 1964)8:26, 40(May 1969)4:52; See Sullivan Management Subseries, JSC History Office.
24. Fred J. Benson to Honorable Olin E. Teague, January 27, 1966, Olin E. Teague Papers, Texas A&M University Archives, Box 224.

Reference Notes

25. Statement of Homer E. Newell, Associate Administrator for Space Science and Applications, NASA, before the Subcommittee on Space Sciences and Applications, Committee on Science and Astronautics, House of Representatives, Olin E. Teague Papers, Texas A&M University Archives, Box 224.
26. Interview, Henry C. Dethloff with Oran Nicks, December 19, 1990; Interview, Henry C. Dethloff with W. Arthur Porter (former Director, Texas Engineering Experiment Station), The Woodlands, Texas, October 23, 1990; Biographical data, David J. Norton and Oran Nicks, Intaglio Files and Records, College Station, Texas.
27. Interview, Dethloff with Nicks, 1990.
28. Interview, Dethloff with Nicks, 1990.
29. Interview, Dethloff with Nicks, 1990.
30. Space Research Center Files and Records, SRC-History, Texas A&M University, College Station, Texas.
31. Space Advisory Board, January 10, 1985, Meeting; Space Board Advisors Meeting, September 27-28, 1985; Proposal to Define a Space Engineering Research Center (October 1984) in Space Research Center Files and Records, David J. Norton-Correspondence, and SRC-History. Texas A&M University faculty and personnel most actively involved included Norton, Nicks, Walter Haisler, Jo Howze, Carl Erdman, Richard Thomas, Robert Merrifield, Duwayne Anderson, and H.H. Richardson; Memorandum, David J. Norton to Henry C. Dethloff (January 3, 1991), Intaglio Files and Records, College Station, Texas.
32. Memorandum, Norton to Dethloff (January 3, 1991), Intaglio Files and Records, College Station, Texas.
33. Space Research Center, Activity Summary, 1985-1990, Presentation to NASA JSC Officials, November 7, 1990, Space Research Center Files and Records; Interview, Dethloff with Huffstetler, 1990; Interview, Dethloff with Nicks, 1991.
34. Interview, Dethloff with Porter, 1990.
35. Interview, Dethloff with Porter, 1990.; Memorandum, Norton to Dethloff (January 3, 1991).
36. University of Houston, Space Vacuum Epitaxy Center, (n.p., n.d.), Intaglio Files and Records, College Station, Texas.
37. University of Houston-Clear Lake, Space Business Research Center, information sheet.
38. See Official Program and Exhibits Guide, 3rd Annual Space Technology Commerce and Communications Conference & Exhibition, November 14-16, 1989, George R. Brown Convention Center, Houston, Texas; Memorandum, Norton to Dethloff (January 3, 1991).
39. Memorandum, Norton to Dethloff (January 3, 1991).
40. Interview, Dethloff with Cohen, 1990.
41. Interview, Dethloff with Aaron Cohen (and Daniel Nebrig), January 3, 1991.
42. National Commission on Space, *Pioneering the Space Frontier* (New York: Bantam Books, 1986), i-4.
43. National Commission on Space, 1986, 8-11.
44. National Commission on Space, 1986, 11-12.
45. National Commission on Space, 1986, 11-12, 180. Commission members included Luis W. Alvarez, Neil A. Armstrong, Paul J. Coleman, George B. Field, William H. Fitch, Charles M. Herzfeld, Jack L. Kerrebrock, Jeane J. Kirkpatrick, Gerard K. O'Neill, Thomas O. Paine (chairman), Bernard A. Schriever, Kathryn D. Sullivan, David C. Webb, Laurel L. Wilkening, and Charles E. Yeager.
46. Ride, 1987, 5.
47. Ride, 1987, 5-11.
48. Ride, 1987, 12.
49. Ride, 1987, 13, 15-63.

Suddenly Tomorrow Came . . .

50. See JSC organization charts, March 1985, March 1986, April 1987, Organizational Subseries, JSC History Office; James C. Fletcher, Summary of Major Actions Taken or in Work in Response to the Recommendations of the NASA Management Study Group, January 7, 1986, Charlesworth Subseries, JSC History Office; JSC Announcement 86-163, December 11, 1986, General Reference Series, JSC History Office.
51. JSC Announcement 86-163.
52. *Houston Chronicle* (September 28, 1988, 24A and October 24, 1988).
53. *Houston Chronicle* (September and October, 1988).
54. *Houston Chronicle* (September 28, 1988), 24A; Roger E. Bilstein, *Orders of Magnitude: A History of the NACA and NASA, 1915-1990* (Washington, D.C.: NASA, 1989), 137-138.

Chapter 17

1. Interview, Henry C. Dethloff with his grandchildren, Brian and Stephen (age 5), Wylie, Texas, July 20, 1990.
2. *Houston Chronicle* (February 19, 1988, April 8, 1989).
3. *Houston Chronicle,* 1989; *Space News Roundup* (hereafter cited *Roundup*) (September 30, 1988).
4. *Roundup* (September 23, 1988).
5. *Houston Chronicle* (April 8, 1989), 8A.
6. Charles Murray and Catherine Bly Cox, *Apollo: The Race to the Moon* (New York: Simon and Schuster, 1989); *New York Times National* (September 9, 1990); and see Alcestis R. Oberg, *Pioneering the Space Frontier: Living on the Next Frontier* (New York: McGraw-Hill, 1986); *Houston Chronicle* (June 25, 1989, September 21, 1988).
7. *New York Times* (February 12, 1987), reprinted in NASA Current News, February 12, 1987.
8. *Houston Chronicle* (September 26, 1988).
9. *Houston Chronicle* (August 4, 1988); The Gallup Poll, *News Service* (July 19, 1989).
10. *Roundup* (October 3, 1988); *Houston Chronicle* (November 15, 1988).
11. *Roundup* (May 6, October 3, 1988); *Clear Lake Area Magazine* (1st Quarter 1990), 7, 9-12; Manned Space Flight Education Foundation, Inc., *Countdown*, I (March 1988)1: 1-4.
12. *Houston Chronicle* (March 26, 1990); Harold Stall presentation at a Partners in Space Symposium, JSC, May 11, 1989.
13. *Houston Chronicle* (December 9, 1988).
14. *Roundup* (October 28, November 11, December 9, 16, 1988 and January 6, 13, 20, 27, 1989).
15. *Roundup* (March 10, 17, 1989).
16. *Roundup* (April 7, 1989).
17. *Houston Chronicle* (April 16, 1989); *Roundup* (October 22-28, 1990).
18. *Roundup* (March 24, June 2, 1989).
19. *Roundup* (June 16, 1989).
20. *Roundup* (May 26, June 2, 16, August 4, 11, 1989); Aaron Cohen, *Report on the 90-Day Study on Human Exploration of the Moon and Mars* (NASA: Washington, D.C., November 1989).
21. *Roundup* (June 16, November 3, 1989).
22. *Roundup* (June 30, 1989).

Reference Notes

23. *Roundup* (June 30, March 31, 1989).
24. *Roundup* (June 30, March 31, 1989).
25. Aaron Cohen, "Project Management: JSC's Heritage and Challenge," *Roundup* (March 31, 1989); Executive Office of the President, *Quality Improvement Prototype Award, 1990,* JSC, NASA.
26. *Roundup* (March 31, 1989).
27. Interview, Henry C. Dethloff with Glenn S. Lunney, Rockwell International, Houston, Texas, October 25, 1990.
28. Note: The author attended and taped the 3-day sessions, which are also available on videotape from the JSC Public Information Office.
29. *Roundup* (July 21, 1989); *Houston Post* (July 21, 1989).
30. *Roundup* (August 18, 25, October 20, November 3, 1989).
31. *Roundup* (September 15, November 3, December 2, 1989).
32. *Roundup* (December 22, 1989).
33. *Roundup* (November 17, 1989, April 13, August 10, 1990).
34. Aaron Cohen, *Report on the 90-Day Study on Human Exploration of the Moon and Mars,* transmittal letter, Aaron Cohen to Vice Admiral Richard H. Truly, and 1-1 to 1-2.
35. *Report on the 90-Day Study,* i.
36. *Report on the 90-Day Study*, 1-1 to 1-9.
37. *Report on the 90-Day Study*, 1-3 to 1-9.
38. *Report on the 90-Day Study*, 2-1 to 9-13.
39. *Report on the 90-Day Study*, transmittal letter; and see NASA, *The Exploration Initiative, a long-range, continuing commitment* (NASA: JSC [Barrios3], March 1990), n.p.
40. *Roundup* (March 9, 1990).
41. *Roundup* (March 23, 30, April 13, 20, 27, May 18, 25, June 1, 8, July 6, 1990).
42. *Roundup* (May 25, June 1, 8, 15, 22, July 6, 13, 1990).
43. Advisory panel members: Laurel L. Wilkening, the Provost and Vice President for Adacemic Affairs at the University of Washington and a former director of the Lunar and Planetary Laboratory; Edward C. "Pete" Aldridge, Jr., President of McDonnell Douglas Electronic Systems was Secretary of the Air Force (1986-1988) and a former astronaut; Joseph P. Allen, President, Space Industries, Inc. of Houston, a former astronaut, holds a doctorate in physics from Yale University; D. James Baker, President of Joint Oceanographic Institutions, Inc. in Washington, D.C., was a distinguished scientist; Congressman Edward P. Boland received the National Science Foundation Distinguished Public Service Award in 1986; Daniel J. Fink, an independent management consultant, was a former vice president of General Electric's Aerospace Group; Don Fuqua, President and General Manager of Aerospace Industries Association, was a former 12-term Congressman from Florida; Retired General Robert T. Herres once served as Commander-in-Chief of the U.S. Space Command and years earlier was the Chief of the Flight Crew Division for the Air Force Manned Orbiting Laboratory project; David T. Kearns was Chairman of Xerox Corporation; Louis J. Lanzerotti a distinguished professor and member of the technical staff of AT&T Bell Laboratory; Thomas O. Paine, NASA's former Administrator. *Roundup* (August 31, September 21, October 12, 19, 1990).
44. *Roundup* (September 28, October 5, October 12, 1990).
45. *Roundup* (October 19, 26, November 9, 16, 30, 1990).
46. *Roundup* (August 3, November 30, 1990).

47. *Report of the Advisory Committee on the Future of the U.S. Space Program* (Washington, D.C.: Government Printing Office, December 1990), 1-59.
48. *Report of the Advisory Committee,* 1990, 1.
49. *Report of the Advisory Committee,* 1990, 2-3.
50. *Report of the Advisory Committee,* 1990, 3-6.
51. *Report of the Advisory Committee,* 1990, 6-9.
52. *Report of the Advisory Committee,* 1990; *Houston Chronicle* (January 27, 1991).
53. *Report of the Advisory Committee,* 1990, 7-9.
54. *Report of the Advisory Committee,* 1990, 42, 46-48
55. *Report of the Advisory Committee,* 1990.
56. *Report of the Advisory Committee,* 1990.
57. Paul E. Purser and William M. Bland, Jr. to Dr. James C. Fletcher, October 10, 1988; Richard H. Truly to Paul E. Purser and William M. Bland, Jr., December 8, 1988; Purser and Bland to Truly, January 6, 1989, in personal papers of Paul E. Purser.
58. H. Hollister Cantus, Associate Administrator for External Relations, to Distribution, October 6, 1988, with NASA Historical Note HHN-161, Crossing the Next Frontier: Precedents for a National Space Program from the History of the United States (September 1988).

Index

A

A. V. Roe Corporation 10, 24, 25, 350
Aaron, John 179, 183, 309, 335
Abbott, Ira H. 32
Acheson, David C. 300
Activation Analysis Laboratory 314
Advanced Missions Program Office 218
Aerojet-General Corporation 143, 152
Aerospace Industries Association 191
Aerospace Summer Intern Program 314
Aerospace Task Force 317, 318
Aerospatiale 296
Agena-D target vehicle 93
Air Force Ballistic Missile Division 22, 23, 122
Air Force Flight Test Center 295
Air Force Missile Test Center 110
Air Force User Committee 239
Akridge, Max 223, 225
Al-Saud, Prince Sultan Salman 296
Aldrich, Arnold D. 183, 325, 335
Aldrich, Edward (Pete) 183, 315, 317
Aldridge, Eleanor 317
Aldrin, Edwin E. (Buzz) 94, 117, 124, 133, 174, 176, 200, 285
Alford, Roy 78
Allen, Joseph P. (Joe) 179, 275, 294, 302
American Aeronautical Society 10
American Association for the Advancement of Science 59
American Institute of Aeronautics and Astronautics 192
American Space Program vii, ix, 11
   and the Cold War 321
   early development 25, 35
   effect on American education 6, 350
   effect on Gulf War 343
   funding vii, 3
   future goals 320-321, 344
   image vii
   institutionalization of 11
   management challenges 107
   management techniques 61, 77, 209
   post-*Challenger* accident 304-305, 320, 326
   public interest in 123, 133, 150, 187, 226
   reaction to Sputnik i, 1-15
   shuttle era 222-225
American Telephone and Telegraph (AT&T) 83, 289, 296
Ames Aeronautical Laboratory (see Lewis Research Center)
Ames Research Center 26, 37, 69, 188, 201, 224, 277, 295
Anders, William A. 171, 332
Anderson, Lyle 265
Anderson, O. A. 220
Anderson, Sen. Clinton P. 12, 189, 190
Anfuso, Victor L. 18
Anik D2 294
Animals, use in space 30
   Able 32
   Baker 32
   Enos 54
   Ham 32, 38, 53, 55
   Laika 1
   Sam and Miss Sam 32
Apollo Applications Programs 108, 192
Apollo Experiments and Training on the Scientific Aspects of the Apollo Program 128
Apollo Extension System 195, 205
Apollo program ix, 31, 53, 56-80, 84-117, 125-180, 185-203, 259-265, 273-289, 309-314, 330, 336-342, 350
   20th anniversary celebrations 252
   and post-Apollo plans 188, 192, 195
   Apollo 4 134, 155
   Apollo 5 155, 173
   Apollo 6 135, 155
   Apollo 7 135, 155, 198, 165
   Apollo 8 156, 169, 171, 172, 224, 333
   Apollo 9 173, 183, 265
   Apollo 10 160, 174, 175, 176-77, 278
   Apollo 11 117, 133, 156, 159, 174-179, 182, 187, 195, 198, 202, 203, 218, 262, 333
   Apollo 12 179, 198
   Apollo 13 160, 173, 182-184, 196-199, 342
   Apollo 14 180, 187, 198, 199
   Apollo 15 180, 198-201, 265
   Apollo 16 180, 199, 202, 214, 265, 278
   Apollo 17 180, 187, 198-203, 214, 262
   Apollo 18 202
   Apollo 19 202
   Apollo AS204 fire vii, 112, 114, 140, 344
   budget 57, 135, 151
   compared to Mercury and Gemini 162-163
   contractor relations 71, 143, 145
   development plan 100, 103
   effect of space program 113-114
   effect on MSC personnel 115
   engineering and design 145-146
   flight tests 53, 169, 173
   historical significance 179
   improvements resulting from 114
   investigation results 113
   lessons from Gemini 64, 162-163, 165

lunar program 79, 137, 141, 169
major elements 142
management conflicts 71, 105-107
management structure 86, 164, 166-168, 170
mission planning and control 103, 178-181
missions 160
origin of name 159
problems 71, 81, 135
program offices 86, 102, 335
public interest 104, 150
public reaction 117
redundancy systems 173
relationships between centers 111
request for contractor proposals 65, 142
review board 113, 114
subcontractors 147-149
technological advances 135, 174
training 127
Apollo spacecraft 73, 82, 90, 93, 103, 107, 108, 114, 118, 132, 134, 139-159, 173-174, 205-212, 221, 279, 311
life support systems 130
specifications 151
Apollo telescope mount 108, 168
Apollo-Soyuz vii, 188, 201, 207, 209, 210, 219-221, 227, 254, 285
Armed Services Committee 1, 2, 3,
Armstrong, Neil 94, 111, 117, 128, 133, 159, 164, 174, 176, 177, 285, 299, 300
Army Ballistic Missile Agency (ABMA) 19, 23, 27, 61, 73, 74, 105, 246, 312, 350
Army Ballistic Missile Program 4
Astronaut Office 118, 124, 127, 212
Astronauts 23, 24, 31, 32, 44, 51, 57, 58, 63, 66, 68, 72, 74, 91-95, 106, 111, 112, 117, 155-157, 172, 178, 184, 187, 195-200, 211-215, 221, 233, 248-254, 261, 265, 278-282, 286-289, 295-300, 307, 309, 332, 336, 342, 344, 350
insurance 122
media attention 117, 122-123
Mercury astronauts 122
privacy 123, 124
recruitment of 57, 82, 127, 248, 265
salary 119
selection criteria 118, 119, 127-128
selection process 117-122, 129, 248
Shuttle astronauts 248, 249
training 122-133, 160-163, 209, 248, 265
women 120, 248, 287
Atlantis 253, 295, 296, 332-334, 337, 338, 343
Atlas rocket 17, 23, 37, 38, 56, 68, 69, 72, 80, 236, 237, 262, 285, 326, 344

Atomic Energy Commission 19, 61, 153, 193-196, 348
Atwood, J. Leland 138, 139, 142, 171
Augerson, William S. 119
Augustine Report 344-46
Augustine, Norman 317, 343, 344
Aurora 7 55
AVCO 143, 147, 152, 189
AVRO Aircraft, Ltd. of Canada 24-26, 63, 64, 80, 165, 249, 350

B

Baker, Ellen 303, 337
Bales, Stephen (Steve) 95, 174, 182
Barnes, Major General G. M. 8
Barrett, H. Ray 265
Barrios Technology Inc. 261, 262, 264-265, 342
Batdorf, Dr. S. A. 19
Baudry, Patrick 290, 296, 302
Bay of Pigs 30, 187
Bayne, James A. 46, 48, 62, 63
Bean, Alan 128, 179, 215
Becker, Harold S. 223
Before Lift Off 118, 292, 293
Beggs, James 308, 310
Bell Aerospace 152
Bell Aerosystems 147
Bell Corporation 9
Bell, David 27
Bell Laboratories 32
Bell Telephone 83, 259
BellComm 32, 83, 107, 152
Bendix 152, 203
Bentsen, Sen. Lloyd 214, 318, 330
Bergen, William D. 114, 179
Berglund, Rene A. 189
Berry, Dr. Charles 93-95, 113, 119
Berry, G. 66, 67
Berry, Ron 171
Beyer, Dr. D. H. 118
Bingman, Charles (Chuck) 62
Bird, John D. 144
Bland, William A. 21, 262
Bluford, Guion 249, 289
Boeing 73, 152, 190, 203, 224, 240, 253, 254, 264, 280, 329
Bogart, General Frank A. 205, 211
Boisjoly, Roger 304
Bolender, Carroll 167
Bond, Aleck 21, 49, 63, 102, 104, 114, 129, 130, 162, 189, 197, 213
Bonestell, Chesley 320

Borman, Frank 44, 93, 111, 114, 128, 133, 167, 171, 172, 213
Boyer, William J. (Bill) 20, 21, 84, 165
Bradley, Ray 224
Brand, Vance D. 128, 210, 220, 302, 303
Brandon, Floyd 45
Braun, Wernher von 4, 6-9, 11, 12, 19, 23, 27, 30, 73, 99, 101, 104-106, 108, 110, 118, 145, 169, 189, 192, 196, 204, 210, 224, 320, 350
   background 7
   conflicts with Gilruth 106
   contributions to American space program 105
   director of Marshall Space Flight Center 27
   management style 106
   role in Shuttle program 320
   testimony at subcommittee hearings 8
Brewer, Jerry 163
Bridges, Sen. Styles 3, 17
British Aerospace Corporation 281
British Interplanetary Society 25
Broadband X-Ray Telescope 342
Brooks, Congressman Jack 319
Brooks, Congressman Overton 18
Broome, Douglas 114
Brown & Root Construction Co. 30, 37, 47, 48, 259
Brown, Clarence 223
Brown, Clinton E. 144
Brown, George R. 30, 37, 41, 260
Brown, Porter 84
Bundy, McGeorge 124
Bush, Dr. Vannevar 4, 5
Bush, President George 330, 332, 333, 334, 337, 342, 343, 345
Butler, Sherwood 23
Byrnes, Martin 39, 41, 42, 46, 48

C

Campagna, I. Edward 37, 41, 45, 46, 48, 62, 63
Canadarm 287, 291, 292, 294, 297
Cape Canaveral 22, 23, 27, 32, 40, 53-55, 70, 74, 77, 82-85, 97, 109-111, 134, 182, 291, 297, 304
   communications problems 83
   launch facilities 110
   Mercury control center 83
   provides backup to MCC 92, 103, 109, 110
   renamed Kennedy Space Center 111
   role in Shuttle program 229
   space allocations and ownership 83
Cape Kennedy (see also Cape Canaveral)
Carpenter, Malcolm S. 122
Carpenter, Scott 54, 55, 56, 68
Carr, Gerald P. 215

Carrying the Fire 117
Carter, President Jimmy 238
Casey, Bob 41, 42, 51
Centaur rockets 25, 97
Center for Space Power 319
Cernan, Eugene A. 93, 109, 174, 203
Chaffee, Roger 111, 112, 140
Challenger ix, 243, 253, 287, 288, 289, 291-296
Challenger accident 113, 298, 301, 307-312, 314, 317, 321, 326, 344
Chamberlin, James A. (Jim) 58, 63-65, 143
   background 79
   Gemini Project Manager 78-81, 82, 87, 165
   management style 81
   reassigned by Gilruth 81
   role in forming STG 80
Chance-Vought Missile Program 4, 143
Chang-Diaz, Franklin 337
Charlesworth, Clifford E. 95, 165, 169, *photo* 171, 172, 178, 198, 305, 325
Charlie Brown 174
Chilton, Robert G. 58, 59, 81
Chimpanzees, used in spaceflight
   Ham 32, 38, 53, 55, 58
   Enos 54
Chrysler Corporation 73, 224
Chu, Dr. Paul 319
Clagett, A. A. 143
Clarke, Arthur C. 25
Clarke, Stuart 42, 61
Clear Lake, Texas 74, 77
Cohen, Aaron vii, 114, 234, 235, 246, 304, 310-312, 316-318, 320, 325, 331, 334, 336-339, 342, 344
Cold War 88, 156, 179, 187, 204, 220, 331
   effect on space program 187
Collins Radio 143, 152, 260
Collins, Michael *photo* 94, 117, 121, 285
Colonna, Richard A. 325, 326, 334
Columbia 253, 275, 279, 280, 282, 285, 287-289, 291, 297, 336, 342-344
Columbia University Division of War Research 17
Comet Kohoutek 216
Command Module 90, 92, 107-114, 117, 135, 138, 142, 144, 145, 146, 152, 165, 169, 173-176, 179, 182-184
Commercialization of space 257-269, 344
   federal support for 257
   small businesses and 257
Committee on Armed Services 1
Committee on Astronautics and Space Exploration 28

Committee on Science and Astronautics 18, 31
Committee on Scientific Training and Facilities 18
Communications Satellite Corporation 300
Communications, worldwide 77, 95, 126, 138, 140, 152, 156
    computer technology 83
    contact with spacecraft 82
    interface problems 83
    light control stations 84-85
    technological advances 84, 86
Compton, William David 153, 252
Computer integrations systems 82
Configuration management 77, 88, 90, 91, 92
Conrad, Charles 93, 94, 133, 179, 214
Construction of MSC 46, 48
    achievements 47-49
    contracts awarded 50
    costs 49
Contracting 77, 80, 81, 91, 102, 122, 137-157
    benefits of 137
    cost-plus-fixed-fee 138, 151
    cost-plus-incentive fee 88, 91, 151
    impact of government regulations 154
    NASA/Contractor relationships 137, 138
    necessity of communications 140
Contractors 36, 48, 50, 60-65, 70-74, 80, 85, 92, 97, 98, 102, 106, 107, 113, 114, 134-155, 164, 174, 178, 184, 197, 203, 205, 209, 214-230, 257
    Apollo program 65, 150-154
    Shuttle program 222, 224
Control Data Corporation 247
Controlled reentry 77, 95
Convair 70-73, 142, 224, 230
Cooper, Henry, Jr. 118, 292, 293
Cooper, Leroy Gordon 68, 69, 70, 82, 93, 122, 133
Corbet, Wayne 223
Corps of Engineers
    role in design and construction of the MSC 46
    supervision of MSC construction 47
Cortright, Edgar M. 99, 104
Covert, Dr. Eugene E. 300
Covey, Richard 326
Covington, Clarke 325
Cox, Catherine Bly 330
Craig, Jerry W. 114
Crippen, Robert 248, 279, 288, 325
Croneis, Carey 37, 41
Cuban Missile Crisis 187
Culbertson, Phil 308
Cunningham, R. Walter 155, 165, 317
Curtis-Wright 4

D

D'Orsey, Leo 122, 123
Daniel, Price 50
Darrow, Whitney, Jr. 95
David Clark Company 174
Davis, Hubert P. and Mary 262, 263, 264
Davis, Jeff 42
Davis, Morgan 41
Debus, Kurt H. 100, 109, 110, 192, 196, 210
Deifenbaker, Prime Minister John 24
Delta rocket 291, 307
Dembling, Paul 37, 124
Denver Aerospace 317
Department of Defense (DoD) 2, 4, 9, 13, 83, 98, 160, 281, 285
    reaction to Sputnik 3
    role in Apollo flights 160, (sidebar 161-162)
    role in Shuttle program 238, 295
Digital Electronics Corporation 247
Discovery 253, 276, 289-297, 326, 329-333, 342
Disher, John H. 104
Dixon Gun Plant 37
Donlan, Charles 17, 23, 25, 60, 120, 227
Doolittle, General James H. 4
Douglas Aircraft 4, 9, 73, 109, 143, 189, 193
Douglas, Donald W. 4
Dryden Flight Research Center 97, 265
Dryden, Dr. Hugh L. 14, 18-19, 23-30, 32, 35-41
Duff, Brian 181
Duke, Charles M., Jr. 202, 254, 265
Duke, Michael B. 316
Dula, Art 262
Dyna-Soar 9, 189, 240

E

Eagle 176-178
Eagle Engineering 257, 261-264
Eagle-Picher Company 152
Earth Observations Aircraft Program Office 218
Earth Radiation Budget Satellite (ERBS) 293
Earth Resources Program 209
Economic impact of MSC 150, 209-214, 223, 230, 257-262
Edwards Air Force Base 14, 97, 282, 291, 295, 296
Eggleston, John 180
Eighth International Astronautical Federation Congress 2
Eisele, Donn 155, 165
Eisenhower, President Dwight D. 1-4, 8, 12, 13, 18-20, 24, 27, 28
Electronic Research Center 194
Ellington Air Force Base 36, 41, 46

Elms, James C. 78, 86, 141
    background 86
    named Deputy Director of MSC 87
    returns to private industry 88
Emergency escape device 17
Emme, E. M. 196
Endeavour 334, 335
Engineering Computation Facility 247
Engle, Joe H. 280
Enterprise 253, 254, 280, 282
Environmental Protection Agency 154
Epitaxy Center 319
Escape tower 22
Eudy, Glenn 304
European Space Agency 266, 281, 285, 289, 290, 297
Evans, Ronald 203
Explorer I 12
Extravehicular activity (EVA) 56, 77, 93, 94, 174, 274, 287
    and Space Lab 215
    lunar 178, 179
    training 131

F

Fabian, John M. 288
Faget, Maxime A. (Max) vii, 10, 14, 17, 19-23, 29, 37, 44, 46, 49, 59-65, 79, 81, 84, 87, 106, 111, 114, 143, 150, 153, 160, 164, 184, 188, 190, 218, 225, *photo* 280
    heads Flight Systems Office 64
    MSC Assistant Director 63
    resigns to work in private sector 262
    role in design of MSC 49
Fairchild Republic 326
Feltz, Charlie 141
Feynman, Dr. Richard P. 300
Fielder, Dennis 78, 82, 84, 85, 164, 165
Fisher, Anna 288, 294
Fisher, William F. 288, 297
Fleming, William (Bill) 104
Fletcher, James C. (Jim) 209, 214, 221, 229, 233, 238, 309-311, 319, 321, 325, 330, 333
Flickinger, Brigadier General Donald D. 120
Flight control 24, 55, 77, 82-85, 103, 163, 164-183, 212, 220, 245, 274-276, 287
Flight endurance 53, 56, 77, 94
Flight simulation systems 86
Flight Systems Office 59, 79
Flory, Donald A. 153
Ford Aerospace 257
Ford, President Gerald 221, 238

Forward, Dr. Robert L. 337
Foster, Willis 153
Freedom 7 32, *photo* 33, 38, 54, 124
Frick, Charles W. 64, 67, 70, 86
Friendship spacecraft 55
Friendswood Development Corporation 260
Frosch, Dr. Robert A. 238
Fullerton, Charles Gordon 128, 282, 290, 302

G

Gagarin, Yuri 29, 32, 36, 38, 64, 74
Galileo spacecraft 334, 337
Gallup Poll 331
Garn, Sen. E. J. (Jake) 290, 295, 302
Garneau, Marc 293, 302
Garrett Corporation 143, 147, 152, 174
Garriott, Owen K. 215
Gavin, James M. 29
Gemini Incentive Task Group 91
Gemini program ix, 31, 32, 53, 55, 61-65, 70-74, 77, 90, 97-111, 115-133, 138, 140, 141, 145, 155, 163-167, 174, 179, 181, 188, 189, 197, 205, 210, 216, 229, 230, 233, 238, 240, 274, 276, 278, 279, 287, 303, 311, 312, 336, 350
    contracts 79
    costs 80
    initiated 56
    lessons from 94
    management requirements 87
    origin of name 79
    reconfigured 74,
    role of private sector 79
    technological achievements 83, 85, 95
Gemini Project Office 63, 64, 67, 79-81, 104, 165
    improvements 92
    management crisis 81
Gemini spacecraft 89, 164
    construction difficulties 141
    Gemini 1 56
    Gemini 2 86
    Gemini 3 86, 92, 93, 134, 279
    Gemini 4 92
    Gemini 5 93
    Gemini 6 93
    Gemini 7/6 58, 93, 163
    Gemini 8 93, 111, 164, 265
    Gemini 9 93, 111
    Gemini 10 94, 95, 111
    Gemini 11 97
    Gemini 12 97
    interior design 93
    operations and mission control 82

specifications 80
two-man 80
unmanned 77
Gemini-Apollo Executive Group 140
General Dynamics 22, 64, 65, 142, 143, 224, 225, 226, 230, 308, 311, 329
General Electric 64, 113, 142, 143, 146, 152, 211, 224, 230, 259, 263, 266, 329
General Motors 139, 148, 151, 152, 174
General Precision 174
General Services Administration 45
Georgia Institute of Technology 211, 279, 281
German rocket scientists 5-9, 17, 105-106, 297, 300
German Rocket Society 6
German Space Operations Center 297
Germany, Daniel M. 334
Getaway Specials 282, 288, 289, 295-298
Gibson, Edward G. 128, 215
Gibson, Robert L. (Hoot) 249, 288, 290, 302, 303
Gillespie, Ben 42, 46
Gilruth, Robert Rowe (Bob) viii, 10, 11, 14, 15, 17, 18-89, 101-114, 119-127, 133, 134, 141-146, 154, 169-172, 180, 181, 184, 189, 190-198, 209, 219-229, 260, 314, 337, 350, *photo* 352
    abolishes Mercury Project Office 74
    and NASA growth pressures 26
    announces cost cutting program 88
    assigns "tiger team" after Apollo fire 114
    commissions JSC self-study 209
    concerns about post-Apollo goals 192, 193
    conflicts with von Braun 106
    contributions to the MSC 58
    creates Apollo Project Office 65
    creates Gemini Project Office 79
    Director of MSC 77, 101
    Director of STG 23
    Gilruth system 53, 58, 60
    influence on space program 59-60, 125
    management style 58-59, 60, 106
    organizes MSC 63-65
    organizes Space Task Group 21, 35, 350
    reaction to Houston location 40-41
    reorganizes MSC 82, 87
    resolves "Gemini Management Crisis" 81
    retirement 198
    role in Apollo flights 169
    role in locating MSC to Houston 37
    transfers to Houston 47
Glenn L. Martin Company 9, 79, 86
Glenn, John 54, 55, 68, 85, 121-125, 128, 295
Glennan, Thomas Keith 15, 19, 20, 23, 27-29, 35, 36, 37, 38, 119

Goddard, Robert H. 6, 7, 26, 350
Goddard Space Flight Center 22, 26, 27, 31, 35-37, 83, 85, 86, 92, 97, 110, 160, 180, 181, 234, 239, 248, 266, 275, 290, 292, 329, 350
Goett Committee 64
Goett, Harry 26, 27, 35, 36, 188
Goetz, Robert C. 198, 308, 310
Goodwin, Burney 42
Goodyear Aircraft Corporation 79
Goodyear Tire and Rubber 260
Gordon, Richard F. 94, 133, 179
Graham, William R. 308-310
Gramm-Rudman-Hollings Deficit Reduction Bill 343
Graves, Barry 84, 87, 88
Gray, Wilber H. 174
Griffin, Gerald 95, 171, 172, 179, 181, 184, 197-199, 295, 307-311, 315, 317, 319, 332, 352
Griffin, Richard L. (Larry) 301, 310
Grillo, Steve 123
Grimwood, James 79
Grissom, Virgil I. 54, *photo* 55, 88, 92, 111, 112, 121-128, 140
Gross, Robert E. 4
Ground-based computer control systems 74
Ground Rules for Manned Lunar Reconnaissance 64
Grumman Aerospace/Boeing 224, 257, 261
Grumman Aircraft Corporation 93, 142, 143, 154, 167, 189, 190, 193, 203, 230, 250, 326
    develops the lunar excursion module 82, 114, 144, 152, 173, 174
Gulfgate Shopping Center 43, 45
Gulfstream II 249
Gustke, Brigadier General Russel F. 37
Guthrie, George 126

H

Hacker, Barton 79
Hage, George 169, 181
Hagerty, James C. 2
Haise, Fred W. 182
Hale, N. Wayne, Jr. 159
Hall, Walter 193
Halley's Comet 297
Hamlin, Jim 86
Hammack, Jerome B. 102, 104, 161, 165, 170, 213, 218, 262, 266
Hamill, Major James P. 8
Hanaway, John 262
Haney, Paul 78, 92, 93, 104, 133, 134, 170, 181
Hannah, David, Jr. 266, 281

Hansen, James E. 14
Hanson, Grant 225
Hartke, Sen. Vance 235
Hartman, Harvey 324, 331
Hauersperger, Karla 282
Hawley, Steven A. 288, 290, 302, 303
Hayden, Sen. Carl 12
Hayes International Corporation 260
Hazard Analysis Group 304
Healy, John P. 114
Hearings 4-5, 11-14
Hello, Bastian 114
Hermes II program 9
Hernandez Engineering 261, 266
Hernandez, Miguel A., Jr. 262, 265, 266
Hess, Dr. Wilmot N. 170, 180, 181
Hicks, Roger 335
High altitude rocket research 7-9
High Speed Flight Center (see Dryden Flight Research Center)
Hilmers, David 326
Hjornevik, Wesley 20, 35, 36, 37, 39, 42, 44, 45, 61- 66, 70, 87, 89, 91, 119, 143, 170, 230
Hobby, William P. 108
Hodge, John D. 64, 83-86, 92, 94, 162-165, 225
Hoffman, Samuel K. 138
Hoften, James Van 290, 292, 297, 302
Holley, I. B., Jr. 105, 106
Holloway, Tommy W. 301
Holmes, D. Brainerd 32, 37, 86, 101
   background 71
   Director 71
   heads Office of Manned Space Flight 51, 107
   reorganizes NASA administrative systems 107
Honeywell, Incorporated 143, 149, 152
Horton, Eugene 42, 125, 134, 254
Horwitz, Solis 2
Hotz, Robert B. 300
Houbolt, Dr. John C. 144, 145
House Committee on Science and Astronautics 20, 31, 38, 39, 41, 42, 50, 102
House Select Committee hearings 14
House Select Committee on Aeronautics and Space Exploration 17, 28
House Subcommittee on Manned Space Flight 187
House, R. O. 20
Houston Advanced Research Center 317
Houston Post 108
Houston, Texas 5, 15
   celebrates opening of MSC 50-51
   chosen by NASA for "spaceflight laboratory" 33, 35-42
   first considered for location of manned spacecraft program 36
   local economy and the space program 150, 257, 259, 260, 350
   support for MSC 42, 257
Hubble Space Telescope 334, 338, 342, 343, 344
Huffstetler, William 312, 313, 324
Hughes Aircraft 4, 143
Humble Oil Company 37, 41
Humphrey, Vice President Hubert 191
Huntoon, Carolyn L. 316, 324, 331
Huntsville, Alabama 17, 19, 23, 27, 145, 169, 215, 246, 350
Hurley, Roy T. 4
Hurricane Carla 41, 42
Hutchinson, Neil 309
Hyland, Lawrence 4

I

IBM 84-86, 146, 152, 230, 247, 248, 257, 259, 326
Indonesian Government 288, 291
Industrial support of space program 103, 137-157
Insects in Flight Motion Study 282
International Geophysical Year 2, 9
International Latex Corporation 152, 174
Intrepid 179
Introduction to Outer Space, An Explanatory Statement 13
Irwin, James 128, 200, 201, 254

J

James, Lee B. 169
Jeffs, George 317
Jet Propulsion Laboratory 7, 9, 97-98, 277, 317
Johns Hopkins Hospital 281
Johnson and Johnson 282
Johnson, Caldwell C. 22, 224
Johnson, Harold I. 126
Johnson, Kimble 25
Johnson, President Lyndon B. ix, 1- 3, 12, 13, 15, 17, *photo* 18, 19, 28, 29, 30, 32, 41, 42, 50, 51, 73, 74, 82, 88, 113, 135, 165, 178, 187, 188, 190-192, 196, 214
   background 5
   compares Sputnik crisis to Pearl Harbor 4
   holds subcommittee hearings 4-5, 11
   signs first space treaty 140
Johnson Space Center (JSC) (see Manned Spacecraft Center)
Johnston, R.E. 262, 263
Johnson, Ray 19
Johnston, Richard S. (Dick) 49, 114

Juno launch vehicle 9, 12, 72
Jupiter rocket program 9, 12, 22, 73, 110, 334, 337, 343

K

Kaplan, Marvin 45
Kapryan, Walter J. 110
Karman, Theodore von 7
Keldysh, Mstislav V. 218
Keller, K. T. 9
Kelly, William R. 211, 324, 331
Kennedy, President John F. ix, 22, 27, 28, 30-33, 39, 41, 54, 123, 124, 145, 160, 176, 178
    announces lunar landing initiative 141, 187
    effect of assassination on space program 74, 111
    relies on Johnson for space leadership 29
    visits MSC 51, 73
Kennedy Space Center (KSC) (see also Cape Canaveral) 74, 85, 86, 97, 98, 101, 103, 107, 109-115, 133, 134, 150, 151, 160, 169, 179, 182, 184, 196, 198, 199, 202, 210, 211, 215, 225, 227, 234, 253, 277, 287, 289, 295, 325, 337, 338
Kentron International 264
Kerr, Sen. Robert 50
Kerwin, Lt. Commander Joseph 128, 129, 215
Killian, Dr. James R. 4, 17
King, Elbert A., Jr. 153
Kinzler, Jack 23, 66, 89
Kissinger, Henry 229
Kleinknecht, Kenneth S. 11, 56, 65, 67, 81, 89, 103, 106, 143, 174, 213
    becomes Manager of Project Mercury 61, 63
    heads Gemini Incentive Task Group 91
    manages Skylab Program Office 211
    moves to Houston 56, 61
Kohl, Chancellor Helmut 291
Kohrs, Richard (Dick) 304, 305, 324, 325, 335
Koppenhaver, James T. 143
Kotanchik, Joseph 114
Kraft, Christopher 10, 21, 55, 58, 59, 60, 63, 65, 66, 81-89, 92, 93, 103, 109, 144, 163, 164, 165, 169, 170, 172, 179, 181, 184, 198, 199, 211-215, 235, 238, *photo* 280
    designs Mercury control center 86
    heads Flight Operations Division 73
    Mission Operations Director 93, 209, *photo* 352
Kranz, Eugene 84, 92, 170, 175, 181, *photo* 280
    background 163
    role in Apollo flights 160, 165, 171-173, 176-178, 183, 184
Kranzberg, Dr. Melvin 281
Kruschev, Nikita 27

Kubasov, Valeriy 220, 221
Kurbjun, Max C. 144
Kusske, Amy 282
Kutyna, Major General Donald J. 300
Kyle, Howard 84, 165

L

LaBerge, Walter 86
Lambert, C. Harold 335
Landing Safety Team 305
Lang, Dave W. 58, 61, 65, 143, 260
Lange, Oswald H. 143
Langley Aeronautical Laboratory 10, 11, 14, 15, 19, 23, 24, 83, 97, 350
Langley Research Center 31, 35, 36, 45, 58, 61, 98, 99, 144, 234, 238, 240-243, 277, 308, 315, 335
Lauderdale, Lloyd 317
Launch Abort and Crew Escape Team 305
Launch vehicle, capabilities 236, 294, 295, 309, 312, 339
    man-rated 130
    reusable 72, 237
Lead Center Management System 233, 346, 273, 311
Leadership Report 321, 325
Lee, Chester M. 181
Lee, Dr. William A. 174
Leonov, Alexei A. 220, 221
Levine, Arnold S. 137, 194
Lewis Flight Propulsion Laboratory 14, 20, 21
Lewis Research Center 17, 32, 35-37, 97, 224, 274, 277, 278, 311, 329
Lewis, George 21
Lichtenberg, Bryon 289, 302
LifeSat 313
Life Sciences Advisory Committee 119
Lilly, William E. 102
Lindburgh, Charles 7
Lindeman, Richard E. 112
Lippmann, Walter 1
Lister, Jack 329
Little Joe launch vehicle 10, 17, 23, 38, 53, 60
Llewellyn, John S. 165
Lockheed 4, 142, 143, 152, 221, 224, 230, 250, 257, 259, 260, 261, 264, 279, 308
Loftus, Joseph P., Jr. 79, 81, 123, 125, 236, 263, 286, 313
Long, Dr. Donlin M. 281
Long Duration Exposure Facility (LDEF) 291, 292, 334, 338
Los Alamos Laboratory 160
Lounge, Mike 290, 302, 303, 326

Lousma, Jack R. 215, 282, 302
Lovelace, Dr. Randolph II 119, 120, 238
Lovell, James A., Jr. 93, 171, 182, 288
Low, George 17, 21, 22, 27-29, 36, 37, 42, 57, 58, 64, 70, 71, 88, 100, 102, 104, 103, 105, 107, 108, 109, 113, 114, 120, 133, 143, 188, 219, 227, 229, 237, 303, 336
    Apollo Program Manager 114, 169
    background 21-22
    head of Manned Lunar Landing Task Group 36
    interface with STG 100
    named Deputy Director of MSC 88
    role in Gemini program 104
    role in Mercury program 73
LTV Corporation (Ling-Temco-Vought) 259, 317
Lucid, Shannon 337
Luna spacecraft 27
Lunar and Mars Exploration Activity 335
Lunar and Planetary Institute 251, 252, 254, 315
Lunar Excursion Module (LEM) 73, 82, 98, 107, 114, 134, 144, 159
Lunar expeditions 179, 187
Lunar landing (see Apollo program) 54, 64, 65, 73, 77, 93, 94, 95, 103, 109, 133, 134, 142, 144, 145, 156, 165, 171-180, 187-202, 205, 331, 333
    20th anniversary 337
    initiative 33, 141
Lunar mission inception and design ix, 31
Lunar orbital missions 169
    training 127
    versus Earth orbital missions 64, 160
    world reaction 157, 175, 179
Lunar module 159, 160, 169, 182, 200, *photo* 200
Lunar probes 192
Lunar Receiving Laboratory 153, 154, 180, 251, 252, 254
Lunar roving vehicle 200
Lunar soil 199
Lunar-Mars initiative 335, 337, 339, 342-345
Lunney, Glynn 95, 165, 178, 179, 182, 184, 220, 278, 307, 337

M

Malfunction procedures 172
Malina, Frank J. 7
Man-rated design criteria 279
Managing NASA in the Apollo Era 137
Manned Maneuvering Unit (MMU) 291, 294
Manned Orbiting Laboratory Program 279
Manned satellite 22, 35
Manned space laboratory 64

Manned space program 17, 18, 20, 26-28, 72, 77, 112, 113
    debate over 30
    early years 28
    economic benefits 31
    funding 31, 36, 58, 75
    personnel and management problems 32, 107, 108
    public reaction 30, 35, 150
Manned Spacecraft Center (MSC) (also Johnson Space Center) vii, ix, 10, 14, 15, 19, 20, 22, 33, 36, 78, 97-115, 273, 329, 350
    aftermath of *Challenger* accident 307, 314, 321, 335
    becomes fully operational 47, 53, 77, 78
    capabilities 312, 338
    computer technology 247-248
    construction cost 38, 40, 49-50, 75
    construction design criteria 47
    contractor relations 71, 257
    division of labor and administrative structure 22
    economic impact on Texas economy 150, 257, 259, 268-269, 350
    effects of layoffs 211-14, 216, 233, 261, 268
    engineering facilities 242, 246, 251
    engineering philosophy 209
    inception 20, 53
    lead center 235-255, 311, 335, 336, 350
    management 60--66, 97-118, 307, 336, 347, 350
    move to permanent quarters 75
    organizational problems 86-87
    organizational review (1972) 205
    personality 99-101, 105
    personnel 63-64, 75, 77
    political influence on new location 41-42, 260
    potential sites 39
    program 103, 105, 107-109, 117, 312, 350
    program offices 63
    relations with NASA Headquarters and other centers 104-106, 307
    relocated to Houston, Texas 37-40, 63, *photo* 40
    renamed Lyndon B. Johnson Space Center 214
    reorganization 87, 89-90, 211, 216
    responsibilities during Apollo 64
    role in research and development 255
    role in Shuttle program 229, 233-255, 260, 311, 319, 350
    senior staff meetings 68-69
    shuttle laboratories 241, 251
    social consciousness 254
    temporary office locations 46, *map* 48
    ties with the academic community 313-318

Manned spacecraft network 84
Maneuverable manned satellite 188
Marine Corps 215, 282
Mark, Hans 290, 295, 308, 332
Marquardt Corporation 143, 148, 152, 174, 188, 189
Mars 12, 13, 192
Mars Mission 313, 320, 321, 330-345, 350
Marshall Space Flight Center 27, 36, 73-75, 97-113, 138-144, 150, 160, 167, 169, 189, 196, 203, 210-215, 221-230, 233, 234, 253, 277, 304, 307, 311, 312, 326, 329
    role in Apollo program 107, 108
    role in Gemini program 107
    role in Space Shuttle program 222, 253
    Spacelab Program office 281
Martin Company 79, 86, 142, 143
Martin, Lou 113
Martin Marietta 64, 65, 114, 203, 211, 224, 263, 317, 343
Massachusetts Institute of Technology 4, 5, 23, 60, 64, 65, 199, 289, 300
Mathews, Charles W. 11, 17, 21, 23, 25, 62, 81-84, 126, 143, 165, 189
    background 81-82
    named head of Gemini Project Office 81, 87
    named head of Spacecraft Research Division 63
Mathews, Fred 84
Mattingly, Thomas K. 128, 182, 202, 203, 282, 290, 302
Mayer, John P. 21, 165, 212
McAuliffe, Christa 293, 290, 299, 303
McCandless, Bruce 128, 291, 302, 303
McCarty, Bill J. 304
McCormack, John W. 12, 17
McCulley, Mike 303, 337
McCurdy, Howard E. 330
McDivitt, Jim 44, 93, 128
McDonnell Aircraft Corporation 23, 63, 79-82, 91-93, 141, 143, 189
    and incentive fee contracts 91, 151
    manufactures Gemini spacecraft 79, 82
    manufactures Mercury spacecraft 8, 80
    suggests Gemini 7/6 rendezvous 93
McDonnell Douglas 211, 223, 224, 230, 257, 261, 266, 282, 293, 317, 329
McElroy, Neil H. 2, 3, 4
McKenzie, Joe 65
McKenzie, Virginia 65
McKinley, C.H. 317
McMullen, Thomas H. 181
Medaris, Major General John B. 9, 110

Medical Operations Office 67, 119, 170
Mendell, Dr. Wendell 252
Merbold, Ulf 289, 302
Mercury astronauts 51
Mercury Mark II 79, 80
Mercury program project ix, 10, 14, 17, 22-32, 36-38, 53-74, 77, 79-87, 89, 94, 95, 97, 98, 100, 103-115, 118-133, 144, 146, 152, 155, 162-165, 179, 188, 216, 224, 229, 230, 233, 287, 303, 311, 312, 331, 336, 350
    completion of 53, 70-71, 82
    costs 28
    management 73, 86
    problems 28, 53
    public reaction 82
    successes 53
Mercury Project Office 48, 55, 56, 61, 63, 74, 211
Mercury spacecraft 23, 53, 54, 80, 87, 107, 109, 133, 138, 189, 197, 205, 274
    flight control 55, 162
    redesign 79
Mercury-Atlas 38, 55, 56, 69, 70, 80
Mercury-Redstone 3 54, 69
Mexico 85, 296, 297
Meyer, Andre 17, 22
Michael, William H. 144
Military role in space program 10-14, 18, 129
Miller, John Q. 304
Miller, Thomas D. 192
Miller, George P. (Rep.) 18, 39, 50
Miller, Philip 37, 39
Minneapolis-Honeywell 152
Minority astronaut candidates 288
Missile and satellite programs 1, 2, 4, 5, 11
Missile Firing Laboratory 110
Missile gap ix, 5, 26, 27, 28
Mission Control Center 85, 86, 103, 117, 131-135, 152, 160, 165, 172, 173, 176-184, 197, 202, 214, 241, 247, 248, 274-278, 288, 292, 294, 326
    at Cape Canaveral 53
    malfunction procedures 164
    press access to 182
    reestablished in Houston 74, 82, 84
    role in Apollo flights 159, 163, 175
Mission Specialists 248, 249, 337
Mississippi Test Facility 278
Mitchell, Edgar Dean 128, 199, 265
Mitchell, George 318
Moore, Jesse W. 308, 310, 311
Morgan, Barbara 293
Morris, Owen 169, 171, 172, 235, 257, 262-264, 273, 278

Morton, Thiokol 301, 326
Motorola 148, 152
Mtekateka, Bishop Josiah 93
Mueller, George E. 71, 74, 99-102, 104, 108, 123, 133, 140, 141, 146, 159, 169, 190-196, 223-227
   background 101-102
   relations with Congress 102
   role in Headquarters administrative changes 101,
Murray, Charles 139, 330
Musgrave, F. Storey 128, 287
Muskie, Edmund 235
Myers, Dale D. 146, 184, 196, 210, 226, 227, 238

N

Nagy, Alex P. 79
National Advisory Committee for Aeronautics (NACA) 4, 6, 9-14, 17-22, 57-59, 61, 70, 81, 86, 97, 98, 101, 105, 114, 123, 137, 224, 234, 262, 339, 347, 350
   appropriations bills 36, 38, 191
   conference on high speed aerodynamics 14
   differences between NACA and NASA 59
   founded by Congress 10
   management 99
   pilotless aircraft research 17
   public relations activities 77
   reconstituted as NASA 10
NASA Advisory Committee 344
NASA Appropriations Bill 36-38
NASA Commercial Centers of Excellence 317
NASA Oversight Committee 190, 259
NASA Planning Steering Group 195, 225
NASA Road One 48
NASA Source Evaluation Board 143
NASA spacecraft center 22
National Academy of Sciences 13
National Advisory Committee for Aeronautics (see NACA)
National Aeronautics and Space Act 11, 14, 18
National Aeronautics and Space Administration (NASA) ix, 4, 10, 14-51, 57-65, 68, 70-75, 77-93, 97-115, 117-146, 150-156, 159-194, 209-211, 214-231, 257-267, 273-282, 285-305, 329-347, 307-328
   budget and personnel 152, 153, 154
   funding 36, 342
   Headquarters 21, 22, 35-38, 41-48, 78, 79, 98-111, 114-127, 133, 134, 138, 140, 143, 152, 153, 169, 181, 189, 193, 195, 198, 205, 209-210, 214, 220, 223-226, 273, 295
   history 97-98
   locates "spaceflight laboratory to Houston" 35
   organizational problems 31, 100
   program offices 32
National Aeronautics and Space Agency 14
National Aeronautics and Space Council 14, 28, 30, 192
National Broadcasting Company (NBC) 78
National Commission on Space 307, 320, 321
National Electronics Corporation 154
National Oceanic and Atmospheric Administration 160
National Science Foundation 3, 13
National Science Teachers Association 282
National Space Council 307, 318, 333, 338, 343, 346
National Space Transportation Program Office 307, 325, 335
Naval Research Laboratory 2, 9, 12, 26, 293
Nebrig, Daniel A. 320
Neil, Roy 78, 331
Nelson, Congressman Bill 298
Nelson, George 326, *photo* 292, 297
Nelson, Todd 282
New Initiatives Office 268, 305, 312-317, 325, 327
New Projects Panel 64
Newell, Homer 32, 153, 195, 196, 210, 225
Nicholson, Leonard 301, 305, 335
Nicks, Oren 28, 29, 102, 315, 316, 318
Nieman, Ed 163
Nixon, President Richard M. 27, 28, 177-179, 196, 203, 204, 214, 218-229, 257-261
   authorizes development of Space Shuttle 206, 235
   goals for Shuttle program 229, 257
North American Aviation 69, 73, 74, 78, 79, 86, 93, 109-114, 137, 138, (sidebar 139), 141-146, 151, 152, 171, 173, 178-190, 222, 250
   and paraglider landing system 75, 79
   constructs launch escape tower 109
   independent reports on Apollo fire 113
   merges with Rockwell 111
   Rocketdyne Division 138
   wins contract for Saturn V booster 74
   wins contract for 3-man spacecraft 73, 143
North American Rockwell Corporation 37, 137, 179, 203, 211, 220-234, 250, 257, 261, 262, 287
North, Warren 17, 19, 22, 78, 83, 104, 127
Northrop Corporation 143, 152
Norton, David J. 315, 316, 317, 318
Notkin and Company 154
Nova launch vehicle 64, 72

O

Oak Ridge National Laboratory 153
Oberth, Hermann 6, 7, 350
Occupational and Safety Act of 1970 154
Occupational Safety and Health Administration 154
Office of Commercial Programs 319
Office of External Relations 198
Office of Management and Budget 194, 235
Office of Manned Space Flight 32, 37, 51, 99, 103, 104, 107, 128, 129, 146, 190, 227
Office of Space and Terrestrial Applications 280, 293
Office of Space Flight Programs 36
Office of Space Science and Applications 98, 128, 129, 152, 282, 293
Office of Space Station 308
Office space 45, 46
    interim facilities, Houston 1962, *map* 48
Omniplan 342
Operation Paperclip 8
Operations Support Contract 342
Orbital maneuvering system 230
Orbiter Project Office 234, 235, 245, 311, 326
Ordnance Guided Missile Center 9
Outer space, development of 12

P

Pacsynski, Al 249
Paine, Thomas O. 169, 184, 196, 209
    head of National Commission on Space 307
    NASA administrator 104, 160, 218, 227
Paraglider landing system 74, 80
PARD 10, 11, 21
Parker, Robert 128
Parker Seal Company 304
Parker, W. A. 42, 45
Parsons, John M. 37, 39
Partners in Recovery 161
Partners in Space 288, 318
Patillo, Pat 46
Paup, John W. 138, 141, 142, 143
Payload Integration Office 245, 278, 282
Payload Operations Control Center 275
Payload Specialists 249, 289, 293, 295, 296, 297
Pearson, Ernest O., Jr. 36
Peck, Robert 42
Pentagon 120, 295
Perkins, Ray 264
Perot, H. Ross 332
Perry, S. O. 4
Pesek, Joan 44

Peterson, Carl 223, 262
Peterson, Donald H. 128, 287
Peterson, John 78
Petrone, Rocco A. 169, 209
Philco 84, 85, 86, 166, 167, 260
Philco-Ford 146, 152, 259
Phillips, Franklin 43
Phillips, Sam 101-3, 104, 107, 132, 134, 169, 181, 189, 257, 309, 311, 319
Phillips, W. Hewitt 58, 59
Physical effects of long-duration spaceflight 95
Piland, Bob 60, 64, 65, 70, 73, 143, 180, 212, 216, 262
Piland, Joseph V. 154, 180, 212, 216, 262
Pilotless Aircraft Research Station (see also Wallops Station) 10, 14, 17, 97, 350
Pitzer, Kenneth 41
Pogue, William R. 128, 215
Pohl, Henry O. 81, 242, 246, 324
Polaris Missile 9, 56
Porter, W. Arthur (Skip) 316, 317, 318
Powers, John A. (Shorty) 41, 42, 77, 78, 122-125, 133, 201
Pratt & Whitney 74, 143, 146, 149, 152, 174
Precision Rubber Products 304
Preparedness Investigating Subcommittee hearings 1-5, 8, 11, 12
President's Scientific Advisory Board 17, 113
Pressure suits 74, 77, 80, 126, 131
Pressurized capsule 17
Preston, G. Merritt 17, 22, 35, 55, 65, 66, 83-89, 108-110
Project HYWARDS 9
Project Mercury (see Mercury Program Project)
Project Offices 64
Project ORDCIT 8
Project Rover 3
Propulsion systems 97, 141-144, 155, 168, 181, 183, 184, 211, 222, 273, 339, 342
Proxmire, Sen. William 191, 235
Public Affairs Office 78, 89, 104, 117, 125, 134, 324
Public relations 77-78, 104, 123, 133
Purser, Paul 10, 11, 17, 19-25, 35, 37, 50, 58, 63, 65, 67, 78, 81, 82, 104-106, 189, 195, 314

Q

Quarantine procedures 199
Quayle, Vice President Dan 317, 333, 343

R

Radiation, Inc. 174

401

Radio Corporation of America (RCA) 71, 83, 152, 174, 257, 263, 297, 290, 311
Raines, Martin L. 304
Ranger 7 171, 173
Ray, Leon 304
Ray, Rex 61, 62
Rayburn, Sam 5, 42
Raytheon 152
Reagan, President Ronald 257, 280, 291, 293, 295, 297, 299, 308, 317, 330
Redding, Ed 37
Redstone Arsenal 4, 9, 73
Redstone rocket 3, 7, 9, 17, 19, 22, 23, 27, 38, 54, 69, 72, 77, 83, 105-107, 110, 130, 262, 163
Redundancy systems 130, 173, 235, 241
Rees, Eberhard 169, 196, 210, 312
Remote Manipulator System 149
Rendezvous and docking 56, 58, 59, 64, 70, 71, 73, 77, 79, 80, 92-95, 103, 142, 144, 145, 160, 164, 165, 174, 178, 189, 209, 211, 219, 221
Rendezvous Committee 144
Reorganizations 88-90, 95, 107
Reusable Ground Launch Vehicle Concept and Development Planning Study 223
Rice University 37, 40, 259, 260
  provides site for spacecraft center 33, 39, 41, 47, 48, 51
Richard, Ludie G. 169
Richardson, Herbert H. 317
Ride Report 344
Ride, Dr. Sally K. 287, 288, 300-304, 311, 321
Roberts, Tecwyn (Tec) 26, 84, 85, 165
Robinson, Emyre Barrios 262, 264
Rockefeller Brothers Fund 12
Rockefeller, Nelson 12
Rocket Research Project 7
Rocketdyne 73, 74, 109, 138, 145, 174, 329
  chosen to design Shuttle engines 230
  prime contractor, Saturn propulsion systems 138
Rockwell International 65, 139, 230, 238, 253, 264, 308, 317, 326, 335, 342
Rockwell Standard 139
Rockwell, Willard F. 137
Rogers Commission 299, 300, 307, 321, 326, 329, 349
Rogers, William P. 299
Roland, Alex 330
Roosa, Stuart Allen 128, 199
Roosevelt, President Franklin 5
Roosevelt, President Theodore 15
Root, Eugene 4
Rose, Rodney G. (Rod) 24, 25, 26, 53, 81, 274

Rosen, Milton 195, 225
Rothrock, Addison M. 36
Rummel, Robert W. 300
Russell, Sen. Richard B. 2
Russian Space Program 17, 29, 32
Rutland, C. H. 223
Ryan Aeronautical Company 79
Ryker, Norman 141

S

Satellites
  ASC-1 290, 297
  Arabsat 290, 296
  ASTRO-1 342
  AUSSAT 290, 297
  GLOMR 296, 297
  Leasat-3 296
  NUSAT 296
  Palapa B 285, 288, 291, 294
  SATCOM Ku-2 297-290
  SOLAR MAX 291, 292
  SYNCOM-3 289, 290, 296, 297
  SYNCOM-4 290, 294, 296, 297
  Telestar 285, 289, 290, 296
  Telesat 288
  Telstar 1 285, 289, 296
  Westar VI 291, 294
Saturn program 27, 54, 64, 107, 108
Saturn rocket 72, 73, 74, 77, 82, 93, 97, 103, 105, 107-115, 134-144, 152, 155, 160, 165, 169, 171, 173, 175, 189, 203, 210, 222, 223, 236
Saudi Arabia 296
Sauter, Linda 44
Scheer, Julian 133
Scherer, Lee R. 210
Schirra, Walter (Wally) 68, 70, 88, 93, 121, 122, 128, 133, 155, 165, 201
Schmitt, Sen. Harrison Hagan (Jack) 128, 203, 252, 295
Schnyer, A. Daniel 223
School of Aviation Medicine 119
Science Advisory Committee 4, 13, 28, 30, 113, 173, 193, 195
Science and Applications Directorate 154, 170, 180, 181, 213, 218, 219, 244, 246, 251
Science Applications International 342
Scott Science & Technology 201, 265, 266
Scott, David R. 93, 111, 200, 201, 265
Scout launch vehicle 72
Scriven, General George P. 10
Scully-Power, Paul 293
Seabrook, Texas 260

Seamans, Robert C., Jr. 32, 79, 101, 124, 141, 144, 145
Security 1, 5, 19, 66, 84, 85, 89, 90, 119, 120, 123, 156, 191, 193, 239, 294, 295, 324, 333
Seddon, Margaret Rhea 288, 290, 302
Select Committee on Astronautics and Space Exploration 12
Selection and Training of Personnel for Space Flight 118
Sells, S. B. 118
Senate Select Committee on Space and Astronautics 17
Shaffer, Philip C. 183
Sharp, Frank 124
Shea, Joseph F. (Joe) 70, 71, 87, 107, 113, 114, 189
Shepard, Alan B., Jr. 32, *photo* 33, 38, 53, 54, 78, 85, 122, 127, 124, 163, 199, 201
Sheppard, Dr. Sallie 318
Sherman, Milton 141, 142
Shuttle Avionics Integration Laboratory (SAIL) 230
Shuttle landing strips 239, 250
Shuttle Management Structure Team 304
Shuttle Mission Simulator (SMS) 248, 250, 278
Shuttle Payload Activities Team 277
Shuttle Payload Integration and Development Program Office 278
Shuttle Program Office 165, 233, 234, 262, 273, 282, 335
Shuttle Program Task Group 227
Silverstein, Abe 17, 18, 19, 23, 26, 31, 36, 57, 79, 102, 104, 120
   background 21-22
   Director of Goddard Space Flight Center 35
   establishes Manned Space Flight Office 21
   interface with STG 100
   leaves NASA 32, 37
   rift with Gilruth 27, 35
   role in Mercury program 71
   role in relocating manned spacecraft program 36, 37
Silviera, Milton A. 230
Simpkinson, Scott 17, 22, 23, 109, 114, 169
Singer Link Division 248, 252, 265, 278
Sisk, Congressman B. F. 18
Sixth International History of Astronautics Symposium 17
Sjoberg, Sigurd A. 198, 209
Skylab Program Office 211
Skylab vii, 127, 178, 188, 190, 197, 203- 205, 207, 209-211, 214-216, 218, 221, 226, 229, 231, 234, 238, 251, 254, 274, 276, 282, 288, 312, 321

Slayton, Donald K. (Deke) 68, 70, 111, 120, 121, 125, 126, 169, 184, 220, 266
   astronaut 122
   Director for Flight Crew Operations 73, 87
   Head of the Astronaut Office 73, 127
   resigns to work in private sector 262
Slayton, Kent 68
Sloop, John 104
Small and Disadvantaged Business Office 268
Small Business Innovation Development Act of 1982 266
Small businesses 150, 154, 257, 267, 268
Sneider, William C. 181
Snoopy 174
Solid Rocket Boosters 238, 253, 279, 285, 301, 326
Sonett, Dr. C. P. 129
Soviet space program 1
Space vii, 159, 196
   mission control 159
   race i, 27, 54, 135, 315
   role in space exploration 209
Space Act 11, 14, 18, 19, 294
   amendment 30
Space Act of 1958 15
Space Adaptation Syndrome 288
Space Advisory Board 317, 318
Space Age 18, 19, 49, 105-106, 196
Space Business Research Center 319
Space Business Roundtable 317
Space Center Houston 331, 332
Space Council 14, 28, 29, 30
Space Environment Simulation Laboratory (SESL) 131
Space Exploration Initiative 337, 338, 342
Spaceflight
   costs 57, 75
   German role in 105-106
   historical significance ix
   human dimensions 118-133
   technical challenges 56-57
Spaceflight program 17
Space Foundation 317, 319
Space Industries, Inc. 264
Space News Roundup 44, 55, 70, 74, 78
Space Research Center 314-319
Space Services Inc. 266-267, 281
Space Shuttle Program ix, 193, 195, 196, 206, 209-231, 233-255, 319, 325, 330, 339
   Canadian-sponsored experiments 293
   challenges 233, 343
   compared with other programs 274
   contractors 222, 224, 230, 260

costs 236-239, 285, 286, 295
design and testing requirements 235
development 274
early concepts 222
effect of *Challenger* disaster 335
experiments 338
financial resources 235, 236-239
first commercial package 282
first shuttle flight 253, 273
flight requirements 285
getaway special 282, 298, 295
laboratories 241, 251
launch vehicles 238
management 224, 234, 308, 312, 319
oceanographic experiments 293
orbiter modifications 326
payload planning 275, 285
post-*Challenger* changes 343
project office 213
public interest 293
public support for 273, 280
purpose 309
remote manipulator (Canadarm) 287, 291-297
retrieve and repair satellite 292
satellites deployed 296-298
scientific knowledge 252, 281
shuttle pallet satellite (SPAS) 287
summary of shuttle flights 302-304
technical challenges 233, 234, 239, 279
technological innovations 251-252
training 293
Space Station *Freedom* vii, 233, 264, 329, 334, 335, 337, 339, 345, 346, 350
Space Station Projects Office 325, 335, 336
Space station vii, ix, 188-196, 204, 205, 210, 221-226, 252, 258, 268, 273, 295, 297, 307-313, 319-321, 325-339, 342-347
Space suit 112, 130, 131, 152, 174, 287, 292
Space Task Group (STG) vii, 11, 15, 20, 21, 25-27, 31, 32, 38-45, 53, 57-59, 61-64, 71, 77-84, 98-101, 105, 106-115, 119, 120, 142, 163, 189, 196, 225, 227, 229, 234, 249, 278, 335
 addresses Mercury technical problems 38
 location 35, 36
 organization memo 21
 origin and development 35
 reaction to relocation to Houston 43
 staffing 23, 24
Space Technology Laboratories 71, 101, 143, 152, 174
Space Transportation System (STS) 159, 193, 196, 205, 209-231, 254, 287, 307, 325, 345

Spacelab 280, 285, 289, 290, 295, 296, 297
Spacelab Program Office 280
Special Committee on Space and Aeronautics 12
Special President's Task Group 196
Sputnik 1-11, 15, 18, 29, 36, 59, 74, 75, 82, 98, 105, 106, 155, 163, 189, 233, 250, 321, 347
Sputnik I 1, 35, 350
 effect on American educational system 6
 effect on American space program 1, 2, 3, 36, 75, 82, 155
 military significance 2, 3
 political repercussions 3
 reaction of scientific community 59-60
 world reaction 1
Sputnik II 17
Stafford, Thomas P. 88, 93, 174, 220, 221, 281, 295, 332
Stall, Harold S. 331, 332
Star Wars 317
Starfire Rocket 266
Stennis Space Center 350
Sterling, William 93
Stevens Institute of Technology 311
Stewart, Bob 291
Stoller, Homer 32
Stone, Ralph W. 144
Storms, Harrison A. (Stormy) 138, 142
Strass, H. Kurt 64
Strategic Defense Initiative 317
Strategic Task Group 10
Subcommittee on International Cooperation 18
Subcommittee on Scientific Research and Development 18
Subcommittee on Space Science and Applications 280
Subcommittee on Space Problems and Life Sciences 18
Sullivan, Kathryn D. 293
Surveyor 3 179, 194
Surveyor 1 171
Survival gear 77, 80
Swigert, John L., Jr. 182
Systems and subsystems management 77
Systems Management American Corp. 342

T

Tapley, Dr. Byron 318
Taylor, General Maxwell 4
Teacher in Space Project 293
Teague, Congressman Olin E. (Tiger) 18, 38, *photo* 41, 42, 43, 51, 77, 102, 108, 109, 113, 127, 146, 150, 187, 190-193, 196, 197, 200, 221, 259, 315

Technological advancements 13, 152, 156
Technology and Commercial Projects Office 268
Television
    role in publicizing space program ix
Teller, Dr. Edward 4, 5
Texas A & M University 259, 234, 262, 311, 314-319
Texas Engineering Experiment Station 314
Texas Instruments 247, 260
Texas Space Grant Consortium 315, 318
Thagard, Dr. Norman 288
The Right Stuff 120
The Space Station 295
Thibodaux, Joseph G. 11
Thiokol 152
Thomas, Sen. Albert 37, 41, 42, 51
Thompson, Bill vii
Thompson, Clarke W. 51
Thompson, Floyd L. 20, 24, 27, 113, 195
Thompson, Robert F. 165, 229, 234, 235
Thor launch vehicle 72
Thor rocket 22, 73
Thorson, Richard A. 335
Tiger Team 114
Tindall, Howard 165, 169
Titan II rocket 72, 74, 77, 79, 88, 107
Toftoy, Colonel H. N. 8
Total Quality Management 336
Tower, Sen. John 50
Tracking and Data Relay Satellite (TDRS) 287, 290
Training 118, 122-133, 151, 163, 164, 172, 175, 180, 198, 211, 214, 220, 248-251, 264-266, 276, 278, 285-287, 292, 293, 301, 312, 315, 338, 342
Training hardware 126
Truly, Lt. Commander Richard H. 128, 129, 280, 289, 319, 333, 334, 338, 343
Truszynski, Gerald 169
TRW 152, 174
Tsiolkovsky, Konstantin E. 6, 7, 350
Turner, Luther 42
Turrentine, Gordon 46

U

Ulmer, Ralph E. 41
Ulysses Probe 334, 343
UniSys 342
United Aircraft 152, 174, 211
University of Houston 39, 46, 264, 268, 314-319
University of Tennessee 288
University of Texas 266, 279, 308, 314-318, 332
Unmanned Moon landing 13
Utah State University 282

V

V-2 missile 7, 8, 9, 105
Van Allen radiation belts 12, 180
Van Allen, James 330
Van Hoften, James D. 292, 297
Van Horn, Richard 317
Vanguard launch vehicle 72
Vanguard rocket program 2, 3, 12, 72
Vanguard satellite program 26, 32
Vavra, Paul 84
Vienna, Austria 17, 22
Vietnam War, effect on space program 108, 113, 135, 137, 156, 178, 187, 188, 192, 193, 197, 205, 206, 226, 261
Viking program 9, 321
Vincent, John 42
Voas, Robert 119, 120, 125, 126
Vostok spacecraft 32, 38
Voyager 337

W

Wainerdi, Richard 317
Walker, Arthur B.C., Jr. 300
Walker, Charles D. 293
Wallops Island Station 22, 23, 32, 36, 53, 97, 326
Walt Disney Imagineering 332
Warrior Constructors, Inc. 154
Washington, D. C. 8, 14, 17, 19, 21, 22, 26, 27, 41, 43-46, 48, 57, 64, 70, 78, 120, 122, 140, 142, 195, 197, 209, 220, 225, 308, 309, 318, 319, 335
Water landing 17, 23, 68, 71, 79, 88, 102-104, 113, 146
Water-landing parachute system 17, 74
Water Transport System 38-40
Waterman, Alan 3
Webb, James (Jim) vii, 24, 29-32, 36, 37, 39-41, 43, 51, 70, 74, 78, 101, 104, 113, 123, 124, 145, 146, 169, 190, 191, 195
    reorganizes NASA management 71
    and NASA management system 99
    retires 160, 197
Webster, Texas 260
Weightless Environment Training Facility (WETF) 151, 249
Weinberger, Caspar 238
Weitz, Paul J. 215, 331
Welch, Louie 260, 261
Welsh, Edward C. 29, 156, 192
West Estate 37, 47
West, Colonel Paul 46, 48
Western Electric 19, 84
Western Union 291

Wheelon, Dr. Albert D. 300
Whirlpool Corporation 211
Whitbeck, Phil 62, 216
White Sands Missile Range 266, 282
White, Dr. Stanley C. 46, 62, 63, 119, 120
White, Edward H. 44, 111, 112, 140
    first spacewalk 93, *photo* 94
Whynot, Charles 265
Wiesner, Jerome B. 29, 30, 124, 145
Williams, Don 337
Williams, Larry 335
Williams, Walter C. 10, 41, 46-48, 51, 60-65, 71, 78, 84-87,
Wilson, Charles E. 2, 3
Winn, Grace 43-45, 68
Wolfe, Tom 120
Women astronaut candidates 120, 288
Worden, Alfred 201, 202
Wright brothers 18
Wright, House Speaker Jim 319
Wyatt, DeMarquis D. 104

Y

Yankee Clipper 179
Yarborough, Sen. Ralph 50
Yardley, John F. 140, 141, 317
Yeager, Brigadier General Charles E. (Chuck) 10, 61
Young, Major John 88, 92, *photo* 94, 128, 174, 179, 201-203, 265, 279, 324, 300-302
Young, R. Wayne 212, 229, 324, 335

Z

Zbanek, Leo T. 62, 66
Zedekar, Raymond 126
Zimmerman, Charles H. 17, 20, 21
Zoerner, Gary 264

The JSC History Series

Reference Works, NASA SP-4000:

Grimwood, James M. *Project Mercury: A Chronology* (NASA SP-4001, 1963).

Grimwood, James M., and Hacker, Barton C., with Vorzimmer, Peter J. *Project Gemini Technology and Operations: A Chronology* (NASA SP-4002, 1969).

Link, Mae Mills. *Space Medicine in Project Mercury* (NASA SP-4003, 1965).

Astronautics and Aeronautics: A Chronology of Science, Technology and Policy (NASA SP-4004 to SP-4025, a series of annual volumes continuing from 1961 to 1985, with an earlier summary volume, *Aeronautics and Astronautics, 1915-1960).*

Ertel, Ivan D., and Morse, Mary Louise. *The Apollo Spacecraft: A Chronology, Volume I, Through November 7, 1962* (NASA SP-4009, 1969).

Morse, Mary Louise, and Bays, Jean Kernahan. *The Apollo Spacecraft: A Chronology, Volume II, November 8, 1962-September 30, 1964* (NASA SP-4009, 1973).

Brooks, Courtney G., and Ertel, Ivan D. *The Apollo Spacecraft: A Chronology, Volume III, October 1, 1964-January 20, 1966* (NASA SP-4009, 1973).

Van Nimmen, Jane, and Bruno, Leonard C., with Rosholt, Robert L. *NASA Historical Data Book, Vol. I: NASA Resources, 1958-1968* (NASA SP-4012, 1976, rep. ed. 1988).

Newkirk, Roland W., and Ertel, Ivan D., with Brooks, Courtney G. *Skylab: A Chronology* (NASA SP-4011, 1977).

Ertel, Ivan D., and Newkirk, Roland W., with Brooks, Courtney G. *The Apollo Spacecraft: A Chronology, Volume IV, January 21, 1966-July 13, 1974* (NASA SP-4009, 1978).

Ezell, Linda Neuman. *NASA Historical Data Book, Vol. II: Programs and Projects, 1958-1968.* (NASA SP-4012, 1988).

Ezell, Linda Neuman. *NASA Historical Data Book, Vol. III: Programs and Projects, 1969-1978* (NASA SP-4012, 1988).

Management Histories, NASA SP-4100:

Rosholt, Robert L. *An Administrative History of NASA, 1958-1963* (NASA SP-4101, 1966).

Levine, Arnold S. *Managing NASA in the Apollo Era* (NASA SP-4102, 1982).

Roland, Alex. *Model Research: The National Advisory Committee for Aeronautics, 1915-1958* (NASA SP-4103, 1985).

Fries, Sylvia D. *NASA Engineers and the Age of Apollo* (NASA SP-4104, 1992).

Project Histories, NASA SP-4200:

Human Space Flight Programs:

Swenson, Loyd S., Jr., Grimwood, James M., and Alexander, Charles C. *This New Ocean: A History of Project Mercury* (NASA SP-4201, 1966).

Hacker, Barton C., and Grimwood, James M. *On Shoulders of Titans: A History of Project Gemini* (NASA SP-4203, 1977).

Benson, Charles D. and Faherty, William Barnaby. *Moonport: A History of Apollo Launch Facilities and Operations* (NASA SP-4204, 1978).

Ezell, Edward Clinton, and Ezell, Linda Neuman. *The Partnership: A History of the Apollo-Soyuz Test Project* (NASA SP-4209, 1978).

Brooks, Courtney G., Grimwood, James M., and Swenson, Loyd S., Jr. *Chariots for Apollo: A History of Manned Lunar Spacecraft* (NASA SP-4205, 1979).

Bilstein, Roger E. *Stages to Saturn: A Technological History of the Apollo/Saturn Launch Vehicles* (NASA SP-4206, 1980).

Compton, W. David, and Benson, Charles D. *Living and Working in Space: A History of Skylab* (NASA SP-4208, 1983).

Compton, W. David. *Where No Man Has Gone Before: A History of Apollo Lunar Exploration Missions* (NASA SP-4214, 1989).

Satellite Space Flight Programs:

Green, Constance McL., and Lomask, Milton. *Vanguard: A History* (NASA SP-4202, 1970; rep. ed. Smithsonian Institution Press, 1971).

Hall, R. Cargill. *Lunar Impact: A History of Project Range* (NASA SP-4210, 1977).

Ezell, Edward Clinton, and Ezell, Linda Neuman. *On Mars: Exploration of the Red Planet, 1958-1978* (NASA SP-4212, 1984).

Scientific Programs:

Newell, Homer E. *Beyond the Atmosphere: Early Years of Space Science* (NASA SP-4211, 1980).

Pitts, John A. *The Human Factor: Biomedicine in the Manned Space Program to 1980* (NASA SP-4213, 1985).

Naugle, John E. *First Among Equals: The Selection of NASA Space Science Experiments* (NASA SP-4215, 1991).

Center Histories, NASA SP-4300:

Hartman, Edwin P. *Adventures in Research: A History of Ames Research Center, 1940-1965* (NASA SP-4302, 1970).

Hallion, Richard P. *On the Frontier: Flight Research at Dryden, 1946-1981* (NASA SP-4303, 1984).

Muenger, Elizabeth A. *Searching the Horizon: A History of Ames Research Center, 1940-1976* (NASA SP-4304, 1985).

Rosenthal, Alfred. *Venture into Space: Early Years of Goddard Space Flight Center* (NASA SP-4301, 1985).

Hansen, James R. *Engineer in Charge: A History of the Langley Aeronautical Laboratory, 1917-1958* (NASA SP-4305, 1987).

Dawson, Virginia P. *Engines and Innovation: Lewis Laboratory and American Propulsion Technology* (NASA SP-4306, 1991).

General Histories, NASA SP-4400:

Corliss, William R. *NASA Sounding Rockets, 1958-1968: A Historical Summary* (NASA SP-4401, 1971).

Wells, Helen T., Whiteley, Susan H., and Karegeannes, Carrie. *Origins of NASA Names* (NASA SP-4402, 1976).

Anderson, Frank W., Jr. *Orders of Magnitude: A History of NACA and NASA, 1915-1980* (NASA SP-4403, 1981).

Sloop, John L. *Liquid Hydrogen as a Propulsion Fuel, 1945-1959* (NASA SP-4404, 1978).

Roland, Alex. *A Spacefaring People: Perspectives on Early Spaceflight* (NASA SP-4405, 1985).

Bilstein, Roger E. *Orders of Magnitude: A History of NACA and NASA, 1915-1990* (NASA SP-4406, 1989).

New Series in NASA History,
Published by The Johns Hopkins University Press:

Cooper, Henry S.F., Jr. *Before Lift-Off: The Making of a Space Shuttle Crew* (1987).

McCurdy, Howard E. *The Space Station Decision: Incremental Politics and Technological Choice* (1990).

Hufbauer, Karl. *Exploring the Sun: Solar Science Since Galileo* (1991).

McCurdy, Howard E. *Inside NASA: High Technology and Organizational Change in the U.S. Space Program* (1993).

www.ingramcontent.com/pod-product-compliance
Lightning Source LLC
Chambersburg PA
CBHW081716170526
45167CB00009B/3593